JACARANDA 7
MATHS QUEST 7

STAGE 4 NSW SYLLABUS | THIRD EDITION

JACARANDA
MATHS QUEST 7

STAGE 4 NSW SYLLABUS | THIRD EDITION

BEVERLY LANGSFORD WILLING

CATHERINE SMITH

Third edition published 2023 by
John Wiley & Sons Australia, Ltd
Level 4, 600 Bourke Street, Melbourne, Vic 3000

First edition published 2011
Second edition published 2014

Typeset in 10.5/13 pt Times LT Std

ISBN: 978-0-730-38611-7

Front cover image: © alexdndz/Shutterstock
© Marish/Shutterstock; © Irina Strelnikova/Shutterstock
© Vector Juice/Shutterstock

Illustrated by various artists, diacriTech and Wiley Composition
Services

Typeset in India by diacriTech

A catalogue record for this
book is available from the
National Library of Australia

The Publishers of this series acknowledge and pay their respects
to Aboriginal Peoples and Torres Strait Islander Peoples as the
traditional custodians of the land on which this resource was
produced.

This suite of resources may include references to (including
names, images, footage or voices of) people of Aboriginal
and/or Torres Strait Islander heritage who are deceased. These
images and references have been included to help Australian
students from all cultural backgrounds develop a better
understanding of Aboriginal and Torres Strait Islander Peoples'
history, culture and lived experience.

It is strongly recommended that teachers examine resources on
topics related to Aboriginal and/or Torres Strait Islander
Cultures and Peoples to assess their suitability for their own
specific class and school context. It is also recommended that
teachers know and follow the guidelines laid down by the
relevant educational authorities and local Elders or community
advisors regarding content about all First Nations Peoples.

All activities in this resource have been written with the safety
of both teacher and student in mind. Some, however, involve
physical activity or the use of equipment or tools. All due care
should be taken when performing such activities. To the
maximum extent permitted by law, the author and publisher
disclaim all responsibility and liability for any injury or loss that
may be sustained when completing activities described in this
resource.

The Publisher acknowledges ongoing discussions related to
gender-based population data. At the time of publishing, there
was insufficient data available to allow for the meaningful
analysis of trends and patterns to broaden our discussion of
demographics beyond male and female gender identification.

Printed in Singapore
M WEP210097 300623

Contents

About this resource

NEW FOR

2024 NSW SYLLABUS

JACARANDA
MATHS QUEST 7

NSW SYLLABUS
THIRD EDITION

Developed by teachers for students

Tried, tested and trusted. The third edition of the *Jacaranda Maths Quest* series, continues to focus on helping teachers achieve learning success for every student — ensuring no student is left behind, and no student is held back.

Because both what and how students learn matter

Learning is personal

Whether students need a challenge or a helping hand, you'll find what you need to create engaging lessons.

Whether in class or at home, students can get unstuck and progress! Scaffolded lessons, with detailed worked examples, are all supported by teacher-led video eLessons. Automatically marked, differentiated question sets are all supported by detailed worked solutions. And Brand-new Quick Quizzes support in-depth skill acquisition.

Learning is effortful

Learning happens when students push themselves. With learnON, Australia's most powerful online learning platform, students can challenge themselves, build confidence and ultimately achieve success.

Learning is rewarding

Through real-time results data, students can track and monitor their own progress and easily identify areas of strength and weakness.

And for teachers, Learning Analytics provide valuable insights to support student growth and drive informed intervention strategies.

Learn online with Australia's most

- Trusted, curriculum-aligned content
- Engaging, rich multimedia
- All the teaching-support resources you need
- Deep insights into progress
- Immediate feedback for students
- Create custom assignments in just a few clicks.

Practical teaching advice and ideas for each lesson provided in teachON

Teaching videos for all lessons

Reading content and rich media including embedded videos and interactivities

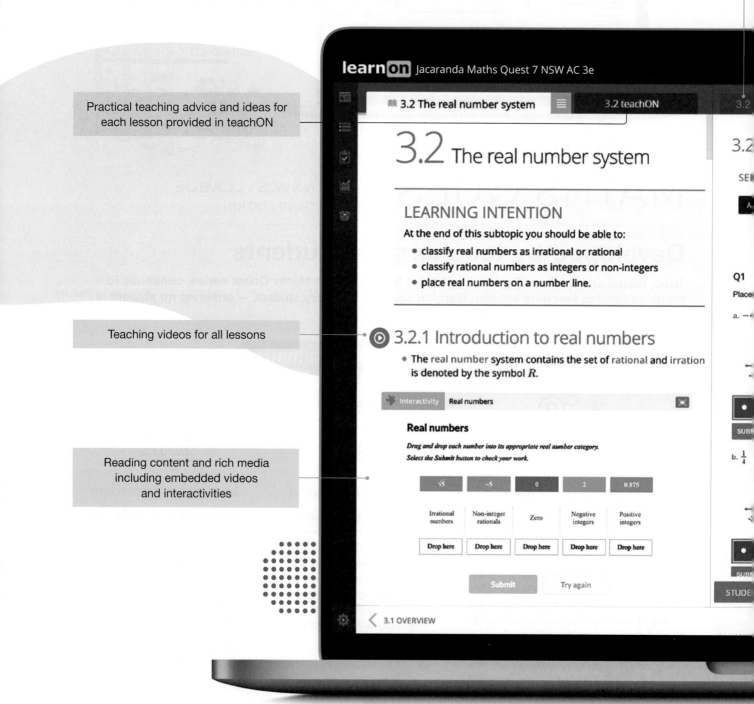

learn**on** Jacaranda Maths Quest 7 NSW AC 3e

📖 3.2 The real number system ☰ 3.2 teachON 3.2

3.2 The real number system

LEARNING INTENTION

At the end of this subtopic you should be able to:
- classify real numbers as irrational or rational
- classify rational numbers as integers or non-integers
- place real numbers on a number line.

◎ 3.2.1 Introduction to real numbers

- The real number system contains the set of rational and irration is denoted by the symbol R.

Interactivity Real numbers

Real numbers

Drag and drop each number into its appropriate real number category.
Select the Submit button to check your work.

$\sqrt{5}$ –5   0   2   0.875

Irrational numbers	Non-integer rationals	Zero	Negative integers	Positive integers
Drop here	Drop here	Drop here	Drop here	Drop here

Submit   Try again

3.1 OVERVIEW

3.2
SE
A
Q1
Place
a. –4
b. $\frac{1}{4}$
SUB
STUDE

powerful learning tool, learnON

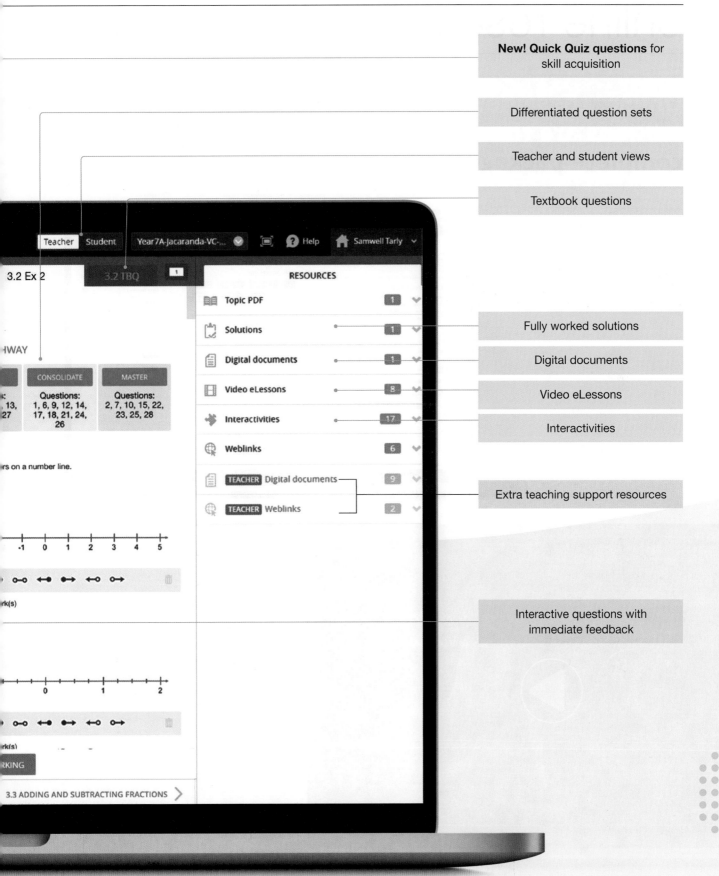

New! **Quick Quiz questions** for skill acquisition

Differentiated question sets

Teacher and student views

Textbook questions

Fully worked solutions

Digital documents

Video eLessons

Interactivities

Extra teaching support resources

Interactive questions with immediate feedback

Teacher Student Year7A-Jacaranda-VC-... Help Samwell Tarly

3.2 Ex 2   3.2 TBQ   1

RESOURCES

Topic PDF   1

Solutions   1

Digital documents   1

Video eLessons   8

Interactivities   17

Weblinks   6

TEACHER Digital documents   9

TEACHER Weblinks   2

HWAY

CONSOLIDATE	MASTER
Questions: 1, 6, 9, 12, 14, 17, 18, 21, 24, 26	Questions: 2, 7, 10, 15, 22, 23, 25, 28

rs on a number line.

3.3 ADDING AND SUBTRACTING FRACTIONS

Get the most from your online resources

Online, these new editions are the complete package

Trusted Jacaranda theory, plus tools to support teaching and make learning more engaging, personalised and visible.

Embedded interactivities and videos enable students to explore concepts and learn deeply by 'doing'.

New teaching videos for every lesson are designed to help students learn concepts by having a 'teacher at home', and are flexible enough to be used for pre- and post-learning, flipped classrooms, class discussions, remediation and more.

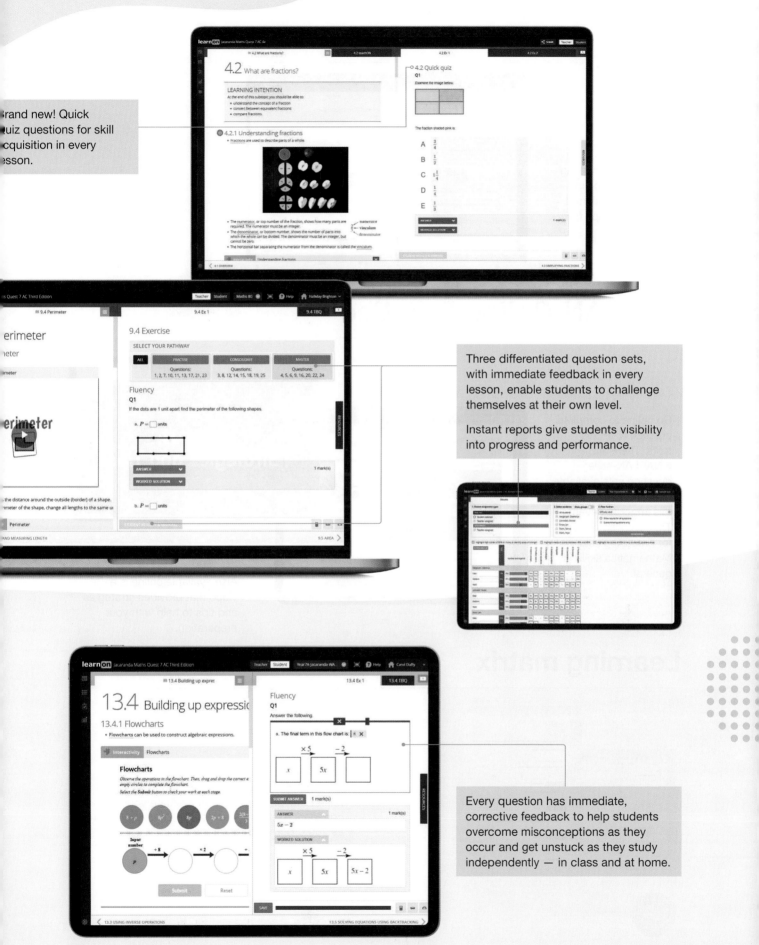

Brand new! Quick quiz questions for skill acquisition in every lesson.

Three differentiated question sets, with immediate feedback in every lesson, enable students to challenge themselves at their own level.

Instant reports give students visibility into progress and performance.

Every question has immediate, corrective feedback to help students overcome misconceptions as they occur and get unstuck as they study independently — in class and at home.

Powering up for Year 7

A six-week 'Powering up for Year 7' online program is designed to plug any gaps from earlier years.

NAPLAN Online Practice

Go online to complete practice NAPLAN tests. There are 6 NAPLAN-style question sets available to help you prepare for this important event. They are also useful for practising your Mathematics skills in general.

Also available online is a video that provides strategies and tips to help with your preparation.

Learning matrix

A Learning matrix in each topic enables you to reflect on your learning and evaluate your confidence level for each of the Learning Intentions as you progress through the topic.

A wealth of teacher resources

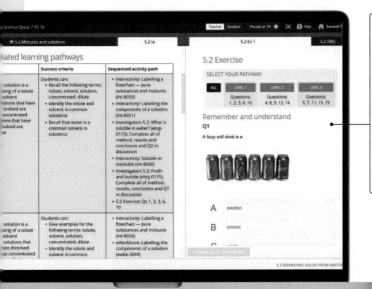

Enhanced teaching-support resources for every lesson, including:

- work programs and curriculum grids
- practical teaching advice
- three levels of differentiated teaching programs
- quarantined topic tests (with solutions)

Customise and assign

An inbuilt testmaker enables you to create custom assignments and tests from the complete bank of thousands of questions for immediate, spaced and mixed practice.

Reports and results

Data analytics and instant reports provide data-driven insights into progress and performance within each lesson and across the entire course.

Show students (and their parents or carers) their own assessment data in fine detail. You can filter their results to identify areas of strength and weakness.

Acknowledgements

The authors and publisher would like to thank the following copyright holders, organisations and individuals for their assistance and for permission to reproduce copyright material in this book.

Subject Outcomes, Objectives and Contents from the NSW Mathematics K–10 Syllabus © Copyright 2022 NSW Education Standards Authority.

Images

• © Jaroslaw Grudzinski/Shutterstock: **371** • © marc - stock.adobe.com: **96** • © elenavah - stock.adobe.com: **334** • © Have a nice day - stock.adobe.c: **329** • © Shutterstock: **5** • © Radu Bercan/Shutterstock: **616** • © Christopher Futcher/Getty Images: **700** • Shutterstock/Gentian Polovina: **338** • Shutterstock/Gringoann: **338** • Shutterstock/Monkey Business I: **317** • Shutterstock/Nerthuz: **339** • Shutterstock/oksana2010: **335** • Shutterstock/Shuang Li: **330** • © Africa Studio/Shutterstock: **344, 375** • © Andrey_Popov/Shutterstock: **646** • © Dragon Images/Shutterstock: **655** • © Dusan Petkovic/Shutterstock: **353** • © Fabrik Bilder/Shutterstock: **661** • © Fresnel/Shutterstock: **345** • © George Rudy/Shutterstock: **643** • © goodluz/Shutterstock: **649** • © hddigital/Shutterstock: **284** • © Hong Vo/Shutterstock: **379** • © Kampol Taepanich/Shutterstock: **352** • © Kostyazar/Shutterstock: **380** • © LightField Studios/Shutterstock: **478** • © Minerva Studio/Shutterstock: **278** • © Monkey Business Images/Shutterstock: **656** • © Murni/Shutterstock: **352** • © naito29/Shutterstock: **345** • © New Africa/Shutterstock: **646** • © nortongo/Shutterstock: **375** • © Pixel-Shot/Shutterstock: **703** • © Pixel-ShotShutterstock: **379** • © rfranca/Shutterstock: **348** • © Shutterstock: **283, 308, 318** • © Shutterstock/Andrey Bayda: **325** • © Shutterstock/Elena Veselova: **318** • © Shutterstock/garagestock: **333** • © Shutterstock/LightField Studio: **318** • © Shutterstock/Madeleine R: **333** • © Shutterstock/Neale Cousland: **323** • © Shutterstock/Radu Bercan: **330** • © Shutterstock/sirtravelalot: **316** • © Shutterstock/Tyler Olson: **337** • © Sudowoodo/Shutterstock: **362** • © Susan Schmitz/Shutterstock: **703** • © Syda Productions/Shutterstock: **648** • © szefei/Shutterstock: **567** • © Tadeusz Wejkszo/Shutterstock: **353** • © Thiti Sukapan/Shutterstock: **702** • © Tomasz Czajkowski/Shutterstock: **486** • © Tyler Olson/ Shutterstock: **695** • © Tyler Olson/Shutterstock: **380** • © Viacheslav Nikolaenko/Shutterstock: **493** • © Vixit/Shutterstock: **486** • © vvvita/Shutterstock: **71** • © zhao jiankang/shutterstock: **487** • © 300 librarians/Shutterstock: **349** • © a. Chones/Shutterstock; b. M. Unal Ozmen/Shutterstock; c. M. Unal Ozmen/Shutterstock: **398** • © a. Raphael Christinat; b.: **121** • © a.Elena Shashki /Shutterstock; b. Tulpahn/Shutterstock: **303** • © Aaron Amat/Shutterstock: **291** • © AbElena/Shutterstock: **355** • © abriendomundo/Shutterstock: **9** • © al7/Shutterstock: **428** • © Alberto Loyo/Shutterstock: **245** • © Alinute Silzeviciute/ Shutterstock: **694** • © ALPA PROD/Shutterstock: **129** • © alterfalter/Shutterstock: **152** • © AmaPhoto/Shutterstock: **246** • © Andrey Armyagov/Shutterstock: **122** • © Andrii_M/Shutterstock: **99** • © Annches/Shutterstock: **392** • © Aptyp_koK/Shutterstock: **461** • © ArliftAtoz2205/Shutterstock: **428** • © auddmin/Shutterstock: **740** • © AustralianCamera/Shutterstock: **232** • © Axel Bueckert/Shutterstock: **260** • © BearFotos/Shutterstock: **208** • © bernashafo/Shutterstock: **304** • © Bilanol/Shutterstock: **40** • © BIRTHPIX/Shutterstock: **645** • © Blend Images/Shutterstock: **33** • © Bloomicon/Shutterstock: **298** • © Brand X Pictures: **513** • © CapturePB/Shutterstock: **730** • © Carlos Gutierrez PhotoShutterstock: **72** • © Carsten Reisinger/Shutterstock: **174** • © ChameleonsEye/Shutterstock: **24** • © clearviewstock/Shutterstock: **745** • © Corbis: **587** • © Corepics VOF/Shutterstock: **23** • © cpaulfell/Shutterstock: **428** • © Damian Pankowiec/Shutterstock: **33** • © Daniel Jedzura/Shutterstock: **214** • © debra hughes/Shutterstock: **115** • © Digital Stock: **513, 514** • © Digital Stock/Corbis Corporation: **88** • © dikkenss/Shutterstock: **548** • © DisobeyArt/Shutterstock: **100** • © Dmytro Zinkevych/Shutterstock: **407** • © dolomite-summits/Shutterstock: **154** • © Dragon Images/Shutterstock: **129** • © Dragon ImagesShutterstock: **83** • © Dudarev Mikhail/Shutterstock: **548** • © eAlisa/Shutterstock: **240** • © Elias/Adobe Stock Photos: **670, 698** • © ESB Professional/Shutterstock: **231** • © ESB Professional/Shutterstock: **653** • © Eva Speshneva/Shutterstock: **359** • © f11photo/Shutterstock: **51** • © fckncg/Shutterstock: **174** • © Fedor Selivanov/Shutterstock: **53** • © Fertas/Shutterstock: **427** • © freevideophotoagency/Shutterstock: **155** • © frenchiestravel/Shutterstock: **57** • © fresher/Shutterstock: **34** • © fritz16/Shutterstock: **8** • © Gino Santa Maria/Shutterstock: **745** • ©

Gjermund/Shutterstock: **447** • © goodluz/Shutterstock: **128** • © GorodenkoffShutterstock: **73** • © Ground Picture/Shutterstock: **100, 122, 177** • © grum_l/Shutterstock: **729** • © gualtiero boffi/Shutterstock: **226** • © Hquality/Shutterstock: **273, 744** • © Hurst Photo/Shutterstock: **514, 534** • © Ilja Generalov/Shutterstock: **459** • © Image Addict: **225** • © Image Disk Photography: **7, 237, 544** • © Image Source: **231** • © Imagenet/Shutterstock: **642** • © ingehogenbijl/Shutterstock: **8** • © Ingo Schulz/Alamy Stock Photo: **241** • © Ingrid Balabanova/Shutterstock: **304** • © Irina Strelnikova/Shutterstock Marish/Shutterstock: **111** • © Irina Strelnikova; © Marish: **1, 65, 257, 343, 389, 435, 565, 715** • © IT StockFree: **587** • © Ivan MarjanovicShutterstock: **77** • © Jan Hopgood/Shutterstock: **390** • © Jasni/Shutterstock: **730** • © Jeni Jenny Jeni/Shutterstock: **164** • © Jennifer Wright/John Wiley & Sons Australia: **175** • © Jennifer Wright/John Wiley & Sons Australia: **175** • © John Wiley & Sons: **574, 589, 621** • © John Wiley & Sons Australia: **33** • © John Wiley & Sons Australia/ Taken by Kari-Ann Tapp: **192** • © John Wiley & Sons Australia/Jo Patterson: **550** • © John Wiley & Sons Australia/Kari-Ann Tapp: **534** • © jokerpro/Shutterstock: **44** • © Joshua Resnick/Shutterstock: **304** • © JR-stock/Shutterstock: **358** • © Kari-Ann Tapp: **732** • © Katilda/Shutterstock: **592** • © Krakenimages.com/Shutterstock: **745** • © l i g h t p o e t/Shutterstock: **33** • © Lance Bellers/Shutterstock: **721** • © Lesya Dolyuk/Shutterstock: **187** • © M. Unal Ozmen/Shutterstock: **58** • © Maks Narodenko/Shutterstock: **185** • © MaraZe/Shutterstock: **45** • © Maridav/Shutterstock: **216** • © mashurov/Shutterstock: **39** • © master1305/Adobe Stock Photos: **678** • © matimix/Shutterstock: **724** • © Maxisport/Shutterstock: **283, 425** • © MCruzUA/Shutterstock: **35** • © Michael Rosskothen/Shutterstock: **461** • © MichaelJayBerlin/Shutterstock: **134** • © michaeljung/Shutterstock: **181** • © Mila Supinskaya Glashchenko/Shutterstock: **39** • © Milagli/Shutterstock: **54** • © Millionstock/Shutterstock: **225** • © Mind Pro Studio/Shutterstock: **229** • © Minerva Studio/ Shutterstock: **665** • © MiniDoodle/Shutterstock: **742** • © mirtmirt/Shutterstock: **726** • © Mitch Hutchinson/Shutterstock: **716** • © MMD Made my dreams/Shutterstock: **100** • © Monkey Business Images/Shutterstock: **118, 187, 203, 220, 221, 241** • © Monkey Business Images/Shutterstock: **40, 397, 403, 654** • © monticello/Shutterstock: **237** • © monticello/Shutterstock.com: **731** • © mooinblack/Shutterstock: **237** • © Natali Glado/Shutterstock: **664** • © Nataliya Peregudova/Shutterstock: **221** • © National Council of Teachers of Mathematics: **736** • © Neale Cousland/Shutterstock: **225** • © Neale Cousland/Shutterstock: **58, 722** • © Neirfy/Shutterstock: **377** • © Nejron PhotoShutterstock: **83** • © New Africa/Shutterstock: **121, 242** • © News Ltd/Newspix: **615** • © Nicholas Piccillo/Shutterstock: **410** • © NicoElNino/Shutterstock: **401** • © nobeastsofierce/Shutterstock: **129** • © Oez/Shutterstock: **624** • © Oleksiy Mark/Shutterstock: **246** • © OlekStock/Shutterstock: **530** • © PA Images/Alamy Stock Photo: **2** • © PA Images/Alamy Stock Photo: **258, 284** • © PANAGIOTIS KARAPANAGIOTIS/Alamy Stock Photo: **54** • © PanyaStudio/Shutterstock: **354** • © Pepgooner/Shutterstock: **88** • © Petrenko Andriy/Shutterstock: **177** • © Phatthanit/Shutterstock: **226** • © Phils Mommy/Shutterstock: **423** • © Photoagriculture/Shutterstock: **416** • © PhotoDisc: **194, 203** • © Photodisc: **514, 529, 530, 535, 588, 605, 614, 621, 648, 650, 677** • © PhotoDisc: Inc, **485** • © PhotoDisc: Katie Dickinson/Shutterstock, Tomasz Pado/Shutterstock, **23** • © photokup/Shutterstock: **423** • © Photozi/Adobe Stock Photos: **666** • © pilipphoto/Shutterstock: **458** • © pio3/Shutterstock: **529** • © PPstock/Shutterstock: **154** • © Pressmaster/Shutterstock: **216** • © Pressmaster/Shutterstock: **353** • © ProStockStudio/Shutterstock: **394** • © Rawpixel.com/Shutterstock: **245** • © Renovacio/Shutterstock: **731** • © Richard Semik/Shutterstock: **416** • © Robyn Mackenzie/Shutterstock: **307** • © Roma Koshel/Shutterstock: **739** • © Roman Samborskyi/Shutterstock: **32** • © Ronald Sumners/Shutterstock: **50** • © Rostislav Král/Shutterstock: **214** • © Ruth Black/Shutterstock: **50** • © safakcakir/Shutterstock: **213** • © Sandarina/Shutterstock: **40** • © science photo/Shutterstock: **545** • © sculpies/Shutterstock: **566** • © Sean Donohue Photo/Shutterstock: **284** • © Serg64/Shutterstock: **301** • © Sergey Novikov/Shutterstock: **231** • © Shutterstock: **272** • © shutterupeire/Shutterstock: **214** • © Shvaygert Ekaterina/Shutterstock: **665** • © Singha Songsak P/Shutterstock: **621** • © sirtravelalot/Shutterstock: **397** • © Sonia Alves-Polidor/Shutterstock: **720** • © Sonulkaster/Shutterstock: **102** • © Source: Reconciliation NSW: **441** • © Source: Wiley art: **437** • © stockcreations/Shutterstock: **88** • © Stocksnapper/Shutterstock: **545** • © Sudowoodo: **359** • © Swapan Photography/Shutterstock: **739** • © szefei/Shutterstock: **589** • © Technotr/E+/Getty Images: **22** • © Tetyana Kokhanets/Shutterstock: **719** • © THPStock/Shutterstock: **725** • © TK Kurikawa/Shutterstock: **301** • © tr3gin/Shutterstock: **215** • © trigga/Getty Images: **72** • © Tum3000/Shutterstock: **549** • © Tyler Olson/Shutterstock.com: **279** • © UnknownLatitude Images/Shutterstock: **7** • © v74/Shutterstock: **720** • © Valentina Kalashnikova/Shutterstock: **112** • © VALUA VITALY/Shutterstock:

150 • © Velychko/Shutterstock: **195** • © Vereshchagin Dmitry/Shutterstock: **649** • © Viktor Fedorenko/Shutterstock: **724, 728** • © Viktoriya/Shutterstock.com: **537** • © VisualArtStudioShutterstock: **22** • © vkilikovShutterstock: **17** • © VladisChern/Shutterstock: **68** • © Volodymyr TVERDOKHLIB/Shutterstock: **118** • © Volt Collection/Shutterstock: **264** • © VovanIvanovich/Shutterstock: **66, 70** • © VP Photo Studio/Shutterstock: **721** • © wavebreakmedia/Shutterstock: **726** • © wolfman57/Shutterstock: **223** • © Yeti Studio/Shutterstock: **736** • © Yevgenij_D/Shutterstock: **436** • © Yorkman/Shutterstock: **545** • © You Touch Pix of EuToch/Shutterstock: **225** • © Zbigniew Guzowski/Shutterstock: **730** • © zbindere/Getty Images: **7** • © zentilia/Shutterstock: **725** • ©Michael D Brown/Shutterstock: **744** • © Andrey Arkusha/Shutterstock: **481** • © Benny Marty/Shutterstock: **481** • © Bruno Herdt/Getty Images: **573** • © chrisdorney/Shutterstock: **503** • © Corbis: **484** • © CurrywurstmitPommes/Shutterstock: **481** • © Digital Stock/Corbis Corporation: **573** • © Digital Vision: **619** • © Dimedrol68/Shutterstock: **576** • © diskoVisnja/Shutterstock: **599** • © diskoVisnja/Shutterstock: **599** • © Eugene Buchko/Shutterstock: **618** • © John Wiley & Sons: **619** • © John Wiley & Sons Australia/Photo by Kari-Ann Tapp: **618** • © Julenochek/Shutterstock: **481** • © Kekyalyaynen/Shutterstock: **503** • © Kitch Bain/Shutterstock.com: **346** • © Kokkai Ng/Getty Images: **484** • © nortongo/Shutterstock: **522** • © Olesia Bilkei/Shutterstock: **576** • © Photo Melon/Shutterstock: **618** • © Photodisc: **542, 574, 610** • © PhotoDisc: Inc, **484** • © PhotoDisc: Inc., **484** • © pornvit_v/Shutterstock: **618** • © Ruud Morijn Photographer/Shutterstock: **600** • © Sheila Fitzgerald/Shutterstock: **349** • © StockphotoVideo/Shutterstock: **574** • © Taras Vyshnya/Shutterstock: **503** • © totajla/Shutterstock: **484** • © urbanbuzz/Shutterstock: **542** • © Zety Akhzar/Shutterstock: **503**

Every effort has been made to trace the ownership of copyright material. Information that will enable the publisher to rectify any error or omission in subsequent reprints will be welcome. In such cases, please contact the Permissions Section of John Wiley & Sons Australia, Ltd.

Powering up for Year 7

We care deeply about ensuring that all students feel confident in knowing they are equipped with the essential core skills in Mathematics as they transition into Year 7.

Powering up for Year 7 is a brand-new online resource developed to fill learning gaps in key mathematical areas to ensure that all students have the building blocks needed to move smoothly into Year 7. The course is organised around key skills over a period of six weeks:

Students will learn key skills such as as place value, operations with whole numbers, mental strategies for calculations, fractions, decimals and percentages.

All questions are automatically marked, with instant feedback and worked solutions. This incredible resource is available online only in our digital learnON platform.

NAPLAN practice

Go online to complete practice NAPLAN tests. There are 6 NAPLAN-style question sets available to help you prepare for this important event. They are also useful for practising your Mathematics skills in general.

Also available online is a video that provides strategies and tips to help with your preparation.

SET A
Calculator allowed

SET B
Non-calculator

SET C
Calculator allowed

SET D
Non-calculator

SET E
Calculator allowed

SET F
Non-calculator

1 Computation with positive integers

LESSON SEQUENCE

LESSON
1.1 Overview

Why learn this?

Positive integers are whole numbers. We use positive integers every day. Addition, subtraction, multiplication and division of positive integers are the building blocks of Mathematics.

When you count the number of runs you make in a game of cricket, you are using integers. When you add your runs to the runs your teammates scored, the total number of runs will be a positive integer. This result is found by the addition of integers.

Many daily activities depend on knowing how to answer simple questions like, 'When you went for that walk, how far did you go?' or 'How many people live in your house?'

Every day we see integers displayed on screens — for example when we look at the weather forecast or check our unread messages. We normally don't even think about integers — they are just the numbers we see all around us.

Understanding integers and their addition, subtraction, multiplication and division is important for everyday life and work. Many jobs, including hospitality, banking, construction, design, engineering, nursing, teaching, finance and medicine all require an understanding of the use of integers.

Hey students! Bring these pages to life online

Watch videos

Engage with interactivities

Answer questions and check solutions

Find all this and MORE in jacPLUS

Reading content and rich media, including interactivities and videos for every concept

Extra learning resources

Differentiated question sets

Questions with immediate feedback, and fully worked solutions to help students get unstuck

Exercise 1.1 Pre-test

1. **MC** Choose the largest of the following numbers.

   A. 5656   B. 5665   C. 5566   D. 5666

2. State how many three-digit numbers you can make from the digits 6, 9 and 1 if you are allowed to repeat digits.

3. Calculate the following additions.
   a. $1366 + 948 + 97$
   b. $654 + 937 + 23 + 68\,941$
   c. $3085 + 38 + 20\,389 + 3000 + 235$

4. Calculate the following subtractions.
   a. $987 - 365$ b. $9432 - 2175$ c. $3001 - 1739$

5. Calculate the following using a mental strategy.
   $2 \times 8 \times 6 \times 5$

6. Evaluate the following.
   $3\overline{)1248}$

7. A large chicken farm sorts its eggs into cartons of 12 eggs. If they have 3852 eggs to sort, calculate the number of complete cartons that can be sorted.

8. A landscape gardener charges an initial fee of $165, plus $52 per hour. Determine how much the landscape gardener will charge for a job that takes 35 hours if a customer has taken advantage of a special offer of $120 off the total.

9. If an order was made for 25 boxes of Fanta and each box contains 24 cans, use factors to calculate the total amount of cans.

10. Round 32 895 to the first digit.

11. Calculate the following.
    a. $48 \div 4$ b. $168 \div 8$ c. $625 \div 25$

12. Complete each statement by placing $<$, $>$ or $=$ in the empty box so that the statement is true.
    a. $2 \times 4 \ \square \ 8 \times 2$ b. $16 \div 2 \ \square \ 8$ c. $16 \times 2 \ \square \ 32 \div 8$

13. Evaluate the following.
    a. $7 \times 6 \div 3 \div 7$ b. $3 \times 8 \div 6 + 4$ c. $7 + [66 - (5 \times 5)] \times 3$

14. Evaluate the following using long division.
    a. $2752 \div 16$ b. $7548 \div 17$ c. $4935 \div 21$

15. Aziz takes 3 minutes to complete one lap of a running course, while Jani takes 4 minutes to complete one lap. Not counting the start of the race, determine when Aziz will next be running beside Jani.

LESSON
1.2 Place value and comparing and ordering positive integers

LEARNING INTENTION

At the end of this lesson you should be able to:
- understand that the value of each digit in a number depends on its position or place value
- recognise and describe the direction and magnitude of integers.

▶ 1.2.1 Place value

eles-3643

- Numbers are made up of the digits 0, 1, 2, 3, 4, 5, 6, 7, 8 and 9.
- The position of a digit in a number gives a different value to the digit. The table below shows the value of the digit 6 in some different numbers.

Number	Name of number	Value of the digit 6
16	Sixteen	6
3562	Three thousand, five hundred and sixty-two	60
18 634	Eighteen thousand, six hundred and thirty-four	600

- Each position in a number has its own **place value**. The number 59 376 can be represented in a place-value table as shown.

Ten thousands	Thousands	Hundreds	Tens	Ones
5	9	3	7	6

WORKED EXAMPLE 1 Expanded form

Write 59 376 in expanded form.

THINK

1. 59 376 is the same as 5 ten thousands, 9 thousands, 3 hundreds, 7 tens and 6 ones.

2. Multiply each of the digits by its place value.

3. Perform each of the multiplications to show how the expanded form is written.

WRITE

Ten thousands	Thousands	Hundreds	Tens	Ones
5	9	3	7	6

$59\,376 = 5 \times 10\,000 + 9 \times 1000 + 3 \times 100 + 7 \times 10 + 6 \times 1$

$= 50\,000 + 9000 + 300 + 70 + 6$

Place-holding zeros

- The 2021 Census of Population recorded the population of Greater Sydney as 5 231 150. In a place-value table, this number appears as follows.

Millions	Hundred thousands	Ten thousands	Thousands	Hundreds	Tens	Ones
5	2	3	1	1	5	0

- The zero (0) in the place-value table means that there are no ones. The zero must be written to hold the place value, otherwise, the number would be written as 523 115 and would no longer have the same value.

Reading numbers

- To make numbers easier to read and name, they are written in groups of three digits with a space between each group. The only exception to this rule is that four-digit numbers are usually written as a group of four digits.

Number	Name of number
4357	Four thousand, three hundred and fifty-seven
12 345	Twelve thousand, three hundred and forty-five
102 345	One hundred and two thousand, three hundred and forty-five
123 456 789	One hundred and twenty-three million, four hundred and fifty-six thousand, seven hundred and eighty-nine

▶ 1.2.2 Ordering and comparing numbers

eles-3644

- Numbers in **ascending order** are placed from smallest to largest, starting on the left. The numbers, 1, 25, 192, 908 and 1115 are in ascending order.

$$1 \quad 25 \quad 192 \quad 908 \quad 1115$$

- Numbers in **descending order** are placed from largest to smallest, starting on the left. The numbers 8532, 934, 105, 53 and 5 are in descending order.

$$8532 \quad 934 \quad 105 \quad 53 \quad 5$$

- Numbers are ordered according to their place values. For whole numbers, the number with the most digits is the greatest in value because the first digit will have the highest place value.
- If two numbers have the same number of digits, then the digits with the highest place value are compared. If they are equal, the next highest place values are compared, and so on.

WORKED EXAMPLE 2 Ordering numbers

Write the following numbers in descending order.
858, 58, 85, 8588, 5888, 855

THINK	WRITE
1. Write the numbers with the most digits.	8588 and 5888
2. There are two numbers with 4 digits. The number with the higher digit in the thousands column is larger. The other number is placed second.	8588, 5888
3. Compare the two numbers with 3 digits. Both have the same hundreds and tens values, so compare the value of the units.	858, 855
4. Compare the two 2-digit numbers.	85, 58
5. Write the answer.	8588, 5888, 858, 855, 85, 58

- The < symbol means **is less than**. The number sentence 3 < 5 is read from left to right like a normal sentence. It is read as 'three is less than five'.
- The > symbol means **is greater than**. The number sentence 5 > 3 is also read from left to right. It is read as 'five is greater than three'.
- When comparing two numbers using the symbols < and >, think of a crocodile's mouth. To work out the right symbol to place between two numbers so the number sentence is correct, just remember that the crocodile always wants to eat the bigger number.

Smaller than 3 < 5

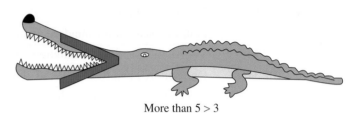

More than 5 > 3

WORKED EXAMPLE 3 Writing number sentences

Copy and complete the following number sentences by placing the < or > symbol in each box.
a. 52 ☐ 160

b. 8764 ☐ 4720

THINK	WRITE
a. 1. Examine the two numbers and determine which number is larger.	**a.** 160 is larger than 52.
2. Recall that the symbols < and > are like a crocodile's mouth. To work out the correct symbol to insert between 52 and 160, remember that the crocodile always wants to eat the larger number.	The crocodile opens its mouth to eat the larger number. 52 < 160
3. The correct number sentence should read '52 is less than 160'. Write the correct statement.	52 < 160
b. 1. Examine the two numbers and determine which number is larger.	**b.** 8764 is larger than 4720.
2. Recall that the symbols < and > are like a crocodile's mouth. To determine the correct symbol to be inserted between 8764 and 4720, remember that the crocodile always wants to eat the larger number.	The crocodile opens its mouth to eat the larger number. 8764 > 4720
3. The correct number sentence should read '8764 is greater than 4720'. Write the correct statement.	8764 > 4720

on Resources

➤ **Interactivity** Place value (int-3921)

Exercise 1.2 Place value and comparing and ordering positive integers

1.2 Quick quiz on	1.2 Exercise

Individual pathways

■ PRACTISE	■ CONSOLIDATE	■ MASTER
1, 4, 7, 9, 10, 13, 16, 17, 19, 22, 25	2, 5, 8, 11, 14, 18, 20, 23, 26	3, 6, 12, 15, 21, 24, 27

Fluency

1. **WE1** Write the following numbers in expanded form.
   a. 925
   b. 28 469
   c. 800 002
   d. 1 080 100

2. Write the following numbers in words.
   a. 765
   b. 9105
   c. 90 450
   d. 100 236

3. Write each of the following as a numeral.
   a. Four hundred and ninety-five
   b. Two thousand, six hundred and seventy
   c. One hundred and nine thousand, six hundred and five

4. Look at the photograph. Using words, write the value of the 4 in the distance to Yancannia.

5. Look at the photograph. Using words, write the value of the 2 in the distance to Esperance.

6. Look at the photograph. Using words, write the value of the 0 in the distance to Katherine.

7. Look at the photograph. Using words, write the value of the 3 in the distance to Coober Pedy.

8. Look at the photograph. Using words, write the value of the 5 in the distance to Alice Springs.

Understanding

9. **MC** Select the largest of the following numbers.

    A. 4884   B. 4488   C. 4848   D. 4888

10. **MC** Select the smallest of the following numbers.

    A. 4884   B. 4488   C. 4848   D. 4888

11. **WE2** Write the following numbers in descending order.
    8569, 742, 48 987, 28, 647

12. Rearrange the following numbers so that they are in descending order.
    47 890, 58 625, 72 167, 12 947, 32 320

13. Organise the following numbers into descending order.
    6477, 7647, 7476, 4776, 6747

14. Change the order of the following numbers to put them in descending order.
    8088, 8800, 8080, 8808, 8008, 8880

15. Write the following numbers in ascending order.
    58, 9, 743, 68 247, 1 258 647

16. Rearrange the following numbers so that they are in ascending order.
    78 645, 58 610, 60 000, 34 108, 84 364

17. Organise the following numbers into ascending order.
    9201, 2910, 1902, 9021, 2019, 1290

18. Change the order of the following numbers so that they are in ascending order.
    211, 221, 212, 1112, 222, 111

19. **WE3** Copy and complete the following number sentences by placing the < or > symbol in each box.
    a. 345 ☐ 567
    b. 89 ☐ 98
    c. 234 596 ☐ 23 459
    d. 7765 ☐ 7756

20. We can use the abbreviation k to represent $1000. For example, $50 000 can be written as $50k. Using this rule, determine what amounts each of the following represent.
    a. $6k
    b. $340k
    c. $58k

21. Write the following using k as an abbreviation to represent $1000.
    a. $430 000
    b. $7000
    c. $800 000

Communicating, reasoning and problem solving

22. Determine the largest five-digit number you can write if each digit must be different and no digit may be prime. Show your working.

23. A new mobile phone company is investigating the habits of phone users. From the following sample of phone calls, determine the most common length of call. Start by rounding each call time to the nearest half-minute (30 seconds).

132 s	10 s	43 s	71 s	243 s	52s
142 s	332 s	6 s	38 s	63 s	32 s
132 s	32 s	43 s	52 s	352 s	101 s
124 s	28 s	153 s	10 s	243 s	34 s

24. Astronomers and other scientists use scientific notation when working with very large or very small numbers. For example, the 'astronomical unit' (AU) is the average distance of the Earth from the Sun; it is equal to150 million km or 1.5×10^{13} cm. Explain why scientists use scientific notation.

25. Determine how many two-digit numbers you can make using the digits 1 and 2 if:
    a. you are not allowed to repeat any digits
    b. you can repeat the digits.

26. Determine how many three-digit numbers you can make using the digits 1, 2 and 3 if:
    a. you are not allowed to repeat any digits
    b. you can repeat the digits.

    Explain whether there is a relationship between the initial number of digits and the number of arrangements when:
    c. repetition is not allowed
    d. repetition is allowed.

27. Without actually counting to one million, determine a way of estimating the time it would take to count out loud from 0 to 1 000 000.

LESSON
1.3 Strategies for adding and subtracting positive integers

LEARNING INTENTION

At the end of this lesson you should be able to:
- add and subtract positive integers using mental strategies.

▶ 1.3.1 Mental strategies for addition

eles-3645

- Mental strategies for addition are different ways to add up large numbers or integers in your head without a calculator.
- There are lots of different methods that can help you add 2- and 3-digit numbers in your head.

Jump strategy for addition

- The jump strategy for addition involves breaking the number being added into smaller parts.

  For example:

  $$535 + 87 =$$

  Break 87 into 80 and 7.

  $$535 + 80 = 615$$

  $$615 + 7 = 622$$

Split strategy for addition

- The split strategy for addition involves splitting both numbers being added into their expanded forms and then adding the parts that have the same place value.

  For example:

  $$165 + 432 =$$

  Split 165 into 1 hundred, 6 tens and 5 ones: $165 = 100 + 60 + 5$
  Split 432 into 4 hundreds, 3 tens and 2 ones: $432 = 400 + 30 + 2$

  $$165 + 432 = 100 + 60 + 5 + 400 + 30 + 2$$

  $$= 100 + 400 + 60 + 30 + 5 + 2$$

  $$= 500 + 90 + 7$$

  $$= 597$$

a. **Calculate $514 + 88$ using the jump strategy.**
b. **Calculate $759 + 412$ using the split strategy.**

THINK	WRITE
a. 1. Write the question.	a. $514 + 88$
2. Break 88 into 80 plus 8.	$514 + 80 + 8$
3. Add 80 to 514 to get 594.	$514 + 80 = 594$
4. Add 8 to 594 to get 602.	$594 + 8 = 602$
5. Write the answer.	$514 + 88 = 602$
b. 1. Write the question.	b. $759 + 412$
2. Split each number up by writing it in expanded form.	$759 = 700 + 50 + 9$ $412 = 400 + 10 + 2$
3. Add the numbers with the same place value together.	$700 + 400 + 50 + 10 + 9 + 2 = 1100 + 60 + 11$ $= 1171$
4. Write the answer.	$759 + 412 = 1171$

Compensation strategy for addition

- The compensation strategy for addition involves making one of the numbers being added larger by rounding up to the nearest ten or hundred, completing the addition, and then subtracting the added amount.

  For example:

  $$243 + 78 =$$

  1. Round 78 up to 80.

  $$243 + 80 = 323$$

  2. Now subtract 2, since 78 was rounded up to 80 by adding 2.

  $$323 - 2 = 321$$

Rearrange strategy for addition

- The rearrange strategy for addition involves rearranging the numbers being added to form numbers that add up to multiples of 10.

  For example:

  $$16 + 17 + 14 =$$

  Swap the order and add 16 and 14 first, since this will result in a multiple of 10.

  $$16 + 14 + 17 =$$

  $$30 + 17 = 47$$

WORKED EXAMPLE 5 Using mental strategies to add numbers

a. Calculate $576 + 86$ using the compensation strategy.

b. Calculate $38 + 13 + 22$ using the rearrange strategy.

THINK	WRITE
a. 1. Write the question.	**a.** $576 + 86$
2. Round 86 up to the nearest 10.	86 becomes 90 when rounded up by adding 4.
3. Add 576 to the rounded-up number (90).	$576 + 90 = 666$
4. Subtract 4, since 86 was rounded up to 90 by adding 4.	$666 - 4 = 662$
5. Write the answer.	$576 + 86 = 662$
b. 1. Write the question.	**b.** $38 + 13 + 22$
2. Rearrange the numbers so that a multiple of 10 can be formed. $38 + 22 = 60$, so add these numbers first.	$= 38 + 22 + 13$ $= 60 + 13$
3. Add 13 to 60.	$= 73$
4. Write the answer.	$38 + 13 + 22 = 73$

WORKED EXAMPLE 6 Using mental strategies to add numbers

Calculate $713 + 143$ using an appropriate mental strategy.

THINK	WRITE
1. Write the question.	$713 + 143$
2. Select a strategy. Since the numbers are both large numbers, it may be easiest to split them up.	Split strategy
3. Use the strategy to add numbers. • Write each number in expanded form. • Add numbers with the same place values. • Complete the final addition.	$713 = 700 + 10 + 3$ $143 = 100 + 40 + 3$ $= 700 + 100 + 10 + 40 + 3 + 3$ $= 800 + 50 + 6$ $= 856$
4. Write the answer.	$713 + 143 = 856$

1.3.2 Mental strategies for subtraction

eles-3646

- Mental strategies for subtraction are different ways to subtract numbers without doing lots of working.
- There are lots of different methods that can help you subtract two- and three-digit numbers in your head.

Jump strategy for subtraction

- The jump strategy for subtraction is like the jump strategy for addition, but you jump backward instead of forward.
- The number being subtracted is broken into smaller parts and then each part is subtracted one by one.

For example:

$$635 - 127 =$$

Break 127 into $100,\ 20$ and 7.

$$635 - 100 = 535$$

$$535 - 20 = 515$$

$$515 - 7 = 508$$

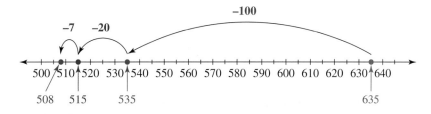

Adding up to find the difference

- Sometimes, even though you are subtracting, it can be helpful to use addition when trying to subtract numbers mentally.
- Start with the number being subtracted and add to that number in parts until you get to the larger number.

For example:

$$672 - 613 =$$
$$613 + 7 = 620$$
$$620 + 50 = 670$$
$$670 + 2 = 672$$

The numbers added are $7,\ 50$ and 2, which give a total of 59.

Therefore $672 - 613 = 59$.

WORKED EXAMPLE 7 Using mental strategies to subtract numbers

Calculate the following.
a. $854 - 84$ using the jump strategy.   **b. $436 - 381$ using the adding strategy.**

THINK	WRITE
a. 1. Write the question.	**a.** $854 - 84$
2. Break 84 into 80 plus 4.	$84 = 80 + 4$
3. Subtract 80 from 854 to get 774.	$854 - 80 = 774$
4. Subtract 4 from 774 to get 770.	$774 - 4 = 770$
5. Write the answer.	$854 - 84 = 770$
b. 1. Write the question.	**b.** $436 - 381$
2. Starting at the smaller number, determine what needs to be added to get to the larger number.	$381 + 9 = 390$ $390 + 10 = 400$
3. Add up all the parts.	$400 + 36 = 436$
4. Write the answer.	$9 + 10 + 36 = 55$ $436 - 381 = 55$

Compensation strategy for subtraction

* The compensation strategy for subtraction is like the compensation strategy for addition, except that you will round up the number being subtracted to the nearest ten or hundred and then add the amount used to do the rounding up.

For example:

$$94 - 49 =$$

Round up 49 to 50 by adding 1.

$$94 - 50 = 44$$

Add 1 to the answer, since 49 was rounded up to 50 by adding 1.

$$44 + 1 = 45$$
$$94 - 49 = 45$$

WORKED EXAMPLE 8 Using the compensation strategy to subtract numbers

Calculate $174 - 38$ using the compensation strategy.

THINK	WRITE
1. Write the question.	$174 - 38$
2. Round 38 up to the nearest 10 by adding 2.	38 becomes 40.
3. Subtract 40 from 174.	$174 - 40 = 134$
4. Add 2 since, 38 was rounded up to 40 by adding 2.	$134 + 2 = 136$
5. Write the answer.	$174 - 38 = 136$

Checking subtraction by adding

Remember, a subtraction can always be checked by adding the two smaller numbers to see if they add up to the larger number.

$$94 - 49 = 45$$

DISCUSSION

What techniques have you used to mentally add numbers? What shortcuts have you used?

Exercise 1.3 Strategies for adding and subtracting positive integers

1.3 Quick quiz on	**1.3 Exercise**

Individual pathways

■ PRACTISE	■ CONSOLIDATE	■ MASTER
1, 2, 5, 7, 12, 13, 16	3, 6, 8, 11, 14, 17	4, 9, 10, 15, 18

Fluency

1. **WE4a** Use the jump strategy to calculate the following.
   a. $143 + 85$
   b. $1537 + 266$

2. **WE4b** Use the split strategy to calculate the following.
   a. $645 + 261$
   b. $370 + 78$

3. **WE5a** Use the compensation strategy to calculate the following.
   a. $471 + 89$
   b. $74 + 28$

4. **WE5b** Use the rearrange strategy to calculate the following.
   a. $27 + 36 + 23$
   b. $14 + 34 + 16$

Understanding

5. State whether the following statement is true or false. Show your working.
   $346 + 451$ calculated using the split strategy becomes $300 + 400 + 40 + 50 + 6 + 1$.

6. **MC** $56 + 12 + 14$ is equal to:
   A. 60
   B. 82
   C. 70
   D. 92

7. **WE6** Calculate $421 + 54 + 372$ using an appropriate mental strategy.

8. **WE7a** Use the jump strategy to calculate the following.
   a. $88 - 43$
   b. $674 - 323$

9. **WE7b** Add up the difference to calculate the following.
   a. $351 - 287$
   b. $79 - 46$

10. **WE8** Use the compensation strategy to calculate the following.
    a. $65 - 48$
    b. $956 - 729$

11. Calculate $953 - 675$ using an appropriate mental strategy.

12. Fill in the gaps for the following sentences.
    a. All the numbers in the 5-times table end in _____ or _____.
    b. The numbers in the _____-times tables all end in 0.

Communicating, reasoning and problem solving

13. A student calculates $145 + 671 - 472$. They give their answer as 344. State whether you agree with their answer. Explain your response.

14. A student has shown the following working on a test.

$$641 + 357 = 600 + 10 + 4 + 300 + 70 + 5$$
$$= 989$$

   a. Determine where the student made an error.
   b. Determine the correct answer.

15. Twenty children were playing in the playground. Eight of the children left for home, then six more children came to the playground. Determine how many children are in the playground now. Show your working.

16. There are 763 students at a senior school campus. Of these, 213 students are currently completing exams. Calculate how many students are not completing exams.

17. Harriet has $243 in savings. She adds $93 to her savings account. The next day she spends $175 from her account. Determine how much is left in her savings account.

18. The sum of two numbers is 19 and their difference is 7. Evaluate the two numbers.

LESSON
1.4 Algorithms for adding and subtracting positive integers

LEARNING INTENTION

At the end of this lesson you should be able to:
- add and subtract positive integers using algorithms.

▶ 1.4.1 Addition of positive integers using algorithms

eles-3647

- To add numbers or positive integers, write them in columns according to place value and then add them, starting at the ones column.
- The **sum** is the result obtained when numbers are added together.

WORKED EXAMPLE 9 Adding integers in columns

Arrange these numbers in columns, then add them.

$$1462 + 78 + 316$$

THINK

1. Set out the numbers in columns according to place value.
2. Add the digits in the ones column in your head $(2 + 8 + 6 = 16)$. Write the 6 in the ones column of your answer and carry the 1 to the tens column, as shown in the plum colour.

3. Add the digits in the tens column $(1 + 6 + 7 + 1 = 15)$. Write the 5 in the tens column of your answer and carry the 1 to the hundreds column, as shown in pink.

WRITE

Thousands	Hundreds	Tens	Ones
1	¹4	¹6	2
		7	8
+	3	1	6
1	8	5	6

4. Add the digits in the hundreds column ($1 + 4 + 3 = 8$).
   Write 8 in the hundreds column of your answer, as
   shown in pink. There is nothing to carry.

5. There is only a 1 in the thousands column. Write 1
   in the thousands column of your answer.

6. State the answer. $1462 + 78 + 316 = 1856$

WORKED EXAMPLE 10 Adding by making a multiple of 10

Perform the addition $27 + 19 + 141 + 73$ by finding suitable pairs of numbers to make multiples of 10.

THINK	WRITE
1. Write the question.	$27 + 19 + 141 + 73$
2. Look for pairs of numbers that the ones digits add to, to make 10. Reorder the sum, pairing these numbers.	$= (27 + 73) + (141 + 19)$
3. Add the number pairs.	$= 100 + 160$
4. Complete the addition.	$= 260$

▶ 1.4.2 Commutative and associative laws for addition

eles-3648

- The **Commutative Law for addition** means that you can add numbers in any order. For example, $3 + 4$ gives the same result as $4 + 3$.

- The **Associative Law for addition**, simply stated, means that the order in which additions are calculated is not important and does not change the result. For example, $2 + 3 + 7$ could be calculated as $5 + 7 = 12$ or $2 + 10 = 12$.
  This is a very useful property when performing mental calculations.

COMMUNICATING — COLLABORATIVE TASK: There's something about Gauss

In the late 1780s, a German teacher gave his class the task of adding the numbers from 1 to 100. One student came up with the correct answer in less than a minute. That student, Johann Carl Friedrich Gauss, used grouping in pairs to work out the sum. He paired the smallest number with the largest number, then the second smallest with the second largest, and so on. Gauss went on to become one of the world's most famous mathematicians.

Try the technique that Gauss used with the numbers from 1 to 10, then 1 to 20, then 1 to 50 and finally 1 to 100. Did you notice any patterns?

▶ 1.4.3 Subtraction of positive integers using algorithms

eles-3649

- Subtraction is the opposite (or 'inverse') of addition. To work out $800 - 360$ you could ask the question, 'What do I need to add to 360 to get 800?' The answer is 440.
- The **difference** is the result when one number is subtracted from another number. For example, the difference between 800 and 360 is 440.
- One of the main distinctions between addition and subtraction is the fact that the order in which you subtract numbers is very important. For example, $6 - 3$ does not give the same result as $3 - 6$.
- This means that subtraction is neither commutative nor associative.

Subtraction of positive integers with no borrowing

- To subtract two numbers, write them in columns according to their place values and then subtract each column in order, starting with the ones column.

WORKED EXAMPLE 11 Subtracting numbers without borrowing

Perform the subtraction $395 - 174$.

THINK	WRITE
1. Set out the difference in columns according to place value.	$\begin{array}{ccc} \text{Hundreds} & \text{Tens} & \text{Ones} \\ 3 & 9 & 5 \\ - \quad 1 & 7 & 4 \\ \hline \end{array}$
2. Subtract the digits in the ones column $(5 - 4 = 1)$. Write the 1 in the ones column of your answer.	$\begin{array}{ccc} 3 & 9 & 5 \\ - \quad 1 & 7 & 4 \\ \hline & & 1 \end{array}$
3. Subtract the digits in the tens column $(9 - 7 = 2)$. Write the 2 in the tens column of your answer.	$\begin{array}{ccc} 3 & 9 & 5 \\ - \quad 1 & 7 & 4 \\ \hline & 2 & 1 \end{array}$
4. Subtract the digits in the hundreds column $(3 - 1 = 2)$. Write the 2 in the hundreds column of your answer.	$\begin{array}{ccc} 3 & 9 & 5 \\ - \quad 1 & 7 & 4 \\ \hline 2 & 2 & 1 \end{array}$
5. State the answer.	The difference between 395 and 174 is 221.

Not all subtraction problems are as simple as the one shown in Worked example 11. For example, working out $32 - 14$ requires a different approach because you can't subtract 4 from 2 in the ones column. A subtraction like this requires borrowing.

$\begin{array}{cc} 3 & 2 \\ - \quad 1 & 4 \\ \hline & ? \end{array}$

Subtraction of positive integers with borrowing

- The most common method of subtraction is to use borrowing, which is called the **decomposition method**.
- It's called 'decomposition' because the larger number is decomposed, or taken apart.
- A visual representation of the decomposition method is shown here to help you understand how it works.

Visual representation of subtracting using decomposition

Evaluate $32 - 14$.

1. Decompose each number by breaking it up into place value parts (tens and ones in this case).

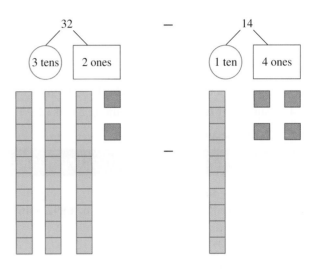

2. Since we can't take 4 ones from 2 ones, we instead borrow a block of 10 and split them into ones.

Borrowing step:

3. Rewrite 32 as 2 tens and 12 ones.
   Note: Effectively we have borrowed a ten from the tens column and given it to the ones column so that we can work out the subtraction.

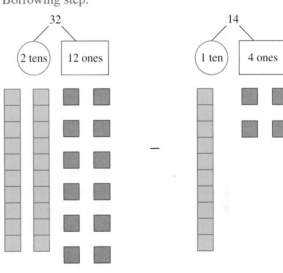

4. We can now complete the subtraction.
   - Subtracting the ones first gives:
     $12 - 4 = 8$ ones
   - Subtracting the tens next gives:
     $2 - 1 = 1$ ten
   - This gives us 1 ten and 8 ones left over after the subtraction.
   - This is the same as 18 individual blocks.

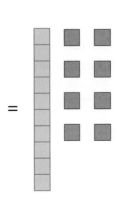

5. State the answer.

$$32 - 14 = 18$$

Algorithmic approach to subtraction using decomposition

The subtraction $32 - 14$ can be solved using the following **algorithm** (set of steps).

1. Set out the difference in columns according to place value.

$$\begin{array}{r} 3\ 2 \\ -\ 1\ 4 \\ \hline \end{array}$$

2. Because 4 can't be subtracted from 2, take one 10 from the tens column of the larger number and add it to the ones column of the same number. So 2 becomes 12, and the 3 tens become 2 tens.

$$\begin{array}{r} {}^2\cancel{3}\ {}^1 2 \\ -\ 1\ 4 \\ \hline 1\ 8 \end{array}$$

3. Subtract 4 units from 12 units ($12 - 4 = 8$).
4. Subtract 1 ten from the remaining 2 tens ($2 - 1 = 1$).
5. Write the answer.

$$32 - 14 = 18$$

WORKED EXAMPLE 12 Subtracting numbers using decomposition

Subtract the following:
a. $154 - 28$ b. $6892 - 467$ c. $3000 - 467$

THINK

a. 1. Since 8 cannot be subtracted from 4, take one 10 from the tens column of the larger number and add it to the ones column of the same number. The 4 ones become 14 ones and the 5 tens become 4 tens.
  2. Subtract 8 ones from the 14 ones ($14 - 8 = 6$).
  3. Subtract 2 tens from the 4 remaining tens ($4 - 2 = 2$).
  4. Subtract 0 hundreds from 1 hundred ($1 - 0 = 1$).
  5. State the answer.

b. 1. Since 7 cannot be subtracted from 2, take one ten from the tens column of the larger number and add it to the units column of the same number. The 2 becomes 12 and the 9 tens become 8 tens.
  2. Subtract the 7 units from the 12 units ($12 - 7 = 5$).
  3. Subtract 6 tens from the 8 remaining tens ($8 - 6 = 2$).
  4. Subtract 4 hundreds from the 8 hundreds ($8 - 4 = 4$).
  5. Subtract 0 thousands from the 6 thousands ($6 - 0 = 6$).
  6. State the answer.

c. 1. Since 7 cannot be taken from 0, 0 needs to become 10.
  2. We cannot take 10 from the tens column because it is also 0. The first column that we can take anything from is the thousands, so 3000 is decomposed to 2 thousands, 9 hundreds, 9 tens and 10 units.
  3. Subtract the units ($10 - 7 = 3$).
  4. Subtract the tens ($9 - 6 = 3$).
  5. Subtract the hundreds ($9 - 4 = 5$).
  6. Subtract the thousands ($2 - 0 = 2$).
  7. State the answer.

WRITE

a.
$$\begin{array}{r} 1\ {}^4\cancel{5}\ {}^1 4 \\ 2\ 8 \\ \hline 1\ 2\ 6 \end{array}$$

$$154 - 28 = 126$$

b.
$$\begin{array}{r} 6\ 8\ {}^8\cancel{9}\ {}^1 2 \\ -\ \ \ 4\ 6\ 7 \\ \hline 6\ 4\ 2\ 5 \end{array}$$

$$6892 - 467 = 6425$$

c.
$$\begin{array}{r} {}^2\cancel{3}\ {}^9\cancel{0}\ {}^9\cancel{0}\ {}^1 0 \\ -\ \ \ 4\ 6\ 7 \\ \hline 2\ 5\ 3\ 3 \end{array}$$

$$3000 - 467 = 2533$$

Digital technology

Scientific calculators can evaluate sums of two or more numbers.

```
[D]            Math ▲
17245+378+936
                18559
```

Scientific calculators can evaluate differences between two numbers.

```
[D]            Math ▲
25487-8954
                16533
```

 Resources

Interactivities Addition of positive integers (int-3922)
Subtraction of positive integers (int-3924)

Exercise 1.4 Algorithms for adding and subtracting positive integers

learnon

| **1.4 Quick quiz** | **1.4 Exercise** |

Individual pathways

■ PRACTISE	■ CONSOLIDATE	■ MASTER
1, 2, 6, 9, 12, 13, 18, 22, 25	3, 5, 7, 14, 15, 17, 19, 23, 26	4, 8, 10, 11, 16, 20, 21, 24, 27, 28

Fluency

1. Answer these questions by doing the working in your head.
   a. $7 + 8$
   b. $20 + 17$
   c. $195 + 15$
   d. $227 + 13$
   e. $1000 + 730$

2. Answer these questions by doing the working in your head.
   a. $17\,000 + 1220$
   b. $125\,000 + 50\,000$
   c. $2 + 8 + 1 + 9$
   d. $6 + 8 + 9 + 3 + 2 + 4 + 1 + 7$
   e. $12 + 5 + 3 + 7 + 15 + 8$

3. Add the following numbers, setting them out in columns as shown. Check your answers using a calculator.

   a.
   ```
      34
   + 65
   ─────
   ```

   b.
   ```
   68 069
      317
        8
   + 4254
   ───────
   ```

   c.
   ```
      399
     1489
     2798
   + 8943
   ───────
   ```

4. **WE9** Arrange the following numbers in columns, then add them.

   a. $137 + 841$
   b. $149 + 562 + 55$
   c. $376 + 948 + 11$
   d. $8 + 12\,972 + 59 + 1423$
   e. $1\,700\,245 + 378 + 930$

   Check your answers with a calculator.

5. **WE10** Mentally perform each of the following additions by pairing suitable numbers together.

   a. $56 + 87 + 24 + 13$
   b. $74 + 189 + 6 + 11$
   c. $98 + 247 + 305 + 3 + 95 + 42$
   d. $180 + 364 + 59 + 141 + 47 + 20 + 16$

6. Answer the following questions without using a calculator.

   a. $11 - 5$
   b. $53 - 30$
   c. $100 - 95$
   d. $150 - 25$
   e. $1100 - 200$

7. Work out the following without using a calculator.

   a. $1700 - 1000$
   b. $100 - 20 - 10$
   c. $1000 - 50 - 300 - 150$
   d. $24 - 3 - 16$
   e. $54 - 28$

8. Calculate the following.

   a. $10 + 8 - 5 + 2 - 11$
   b. $40 + 15 - 35$
   c. $120 - 40 - 25$
   d. $15 + 45 + 25 - 85$
   e. $100 - 70 + 43$
   f. $1000 - 400 + 250 + 150 + 150$

9. **WE11&12** Work out the following subtractions.

   a. $167 - 132$
   b. $47\,836 - 12\,713$
   c. $642\,803 - 58\,204$

Understanding

10. Solve the following subtractions.

    a. $664 - 397$
    b. $12\,900 - 8487$
    c. $69\,000 - 3561$
    d. $2683 - 49$

    Check your answers using a calculator.

11. Work out the following subtractions.

    a. $70\,400 - 1003$
    b. $27\,321 - 25\,768$
    c. $812\,741 - 462\,923$
    d. $23\,718\,482 - 4\,629\,738$

    Check your answers using a calculator.

12. Hella was performing in a ballet and needed to buy a tutu, ballet shoes and white tights. Based on the prices shown in the photograph, calculate how much she spent in total on her costume.

$150

$12

$85

13. A print dictionary is split into two volumes. There are 1544 pages in the A–K volume and 1488 pages in the L–Z volume. Calculate how many pages the dictionary has in total.

14. Nathan has taken his sister and her friends to lunch for her birthday. Look at the photograph. Determine what the total cost of the items shown will be.

15. Of all the world's rivers, the Amazon in South America and the Nile in Africa are the two longest. The Amazon is 6437 kilometres in length and the Nile is 233 kilometres longer than the Amazon. Calculate the length of the Nile.

16. An **arithmagon** is a triangular figure in which the two numbers at the end of each line add to the number along the line, like in the example shown.

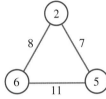

Solve each of these arithmagons.

a.

b.

c.

17. The following photographs show three of the highest waterfalls in the world. Determine how much higher:
    a. Victoria Falls is than Iguazu Falls
    b. Iguazu Falls is than Niagara Falls
    c. Victoria Falls is than Niagara Falls.

18. Prithvi received a box of 36 chocolates. They ate 3 on Monday, 11 on Tuesday and gave 7 away on Wednesday. Calculate how many they had left.

19. A crowd of 24 083 attended an NRL match between Canterbury-Bankstown Bulldogs and St George Illawarra Dragons. If 14 492 people supported the Bulldogs and the rest supported the Dragons, calculate how many supporters the Dragons had.

20. A school bus left Laurel High School with 31 students aboard. Thirteen of these passengers disembarked at Hardy Railway Station. The bus collected 24 more students at Hardy High School and a further 11 students disembarked at Laurel Swimming Pool. Calculate how many students were still on the bus.

21. Shu-Ling and Ty were driving from Melbourne to Sydney for a holiday. The distance between Melbourne and Sydney via the Hume Highway is 867 kilometres, but they chose the more scenic Princes Highway even though the distance is 1039 kilometres. They drove to Lakes Entrance on the first day (339 kilometres), a further 347 kilometres to Narooma on the second day, and arrived in Sydney on the third day.

    a. Calculate how much further it is from Melbourne to Sydney via the Princes Highway than via the Hume Highway.
    b. Determine how far Shu-Ling and Ty travelled on the third day.

Communicating, reasoning and problem solving

22. Fill in the missing numbers. The * can represent any digit.

    a.
    $$\begin{array}{r} 6*8*2* \\ -\,488417 \\ \hline *499*4 \\ \hline \end{array}$$

    b.
    $$\begin{array}{r} 3*9* \\ -\,*6*5 \\ \hline 1*07 \\ \hline \end{array}$$

23. In less than 10 seconds, without using a calculator, calculate the answer to $6\,849\,317 - 999\,999$. Explain how you reached your answer.

24. A beetle has fallen into a hole that is 15 metres deep. It is able to climb a distance of 3 metres during the day, but at night the beetle is tired and must rest. However, during the night it slides back 1 metre. How many days will it take the beetle to reach the top of the hole to freedom? Explain your answer.

25. Five termites start munching through a log.
    Pixie is 19 mm ahead of Bitsie.
    Trixie is 6 mm ahead of Mixie.
    Mixie is twice as far ahead as Itsie.
    Itsie is 10 mm behind Bitsie.
    Mixie has eaten through 52 mm.

    Determine which termite has eaten the most. Show your working.

26. In a class of 22 students, each attends swimming practice or lifesaving, or both. If 16 students attend swimming practice and 9 attend lifesaving, work out how many attend both.

27. Zaina works in a restaurant where people work from Monday to Friday, on weekends only, or from Monday to Sunday. There are 15 employees at the restaurant. If 13 employees work every Monday and 14 employees work every Sunday, evaluate how many employees do not work from Monday to Sunday.

28. There are 20 chocolate bars on the table all containing only fruit, only nuts, or fruit and nuts together. If 12 chocolate bars contain nuts and 14 chocolate bars contain fruit, evaluate the number of chocolate bars that contain both fruit and nuts.

LESSON
1.5 Multiplying positive integers

LEARNING INTENTION

At the end of this lesson you should be able to:
- multiply positive integers using mental strategies and algorithms.

▶ 1.5.1 Multiplying positive integers

eles-3650

- **Multiplication** is a short way of performing repeated addition of the same number.
  For example, $3 + 3 + 3 + 3 + 3$ represents 5 groups of 3 and can be written as 5×3.
- The **product** is the answer you get when numbers are multiplied.
  For example, the product of 5 and 3 is 15 because $5 \times 3 = 15$.
- Multiplication is commutative because the order in which numbers are multiplied is not important.
  For example, $3 \times 8 = 8 \times 3$.

Multiplying using diagrams

- The following diagram shows 2 groups of balls with 4 balls in each group. There are 8 balls in total.
  This can be written as a multiplication.

$$2 \times 4 = 8$$

Group 1   Group 2

- Instead, what if we looked at the picture as 4 groups of balls with 2 balls in each group? Would we get the same answer? Count the balls and see. There are still 8 balls in total.

$$4 \times 2 = 8$$

Group 1   Group 2   Group 3   Group 4

Multiplying two numbers

When two numbers are multiplied, the order does not matter; the answer will be the same.

WORKED EXAMPLE 13 Using a diagram to multiply numbers

a. **Draw a diagram to show 2×6.**   b. **State what 2×6 is equal to.**

THINK

a. Draw a diagram with 2 groups of 6.

b. Count how many circles there are altogether.
Write the answer.

WRITE/DRAW

a. Group 1   Group 2

b. There are 12 circles.
$2 \times 6 = 12$

- To multiply small numbers, we learn our multiplication facts and families. We can also use diagrams.
- To multiply larger numbers, we need to use other methods.

Multiplying using expanded form

- In this method of multiplication the larger number being multiplied is broken up into expanded form first, then each component is multiplied by the smaller number.
  For example, here is how the expanded form can be used to calculate 215×4.
  The expanded form of $215 = 200 + 10 + 5$

$$215 \times 4 = (200 \times 4) + (10 \times 4) + (5 \times 4)$$
$$= 800 + 40 + 20$$
$$= 860$$

Multiplying using the area model

- The area model of multiplication involves using the expanded form of the larger number and constructing rectangles for each component, arranged according to place value.
  This diagram shows how to calculate 215×4 using an area model. (*Note:* This graph is not drawn to scale.)

▶ 1.5.2 The short multiplication algorithm

eles-3651

- The short multiplication algorithm can be used when multiplying a large number by a single digit number.
- Start by lining up the numbers in columns, with one number under the other number. Place the larger number on top.
- Next, multiply each place value in the larger number by the smaller number, starting with the units column.
- Write the answer to each multiplication underneath the smaller number in the appropriate column.
- If an answer has two place values, carry the larger place value to the next column on the left and add it to the result of the next multiplication.

$$
\begin{array}{r}
2 \\
215 \\
\times\ \ 4 \\
\hline
860 \\
\hline
\end{array}
$$

1. $5 \times 4 = 20$
2. Write the 0 in the units column and carry the 2 over to the tens column.
3. Perform the next multiplication and add the carried number to the result.
   $1 \times 4 = 4;\ 4 + 2 = 6$
4. Write the 6 in the tens column.
5. $2 \times 4 = 8$
6. Write the 8 in the hundreds column.

WORKED EXAMPLE 14 Multiplying using short multiplication

Calculate 1456×5.

THINK

1. Write the numbers in columns according to place value, with the larger number on top.

2. Multiply the units by 5 ($5 \times 6 = 30$). Write down the 0 and carry the 3 to the tens column.

3. Multiply the tens by 5 and add the carried number ($5 \times 5 = 25; 25 + 3 = 28$). Write the 8 in the tens column and carry the 2 to the hundreds column.

4. Multiply the hundreds by 5 and add the carried number ($5 \times 4 = 20; 20 + 2 = 22$). Write the last 2 in the hundreds column and carry the other 2 to the thousands column.

5. Multiply the thousands by 5 and add the carried number ($5 \times 1 = 5;\ 5 + 2 = 7$). Write 7 in the thousands column of the answer.

6. Write the answer.

WRITE

$$
\begin{array}{r}
1\ 4\ 5\ 6 \\
\times\quad 5 \\
\hline
 \\
\hline
\end{array}
$$

$$
\begin{array}{r}
{}^2 1\,{}^2 4\,{}^3 5\ 6 \\
\times\qquad 5 \\
\hline
7\ 2\ 8\ 0 \\
\hline
\end{array}
$$

$1456 \times 5 = 7280$

▶ 1.5.3 The long multiplication algorithm

eles-3652

- The long multiplication algorithm is used to multiply numbers that both have more than one digit. The process is the same as that for short multiplication, but it is repeated for each additional digit.
- When multiplying two numbers that both have more than one digit, a 0 is added in the second row.

WORKED EXAMPLE 15 Multiplying whole numbers

Evaluate the following.
a. 547×6
b. 35×62

THINK	WRITE
a. 1. Set up the multiplication in columns, with the smaller number on the bottom.	a. $\begin{array}{r} 547 \\ \times\ 6 \end{array}$
2. Multiply each digit in 547 by 6. • $7 \times 6 = 42$. Write 2 and carry 4. • $4 \times 6 + 4 = 28$. Write 8 and carry 2. • $5 \times 6 + 2 = 32$. Write 32.	$\begin{array}{r} {}^2 5 {}^4 4\ 7 \\ \times\ 6 \\ \hline 3\ 2\ 8\ 2 \end{array}$
3. Write the answer.	$547 \times 6 = 3282$
b. 1. Set up the multiplication in columns.	b. $\begin{array}{r} 35 \\ \times\ 62 \end{array}$
2. Multiply 35 by 2. • $5 \times 2 = 10$. Write 0 and carry 1. • $3 \times 2 + 1 = 7$. $6 + 1 = 7$. Write 7.	$\begin{array}{r} {}^1 3\ 5 \\ \times\ 6\ 2 \\ \hline 7\ 0 \end{array}$
3. Add a 0 to the second row because you are multiplying by a number in the tens column.	$\begin{array}{r} {}^3 3\ 5 \\ \times\ 6\ 2 \\ \hline 7\ 0 \\ 0 \end{array}$
4. Multiply 35 by 6. • $5 \times 6 = 30$. Write 0 and carry 3. • $3 \times 6 + 3 = 21$. Write 21. 5. Add 70 and 2100 to get the answer.	$\begin{array}{r} {}^3 3\ 5 \\ \times\ 6\ 2 \\ \hline 7\ 0 \\ +2\ 1\ 0\ 0 \\ \hline 2\ 1\ 7\ 0 \end{array}$
6. Write the answer.	$35 \times 62 = 2170$

WORKED EXAMPLE 16 Multiplying using long multiplication

Calculate 1456×132 using long multiplication.

THINK	WRITE
1. Set the product up in columns according to place value, writing the larger number on top.	$\begin{array}{r} 1456 \\ \times\ \ 132 \\ \hline \end{array}$
2. Multiply the larger number by the units digit in the smaller number using short multiplication ($1456 \times 2 = 2912$). Write the answer directly below the problem.	$\begin{array}{r} 1\,^14\,^156 \\ \times\ \ 132 \\ \hline 2912 \end{array}$
3. Place a zero in the units column, as shown in pink, when multiplying the larger number by the tens digit of the smaller number ($1456 \times 3 = 4368$) This is because you are really working out $1456 \times 30 = 43\,680$. Write the answer directly below the previous answer.	$\begin{array}{r} ^11\,^14\,^156 \\ \times\ \ 132 \\ \hline 2912 \\ 43680 \end{array}$
4. Place zeros in the units and tens columns, as shown in pink, when multiplying the larger number by the hundreds digit of the smaller number ($1456 \times 1 = 1456$). This is because you are really working out $1456 \times 100 = 145\,600$. Write the answer directly below the previous answer.	$\begin{array}{r} 1456 \\ \times\ \ 132 \\ \hline 2912 \\ 43680 \\ +145600 \\ \hline 192192 \end{array}$
5. Add the numbers in each column of the three rows to determine the answer.	
6. Write the answer.	$1456 \times 132 = 192\,192$

- The **Distributive Law** applies only to multiplication. It states that $a(b + c) = ab + ac$. The Distributive Law helps with mental calculation because it means, for example, that 8×13 can be thought of as:

$$\begin{aligned} 8 \times 13 &= 8 \times (10 + 3) \\ &= 8 \times 10 + 8 \times 3 \\ &= 80 + 24 \\ &= 104 \end{aligned}$$

Digital technology

Scientific calculators can evaluate the multiplication of large numbers.

Note: Some scientific calculators display * for multiplication.

```
57835×2074              D    Math ▲

                    119949790
```

⊙ 1.5.4 Mental strategies for multiplication
eles-4838

- In many cases it is not practical to use pen and paper (or even a calculator) to perform multiplication.
- There are mental strategies that can make multiplication easier.
- Multiplication is associative — this means smaller numbers can be paired up for multiplication.
  For example, $2 \times 17 \times 5 = 17 \times (2 \times 5)$
  $$\begin{aligned} &= 17 \times 10 \\ &= 170 \end{aligned}$$
  This calculation has been simplified by finding the pair of numbers that multiply to equal 10.

WORKED EXAMPLE 17 Using mental strategies for multiplication

Use mental strategies to calculate $4 \times 23 \times 25$.

THINK	WRITE
1. Write the question.	$4 \times 23 \times 25$
2. Look for a number pair that makes a simpler multiplication and rearrange. $4 \times 25 = 100$, and multiplying by 100 is simpler than multiplying by either 4 or 25.	$= 23 \times (4 \times 25)$
3. Mentally calculate 4×25.	$= 23 \times 100$
4. Mentally calculate the final answer.	$= 2300$

WORKED EXAMPLE 18 Using mental strategies for multiplication

Use a mental strategy to calculate 34×200.

THINK	WRITE
1. Write the question.	34×200
2. Write 200 as 2×100.	$= 34 \times 2 \times 100$
3. Mentally calculate 34×2.	$= 68 \times 100$
4. Mentally calculate 68×100.	$= 6800$

Multiplying multiples of 10 or 100

If both numbers are multiples of 10, 100 and so on, ignore the zeros, multiply the remaining numbers, then add the total number of zeros to the answer.

For example, $900 \times 6000 = 5\,400\,000$.

Multiply 9 by 6 to get 54, then add the zeros after.

- Consider the multiplication 9×58. This multiplication can be regarded as $10 \times 58 - 1 \times 58$. Using this way of writing the multiplication, the answer can be mentally calculated by multiplying 58 by 10 and then subtracting 58 from the answer. This can be thought of as the 'multiply by 10' strategy.

WORKED EXAMPLE 19 Using mental strategies for multiplication

Use a mental strategy to calculate 77×9.

THINK	WRITE
1. Write the question.	77×9
2. Use the strategy of 'multiply by 10'.	$= 77 \times 10 - 77 \times 1$
3. Calculate 77×10 and then subtract 77.	$= 770 - 77$
	$= 693$

Exercise 1.5 Multiplying positive integers

learnon

1.5 Quick quiz	1.5 Exercise

Individual pathways

■ PRACTISE	■ CONSOLIDATE	■ MASTER
1, 2, 6, 8, 11, 13, 16, 17, 22, 23, 24, 27, 30, 33	3, 4, 9, 12, 14, 18, 19, 25, 26, 31, 34	5, 7, 10, 15, 20, 21, 28, 29, 32, 35, 36

Fluency

1. **WE13** Represent each of the following as a diagram and then use the diagram to calculate the product.
   a. 8×3 b. 6×9 c. 12×5 d. 13×2 e. 15×3

2. Calculate the product of each of the following without using a calculator.
   a. 25×4 b. 45×2 c. 16×2 d. 21×3 e. 54×2

3. Calculate the product of each of the following without using a calculator.
   a. 25×3 b. $3 \times 4 \times 6$ c. $3 \times 3 \times 3$ d. $5 \times 6 \times 3$ e. $8 \times 5 \times 2$

4. Calculate 267×3 by rewriting the larger number in expanded form.

5. Use an area model to:
   a. show 523×4
   b. calculate 523×4.

6. **WE14** Calculate the following using short multiplication.
   a. 16×8 b. 137×9 c. 857×3 d. 4920×5

7. Calculate the following using short multiplication.
   a. 7888×8 b. 2015×8 c. $10\,597 \times 6$ d. $41\,060 \times 12$
   Check your answers using a calculator.

8. **WE16** Calculate the following using long multiplication.
   a. 52×44 b. 97×31 c. 59×28

9. Calculate the following using long multiplication.
   a. 16×57 b. 850×76 c. 407×53

10. Calculate the following using long multiplication.
    a. $80\,055 \times 27$ b. $57\,835 \times 476$ c. 8027×215
    Check your answers using a calculator.

11. **WE17** Use mental strategies to calculate the following.
    a. $2 \times 8 \times 5$ b. $4 \times 19 \times 25$ c. $50 \times 45 \times 2$ d. $4 \times 67 \times 250$

12. **WE18** Use mental strategies to calculate the following.

    a. 45×20 b. 62×50 c. 84×200 d. 86×2000

13. Calculate each of the following.

    a. 200×40 b. 600×800 c. 1100×5000 d. $900\,000 \times 7000$

14. Calculate each of the following.

    a. 90×80 b. 800×7000 c. 9000×6000 d. $12\,000 \times 1100$

15. **WE19** Use mental strategies to calculate each of the following.

    a. 34×9 b. 628×9 c. 75×99 d. 26×8

16. a. Calculate 56×100.
    b. Calculate 56×10.
    c. Use your answers to parts a and b to calculate the answer to 56×90.

17. Use the method demonstrated in question 16 to calculate each of the following.

    a. 48×90 b. 125×90 c. 32×900

18. a. Calculate 25×6.
    b. Multiply your answer to part a by 2.
    c. Now calculate 25×12.

19. Use one of the methods demonstrated in this lesson to mentally calculate the value of each of the following.

    a. 15×12 b. 70×12 c. 40×16 d. 34×20

20. Answer the following questions.
    a. Calculate the value of 9×10.
    b. Calculate the value of 9×3.
    c. Calculate the value of 9×13.

21. Use one of the methods demonstrated in this lesson to mentally calculate the value of each of the following.

    a. 25×13 b. 24×13

Understanding

22. A school has eight Year 7 classes with 26 students in each class. Calculate how many students there are in Year 7 in total.

23. A shop owner sells 84 bananas each day. Calculate how many bananas are sold in 2 weeks.

24. Ezra wants to make a telephone call to their friend Aasuka, who lives in San Francisco. The call will cost \$3 per minute. If Ezra speaks to Aasuka for 24 minutes:

    a. calculate what the call will cost
    b. calculate what Ezra would pay if they made this call every month for 2 years.

25. Santilla is buying some generators. The generators cost \$12\,000 each. She needs 11 of them. Calculate how much they will cost her in total.

26. Julie was saving money to buy a digital camera. She was able to save $75 each month.

   a. Calculate how much she saved over 9 months.
   b. Calculate how much she saved over 16 months.
   c. If Julie continues to save money at the same rate, calculate how much she will save over a period of 3 years.

27. A car can travel 14 kilometres using 1 litre of fuel. Calculate how far it could travel with 35 litres of fuel.

28. In 1995 a team of British soldiers in Hamelin, Germany, constructed a bridge in the fastest time ever. The bridge spanned an 8-metre gap and took the soldiers 8 minutes and 44 seconds to build. Determine how many seconds in total it took to build the bridge.

29. Narissa does a paper round each morning before school. She travels 2 kilometres each morning on her bicycle, delivers 80 papers and is paid $35. She does her round each weekday.

   a. Calculate how far she travels in 1 week.
   b. Determine how much she gets paid in 1 week.
   c. Calculate how far she travels in 12 weeks.
   d. Calculate how much she would be paid over 52 weeks.
   e. Calculate how many papers she would deliver in 1 week.
   f. Determine how many papers she would deliver in 52 weeks.

Communicating, reasoning and problem solving

30. Jake says the product of 42 and 35 is 1428. His friend Ben worked out the answer to be 1470. State who is correct, Jake or Ben. Explain your answer.

31. Explain whether the mathematical statement $23 \times 4 > 20 \times 7$ is true.

32. Imagine a simplified form of a car numberplate that consists of one letter followed by one number, from A0 up to Z9. Determine how many numberplates could be issued under this system.
Remember that all numberplates issued have to be different.

33. Evaluate the product of 4, 16 and 9 using expanded form.

34. Consider numbers with two identical digits multiplied by 99. Determine:

$$11 \times 99, \ 22 \times 99, \ 33 \times 99$$

Can you see a pattern? Explain it.
Without using a calculator or long multiplication, write down the answers to the following.

   a. 44×99 b. 55×99 c. 66×99
   d. 77×99 e. 88×99 f. 99×99

35. Theo eats one of the strawberries in a basket, then he and his two friends split them evenly. If Theo's friends get 15 strawberries each, determine how many there were in the basket to begin with.

36. The product of two consecutive numbers is 1190. Evaluate the two numbers and their sum.

LESSON
1.6 Dividing positive integers

LEARNING INTENTION

At the end of this lesson you should be able to:
* divide positive integers using mental strategies and the short division algorithm.

▶ 1.6.1 Dividing positive integers

eles-3653

* **Division** is the process of sharing (or dividing) a number into equal parts. If 10 is divided into five parts, you can write this as $10 \div 5 = 2$.
* Division is the opposite (or inverse) of multiplication. To evaluate $56 \div 8$, you could ask the question, 'By which number do I multiply 8 to get 56?' $8 \times 7 = 56$, so the answer is 7.
* The commutative law does not apply to division. This means the order in which you divide numbers is important. For example, $10 \div 5$ is not the same as $5 \div 10$. 10 people in 5 groups is not the same as 5 people in 10 groups.
* A division can be written in a variety of ways, as shown.

$$10 \div 5 = 2$$

$$5 \overline{)10} \quad \text{(with } 2 \text{ above)}$$

The vinculum represents the division symbol. $\longrightarrow \dfrac{10}{5} = 2$

* Consider a box of chocolates that is going to be shared between a group of friends.

There are 20 chocolates in total and 4 people sharing the box. How can the chocolates be divided evenly between the 4 friends?

20 chocolates divided by 4 people means they will get 5 chocolates each.

We can write this mathematically as:

$20 \div 4 = 5$

Checking division by multiplying

A division can always be checked by multiplying.

If $20 \div 4 = 5$, then $4 \times 5 = 20$.

Dividing using expanded form

- One way to divide numbers is to first break the dividend (the number being divided) into expanded form and then divide each part by the divisor (the number of parts that the dividend is being divided into).

For example, calculate $636 \div 6$.
1. Write the dividend (636) in expanded form: $636 = 600 + 30 + 6$.
2. Divide each part by the divisor (6).
3. Add the results.

$$600 \div 6 = 100$$
$$30 \div 6 = 5$$
$$6 \div 6 = 1$$

So $636 \div 6 = 100 + 5 + 1 = 106$.

1.6.2 The short division algorithm

eles-3654

- Another way to divide numbers is to use an algorithm known as short division.
- Here is a demonstration of how to use this algorithm to calculate 636 divided by 6.

$$6\overline{)636}$$

$$6\overline{)6\boxed{3}6}^{\,1}$$
Divide 6 into 6. $6 \div 6 = 1$.
Write 1 on top.

$$6\overline{)6\boxed{3}^{3}6}^{\,10}$$
Divide 6 into 3.
6 cannot divide into 3.
Therefore 0 is placed on top.
3 is carried over to the next digit (6).

$$6\overline{)6\boxed{3}^{3}6}^{\,10\ 6}$$
When 3 is carried over to 6,
the new number formed is 36.
Divide 6 into 36. $36 \div 6 = 6$.
Write 6 on top.

- Sometimes there is a number left over after numbers have been divided. This number is called a **remainder**. The letter r is used to show that a number is a remainder.
A box of 9 chocolates needs to be shared evenly between 2 people. Because 9 cannot be divided exactly into 2 groups, each person will get 4 chocolates, with 1 chocolate left over.

$$9 \div 2 = 4 \text{ remainder } 1$$
$$\text{or}$$
$$9 \div 2 = 4 \text{ r } 1$$

WORKED EXAMPLE 20 Dividing using short division

Evaluate the following.

a. $834 \div 2$

b. $466 \div 3$

THINK

WRITE

a. 1. Write the equation as a short division.

a. $2\overline{)834}$

  2. Divide each digit by 2.
- $8 \div 2 = 4$. Write 4 on top.
- $3 \div 2 = 1$ r 1. Write 1 on top and carry the remainder of 1 to the next digit. The next digit becomes 14.
- $14 \div 2 = 7$. Write 7 on top.

$\begin{array}{r} 417 \\ 2\overline{)83^14} \end{array}$

  3. Write the answer.

$834 \div 2 = 417$

b. 1. Write the equation as a short division.

b. $3\overline{)466}$

  2. Divide each digit by 3.
- $4 \div 3 = 1$ r 1. Write 1 on top and carry the remainder of 1 to the next digit. The next digit becomes 16.
- $16 \div 3 = 5$ r 1. Write 5 on top and carry the remainder of 1 over to the next digit. The next digit becomes 16.
- $16 \div 3 = 5$ r 1. Write 5 r 1 on top.

$\begin{array}{r} 1\ 5\ 5\ r\ 1 \\ 3\overline{)4^16^16^1} \end{array}$

  3. Write the answer.

$466 \div 3 = 155\,r\,1$

WORKED EXAMPLE 21 Dividing using short division

Calculate $89\,656 \div 8$ using short division.

THINK

WRITE

1. Write the equation as a short division.

$8\overline{)89\,656}$

2. Divide 8 into the first digit and carry the remainder to the next digit. 8 divides into 8 once. Write 1 above the 8 as shown. There is no remainder.

$\begin{array}{r} 1\ 1\ 2\ 0\ 7 \\ 8\overline{)89^16\,5^56} \end{array}$

3. Divide 8 into the second digit and carry the remainder to the next digit. 8 divides into 9 once, with 1 left over. Write 1 above the 9 and carry 1 to the hundreds column.

4. Divide 8 into the third digit and carry the remainder to the next digit. 8 divides into 16 twice, with no remainder. Write 2 above the 6.

5. Divide 8 into the fourth digit and carry the remainder to the next digit. 8 doesn't divide into 5. Write 0 above the 5. Carry 5 to the next digit.

6. Divide 8 into 56. 8 divides into 56 seven times. Write 7 above the 6. There is no remainder.

7. Write the answer.

$89\,656 \div 8 = 11\,207$

⏵ 1.6.3 Dividing numbers that are multiples of 10

eles-3655

- When we are dividing numbers that are multiples of 10, an equivalent number of zeros can be cancelled from both the **dividend (numerator)** and the **divisor (denominator)**.
- The dividend is the quantity to be divided.
- The divisor is the number by which the dividend is to be divided.
- The **quotient** is the answer obtained when one number is divided by another number.

WORKED EXAMPLE 22 Dividing numbers that are multiples of 10

Calculate $48\,000 \div 600$. Note that both numbers are multiples of 10.

THINK	WRITE
1. Write the question.	$48\,000 \div 600$
2. Write the question as a fraction.	$= \dfrac{48\,0\cancel{0}\cancel{0}}{6\cancel{0}\cancel{0}}$
3. Cancel as many zeros as possible, crossing off the same number of zeros in both numerator and denominator.	$= \dfrac{480}{6}$
4. Perform the division.	$\begin{array}{r} 0\,8\,0 \\ 6\overline{)4^{4}80} \end{array}$
5. Write the answer.	$48\,000 \div 600 = 80$

Note: The principle of associativity is not true for division. This means that, for example, $(80 \div 8) \div 2$ is not equivalent to $80 \div (8 \div 2)$. This is shown by the following.

$$(80 \div 8) \div 2 = 10 \div 2 \qquad\qquad 80 \div (8 \div 2) = 80 \div 4$$
$$= 5 \qquad\qquad\qquad\qquad\qquad\qquad = 20$$

Digital technology

Scientific calculators can evaluate the division of large numbers.

```
27768÷24        D    Math ▲

                      1157
```

COMMUNICATING — COLLABORATIVE TASK: Magic division

1. a. Choose a digit from 2 to 9. Write it 6 times to form a 6-digit number. For example, if you choose 4 your 6-digit number will be 444 444.
   b. Divide your 6-digit number by 33.
   c. Divide your result from part **b** by 37.
   d. Divide your result from part **c** by 91.
   e. What do you notice about your result from part **d**?
2. Repeat question **1** with a different 6-digit number and explain how the 'magic division' works.

Exercise 1.6 Dividing positive integers

learn on

1.6 Quick quiz on	1.6 Exercise

Individual pathways

■ PRACTISE	■ CONSOLIDATE	■ MASTER
1, 6, 8, 12, 15, 18	2, 4, 7, 9, 13, 14, 16, 19	3, 5, 10, 11, 17, 20

Fluency

1. Calculate these divisions without using a calculator. There should be no remainder.

   a. $24 \div 6$　　　　　b. $36 \div 9$　　　　　c. $96 \div 12$
   d. $56 \div 7$　　　　　e. $26 \div 2$　　　　　f. $45 \div 15$

2. Calculate these divisions without using a calculator. There should be no remainder.

   a. $27 \div 3 \div 3$　　　　b. $96 \div 8 \div 6$　　　　c. $48 \div 12 \div 2$
   d. $56 \div 7 \div 4$　　　　e. $100 \div 2 \div 10$　　　f. $90 \div 3 \div 2$

   Check your answers using multiplication.

3. Perform the following calculations, which involve a combination of multiplication and division. Remember to work from left to right.

   a. $4 \times 5 \div 2$　　　　b. $80 \div 10 \times 7$　　　　c. $144 \div 12 \times 7$
   d. $120 \div 10 \times 5$　　　e. $121 \div 11 \times 4$

4. **WE20** Calculate the following using short division.

   a. $3\overline{)1455}$　　　b. $7\overline{)43\,456}$　　　c. $11\overline{)30\,371}$　　　d. $8\overline{)640\,360}$

5. **WE21** Calculate the following using short division.

   a. $3\overline{)255\,194}$　　　b. $6\overline{)516\,285}$　　　c. $7\overline{)6\,328\,520}$　　　d. $8\overline{)480\,594}$

   Check your answers using a calculator.

6. Divide the following numbers.

   a. $4200 \div 6$　　　　b. $210 \div 30$　　　　c. $720\,000 \div 800$

7. Divide the following numbers, which are multiples of 10.

   a. $4\,000\,000 \div 8000$　　　b. $600\,000 \div 120$　　　c. $480\,000 \div 600$

8. Calculate $144 \div 2$ by breaking 144 up into expanded form.

9. Calculate $642 \div 3$ by breaking 642 up into expanded form.

10. Calculate $4256 \div 8$ by breaking 4256 up into expanded form.

Understanding

11. Kelly works part time at the local pet shop. Last year she earned $2496.

    a. Calculate how much Kelly earned each month.
    b. Calculate how much Kelly earned each week.

12. At the milk processing plant, the engineer asked Farid how many cows he had to milk each day. Farid said he milked 192 cows, because he had 1674 litres of milk at the end of each day and each cow produced 9 litres. Determine whether Farid really milks 192 cows each day. If not, calculate how many cows he does milk.

13. When Juan caters for a celebration like a wedding, he confirms the arrangements with the client by filling out a form. Juan has been called to answer the phone, so it has been left to you to fill in the missing details. Copy and complete this planning form.

Celebration type	Wedding
Number of guests	152
Number of people per table	8
Number of tables required	
Number of courses for each guest	4
Total number of courses to be served	
Number of courses each waiter can serve	80
Number of waiters required	
Charge per guest	$55
Total charge for catering	

14. Janet is a land developer. She has bought 10 450 square metres of land. She intends to subdivide the land into 11 separate blocks.

    a. Calculate how many square metres each block will be.
    b. If she sells each block for $72 250, determine how much she will receive for the subdivided land.

Communicating, reasoning and problem solving

15. Kayla divides 243 by 9 and works out the answer to be 27. Her friend Tristan calculates the answer to be 26. State whether either Kayla or Tristan calculated the correct answer.

16. Explain whether this statement is true: $1148 \div 4 = 282$. Justify your answer by breaking 1148 up into expanded form.

17. Some people like to keep mice as pets. Female mice give birth to about six litters of babies per year, with about six babies in each litter.

  a. If you own a female mouse, how many pet mice will you have at the end of one year if none of the female's babies have any litters of their own? Explain how you reached your answer.

  b. Explain how many female mice you had at the start of the year if, at the end of the year, you had 111 mice and none of your females' babies had any litters of their own. Show your working.

18. Danh has booked a beach house for a week over the summer period for a group of 12 friends. The house costs $1344 for the week. If all 12 people stayed for 7 nights, determine how much each person will pay per night.

19. A total of 62 people have been invited to a dinner party. Each table can seat 8 people. Determine how many tables are needed.

20. List the first four whole numbers that will have a remainder of 2 when you divide by 8. Describe the pattern and explain how you found these numbers.

LESSON
1.7 Rounding and estimating numbers

▶ 1.7.1 Rounding integers

eles-3656

- Numbers or integers can be rounded to different degrees of accuracy.

Rounding to the nearest 10

- To round to the nearest 10, think about which multiple of 10 the number is closest to. For example, if 34 is rounded to the nearest 10, the result is 30 because 34 is closer to 30 than it is to 40.

34 ≈ 30

(Note: The symbol ≈ means 'is approximately equal to'.)

- When the number is halfway between, such as 35 in the example above, we always round up. 35 would round to 40 in that case.

Rounding to the nearest 100

- To round to the nearest 100, think about which multiple of 100 the number is closest to. For example, if 177 is rounded to the nearest 100, the result is 200 because 177 is closer to 200 than it is to 100.

Rounding to the first (or leading) digit

- To round to the first (or leading) digit, use the following guidelines:
  - Consider the digit after the leading one (i.e. the second digit).
  - If the second digit is 0, 1, 2, 3 or 4, the first digit stays the same and all the following digits are replaced with zeros.
  - If the second digit is 5, 6, 7, 8 or 9, the first digit is raised by 1 (rounded up) and all the following digits are replaced with zeros.
- For example, if 2345 is rounded to the first digit, the result is 2000 because 2345 is closer to 2000 than it is to 3000.

2345 ≈ 2000

Round the following numbers to the first (or leading) digit.

a. **2371**

b. **872**

THINK

a. Since the second digit (3) is less than 5, leave the leading digit unchanged and replace all other digits with zeros.

b. Since the second digit (7) is greater than 5, add 1 to the leading digit and replace all other digits with zeros.

WRITE

a. $2371 \approx 2000$

b. $872 \approx 900$

1.7.2 Estimation

eles-3657

- An estimate is not the same as a guess, because an estimate is based on information.
- When you do not need to know an exact amount, an estimate or approximation is enough.
- To estimate the answer to a mathematical problem, round the numbers to the first digit and find an approximate answer.
- Estimations can be made when multiplying, dividing, adding or subtracting. They can also be used when there is more than one operation in the same question.

Estimate $48\,921 \times 823$ by rounding each number to the first digit.

THINK

1. Write the question.

2. Round each part of the question to the first digit.

3. Multiply

4. Write the answer.

WRITE

$48\,921 \times 823$

$\approx 50\,000 \times 800$

$\approx 40\,000\,000$

$48\,921 \times 823 \approx 40\,000\,000$

DISCUSSION

Is the estimated answer to the multiplication in Worked example 24 higher or lower than the actual answer? Justify your response.

 Resources

 Interactivity Rounding (int-3932)

Exercise 1.7 Rounding and estimating numbers

1.7 Quick quiz on	1.7 Exercise

Individual pathways

■ PRACTISE	■ CONSOLIDATE	■ MASTER
1, 4, 5, 6, 15, 18	2, 7, 8, 13, 14, 16, 19	3, 9, 10, 11, 12, 17, 20

Fluency

1. **WE23** Round each of the following to the first (or leading) digit.

   a. 6
   b. 45
   c. 1368
   d. 12 145
   e. 168 879
   f. 4 985 452

2. **WE24** Estimate $67\,451 \times 432$ by rounding each number to the first digit.

3. Copy and complete the following table by rounding the numbers to the first digit. The first row has been completed as an example.
   - In the 'Estimate' column, round each number to the first digit.
   - In the 'Estimated answer' column, calculate the answer.
   - In the 'Prediction' column, guess whether the actual answer will be higher or lower than your estimate.
   - Use a calculator to work out the actual answer and record it in the 'Calculation' column to determine whether it is higher or lower than your estimate.

		Estimate	Estimated answer	Prediction	Calculation
Example	$4129 \div 246$	$4000 \div 200$	20	Lower	16.784 553, so lower
a.	$487 + 962$				
b.	$33\,041 + 82\,629$				
c.	$184\,029 + 723\,419$				
d.	$93\,261 - 37\,381$				
e.	$321 - 194$				
f.	$468\,011 - 171\,962$				
g.	36×198				
h.	$623 \times 12\,671$				
i.	$29\,486 \times 39$				
j.	$31\,690 \div 963$				
k.	$63\,003 \div 2590$				
l.	$69\,241 \div 1297$				

4. **MC** Select which of the following is the best estimate of $4372 + 2587$.

   A. 1000
   B. 5527
   C. 6000
   D. 7000

5. **MC** Choose which of the following is the best estimate of 672×54.

   A. 30 000
   B. 35 000
   C. 36 000
   D. 40 000

6. **MC** Select which of the following is the best estimate of $67\,843 \div 365$.

   A. 150
   B. 175
   C. 200
   D. 230

Understanding

7. Estimate the answers to each of the following by rounding to the first digit.
   - a. $5961 + 1768$
   - b. $48\,022 \div 538$
   - c. 9701×37
   - d. $98\,631 + 608\,897$
   - e. $6501 + 3790$

8. Estimate the answers to each of the following by rounding to the first digit.
   - a. $11\,890 - 3642$
   - b. $112\,000 \times 83$
   - c. $66\,501 \div 738$
   - d. $392 \times 113\,486$
   - e. $12\,476 \div 24$

Questions 9 to 12 relate to the following information.

Su-Lin was using her calculator to answer some mathematical questions, but she got a different answer each time she performed the same calculation. Using your estimation skills, predict which of Su-Lin's answers is most likely to be correct.

9. **MC** 217×489
   - A. 706
   - B. 106 113
   - C. 13 203
   - D. 105 203

10. **MC** $89\,344 \div 256$
    - A. 39
    - B. 1595
    - C. 89 088
    - D. 349

11. **MC** 78×6703
    - A. 522 834
    - B. 52 260
    - C. 6781
    - D. 56 732 501

12. **MC** $53\,669 \div 451$
    - A. 10
    - B. 1076
    - C. 53 218
    - D. 119

13. Kody is selling tickets for his school's theatre production of *South Pacific*. So far he has sold 439 tickets for Thursday night's performance, 529 for Friday's and 587 for Saturday's. The costs of the tickets are $9.80 for adults and $4.90 for students.
    - a. Round the figures to the first digit to estimate the number of tickets Kody has sold so far.
    - b. If approximately half the tickets sold were adult tickets and the other half were student tickets, estimate how much money has been received so far by rounding the cost of the tickets to the first digit.

14. For the following calculations:
    - i. estimate the answer by rounding to the first digit
    - ii. calculate the answer using a written method for parts a and b and a calculator for part c.
    - a. $6650 - 1310$
    - b. 36×223
    - c. $18\,251 \div 391$

Communicating, reasoning and problem solving

15. Ava wants to buy three items that cost $5.58, $15.92 and $7.22. Estimate the total cost of the three items by rounding each number to the nearest whole number, and then adding.

16. a. Explain how you can estimate the number of graduating students shown in the photograph. (Do not count all of the students in the photograph.)
    b. If the hall where they are graduating holds 12 times this number, estimate the total capacity of the hall.

17. At Kooboora Secondary College there are 127 students in Year 7, 152 students in Year 8 and 319 students in Years 9 and 10 combined. Estimate the total number of students in Years 7–10.

18. Describe a situation in everyday life where it is *not* appropriate to estimate values.

19. Shari bought a car five years ago for $21 798. The car's value has decreased to one-third of its initial value. Determine the approximate value of the car today.

20. During the intermission of the school production of *South Pacific*, Jia is planning to run a stall selling hamburgers to raise money for the school. She has priced the items she needs and made a list in order to estimate her expenses.
    a. By rounding the item price to the first digit, use the following table to estimate how much each item will cost Jia for the quantity she requires.
    b. Estimate what Jia's total shopping bill will be.
    c. If Jia sells 300 hamburgers over the 3 nights for $2 each, determine how much money she will receive for the hamburgers.
    d. Determine approximately how much money Jia will raise through selling hamburgers over the 3 nights.

Item	Item price	Quantity required	Estimated cost
Bread rolls	$2.90/dozen	25 packets of 12	
Hamburgers	$2.40/dozen	25 packets of 12	
Tomato sauce	$1.80/litre	2 litres	
Margarine	$2.20/tub	2 tubs	
Onions	$1.85/kilogram	2 kilograms	
Tomatoes	$3.50/kilogram	2 kilograms	
Lettuce	$1.10 each	5 lettuces	

LESSON
1.8 Order of operations

LEARNING INTENTION

At the end of this lesson you should be able to:
- understand how the order of operations is to be applied
- apply the order of operations to solve problems.

▶ 1.8.1 BIDMAS

eles-3658

- There are rules in mathematics that ensure everyone has the same understanding of mathematical operations. These rules are known as the **order of operations**.
- The acronym **BIDMAS** (**B**rackets, **I**ndices or roots, **D**ivision and **M**ultiplication, **A**ddition and **S**ubtraction) can be used to remember the correct order of operations.

Note: Division and multiplication have the same order as each other so are to be performed left to right. Addition and subtraction have the same order as each other and also should be performed left to right.

BIDMAS

BIDMAS helps us to remember the correct order in which we perform the various operations, working from left to right.

- It is important that you follow BIDMAS to get the correct answer.
  Take the example of $15 - 10 \div 5$:

Following BIDMAS:	Following left to right:
$15 - 10 \div 5$ $15 - 2$ $13 \checkmark$	$15 - 10 \div 5$ $5 \div 5$ $1 \times$

WORKED EXAMPLE 25 Applying the order of operations

Calculate $6 + 12 \div 4$.

THINK

1. Write the question.

2. Follow BIDMAS.
   - Brackets: there are none.
   - Indices or roots: there are none.
   - Division and multiplication: there is one division.
   - Addition and subtraction: there is one addition.

3. Apply BIDMAS by working out the division before the addition.

4. Calculate the answer.

WRITE

$6 + 12 \div 4$

$6 + 12 \div 4 = 6 + 3$

$= 9$

WORKED EXAMPLE 26 Applying the order of operations

Calculate the following.

a. $12 \div 2 + 4 \times (4 + 6)$

b. $80 \div [((11 - 2) \times 2) + 2]$

THINK

a. 1. Write the question.

2. Apply BIDMAS. Remove the brackets by working out the addition inside.

3. Perform the division and multiplication next, doing whichever occurs first as you work from left to right. Work out $12 \div 2$ first, then work out 4×10.

4. Complete the addition last and calculate the answer.

WRITE

a. $12 \div 2 + 4 \times (4 + 6)$

$= 12 \div 2 + 4 \times 10$

$= 12 \div 2 + 4 \times 10$
$= 6 + 40$

$= 46$

b. 1. Write the question.

2. Remove the innermost brackets first by working out the subtraction inside them.

3. Remove the next pair of brackets by working out the multiplication inside them.

4. Remove the final pair of brackets by working out the addition inside them.

5. Perform the division last and calculate the answer.

b. $80 \div [((11 - 2) \times 2) + 2]$

$= 80 \div [(9 \times 2) + 2]$

$= 80 \div [18 + 2]$

$= 80 \div 20$

$= 4$

COMMUNICATING — COLLABORATIVE TASK: Order of operations

Using the expressions from Worked example 26, investigate whether the order of operations is applied on different devices such as your mobile phone or a computer. Do you have to take any special steps to ensure that the order of operations is followed?

WORKED EXAMPLE 27 Inserting brackets to make a statement true

Insert one set of brackets in the appropriate place to make the following statement true.
$3 \times 10 - 8 \div 2 + 4 = 7$

THINK	WRITE
1. Write the left-hand side of the equation.	$3 \times 10 - 8 \div 2 + 4$
2. Place one set of brackets around the first two values.	$= (3 \times 10) - 8 \div 2 + 4$
3. Perform the multiplication inside the bracket.	$= 30 - 8 \div 2 + 4$
4. Perform the division.	$= 30 - 4 + 4$
5. Perform the subtraction and addition working from left to right. *Note*: Since this is not the answer, the above process must be repeated.	$= 26 + 4$ $= 30$ Since this is not equal to 7 we must place the brackets in a different position.
6. Place one set of brackets around the second and third values.	$3 \times 10 - 8 \div 2 + 4$ $= 3 \times (10 - 8) \div 2 + 4$
7. Perform the subtraction inside the bracket.	$= 3 \times 2 \div 2 + 4$
8. Perform the multiplication and division working from left to right.	$= 6 \div 2 + 4$ $= 3 + 4$
9. Perform the addition last and calculate the answer.	$= 7$
10. Write the answer showing the bracket placement that makes the statement true.	$3 \times (10 - 8) \div 2 + 4 = 7$

WORKED EXAMPLE 28 Using the order of operations

Calculate $58 - (2 \times 8 + 3^2)$ using the order of operations.

THINK	WRITE
1. Write the question.	$58 - (2 \times 8 + 3^2)$
2. Apply BIDMAS.	
• Remove the brackets first.	
• Calculate the indices inside the brackets.	$= 58 - (2 \times 8 + 9)$
• Perform the multiplication inside the brackets.	$= 58 - (16 + 9)$
• Perform the addition inside the brackets.	$= 58 - (25)$
• Perform the subtraction.	
3. Write the answer.	$= 33$

Digital technology

Scientific calculators can evaluate the expression using the order of operations.

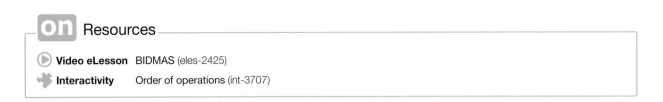

DISCUSSION

Use a calculator to work out the answer to Worked example 28. Do you think it is easier or harder to perform the calculation using a calculator? What do you think the advantages and disadvantages are of using a calculator compared to calculating using mental strategies?

on Resources

▶ **Video eLesson** BIDMAS (eles-2425)

Interactivity Order of operations (int-3707)

Exercise 1.8 Order of operations

1.8 Quick quiz on	1.8 Exercise

Individual pathways

■ PRACTISE	■ CONSOLIDATE	■ MASTER
1, 2, 6, 9, 14, 17, 20, 21	3, 7, 10, 12, 13, 16, 18, 22, 23	4, 5, 8, 11, 15, 19, 24, 25

Fluency

1. **WE25 & 26** Follow the order of operations rules and calculate each of the following.

   a. $3 + 4 \div 2$ b. $8 + 1 \times 1$ c. $24 \div (12 - 4)$

   d. $15 \times (17 - 15)$ e. $11 + 6 \times 8$

2. Follow the order of operations rules and calculate each of the following.

   a. $30 - 45 \div 9$ b. $56 \div (7 + 1)$ c. $12 \times (20 - 12)$ d. $(7 + 5) - (10 + 2)$

3. Follow the order of operations rules and calculate each of the following.

   a. $3 \times 4 + 23 - 10 - 5 \times 2$ b. $42 \div 7 \times 8 - 8 \times 3$

   c. $10 + 40 \div 5 + 14$ d. $81 \div 9 + 108 \div 12$

   e. $16 + 12 \div 2 \times 10$ f. $(18 - 15) \div 3 \times 27$

4. Follow the order of operations rules and calculate each of the following.

   a. $4 + (6 + 3 \times 9) - 11$ b. $52 \div 13 + 75 \div 25$

   c. $88 \div (24 - 13) \times 12$ d. $(4 + 5) \times (20 - 14) \div 2$

   e. $\{[(16 + 4) \div 4] - 2\} \times 6$ f. $60 \div \{[(12 - 3) \times 2] + 2\}$

Understanding

5. **WE27** Insert one set of brackets in the appropriate place to make these statements true.

   a. $12 - 8 \div 4 = 1$ b. $4 + 8 \times 5 - 4 \times 5 = 40$

   c. $3 \times 10 - 2 \div 4 + 4 = 10$ d. $12 \times 4 + 2 - 12 = 60$

   e. $10 \div 5 + 5 \times 9 \times 9 = 81$ f. $18 - 3 \times 3 \div 5 = 9$

6. Calculate the following, using the order of operations.

   a. $6 + 3 \times 4$ b. $18 - 12 \div 3$ c. $2 - 4 - 6$ d. $17 - 3 + 8$

7. Calculate the following, using the order of operations.

   a. $6 \times 3 \div 9$ b. $72 \div 8 \times 3$ c. $7 - (3 + 4)$ d. $(6 + 3) \div 9$

8. Calculate the following, using the order of operations.

   a. $3 \times 2 - 3 \times 1$ b. $6 \times 5 - 2 \times 6$ c. $4 \times (6 + 4)$ d. $(8 + 3) \times 7$

9. **WE28** Evaluate each of the following.

   a. $4 + 7 \times 3 - 2$ b. $6 - 4 + (4)^2$

   c. $(2)^3 - 3 \times 2$ d. $3 + (8 - 2) + 6$

10. Evaluate each of the following.

    a. $8 \div 2 + (2)^2$ b. $4 \times 8 - \left[2 + (3)^2\right]$

    c. $(7 + 5) + 24 \div 6$ d. $30 \div (5 - 2) - 10$

11. Evaluate each of the following.

    a. $54 \div 6 + 8 \times 9 \div 4$ b. $(9 + 6) \div 5 - 8 \times 0$

    c. $7 + 7 \div 7 \times 7 + 7$ d. $9 \times 5 - (3 + 2) - 48 \div 6$

12. **MC** $20 - 6 \times 3 + 28 \div 7$ is equal to:

A. 46   B. 10   C. 6   D. 4

13. **MC** Choose which of the following is the correct order of the two signs that can replace * in the equation $7*2*4 - 3 = 12$.

A. $-, +$ B. $+, \times$ C. $=, \div$ D. $\div, \times$

14. **MC** Select which of the following the expression $6 - 2 \times 5 + 10 \div 2$ is equal to:

A. 25   B. 40   C. 30   D. 1

15. Insert brackets (if necessary) to make each statement true.

a. $6 + 2 \times 4 - 3 \times 2 = 10$ b. $6 + 2 \times 4 - 3 \times 2 = 26$

c. $6 + 2 \times 4 - 3 \times 2 = 16$ d. $6 + 2 \times 4 - 3 \times 2 = 8$

16. Model each situation with integers, then calculate the result.

a. Jemma starts out with $274 in the bank. She makes 2 withdrawals of $68 each, then 3 deposits of $50 each. Calculate how much money Jemma now has in the bank.

b. If 200 boxes of apples were each 3 short of the stated number of 40 apples, calculate the overall shortfall in the number of apples.

c. A person with a mass of 108 kg wants to reduce his mass to 84 kg in 3 months. Determine what average mass reduction is needed per month.

Communicating, reasoning and problem solving

17. Discuss a real-life situation in which performing calculations in the wrong order could cause problems.

18. Keenan and Amona discovered that they had different answers to the same question, which was to calculate $6 + 6 \div 3$. Keenan thought the answer was 8. Amona thought the answer was 4. State who was correct, Keenan or Amona. Explain who made the error and why.

19. I have a bag of pebbles. Arranging them in bags of 7 leaves 1 extra pebble, and arranging them in bags of 5 leaves 3 extra pebbles. If there are more than 10 pebbles in my bag, determine the smallest number of pebbles possible for my bag to have in it. Show your working.

20. The student lockers at Jacaranda College are to be numbered consecutively from 1 to 800 using plastic digits. Each digit costs 10c. Calculate the total cost of all the digits.

21. Xanthe bakes and sells cupcakes. A customer ordered 3 lemon cupcakes at $4.25 each, 3 chocolate cupcakes at $4.75 each and 2 original cupcakes at $3.75 each.

a. Write an expression to help Xanthe calculate the amount the customer has to pay.

b. Use this expression to calculate the amount that the customer has to pay.

22. Local time in Sydney is 3 hours ahead of Singapore time, which is 5 hours behind Auckland (NZ) time. Auckland is 11 hours ahead of Berlin (Germany) time. Determine the time difference between:

   a. Sydney and Berlin
   b. Singapore and Berlin.

23. Minh was asked to buy enough ice creams for everyone in his class. There are 31 people in the class. At the supermarket, ice creams come in packets of 3 or packets of 5. Minh is told that the cost of one ice cream is the same from either packet. Determine how many packets (or what combination of packets) Minh should buy.

24. Use the digits 1, 2, 3 and 4 and the operators $+, -, \times$ and $\div$ to construct equations that result in the numbers 1 to 5 (the numbers 2 and 4 are already done for you). You must use each digit in each expression, and you may not use the digits more than once. You may combine the digits to form larger numbers (such as 21). You may also use brackets to make sure the operations are done in the correct order.

$$1 =$$
$$2 = 4 - 3 + 2 - 1$$
$$3 =$$
$$4 = 4 - 2 + 3 - 1$$
$$5 =$$

25. Use the digits 1, 2, 3 and 4 and the operators $+, -, \times$ and $\div$ to construct equations that result in the numbers 5 to 10 (the numbers 7 and 10 have already been done for you). You must use each digit in each expression and you may not use the digits more than once. You may combine the digits to form larger numbers (such as 21). You may also use brackets to make sure the operations are done in the correct order.

$$5 =$$
$$6 =$$
$$7 = 24 \div 3 - 1$$
$$8 =$$
$$9 =$$
$$10 = 1 + 2 + 3 + 4$$

LESSON
1.9 Review

1.9.1 Topic summary

Place value

The position of a digit in a number gives a different value to the digit.

For example, the place value of each digit in the number 613 452 is:

Hundred thousands	Ten thousands	Thousands	Hundreds	Tens	Ones
6	1	3	4	5	2

- The number 613 452 is read as *six hundred and thirteen thousand, four hundred and fifty-two*.

Rounding and estimating

To round to the first digit, use the following guidelines:
- If the second digit is 0, 1, 2, 3 or 4, the first digit stays the same and all the following digits are replaced with zeros.
- If the second digit is 5, 6, 7, 8 or 9, the first digit is increased by 1 (rounded up) and all the following digits are replaced with zeros.

COMPUTATION WITH POSITIVE INTEGERS

Positive integers

Positive integers are whole numbers including 0 without decimals or fractions.
- The integers 1, 2, 3, … are called positive integers.
- 0 is neither positive nor negative.

Addition of positive integers

Some mental strategies and algorithms used to add up numbers without a calculator are:
- jump strategy
- split strategy
- compensation strategy
- rearrange strategy
- algorithm to add positive integers.

e.g. 462 + 78 + 316

$$\begin{array}{r} {}^{1}4\ \ {}^{1}6\ \ 2 \\ 7\ \ 8 \\ 3\ \ 1\ \ 6 \\ \hline 8\ \ 5\ \ 6 \end{array}$$

Subtraction of positive integers

Some mental strategies and algorithms used to subtract numbers without a calculator are:
- jump strategy
- adding up to find the difference
- compensation strategy
- algorithm to subtract positive integers.

e.g. 892 – 467

$$\begin{array}{r} 8^{8}\ \ \cancel{9}^{1}\ \ 2 \\ 4\ \ 6\ \ 7 \\ \hline 4\ \ 2\ \ 5 \end{array}$$

Multiplication of positive integers

- A short way of performing repeated addition of the same number can be used.
- Expanded form, area model, short and long multiplication algorithms are used to multiply.

e.g.

Short multiplication

$$\begin{array}{r} {}^{2}1{}^{2}456 \\ \times\ \ \ \ 5 \\ \hline 7\ 2\ 8\ 0 \end{array}$$

Long multiplication

$$\begin{array}{r} 1\ 4\ 5\ 6 \\ \times\ \ \ 1\ 3\ 2 \\ \hline 2\ 9\ 1\ 2 \\ 4\ 3\ 6\ 8\ 0 \\ 1\ 4\ 5\ 6\ 0\ 0 \\ \hline 1\ 9\ 2\ 1\ 9\ 2 \end{array}$$

Division of positive integers

- A number can be shared or divided into equal parts.
- Expanded form and short division algorithm are used to divide.

e.g.

$$\begin{array}{r} 417 \\ 2\overline{)83^{1}4} \end{array}$$

Quotient

$$5\overline{)10}^{\ 2}$$

Divisor   Dividend

Order of operations

- The order of operations is a set of rules we must follow so that we all have a common understanding of mathematical operations.
- The set order in which we calculate problems is:
  1. **Brackets**   ( ) or [ ]
  2. **Indices or roots** a^x or $\sqrt[n]{x}$
  3. **Division and Multiplication** (working left to right)   ÷ or ×
  4. **Addition and Subtraction** (working left to right)   + or −
- The acronym **BIDMAS** can be used to remember the correct order of operations.

1.9.2 Project

Ancient number systems

All ancient civilisations developed methods to count and use numbers. The methods they used to represent numbers did not involve the ten digits 1, 2, 3, 4, 5, 6, 7, 8, 9 and 0 that we use today. Instead they involved different systems like letters, other symbols, pebbles, tying knots in ropes, or cutting notches in sticks. As the need to use and represent larger numbers became more and more important, many old systems fell out of use.

Egyptian numbers

The ancient Egyptians were one of the oldest known civilisations to have a recorded number system. In about 5000 BCE, Egyptians used a system of symbols called hieroglyphs. An example of hieroglyphs can be seen in the photograph.

The hieroglyphs used for numbers are shown in this table.

Number	1	10	100	1000	10 000	100 000	1 000 000	10 000 000
Symbol	ı	∩	℮	⬆	⟩	⟅	𓀀	☻

The following example shows how the symbols are used to represent numbers. As the numbers get larger in value, the symbols are simply repeated. This set of symbols represents 241 513.

$$\text{ꟼꟼ}\,\text{ııı}\,\text{ıı}\,\text{⬆}\,\text{℮℮℮℮℮}\,\text{∩ıııı}$$

1. What number is represented by the following Egyptian hieroglyphs?

2. Show how the ancient Egyptians would have represented the number 2147.
   The ancient Egyptian system of numbers was not limited to simply representing numbers. Addition and subtraction could also be performed when it was required. Perform the following additions and subtractions, giving your answers as Egyptian numbers.

3. $$\text{∩∩∩∩}\,\text{ıııı}\,\text{∩∩∩}\,\text{ıııı} + \text{℮}\,\text{∩∩∩}\,\text{ıııı}\,\text{∩∩∩}\,\text{ıııı}$$

4. $$\text{⬆}\,\text{℮℮℮}\,\text{∩∩∩}\,\text{℮℮℮}\,\text{∩∩∩}\,\text{ııı} + \text{ıı⬆⬆}\,\text{∩}\,\text{ıııı}$$

5. $$\text{∩∩∩∩}\,\text{ııı}\,\text{∩∩∩}\,\text{ııı} - \text{∩∩∩}\,\text{ıııı}\,\text{∩}\,\text{ııı}$$

6. ⬆ℯℯ⌒⌒⌒∩∩|| − ℯℯℯℯ ∩∩ |||| ℯℯℯℯ ∩∩ ||||

7. Comment on any difficulties you encountered when performing the calculations using Egyptian hieroglyphs.

Greek numbers

The ancient Greeks used an alphabet that had 27 letters (the modern Greek alphabet has only 24 letters). These letters were also used to represent numbers.

The first nine letters of the alphabet represented the numbers 1 to 9. The middle nine letters of the alphabet represented the tens from 10 to 90. The last nine letters of the alphabet represented the hundreds from 100 to 900. Numbers were written by combining these letters.

8. Search online to find the 27 symbols used in the ancient Greek number system and the number that each symbol represented.

9. How did the ancient Greeks represent numbers greater than or equal to 1000?

10. Search online to find out about number systems used by other ancient civilisations. Try to find out what symbols were used to represent numbers and how those symbols were used in calculations. Discuss any advantages of these systems and suggest why the system may have fallen out of use. Present your findings on a separate sheet of paper.

Maya Numbers

0	1	2	3	4	5	6
7	8	9	10	11	12	13
14	15	16	17	18	19	20

on Resources

  Interactivities Crossword (int-2586)

Sudoku puzzle (int-3163)

Exercise 1.9 Review questions

Fluency

1. State the place value of the digit shown in pink in each of the following.
   a. 74 037   **b.** 541 910   **c.** 1 904 000   **d.** 290

2. Write each of the following numbers in expanded form.
   a. 392   **b.** 4109   **c.** 42 001   **d.** 120 000

3. List the numbers 394, 349, 943, 934, 3994, 3499 in ascending order.

4. List the numbers 1011, 101, 110, 1100, 1101 in descending order.

5. Calculate the sum of these numbers.
   a. $43 + 84$ **b.** $139 + 3048$
   c. $3488 + 91 + 4062$ **d.** $3\,486\,208 + 38\,645 + 692\,803$

6. Calculate each of the following.
   a. $20 - 12 + 8 - 14$ **b.** $35 + 15 + 5 - 20$
   c. $300 - 170 + 20$ **d.** $18 + 10 - 3 - 11$

7. Calculate the difference between the following numbers.
   a. $688 - 273$ **b.** $400 - 183$
   c. $68\,348 - 8026$ **d.** $46\,234 - 8476$
   e. $286\,005 - 193\,048$

8. Use mental strategies to multiply each of the following.
   a. $2 \times 15 \times 5$ **b.** $4 \times 84 \times 25$ **c.** 62×20
   d. 67×9 **e.** 31×19

9. Evaluate the following multiplications using a diagram.
   a. 3×6 **b.** 5×4 **c.** 2×7

10. Calculate the following using expanded form.
    a. 24×4 **b.** 134×8 **c.** 642×5

11. Calculate the following using an area model or a method of your choice.
    a. 236×6 **b.** 347×4 **c.** 675×5

12. Calculate the following using the short multiplication algorithm.
    a. 128×9
    b. 762×3
    c. 3840×5

13. Calculate the following using the long multiplication algorithm.
    a. 64×72 **b.** 542×57 **c.** $50\,426 \times 364$

14. Calculate each of the following using expanded form.
    a. $645 \div 5$ **b.** $1040 \div 4$ **c.** $15\,063 \div 3$

▶

15. Calculate each of the following.

    a. $6 \times 4 \div 3$ **b.** $4 \times 9 \div 12$ **c.** $49 \div 7 \times 12$

    d. $81 \div 9 \times 5$ **e.** $6 \times 3 \div 9 \div 2$

16. Divide the following multiples of 10.

    a. $84\,000 \div 120$ **b.** $4900 \div 700$ **c.** $12\,300 \div 30$

17. Evaluate the following using short division.

    a. $4704 \div 3$ **b.** $6867 \div 7$ **c.** $3762 \div 7$

    d. $5702 \div 17$ **e.** $11\,852 \div 13$

18. By rounding each number to its first digit, estimate the answer to each of the following calculations.

    a. $6802 + 7486$ **b.** $8914 - 3571$

    c. $5304 \div 143$ **d.** 5706×68

    e. $49\,581 + 73\,258$ **f.** 289×671

19. **MC** Choose which of the following is an integer.

    A. $2\dfrac{1}{2}$ **B.** 0.45 **C.** 0 **D.** 201.50

Understanding

20. Complete each statement by inserting $=, <$ or $>$ in the box to make the statement correct.

    a. $6 - 4 \,\square\, 3 + 2$ **b.** $1 \,\square\, 7 \div 7$

    c. $5 - 4 \,\square\, 0 \times 5$ **d.** $100 - 50 \,\square\, 9 \times 5$

21. Calculate the following using a mental strategy.

    a. $315 + 12$ **b.** $1010 + 80$

    c. $735 + 54$ **d.** $154 + 62$

22. Calculate the following using a mental strategy.

    a. $152 - 25$ **b.** $135 - 54$

    c. $745 - 34$ **d.** $354 - 76$

23. Evaluate the following.

    a. $(6 - 2) \times 7$ **b.** $4 \times (4 + 2)$

    c. 2×5 **d.** $(10 - 6) \times 2$

    e. $7 + 8 \div 2$ **f.** $3 \times 5 + (6 \times 2)$

24. Calculate the following.

    a. $36 \div 3$ **b.** $210 \div 7$

    c. $45 \div 9$ **d.** $1800 \div 2$

    e. $672 \div 48$ **f.** $625 \div 25$

25. Complete the following by inserting $=, <$ or $>$ in the box to make the statement correct.

    a. $3.5 \,\square\, 3.75$ **b.** $22 \times 2 \,\square\, 44$

    c. $4 \times 3 \,\square\, 8 \times 0.5$ **d.** $5 \times 2 \,\square\, 15$

26. Complete the following by inserting $=, <$ or $>$ in the box to make the statement correct.

    a. $12 \div -4 \,\square\, 3$ **b.** $5 \times (2 + 10) \div 4 \,\square\, 25$

    c. $10 \times 6.5 \,\square\, 50$ **d.** $5 \times 2 - 6 \div 4 \,\square\, 2.5$

27. The lowest temperature recorded on a particular day in Melbourne was 4 °C and the highest was 27.5 °C. Calculate the difference between these two temperatures.

28. Follow the rules for the order of operations to calculate each of the following.
   a. $35 \div (12 - 5)$
   b. $11 \times 3 + 5$
   c. $8 \times 3 \div 4$
   d. $5 \times 12 - 11 \times 5$
   e. $(6 + 4) \times 7$
   f. $5 + [21 - (5 \times 3)] \times 4$

Communicating, reasoning and problem solving

29. Uluru is a sacred Aboriginal site. The following map shows some roads between Uluru and Alice Springs. The distances (in kilometres) along particular sections of road are indicated.
   a. Calculate how far Kings Canyon Resort is from Ayers Rock Resort near Uluru.
   b. Determine the shortest distance by road from Kings Canyon Resort to Alice Springs.
   c. If you are travelling in a hire car in this part of Australia, you must travel only on sealed roads. Work out the distance you need to travel if you are driving in a hire car from Kings Canyon Resort to Alice Springs.

30. In summer, an ice cream factory operates 16 hours a day and makes 28 ice creams an hour.

   a. Calculate how many ice creams are made each day.
   b. If the factory operates 7 days a week, determine how many ice creams are made in one week.
   c. If there are 32 staff who run the machines over a week, calculate how many ice creams each person would make.

31. When you add two even positive integers, the answer is even. The sum of an even and an odd positive integer is an odd integer. Two odd positive integers add to give an even integer.
    Determine what happens if you perform multiplication on these types of integers. Give examples to support your answers. Write a general statement to summarise what would happen if you multiplied more than two positive integers.

32. Complete the following subtraction by adding the missing digits to the boxes.

$$
\begin{array}{ccccc}
 & 8 & 2 & \square & 0 \\
- & 3 & \square & 8 & \square \\
\hline
 & \square & 5 & 5 & 9 \\
\hline
\end{array}
$$

33. A lolly factory machine dispenses 760 lollies into 8 jars. Assuming that an equal number of lollies is dispensed into each jar, evaluate how many lollies there are in each jar.

34. Julie sells 8 bottles of soft drink for $3 each and 12 bottles of water for $2 each.
   a. Write a calculation that will find the total value of Julie's sales.
   b. Calculate the total value of Julie's sales.

35. Dara estimates the crowd at a football match to be 20 000 people. If Dara's estimate is correct to the nearest 1000 people, evaluate the greatest number of people that could possibly be at the match.

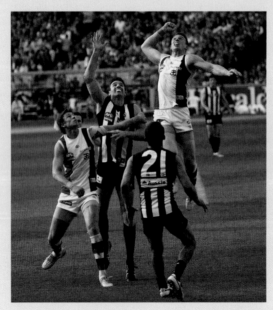

36. a. Complete the addition table below.

+	6		3
	11	15	
12			
	15		

b. Complete the multiplication table below.

×	4		6
	12	24	18
5			
	8		

37. Minimum and maximum temperatures are recorded at many locations around the world.

a. In Melbourne the temperature often changes rapidly during the day. On one very hot day the temperature reached 42 °C. A cool change arrived during the afternoon and the temperature dropped by 18 °C. Determine the temperature after the cool change.

b. In Alaska the weather often changes suddenly as storms sweep across the frozen plains. On one day, the temperature was 3 °C, but a storm caused the temperature to drop by 24 °C. Determine the temperature during the storm.

38. You have 30 questions to answer on a multiple choice test. Each correct answer scores one mark (+1). Each incorrect answer loses two marks (−2). Any unmarked questions will be counted as incorrect. Evaluate the lowest number of questions that you can answer correctly and still obtain a score greater than 0. Determine how many marks you would earn. Explain your reasoning.

39. Place the digits 1–9 (with no repeats) in the crossword shown so that the equations reading across and down are true.

40. The following table lists a business's profit over the past four years. Evaluate which two consecutive years had the greatest difference. Show your working.

Year	2016	2017	2018	2019
Profit/loss ($)	1020	2620	3808	8222

on To test your understanding and knowledge of this topic, go to your learnON title at www.jacplus.com.au and complete the **post-test**.

Answers

Topic 1 Computation with positive integers

1.1 Pre-test

1. D
2. 27
3. a. 2411   b. 70 555   c. 26 747
4. a. 622   b. 7257   c. 1262
5. 480
6. 416
7. 321
8. $1865
9. 600
10. 30 000
11. a. 12   b. 21   c. 25
12. a. <   b. =   c. >
13. a. 2   b. 8   c. 130
14. a. 172   b. 444   c. 235
15. 12 minutes after the race has started.

1.2 Place value and comparing and ordering positive integers

1. a. $900 + 20 + 5$
   b. $20 000 + 8000 + 400 + 60 + 9$
   c. $800 000 + 2$
   d. $1 000 000 + 80 000 + 100$

2. a. Seven hundred and sixty-five
   b. Nine thousand, one hundred and five
   c. Ninety thousand, four hundred and fifty
   d. One hundred thousand, two hundred and thirty-six

3. a. 495   b. 2670   c. 109 605
4. Forty
5. Two hundred
6. Zero
7. Three hundred
8. Fifty
9. D
10. B
11. 48 987, 8569, 742, 647, 28
12. 72 167, 58 625, 47 890, 32 320, 12 947
13. 7647, 7476, 6747, 6477, 4776
14. 8880, 8808, 8800, 8088, 8080, 8008
15. 9, 58, 743, 68 247, 1 258 647
16. 34 108, 58 610, 60 000, 78 645, 84 364
17. 1290, 1902, 2019, 2910, 9021, 9201
18. 111, 211, 212, 221, 222, 1112
19. a. $345 < 567$ b. $89 < 98$ c. $234 596 > 23 459$
    d. $7765 > 7756$
20. a. $6000   b. $340 000   c. $58 000

21. a. $430k   b. $7k   c. $800k
22. 98 641
23. Half a minute (30 s)
24. Astronomers and other scientists work with very small or very large numbers. Scientific notation is a system that makes working with very small or very large numbers easier.
25. a. 2
    b. 4
26. a. 6
    b. 27
    c. In part a, the number of two-digit numbers is the same as the number of digits. In part b, the number of three-digit numbers is double the number of digits because the number can begin with any one of the three digits, and then there are two digits to choose from for the second place and one digit for the third place; this makes six possible numbers.
    d. In part a, the number of two-digit numbers is 2^2. In part b the number of three-digit numbers is 3^3.
27. About 12 days at a rate of 1 number per second

1.3 Strategies for adding and subtracting positive integers

1. a. 228   b. 1803
2. a. 906   b. 448
3. a. 560   b. 102
4. a. 86   b. 64
5. True
6. B
7. 847
8. a. 45   b. 351
9. a. 64   b. 33
10. a. 17   b. 227
11. 278
12. a. 5 or 0   b. 10
13. Yes. Working shown:
    $145 + 671$: Break into 600, 70 and 1
    $145 + 600 = 745$
    $745 + 70 = 815$
    $815 + 1 = 816$
    $816 - 472$: Break into 400, 70 and 2
    $816 - 400 = 416$
    $416 - 70 = 346$
    $346 - 2 = 344$
14. a. The student has not listed the tens and ones correctly.
    b. 998
15. There are 18 children on the playground now.
16. 550 students are not completing exams.
17. Harriet has $161 left in her savings account.
18. The two numbers are 13 and 6.

1.4 Algorithms for adding and subtracting positive integers

1. a. 15　　b. 37　　c. 210
   d. 240　　e. 1730

2. a. 18 220　　b. 175 000　　c. 20
   d. 40　　e. 50

3. a. 99　　b. 72 648　　c. 13 629

4. a. 978　　b. 766　　c. 1335
   d. 14 462　　e. 1 701 553

5. a. 180　　b. 280　　c. 790
   d. 827

6. a. 6　　b. 23　　c. 5
   d. 125　　e. 900

7. a. 700　　b. 70　　c. 500
   d. 5　　e. 26

8. a. 4　　b. 20　　c. 55
   d. 0　　e. 73　　f. 1150

9. a. 35　　b. 35 123　　c. 584 599

10. a. 267　　b. 4413　　c. 65 439
    d. 2634

11. a. 69 397　　b. 1553　　c. 349 818
    d. 19 088 744

12. $247

13. 3032 pages

14. $71

15. 6670 km

16. a. 7, 6, 14　　b. 15, 10, 15　　c. 8, 20, 36

17. a. 26 m　　b. 26 m　　c. 52 m

18. 15

19. 12 149

20. 31

21. a. 172 km　　b. 353 km

22. a. 638 321 − 488 417 = 149 904
    b. 3492 − 1685 = 1807 (other answers possible)

23. 5 849 318

24. 7 days

25. Trixie

26. Three students attend both swimming practice and lifesaving.

27. Three people do not work Monday to Sunday.

28. 6

1.5 Multiplying positive integers

1. a. 24　　b. 54　　c. 60
   d. 26　　e. 45

2. a. 100　　b. 90　　c. 32
   d. 63　　e. 108

3. a. 75　　b. 72　　c. 27
   d. 90　　e. 80

4. 801

5. a.

2000 + 80 + 12 = 2092

   b. 2092

6. a. 128　　　　b. 1233
   c. 2571　　　　d. 24 600

7. a. 63 104　　　　b. 16 120
   c. 63 582　　　　d. 492 720

8. a. 2288　　b. 3007　　c. 1652

9. a. 912　　b. 64 600　　c. 21 571

10. a. 2 161 485　　b. 27 529 460　　c. 1 725 805

11. a. 80　　　　b. 1900
    c. 4500　　　　d. 67 000

12. a. 900　　　　b. 3100
    c. 16 800　　　　d. 172 000

13. a 8000　　　　b 480 000
    c 5 500 000　　　　d 6 300 000 000

14. a. 7200　　　　b. 5 600 000
    c. 54 000 000　　　　d. 13 200 000

15. a. 306　　　　b. 5652
    c. 7425　　　　d. 208

16. a. 5600　　b. 560　　c. 5040

17. a 4320　　b 11 250　　c 28 800

18. a. 150　　b. 300　　c. 300

19. a. 180　　b. 840　　c. 640　　d. 680

20. a. 90　　b. 27　　c. 117

21. a. 325　　b. 312

22. 208

23. 1176

24. a. $72　　b. $1728

25. $132 000

26. a. $675　　b. $1200　　c. $2700

27. 490 km

28. 524 seconds

29. a. 10 km　　b. $175　　c. 120 km
    d. $9100　　e. 400 papers　　f. 20 800 papers

30. The correct answer is 1470 so Ben is correct.

31. 92 > 140 False

32. 260

33. 576

34. 1089; 2178; 3267
    a. 4356　　b. 5445　　c. 6534
    d. 7623　　e. 8712　　f. 9801

35. 46

36. The two numbers are 34 and 35. Their sum is 69.

1.6 Dividing positive integers

1. a. 4   b. 4   c. 8
   d. 8   e. 13   f. 3

2. a. 3   b. 2   c. 2
   d. 2   e. 5   f. 15

3. a. 10   b. 56   c. 84
   d. 60   e. 44

4. a. 485   b. 6208
   c. 2761   d. 80 045

5. a. 85 064 remainder 2   b. 86 047 remainder 3
   c. 904 075 remainder 5   d. 60 074 remainder 2

6. a. 700   b. 7   c. 900

7. a. 500   b. 5000   c. 800

8. 72

9. 214

10. 532

11. a. $208   b. $48

12. 186 cows

13. Number of tables required: 19; total number of courses to be served: 608; number of waiters required: 8; total charge for catering: $8360

14. a. 950 m^2 b. $794 750

15. Kayla: $243 \div 9 = 27$

    $9\overline{)2^2 4^6 3}$ = 027

16. The statement is false.

17. a. 37   b. 3

18. $16 per night

19. There needs to be 8 tables.

20. 10, 18, 26, 34. Students should mention that the pattern in the solution differs by the divisor and that it is found by adding the remainder to each of the multiples.

1.7 Rounding and estimating numbers

1. a. 10   b. 50   c. 1000
   d. 10 000   e. 200 000   f. 5 000 000

2. 28 000 000

3. Estimation table:

	Estimate	Estimated answer	Actual answer
a.	500 + 1000	1500	1449
b.	30 000 + 80 000	110 000	115 670
c.	200 000 + 700 000	900 000	907 448
d.	90 000 − 40 000	50 000	55 880
e.	300 − 200	100	127
f.	500 000 − 200 000	300 000	296 049
g.	40 × 200	8000	7128
h.	600 × 10 000	6 000 000	7 894 033
i.	30 000 × 40	1 200 000	1 149 954
j.	30 000 ÷ 1000	30	32.907 58
k.	60 000 ÷ 3000	20	24.325 483
l.	70 000 ÷ 1000	70	53.385 505

4. D

5. B

6. B

7. a. 8000   b. 100   c. 400 000
   d. 700 000   e. 11 000

8. a. 6000   b. 8 000 000   c. 100
   d. 40 000 000   e. 500

9. D

10. D

11. A

12. D

13. a. 1500 tickets   b. $11 250

14. a.   i. 6000   ii. 5340
    b.   i. 8000   ii. 8028
    c.   i. 50   ii. 46.677 749

15. $29

16. a. One way to estimate the number of students is to count (to your best ability) the number of students in 1 row and in 1 column and multiply. It appears to be about $15 \times 20 = 300$ students in the photograph.
    b. If the hall holds 12 times this number, the capacity of the hall can be estimated as $30 \times 12 = 3600$.

17. 600 students

18. It is important not to estimate calculations by rounding when a very accurate answer is needed.

19. $7270

20. a.

Item	Estimated cost
Bread rolls	$75
Hamburgers	$50
Tomato sauce	$4
Margarine	$4
Onions	$4
Tomatoes	$8
Lettuce	$5

   b. $150   c. $600   d. $450

1.8 Order of operations

1. a. 5   b. 9   c. 3
   d. 30   e. 59

2. a. 25   b. 7   c. 96
   d. 0

3. a. 15   **b.** 24   **c.** 32
    d. 18   **e.** 76   **f.** 27

4. a. 26   **b.** 7   **c.** 96
    d. 27   **e.** 18   **f.** 3

5. a. $(12 - 8) \div 4 = 1$
    b. $(4 + 8) \times 5 - 4 \times 5 = 40$
    c. $3 \times (10 - 2) \div 4 + 4 = 10$
    d. $12 \times (4 + 2) - 12 = 60$
    e. $10 \div (5 + 5) \times 9 \times 9 = 81$
    f. $(18 - 3) \times 3 \div 5 = 9$

6. a. 18   **b.** 14   **c.** -8 **d.** 22

7. a. 2   **b.** 27   **c.** 0   **d.** 1

8. a. 3   **b.** 18   **c.** 40   **d.** 77

9. a. 23   **b.** 18   **c.** 2   **d.** 15

10. a. 8   **b.** 21   **c.** 16   **d.** 0

11. a. 27   **b.** 3   **c.** 21   **d.** 32

12. C

13. B

14. D

15. a. $6 + 2 \times (4 - 3) \times 2 = 10$
    b. $(6 + 2) \times 4 - 3 \times 2 = 26$
    c. $6 + (2 \times 4 - 3) \times 2 = 16$
    d. $6 + 2 \times 4 - 3 \times 2 = 8$, no brackets required

16. a. $274 + 2 \times -68 + 3 \times 50 = \$288$
    b. 600 apples
    c. 8 kg

17. When dealing with money.

18. Keenan ($6 + 6 \div 3 = 6 + 2 = 8$)

19. 43 pebbles

20. $229.20

21. a. $3 \times (4.25 + 4.75) + 2 \times 3.75$
    b. $34.50

22. a. Sydney 9h ahead
    b. Singapore 6h ahead

23. Two possible solutions: 2 packets with 3 ice creams and 5 packets with 5 ice creams, or 7 packets with 3 ice creams and 2 packets with 5 ice creams.
With further investigation of the number pattern obtained by summing 3s and 5s, it appears that there are combinations of 3 packs and 5 packs that will amount to any quantity of ice creams equal to or larger than 8.

24. An example is shown below.
$1 = (4 - 3) \div (2 - 1)$
$3 = 21 \div (3 + 4)$
$5 = 3 + 4 - (2 \times 1)$

25. An example is shown below.
$5 = 4 + 3 - (2 \times 1)$
$6 = 3 \times 4 \div (2 \times 1)$
$8 = 4 \times (3 - 2 + 1)$
$9 = 4 + 3 + 2 \div 1$

Project

1. 1 530 430

2. ⇑⇑℮∩∩ |||| |||

3. ℮℮∩∩∩∩||

4.)|⇑⇑⇑ ℮℮℮ ∩∩∩∩ ||

5. ∩∩||||||

6. ℮℮℮℮∩∩∩∩||||

7. List any difficulties you encountered.

8.

Symbol	Value	Symbol	Value
α	1	ξ	60
β	2	o	70
γ	3	π	80
δ	4	φ or ζ	90
ε	5	ρ	100
ζ or F	6	σ	200
ξ	7	τ	300
η	8	υ	400
θ	9	ϕ	500
ι	10	χ	600
κ	20	ψ	700
λ	30	ω	800
μ	40	$\bar{\mathfrak{z}}$	900
ν	50		

9. Adding a subscript or superscript *iota* (Greek letter for 10) with the symbols used for 1 to 9 represented the numbers 1000 to 9000. *M* was used to represent 10 000. When symbols were placed on top of the *M*, this meant that the value of the symbols was multiplied by 10 000. This enabled the ancient Greeks to represent larger numbers.

10. Sample response can be found in the worked solutions in the online resources.

1.9 Review questions

1. a. 7   **b.** 40 000
    c. 1 000 000   **d.** 90

2. a. $300 + 90 + 2$ **b.** $4000 + 100 + 9$
    c. $40 000 + 2000 + 1$ **d.** $100 000 + 20 000$

3. 349, 394, 934, 943, 3499, 3994

4. 1101, 1100, 1011, 110, 101

5. a. 127   **b.** 3187
    c. 7641   **d.** 4 217 656

6. a. 2   **b.** 35   **c.** 150   **d.** 14

7. a. 415   **b.** 217   **c.** 60 322
    d. 37 758   **e.** 92 957

8. a. 150   **b.** 8400   **c.** 1240
    d. 603   **e.** 589

9. a. 3×6

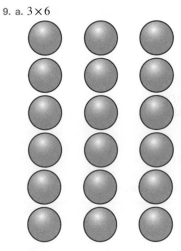

There are 18 balls = 18

b. 5×4

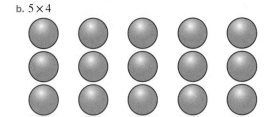

There are 20 balls = 20

c. 2×7

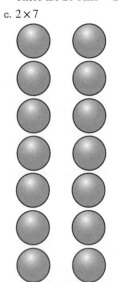

There are 14 balls = 14

10. a. 96   b. 1072   c. 3210

11. a. 1416   b. 1388   c. 3375

12. a. 1152   b. 2286   c. 19 200

13. a. 4608   b. 30 894   c. 18 355 064

14. a. 129   b. 260   c. 5021

15. a. 8   b. 3   c. 84   d. 45   e. 1

16. a. 700   b. 7   c. 410

17. a. 1568   b. 981   c. 537 r 3
    d. 335 r 7   e. 911 r 9

18. a. 14 000   b. 5000   c. 50
    d. 420 000   e. 120 000   f. 210 000

19. C

20. a. $<$ b. $=$ c. $>$ d. $>$

21. a. 327   b. 1090   c. 789   d. 216

22. a. 127   b. 81   c. 711   d. 278

23. a. 28   b. 24   c. 10   d. 8   e. 11   f. 27

24. a. 12   b. 30   c. 5
    d. 900   e. 14   f. 25

25. a. $<$ b. $=$ c. $>$ d. $<$

26. a. $<$ b. $<$ c. $>$ d. $>$

27. 23.5

28. a. 5   b. 38   c. 6   d. 5   e. 70   f. 29

29. a. 306 km   b. 322 km   c. 482 km

30. a. 448   b. 3136   c. 98

31. As long as one of the factors in a multiplication is even, the result will be even. Otherwise, the answer will be odd.

32.
$$\begin{array}{r} 8240 \\ -3681 \\ \hline 4559 \end{array}$$

33. 95

34. a. $8 \times 3 + 12 \times 2$ b. $48

35. 20 499

36. a.

+	6	10	3
5	11	15	8
12	18	22	15
9	15	19	12

b.

×	4	8	6
3	12	24	18
5	20	40	30
2	8	16	12

37. a. 24 °C   b. -21 °C

38. 21 questions for 3 marks

39.

8	+	4	×	6	=	32
+		×		×		
3	−	5	+	9	=	7
×		÷		÷		
7	−	2	−	1	=	4
=		=		=		
29		10		15		

40. 2019 and 2018

2 Computation with positive and negative integers

LESSON SEQUENCE

LESSON
2.1 Overview

Why learn this?

Integers are whole numbers. They can be positive, negative or equal to zero. Positive integers have values above zero and negative integers have values below zero.

If you live in the southern parts of Australia, overnight temperatures often fall below zero in winter. These very cold temperatures are written as negative integers. In summer, the generally very hot temperatures are written as positive integers.

Using your understanding of the basic operations of addition, subtraction, multiplication and division, you can measure the temperature changes throughout a single day and the average daily temperature over a month, and use those measurements to compare your home town to other parts of Australia, or other parts of the world.

In finance, positive numbers are used to represent the amount of money in someone's bank account, while negative numbers are used to represent how much money someone owes — for example, how much they have to pay back after borrowing money or taking out a loan from a bank.

Hey students! Bring these pages to life online

Watch videos

Engage with interactivities

Answer questions and check solutions

Find all this and MORE in jacPLUS

Reading content and rich media, including interactivities and videos for every concept

Extra learning resources

Differentiated question sets

Questions with immediate feedback, and fully worked solutions to help students get unstuck

1. Write an integer to express the following.
   a. A debt of $250
   b. 15 metres below sea level
   c. The temperature is $20\,°C$

2. Complete each statement by placing $<, >$ or $=$ in the empty box so that the statement is true.
   a. $-3\,\square\,2$ b. $-5\,\square\,-7$ c. $-21\,\square\,-17$ d. $0\,\square\,-8$

3. **MC** Select which of the following options has the numbers $-5, 8, 2, 0, -1, -8, 3$ arranged in ascending order.
   A. $8, -5, -1, 0, 2, 3, -8$ B. $-8, -5, -1, 0, 2, 3, 8$
   C. $0, -1, 2, 3, -5, -8, 8$ D. $-8, 5, -1, 0, 2, 3, 8$

4. Determine the coordinates of point A in the graph.

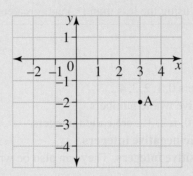

5. **MC** Select the quadrant in which the point $(-7, -1)$ is located.
   A. First   B. Second   C. Third   D. Fourth

6. **MC** If the points $(-2, -2)$, $(-2, 2)$, $(2, 2)$ and $(2, -2)$ were plotted on a Cartesian plane and joined by straight lines, select the shape this would make.
   A. Rectangle   B. Circle   C. Square   D. Trapezium

7. Calculate the following.
   a. $3 + -5 =$ b. $7 - -6 =$ c. $-7 + -12 =$ d. $-16 + 5 =$

8. Write the number that is 12 more than -5.

9. Narelle went for a hike at Mount Baw Baw. From her starting point, she climbed up $72\,m$ and then descended $15\,m$, before ascending a further $21m$. Determine how far above her starting point she is now.

10. Evaluate -8×-6.

11. Evaluate the following.
    a. $7 \times 6 \div 3 \div 7$ b. $3 \times 8 \div 6 + 4$ c. $7 + [66 - (5 \times 5)] \times 3$

12. **MC** Select the simplified form of the algebraic expression $5g \times -3h \times -6k$.
    A. $-15ghk$ B. $18ghk$ C. $-90ghk$ D. $90ghk$

13. If $a = -5$, $b = 4$ and $c = -2$, evaluate $4(a - b) \div 3c$.

14. Hang has $340 in the bank. She makes a deposit of $75 each month for the next three months, before withdrawing $185 to buy a new dress. Determine the amount of money Hang has left in her account.

15. To improve goal kicking a new training drill was developed. In this training run, kicking a goal scores you 6 points, but missing a goal loses you 2 points. If a team had 30 shots at goal and kicked 21 goals, determine how many points they scored.

LESSON
2.2 Integers on the number line

LEARNING INTENTION

At the end of this lesson you should be able to:
- identify and represent integers on a number line
- compare the value of integers using the less than (<) and greater than (>) symbols.

▶ 2.2.1 Integers

eles-3695

- **Whole numbers** are the numbers $0, 1, 2, 3, 4 \ldots$
- **Integers** are positive and negative whole numbers.
  - The numbers $\ldots -3, -2, -1, 0, 1, 2, 3 \ldots$ are called integers.
  - The integers $-1, -2, -3 \ldots$ are **negative integers**.
  - The integers $1, 2, 3 \ldots$ are **positive integers**. They can also be written as $+1, +2, +3 \ldots$
  - 0 is an integer. It is neither positive nor negative.
  - The direction of an integer indicates whether it is positive or negative. The $+$ sign can be used to represent either an addition (if it appears between two quantities) or the direction of an integer (if it appears in front of an integer). In the same way, the $-$ sign can be used to represent either a subtraction or the direction of an integer.

DISCUSSION

Can you think of an example of negative numbers that we encounter in everyday life?

WORKED EXAMPLE 1 Converting a description into an integer

Write the integer that is suggested by the following descriptions.
a. **My flat is on the 5th floor.**
b. **This town is 20 m below sea level.**

THINK	WRITE
a. Numbers above 0 are positive, so place a + sign before the number.	a. $+5$ (or 5)
b. Numbers below 0 are negative, so place a – sign before the number.	b. -20

▶ 2.2.2 The number line

eles-3696

- Integers can be represented on a number line. On a number line, positive integers are written to the right of 0, while negative integers are written to the left of 0.

- The **magnitude** of an integer shows the size of the integer in terms of its distance from 0. For example, the number −3 is negative and has a magnitude of 3. This means it is a distance of 3 units from 0.

- The further a number is to the right of any number on a number line, the larger it is. For example, +5 is greater than +2, while + 2 is greater than 0 and 0 is greater than −5.

- **Opposite integers** are the same distance from 0, but on opposite sides of 0. For example, 3 and −3 are opposite integers because they are both 3 units from 0.

WORKED EXAMPLE 2 Placing integers on a number line

Place the integers 9, − 2, − 7, 5 on a number line.

THINK	WRITE
1. Draw a number line from −10 to 10 so that you can include all of the numbers.	
2. Mark the values 9, −2, −7, 5 on the number line.	

WORKED EXAMPLE 3 Writing integers in numerical order

Use a number line to help list the integers −4, 2, −3, 1 in numerical order.

THINK	WRITE
1. Draw a number line from −5 to 5 so that you can include all of the numbers.	
2. Mark the values. As you move from left to right on the number line, the numbers become larger.	In numerical order: −4, −3, 1, 2

Write opposites for the following integers.

a. −2   b. 3   c. 4

THINK	WRITE
a. An opposite integer is the same distance from 0, but has the opposite sign.	a. The opposite of −2 is 2.
b. An opposite integer is the same distance from 0, but has the opposite sign.	b. The opposite of 3 is −3.
c. An opposite integer is the same distance from 0, but has the opposite sign.	c. The opposite of 4 is −4.

Comparing integers

- The following inequality symbols can be used when comparing integers.
    - $>$ means 'greater than'
    - $<$ means 'less than'
    - $\geq$ means 'greater than or equal to'
    - $\leq$ means 'less than or equal to'
- Remember that inequality signs look a little bit like a crocodile's mouth. When placing an inequality symbol between two numbers, the crocodile's mouth always opens up to eat the larger number.

Complete each statement by inserting the correct symbol: < or >.

a. 2 ☐ 5   b. −4 ☐ −1   c. 0 ☐ −3   d. 6 ☐ −2

THINK	WRITE
a. 2 sits to the left of 5 on the number line, so 2 is smaller.	a. $2 < 5$
b. −4 sits to the left of −1 on the number line, so −4 is smaller.	b. $-4 < -1$
c. 0 is to the right of −3 on the number line, so 0 is larger.	c. $0 > -3$
d. 6 is to the right of −2 on the number line, so 6 is larger.	d. $6 > -2$

▶ 2.2.3 Using positive and negative numbers in daily life

eles-3697

- Because positive and negative numbers show both their size and their direction away from zero, they are often referred to as **directed numbers**.

- Directed numbers are often used to measure objects from a reference point, for example, the temperature above or below zero, the height above or below sea level, or archaeological measurements that show the depth beneath the Earth's surface. Directed numbers are also used when describing banking transactions. Withdrawals of money from a bank account are shown using negative signs while deposits into a bank account are shown using positive signs.

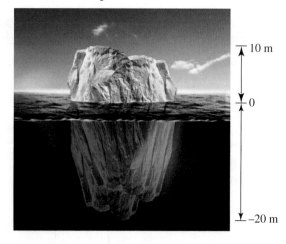

WORKED EXAMPLE 6 Converting a statement into a directed number

Write the integer that is suggested by each of the following situations.

a. The temperature was 1 degree below zero.   b. Romie has $500 in her bank account.

THINK	WRITE
a. 'Below zero' means negative.	a. -1
b. Romie has money in her account, so the integer is positive.	b. $+500$

COMMUNICATING — COLLABORATIVE TASK: Temperatures in local Aboriginal Nations or various cities around the world

Equipment: paper, pen, whiteboard, whiteboard marker, internet and device.

Map of Indigenous Australia: aiatsis.gov.au/explore/map-indigenous-australia

1. Assign different local Aboriginal Nations or cities around the world to students in pairs.
2. Research the current temperature in that location.
3. On the whiteboard, draw one large number line.
4. Ask groups to record the researched temperatures on the number line.
5. Discuss the results as a group.

 Resources

Interactivities   Number line (int-4033)

Exercise 2.2 Integers on the number line   learn on

2.2 Quick quiz on	2.2 Exercise

Individual pathways

■ PRACTISE	■ CONSOLIDATE	■ MASTER
1, 4, 7, 10	2, 5, 8, 11	3, 6, 9, 12

Fluency

1. **WE1&6** Write the integer suggested by each of the following descriptions.

   a. The temperature today is 7 degrees below zero.
   b. The penthouse is on the 10th floor.
   c. The Dead Sea is 422 m below sea level.
   d. Mount Kosciuszko is 2228 m above sea level.
   e. The Dow Jones dropped 3 points overnight.
   f. Chloe lost $30.

2. **WE2** Draw a number line from −10 to 10. Mark the following values on your number line.

   a. 2
   e. 5

   b. −4
   f. −1

   c. −9
   g. 0

   d. 7
   h. −3

Understanding

3. **WE3** Use a number line to place the following numbers in numerical order.

   a. $5, -2, 3, -6, 7$
   b. $0, -1, 3, -4, -6$
   c. $-5, 7, -2, -3, 1$
   d. $-4, 2, 1, -3, 5$

4. **WE4** Write opposite integers for the following.

   a. −6
   b. 7
   c. 1
   d. −8

5. **WE5** Complete the following by inserting < or > in the box.

   a. $2 \square -5$
   d. $6 \square -2$

   b. $3 \square 7$
   e. $-1 \square 3$

   c. $-2 \square -6$
   f. $-10 \square -6$

6. From the following lists, select:

   i. the smallest number

   ii. the largest number.

   a. $-3, 7, -5$
   b. $2, -4, 5, 3, -2$
   c. $7, -10, 5, -2, -4$
   d. $-4, -1, 3, 0, -2$

Communicating, reasoning and problem solving

7. The temperature in Oslo dropped to −6 °C overnight. During the day, the temperature rose by 4 °C. Determine the maximum temperature on that day. Show full working.

8. A new apartment building is being built. It will include a ground floor reception area. If there will be 10 floors of apartments and 3 floors of underground parking, how many buttons will be needed in the lift? Explain your answer.

9. Lake Mungo, part of Mungo National Park in New South Wales, is a World Heritage listed area. Archaeologists estimate that the lake dried out 10 000 years ago and, before that, Aboriginal people lived on the shores of the lake. Lake Mungo became one of the world's most significant archaeological sites following the discovery there of some of the oldest remains of human civilisation on Earth.
   Explain what −60 000 could mean in terms of human occupation in the area around Lake Mungo.

10. Raj and Vanessa are going home from school in opposite directions. Raj lives 1.2 km south of the school and Vanessa lives 1.6 km north of the school.

    a. Draw a number line to represent the three locations: Raj's house, Vanessa's house and the school. Label the three locations with directed numbers.
    b. Determine how far apart Raj and Vanessa live from each other.

11. An aeroplane flying at 12 km above sea level flies directly over a submarine travelling at 500 m below sea level.

    a. Draw a number line to represent the three locations: the aeroplane, the submarine and the ocean level. Label the three locations with directed numbers.
    b. Evaluate the distance between the submarine and the aeroplane.

12. **a.** Historians use BCE (Before Common Era) and CE (Common Era) when discussing dates. Explain what is meant by 2000 CE.

b. If the millennium year 2000 CE (+2000) was considered to be zero, determine the:

    i. integer value that the year 2010 would represent

    ii. integer value that the year 1894 would represent.

c. Human exploration of space started in 1957, which led to new discoveries. The successful orbit of Earth by Sputnik 1 gave hope to humans that one day they would enter space and even walk on the Moon. If we consider 1957 as being zero, state the directed numbers that could be used to describe the following information.

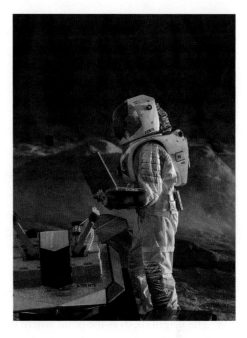

    i. Cosmonaut Yuri Gagarin was the first man to enter space — 1961

    ii. Neil Armstrong and Edwin 'Buzz' Aldrin of Apollo 11 were the first humans to walk on the Moon — 1969

    iii. Sally Ride was the first American woman in space — 1983

    iv. *Discovery* was the first shuttle to dock with the International Space Station — 1999

    v. *Rosetta* was the first man-made probe to make a soft landing on a comet — 2014

d. Draw a number line from −10 to 60 to show your answers to part **c**.

LESSON
2.3 Integers on the number plane

LEARNING INTENTION

At the end of this lesson you should be able to:
- understand the number plane (also called the Cartesian plane), origin and coordinates
- infer the location of a point described by its coordinates
- plot one or more points using their coordinates.

2.3.1 Positive integers and zero on the number plane

eles-3698

- The **number plane**, also called the Cartesian plane, has two axes: the horizontal axis, also called the *x*-axis, and the vertical axis, also called the *y*-axis.
- Every point on the number plane can be described by its position relative to the *x*- and *y*-axes.
- A pair of coordinates, also called an ordered pair (x, y), fixes the position of a point. The *x* value shows the point's distance along the *x*-axis, while the *y* value shows the point's distance along the *y*-axis.

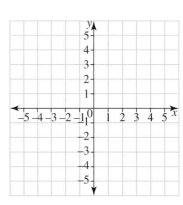

Determine the coordinates of the following points in the number plane shown.

a. A   **b. B**

THINK

a. Point A is 1 unit to the right of zero along the x-axis and 4 units above zero along the y-axis.

b. Point B lies on the y-axis, so it is 0 units to the right of zero along the x-axis. It is 2 units above zero along the y-axis.

WRITE

a. A $(1, 4)$

b. B $(0, 2)$

⏵ 2.3.2 Integers on the number plane

eles-3699

- The origin $(0, 0)$ is the point where the x- and y-axes intersect to divide the number plane into 4 quadrants, as shown. The origin is labelled as '0'.
- The first quadrant has positive x and positive y values.
- The second quadrant has negative x and positive y values.
- The third quadrant has negative x and negative y values.
- The fourth quadrant has positive x and negative y values.
- The illustration demonstrates the position of points on the number plane. Starting at the origin:
  - A $(3, 2)$ — 3 right and 2 up
  - B $(-3, 2)$ — 3 left and 2 up
  - C $(-3, -2)$ — 3 left and 2 down
  - D $(3, -2)$ — 3 right and 2 down

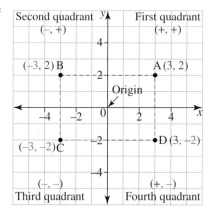

Write the coordinates and state the quadrant or axis of each point on this number plane.

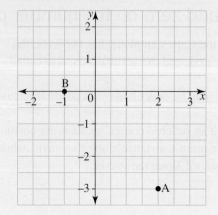

THINK

a. 1. A is 2 units to the right of the origin along the x-axis and 3 units down from the origin along the y-axis.

   2. A is in the lower right-hand corner.

b. 1. B is 1 unit to the left of the origin along the x-axis and 0 units up or down from the origin along the y-axis.

   2. B is on the x-axis.

WRITE

a. A $(2, -3)$

   A is in the fourth quadrant.

b. B $(-1, 0)$

   B is on the x-axis.

 Resources

Interactivities   Integers on the number plane (int-4035)

                  Positive integers and zero on the number plane (int-4034)

Exercise 2.3 Integers on the number plane   **learn** on

| 2.3 Quick quiz   on | 2.3 Exercise |

Individual pathways

■ PRACTISE	■ CONSOLIDATE	■ MASTER
1, 2, 4, 7, 10, 13, 16, 19	3, 6, 8, 12, 15, 17, 20	5, 9, 11, 14, 18, 21

Fluency

Questions **1** and **2** refer to the graph shown.

1. **WE7** Determine the coordinates of the following.

   a. A   b. B   c. C   d. D   e. E

2. Match a letter to the points with the following coordinates.

   a. $(0, 0)$ b. $(6, 6)$ c. $(8, 1)$ d. $(1, 8)$

3. **MC** Select the quadrant in which all the points have negative x and negative y coordinates.

   A. Quadrant 1   B. Quadrant 2
   C. Quadrant 3   D. Quadrant 4

4. **MC** Select the quadrant in which all the points have positive x and negative y coordinates.

   A. Quadrant 1   B. Quadrant 2
   C. Quadrant 3   D. Quadrant 4

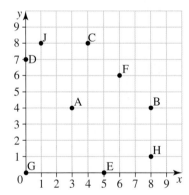

5. If the x-axis represents the direction from west to east, and the y-axis represents the direction from south to north, calculate the coordinates if a person starts at the origin and walks 5 m west and then 2 m north.

Understanding

Use the graph shown for questions **6** through **10**.

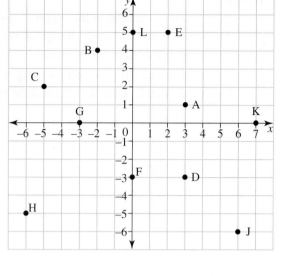

6. **WE8** Write the coordinates and state the quadrant or axis of each of the following points.

    a. A
    b. B
    c. H
    d. F
    e. J

7. Match the following points to a letter and state their quadrant or axis.

    a. $(-5, 2)$
    b. $(0, 5)$
    c. $(3, -3)$
    d. $(2, 5)$
    e. $(-3, 0)$

8. Determine the *x*-coordinate of the following points.

    a. A
    b. D
    c. K
    d. L

9. Determine the *y*-coordinate of the following points.

    a. C
    b. J
    c. G
    d. F

10. State all the points lying in the third quadrant in the number plane.

11. Determine whether the following statements are true or false.

    a. The origin has coordinates $(0, 0)$.
    b. The point at $(3, 5)$ is the same point as at $(5, 3)$.
    c. The point at $(-5, 4)$ is in the third quadrant.
    d. The point at $(0, 2)$ must lie on the *y*-axis.

12. **MC** The point $(5, -2)$ lies:

    A. in the first quadrant.
    B. in the second quadrant.
    C. in the third quadrant.
    D. in the fourth quadrant.

13. **MC** The point $(0, 4)$ lies:

    A. in the first quadrant.
    B. on the *y*-axis.
    C. in the third quadrant.
    D. in the fourth quadrant.

14. **MC** The point $(-4, 5)$ lies:

    A. in the first quadrant.
    B. in the second quadrant.
    C. in the third quadrant.
    D. in the fourth quadrant.

15. Draw a number plane with both axes scaled from -6 to 6. Plot the points listed and join them with straight lines in the order given. State the name for each completed shape.

    a. $(5, 5), (3, 2), (-2, 2), (0, 5), (5, 5)$
    b. $(4, -1), (4, -5), (-1, -3), (4, -1)$
    c. $(-4, 4), (-2, 1), (-4, -5), (-6, 1), (-4, 4)$
    d. $(-2, 1), (1, 1), (1, -2), (-2, -2), (-2, 1)$

Communicating, reasoning and problem solving

16. Consider the graph relating to question **6** and explain whether the points F and D have the same *y*-coordinate, and whether points A and D have the same *x*-coordinate.

17. Consider the graph shown.

   a. Determine the coordinates of a point, C, so that ABCD is a parallelogram.
   b. Determine the coordinates of a point, F, so that DBEF is a kite shape.
   c. Show that the point $(4, -1)$ lies on the line that passes through D and the origin.
   d. List 2 points on the line joining D to E.
   e. Give the coordinates of a point, T, in the third quadrant that would complete the isosceles triangle ADT.

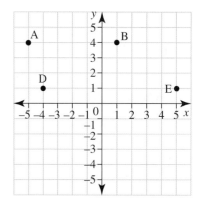

18. a. If you were moving east and then you did a U-turn, determine the direction you would then be moving in. Assume that a U-turn is a 180-degree turn, or going back the way you came and is the same as putting a negative sign in front of you.
   b. Determine the direction you would be facing if you had two negative signs in front of you.
   c. Determine the direction you would be facing if you had three negative signs in front of you.
   d. Determine the direction you would be facing if you had ten negative signs in front of you.
   e. Determine the direction you would be facing if you had 101 negative signs in front of you. Explain your answer.
   f. Explain what would happen if you were moving north and then you made a U-turn (which would mean putting a negative sign in front of you). Determine the direction you would be moving in after you did the U-turn.
   g. If you were moving north, explain which direction you would be facing after an odd number of negative signs. Explain which direction you would be facing after an even number of negative signs.

19. Eugenia uses a remote control to move a robot car on a grid map. The robot car moves from point $(7, 5)$ to point $(7, -4)$. Determine the distance travelled by the robot car.

20. The tennis court shown is enclosed in an area that has a post in each corner.

   a. Draw a number plane with the origin in the centre of the court.
   b. Determine the coordinates of the four posts on the number plane drawn in part a.

21. If you swap the x- and y-coordinates of a point, the position of the point may change to another quadrant. Determine what happens to points from each quadrant if you swap their coordinates.

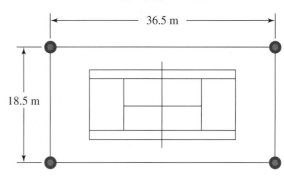

LESSON
2.4 Adding integers

2.4.1 Addition of integers

eles-3700

- A number line can be used to add integers.
- When adding a positive integer, move right along the number line.
- When adding a negative integer, move left along the number line.

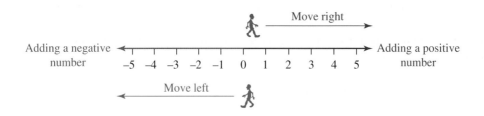

- Consider the addition of $-1 + (+4)$.

Symbol	Meaning	Action to take
-1	Starting point	Start at -1.
$+$	Direction to face on the number line	Face right for addition $+$.
$(+)$	Direction to move on the number line	Move forward for positive $(+)$.
(4)	Number of steps to take from the starting point	Move 4 places from -1.
$=$	Finishing point	Stop at $+3$.

Therefore, $-1 + (+4) = +3$.

Calculate the value of each of the following, using a number line.

a. $-5 + (+3)$

b. $-5 + (-3)$

THINK

a. 1. Start at -5 and face right (because this is an addition). Move 3 steps forward (because this number is positive). The finishing point is -2.

WRITE

a.

2. Write the answer.

$-5 + (+3) = -2$

b. 1. Start at -5 and face right (because this is an addition). Take 3 steps backwards (because this number is negative). The finishing point is -8.

b.

2. Write the answer.

$-5 + (-3) = -8$

Write number sentences to show the addition problems in the following diagrams.

a.

b.

THINK

a. 1. The first integer is 2 units to the left of zero, so start at -2. The second integer is 4 units to the left of -2, finishing at -6.

WRITE

a.

2. Write the number sentence.

$-2 + (-4) = -6$

b. 1. The first integer is 2 units to the left of zero, so start at -2. The second integer is 5 units to the right of -2, finishing at $+3$.

b.

2. Write the number sentence.

$-2 + (+5) = +3$

Addition of integers using the sign model

- In the sign model, a ⊖ sign and a ⊕ sign cancel each other out. This model can be used to add integers, as shown in Worked example 11.

WORKED EXAMPLE 11 Adding integers using the sign model

Use the sign model to calculate each of the following.

a. $-4 + (+1)$ b. $+5 + (-3)$

THINK	WRITE
a. 1. Show -4 as four negative signs.	a.
2. Show $+1$ as one positive sign.	
3. Cancel out the opposite signs. There are 3 negative signs left over.	
4. Write the answer.	$-4 + (+1) = -3$
b. 1. Show $+5$ as 5 positive signs.	b.
2. Show -3 as 3 negative signs.	
3. Cancel out the opposite signs. There are 2 positive signs left over.	
4. Write the answer.	$+5 + (-3) = +2$

Digital technology

Scientific calculators can evaluate the addition of positive and negative numbers.

```
                        D        Math ▲
-350+82+(-156)

                          -424
```

---

on Resources

⬧ **Interactivities** Addition of integers (int-4036)

2.4 Quick quiz on	2.4 Exercise

Individual pathways

■ PRACTISE	■ CONSOLIDATE	■ MASTER
1, 3, 6, 9, 12, 15	2, 4, 7, 10, 13, 16	5, 8, 11, 14, 17, 18, 19

Fluency

1. **WE9** Calculate the value of each of the following.
   a. $-25 + (+10)$ b. $-7 + (-3)$ c. $6 + (-7)$
   d. $-8 + (-5)$ e. $13 + (+6)$ f. $16 + (-16)$

2. **WE10** Write number sentences to show the addition problems in the following diagrams.

   a.   b.

   c.   d.

   e.   f.

3. **WE11** Use the sign model to calculate each of the following.
   a. $+2 + (-3)$ b. $+3 + (-4)$ c. $+4 + (-4)$
   d. $+3 + (-2)$ e. $-4 + (+2)$

4. Calculate each of the following using number sentences. (Draw a number line or other model if you wish.)
   a. $5 + (-2)$ b. $-3 + (-4)$ c. $-2 + 2$
   d. $6 + (-5)$ e. $-5 + 5$

5. Calculate each of the following using number sentences. (Draw a number line or other model if you wish.)
   a. $4 + (-6)$ b. $-5 + 7$ c. $6 + (-9)$
   d. $3 + (-3)$ e. $0 + (-6)$

6. Write the answer for each of the following.
   a. $-5 + (-2)$ b. $-6 + 4$ c. $-8 + 8$
   d. $3 + (-7)$ e. $-3 + 7$ f. $-3 + (-7)$

7. Write the answer for each of the following.
   a. $-8 + 12$ b. $19 + (-22)$ c. $-64 + (-36)$
   d. $-80 + 90$ e. $-2 + 4$ f. $-15 + (-7)$

8. Copy and complete this addition table.

+	−13	5
21		
−18		

Understanding

9. Determine the missing number in each of these incomplete number sentences.

   a. $8 + \underline{\quad} = 0$

   b. $-2 + \underline{\quad} = -8$

   c. $\underline{\quad} + (-6) = 4$

   d. $\underline{\quad} + 5 = -2$

   e. $\underline{\quad} + (-5) = -2$

10. a. Copy and complete the addition table shown.

    b. i. Determine the pattern shown along the leading (dotted) diagonal.

    ii. Determine the pattern shown along the other (unmarked) diagonal.

    iii. State whether the chart is symmetrical about the leading diagonal.

+	-2	-1	0	1	2
-2		-3			
-1				0	
0	-2				
1					
2					

11. Model each situation with an integer number sentence that shows the result of the following.

    a. Starting at ground level, a lift went down 2 floors, then up another 3 floors.
    b. Starting at ground level, a lift went down 3 floors, then up 5 floors.
    c. In cricket, Australia was 50 runs behind, then made another 63 runs.
    d. A submarine at sea level, dived 50 metres then rose 26 m.
    e. An account with a balance of $200 had $350 withdrawn from it.

Communicating, reasoning and problem solving

12. Write a real-life scenario to fit each of the following number sentences.

    a. $-3 + (-2) = -5$

    b. $-10 + (-40) = -50$

    c. $2 + (-6) = -4$

    d. $-20 + 20 = 0$

    e. $-8 + 10 = 2$

13. a. Write a rule starting with 'When adding two negative numbers…'
    b. Write a rule starting with 'When adding a negative number and a positive number…'

14. Arrange the numbers in this magic square so that all rows, columns and diagonals add up to the same value. No numbers can be used twice.

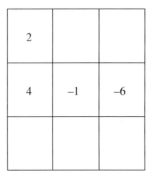

15. Kuni has a sporting equipment business. Due to a financial crisis, she lost $2500 after six months and then lost $3075 by the end of the year. Calculate the total amount that Kuni lost by the end of the year.

16. In an AFL match, Dustin Martin scored two goals in the first quarter and was awarded six points for each. He scored three behinds in the fourth quarter and was awarded 1 point for each. Calculate the total points gained by Dustin for his team.

17. Avery parks her car in a car park two floors below ground level. She takes a lift to the shopping centre and goes up seven floors. Then she takes a lift down three floors to catch up with her friend Lucy. Determine the floor at which Avery gets off to meet Lucy.

18. Determine the effect of adding a number to its opposite integer.

19. Sydney time is 3 hours ahead of Perth time. A flight from Perth to Sydney takes 4 hours. If a flight leaves Perth for Sydney at 9 am, determine what time it lands in Sydney.

LESSON
2.5 Subtracting integers

LEARNING INTENTION

At the end of this lesson you should be able to:
- understand how a number line can be used to subtract integers
- use a number line to subtract integers
- subtract integers without the aid of a number line.

▶ 2.5.1 Subtraction of integers

eles-3701
- A number line can be used to subtract integers.
- Subtracting a number gives the same result as adding its opposite integer.
- When subtracting a positive integer, move left along the number line.
- When subtracting a negative integer, move right along the number line.

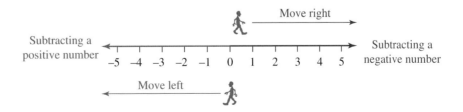

- Consider the subtraction of $12 - (+4)$.

Symbol	Meaning	Action to take
12	Starting point	Start at 12.
$-$	Direction to face on the number line	Face left for subtraction $-$.
$(+)$	Direction to move on the number line	Move forward for positive $(+)$.
(4)	Number of steps to take from the starting point	Move 4 places from 12.
$=$	Finishing point	Stop at +8.

Therefore, $12 - (+4) = +8$.

WORKED EXAMPLE 12 Subtracting integers using a number line

Calculate the value of each of the following, using a number line.
a. $-2 - (+3)$
b. $-2 - (-3)$

THINK

a. 1. Start at -2 and face left (because this is subtraction). Take 3 steps forwards (because the number being subtracted is positive). The finishing point is -5.

 2. Write the answer.

b. 1. Start at -2 and face left (because this is subtraction). Take 3 steps backwards (because the number being subtracted is negative). The finishing point is $+1$.

 2. Write the answer.

WRITE

a.

$-2 - (+3) = -5$

b.

$-2 - (-3) = +1$

eles-4091

▶ 2.5.2 Subtracting integers by adding opposites

- By developing and extending a pattern, we can show that subtracting negatives has the same effect as adding a positive. This means that subtracting a negative integer is the same as adding the opposite integer. For example, as demonstrated in part **b** of Worked example 13:

$$-2 - (-3) = -2 + (+3)$$
$$= -2 + 3$$
$$= +1$$

WORKED EXAMPLE 13 Subtracting integers

Calculate the following.

a. $2 - 5$ b. $-3 - 6$ c. $5 - (-3)$ d. $-5 - (-4)$

THINK

a. 1. Write the question.
   2. Rewrite the equation, changing subtraction to addition of the opposite integer.
   3. Add, using the addition rule for addition of integers.
      - Start at +2.
      - Face right for addition.
      - Move backwards 5 units.

b. 1. Write the question.
   2. Rewrite the equation, changing subtraction to addition of the opposite integer.
   3. Add, using the addition rule for addition of integers.
      - Start at −3.
      - Face right for addition.
      - Move backwards 6 units.

c. 1. Write the question.
   2. Rewrite the equation, changing subtraction to addition of the opposite integer because we are subtracting a negative.
   3. Add, using the addition rule for addition of integers.
      - Start at +5.
      - Face right for addition.
      - Move forward 3 units.

d. 1. Write the question.
   2. Rewrite the equation, changing subtraction to addition of the opposite integer.
   3. Add, using the addition rule for addition of integers.
      - Start at −5.
      - Face right for addition.
      - Move forward 4 units.

WRITE

a. $2 - 5 = 2 - (+5)$
   $= 2 + (-5)$
   $= -3$

b. $-3 - 6 = -3 - (+6)$
   $= -3 + (-6)$
   $= -9$

c. $5 - (-3) = 5 + (+3)$
   $= 5 + 3$
   $= 8$

d. $-5 - (-4) = -5 + (+4)$
   $= -5 + 4$
   $= -1$

A news flash in Freezonia announced that there had been a record drop in temperature overnight. At 6 pm the temperature was 10 °C and by 4 am it had fallen by 25 °C. What was the temperature at 4 am?

THINK	WRITE
1. The original temperature is 10 °C, so write the number 10. The temperature fell by 25 °C, so write this as 10 − 25. Write the number sentence. Rewrite as the addition of the opposite.	$10 - 25$ $= 10 + (-25)$
2. Complete the addition by starting at +10, facing right, then moving backwards by 25 units.	$= -15$
3. Write the answer in a sentence.	The temperature in Freezonia at 4 am was −15 °C.

Digital technology

Scientific calculators can evaluate subtraction of positive and negative numbers.

```
                          D        Math ▲
247-57-(-68)

                            258
```

COMMUNICATING — COLLABORATIVE TASK: Radial diagrams

In pairs, insert the integers from −6 to +2 into the circles on the diagram so that each straight line of three circles has each of the following totals.

a. −6
b. −3
c. −9

 Resources

▶ **Video eLesson** Addition and subtraction of positive and negative numbers (eles-1869)

Interactivities Subtraction of integers (int-4037)

Exercise 2.5 Subtracting integers

2.5 Quick quiz on	2.5 Exercise

Individual pathways

■ PRACTISE	■ CONSOLIDATE	■ MASTER
1, 5, 9, 12, 14, 17	2, 4, 6, 8, 11, 15, 18	3, 7, 10, 13, 16, 19, 20

Fluency

1. **WE12** Calculate the value of each of the following.

   a. $7 - (+2)$
   b. $-18 - (+6)$
   c. $3 - (+8)$
   d. $17 - (-9)$
   e. $-28 - (-12)$
   f. $-17 - (-28)$

2. **WE13** Calculate the value of each of the following.

   a. $7 - 5$
   b. $8 - (-2)$
   c. $-4 - 6$
   d. $-6 - (-8)$
   e. $1 - 10$

3. Calculate the value of each of the following.

   a. $-5 - 5$
   b. $-8 - (-8)$
   c. $0 - 4$
   d. $0 - (-3)$
   e. $-10 - (-20)$
   f. $-5 - (-5)$

4. Write number sentences to show the subtraction problems in the following diagrams.

   a.

   b.

   c.

   d.

Understanding

5. Mentally calculate the following and write the answer.

   a. $-7 - 3$
   b. $8 - (-5)$
   c. $-6 - (-9)$
   d. $-0 - (-12)$

6. Mentally calculate the following and write the answer.

   a. $-8 - 8$
   b. $3 - 20$
   c. $20 - (-3)$
   d. $-4 - 8$

7. Determine the missing number in these incomplete number sentences.

   a. _____ $- 7 = -6$
   b. $-8 - $_____ $= -17$
   c. $-8 - $_____ $= 17$
   d. _____ $- (-2) = 7$

8. Write each of the numbers that can be described by the following.

   a. 6 less than -2
   b. 5 less than -8
   c. $8\,°C$ below $-1\,°C$
   d. $3\,°C$ below $2\,°C$
   e. 3 to the left of $+4$
   f. 4 to the left of -3

9. **MC** Select which of the following is equal to $7 + (-4) - (-2)$.

   A. 9
   B. 1
   C. 13
   D. 5

10. Model each situation with an integer number sentence that shows the result of the following.

    a. From ground level, a lift went down 2 floors, then down another 3 floors.

    b. An Olympian dived down 5 metres from a board 3 metres above water level.

    c. At 5:00 pm in Falls Creek the temperature was 1 °C. It then fell 6 °C by 11:00 pm.

    d. A bank account with a balance of $500 had $150 withdrawn from it.

11. **MC** From ground level, a lift went down 2 floors, then down another 3 floors to a level 5 floors below the ground. Select which of the following number sentences describes this situation.

    A. $2 + 3 = 5$ **B.** $-2 + (-3) = -5$

    C. $-2 + 3 = 1$ **D.** $2 + (-3) = -1$

12. State whether the following number sentences are true or false. Show your full working.

    a. $7 - 9 = 7 + -9$ **b.** $0 - 8 = -8 - 0$ **c.** $8 - 12 = -12 + 8$

13. **WE14** After lunch, Chen took a tub of ice-cream out of the freezer to get a few scoops, and went out without putting the tub back into the freezer. The tub of ice-cream was initially at -18 °C, and its temperature rose by 16 °C by the time it was found on the kitchen counter. Determine the temperature of the ice-cream tub when it was found.

Communicating, reasoning and problem solving

14. Nihal has to hand in an essay for English class with a maximum of 400 words. Currently their essay has 520 words. Calculate the number of words Nihal has to cut out of their essay.

15. Jill is climbing up a steep and slippery path to fetch a bucket of water. When she is 6 m above her starting point, she slips back 1 m, grabs onto some bushes by the side of the path and climbs 7 m more to a flat section of the path. Determine how far above her starting point Jill is when she reaches the flat section. Show your working.

16. Sharyn opened a bank account when she started working at her local fish and chip shop. Below is a week-by-week breakdown of her transactions.
Week 1: deposited $56.00, then withdrew $45.00
Week 2: deposited $44.00, then withdrew $75.00
Week 3: deposited $52.80, then withdrew $22.00
Week 4: deposited $39.20, then withdrew $50.00
Sharyn's bank account was opened with a $10 deposit.

    a. Determine how much money was in Sharyn's account at the end of the month.

    b. Show full working to determine in which week the balance of Sharyn's account was negative.

17. Stuart is diving in the ocean. They start at a depth of 5 m below the water's surface, then rises 2 m. Stuart then dives a further 12 m down. Calculate the final depth that Stuart reached.

18. A hospital building has 6 floors above ground level and 4 floors below ground level. Over a 10-minute period one busy morning, the lift started from ground level (reception) and made the following moves: up 2 floors, up 1 floor, down 4 floors, down 2 floors, up 3 floors, down 4 floors and up 6 floors.

    a. Determine the floor on which the lift finished.
    b. Determine the number of floors the lift travelled in the whole 10 minutes.

19. Rendah has a maths trick. For her trick she asks people to do the following mental maths steps:

    1. Pick a number between 1 and 10.
    2. Multiply the number by 9.
    3. Add the digits of the number. For example, if you chose the number 25 you would add 2 and 5 to get 7.
    4. _____
    5. Match your number to a letter of the alphabet; for example, 1 is A, 2 is B, 3 is C and so on.
    6. Think of an animal that starts with your letter.
       When the trick is done, Rendah asks people if they are thinking of a dog, and they almost always say yes. If they say no, it's usually because they picked a donkey, a dolphin or a dugong. Determine the missing step in Rendah's trick.

20. Determine the effect of subtracting a number from its opposite integer.

LESSON
2.6 Multiplying and dividing integers

> **LEARNING INTENTION**
>
> At the end of this lesson you should be able to:
> - represent multiples of integers as repeated addition on a number line
> - understand the patterns for multiplying and dividing integers
> - multiply positive and negative integers
> - divide positive and negative integers.

▶ 2.6.1 Multiplication of integers as repeated addition

eles-6131

- Multiplication can be thought of as repeated addition.
  For example, 4×3 is 'four groups of three' or the same as 'adding three, four times'.

$$4 \times 3 = 3 + 3 + 3 + 3$$
$$= 12$$

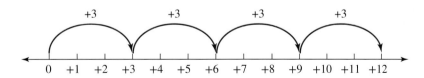

Repeated addition with one negative number

- Multiplication of negative numbers can also be shown using repeated addition.
  For example, 4×-3 is 'four groups of negative three' or the same as 'adding negative three, four times'.

$$4 \times -3 = (-3) + (-3) + (-3) + (-3)$$
$$= -12$$

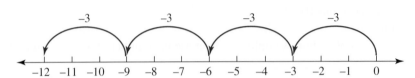

- We can consider -4×3 the same as $-(4 \times 3)$, read as 'four groups of three in the opposite direction', giving the same result of -12.

Repeated addition with two negative numbers

- When a number sentence contains two negative numbers such as -4×-3, we can combine what we have learnt above.
- For example, -4×-3 or $-(4 \times -3)$ can be read as 'four groups of negative three in the opposite direction'. Firstly, we know that 'four groups of negative three' gives the result of -12. Because of the extra negative, this means it would result in the opposite direction equaling positive 12.

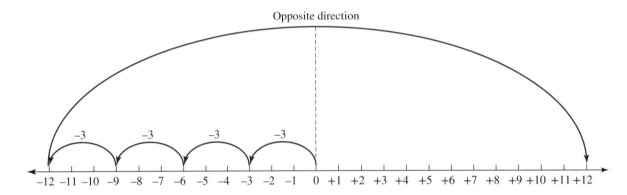

WORKED EXAMPLE 15 Multiplying using a number line

Use a number line to model and evaluate:
a. 3×-2   **b. -3×5**   **c. -3×-2**

THINK

a. 1. 3×-2 is read as 'three groups of negative two'.

2. Draw a number line to model 3×-2.

3. State the answer.

b. 1. -3×5 or $-(3 \times 5)$ is read as 'three groups of five in the negative direction'.

WRITE

a. 3×-2

$3 \times -2 = -6$

b. -3×5

---

2. Draw a number line to model -3×5.

3. State the answer. $-3 \times 5 = -15$

c. 1. -3×-2 is read as 'three groups of negative two in the opposite direction'.   **c.** -3×-2

2. Draw a number line to model 3×-2.

3. Flip to the opposite direction from zero.

4. State the answer. $-3 \times -2 = 6$

▶ 2.6.2 Recognising patterns for multiplication of integers

eles-3702

- Patterns from multiplication tables can also be used to determine the product when two directed numbers are multiplied.
- Consider the following multiplication table — we already know how to complete part of the table.
  - Notice that multiplying by 0 always results in 0.
  - Notice that multiplying by 1 does not change the value of the number.

×	−3	−2	−1	0	1	2	3
3				0	3	6	9
2				0	2	4	6
1				0	1	2	3
0				0	0	0	0
−1							
−2							
−3							

- Using a calculator to help with the multiplication of negative numbers, we can complete the table, as shown here.

×	−3	−2	−1	0	1	2	3
3	−9	−6	−3	0	3	6	9
2	−6	−4	−2	0	2	4	6
1	−3	−2	−1	0	1	2	3
0	0	0	0	0	0	0	0
−1	3	2	1	0	−1	−2	−3
−2	6	4	2	0	−2	−4	−6
−3	9	6	3	0	−3	−6	−9

- From this table, we can develop the following rules for multiplying integers.

Rules for multiplying integers

- **When multiplying two integers with the *same* sign, the answer is always *positive*.**

$$+ \times + = +$$
or
$$- \times - = +$$

- **When multiplying two integers with *different* signs, the answer is always *negative*.**

$$+ \times - = -$$
or
$$- \times + = -$$

WORKED EXAMPLE 16 Multiplying integers

Evaluate:

a. $-5 \times +2$

b. -4×-6

THINK	WRITE
a. 1. Write the question.	a. $-5 \times +2$
2. Negative $\times$ positive = negative.	$= -10$
b. 1. Write the question.	b. -4×-6
2. Negative $\times$ negative = positive.	$= 24$

- Brackets can also be used to represent multiplication. For example, the expression $(-3)(5)$ is the same as -3×5. Likewise, the expression $(2-4)(11-3)$ is the same as $(2-4) \times (11-3)$.

⏵ 2.6.3 Division of integers

eles-3703

- Division is the inverse, or opposite, of multiplication.
- The multiplication rules for directed numbers can be used to work out the division rules for directed numbers.
  Since $3 \times 2 = 6$ then $6 \div 3 = 2$ and $6 \div 2 = 3$.
  Since $3 \times -2 = -6$ then $-6 \div 3 = -2$ and $-6 \div -2 = 3$
  Since $-2 \times -3 = 6$ then $6 \div -3 = -2$ and $6 \div -2 = -3$.

From this, we can develop the following rules for dividing integers.

Rules for dividing integers

- **When dividing two integers with the *same* sign, the answer is always *positive*.**

$$+ \div + = +$$
or
$$- \div - = +$$

- **When dividing two integers with *different* signs, the answer is *negative*.**

$$+ \div - = -$$
or
$$- \div + = -$$

WORKED EXAMPLE 17 Dividing integers

Evaluate the following.

a. $10 \div -2$ b. $-12 \div 4$ c. $-20 \div -5$

THINK	WRITE
a. 1. Write the question.	a. $10 \div -2$
2. The two integers have different signs, so the answer will be negative.	$= -5$
b. 1. Write the question.	b. $-12 \div 4$
2. The two integers have different signs, so the answer will be negative.	$= -3$
c. 1. Write the question.	c. $-20 \div -5$
2. Both numbers have the same sign, so the answer will be positive. *Note:* We could also think of $-20 \div -5$ as 'How many groups of -5 are in -20?'	$= 4$

WORKED EXAMPLE 18 Dividing integers in fraction form

Evaluate:

a. $\dfrac{-16}{+2}$ b. $\dfrac{-45}{-5}$

THINK	WRITE
a. 1. Write the question. *Note:* $\dfrac{-16}{+2}$ is the same as $-16 \div +2$.	a. $\dfrac{-16}{+2}$
2. Evaluate the expression.	$\dfrac{-16}{+2} = -8$
b. 1. Write the question.	b. $\dfrac{-45}{-5}$
2. Evaluate the expression. *Note:* Negative ÷ negative = positive	$\dfrac{-45}{-5} = 9$

Digital technology

Scientific calculators can evaluate multiplication of positive and negative numbers.

```
           D        Math ▲
(2-4)×(11-3)

                  -16
```

Scientific calculators can evaluate division of positive and negative numbers.

$$-15 \div 3$$

$$-5$$

 Resources

 Interactivities Multiplication of integers (int-4038)

Division of integers (int-4039)

Exercise 2.6 Multiplying and dividing integers

learnon

2.6 Quick quiz on	2.6 Exercise

Individual pathways

■ PRACTISE	■ CONSOLIDATE	■ MASTER
1, 2, 5, 10, 15, 16, 19	3, 6, 8, 9, 11, 12, 17, 20	4, 7, 13, 14, 18, 21

Fluency

1. **WE15** Use a number line to model and evaluate:

   a. 4×-2 b. -5×2 c. -2×-4

2. **WE16** Calculate the following.

   a. 2×-5 b. -6×3 c. -7×9

   d. 6×-5 e. -2×-3

3. Evaluate the following.

   a. -6×-3 b. -5×-5 c. 0×-7 d. -8×-1

4. Calculate the following.

   a. 10×-1 b. -15×2 c. -20×-10 d. -6×-6

5. **WE17** Evaluate the following.

   a. $-8 \div -2$ b. $-8 \div 2$ c. $12 \div -3$

   d. $-16 \div -8$ e. $-90 \div -10$

6. Calculate the following.

   a. $88 \div -11$ b. $-6 \div 1$ c. $0 \div -4$

   d. $-84 \div 4$ e. $-184 \div 2$

7. Evaluate the following.

   a. $-125 \div -5$ b. $-67 \div -1$ c. $129 \div -3$

   d. $-284 \div 4$ e. $336 \div -6$

Understanding

8. **WE18** Calculate the following.

    a. $\dfrac{-6}{2}$

    b. $\dfrac{-24}{-8}$

    c. $\dfrac{8}{-8}$

9. Calculate the value of each of the following.

    a. $2 \times -3 \times 4$

    b. $-4 \times -3 \times 3$

    c. $-8 \times 9 \times -2$

10. Fill in the missing numbers in these number sentences.

    a. $6 \times \underline{\hspace{1cm}} = -18$

    b. $\underline{\hspace{1cm}} \times 3 = -18$

    c. $-8 \times \underline{\hspace{1cm}} = -8$

    d. $-8 \times \underline{\hspace{1cm}} = 8$

    e. $-8 \times \underline{\hspace{1cm}} = 0$

11. Fill in the missing numbers in these number sentences.

    a. $-21 \div \underline{\hspace{1cm}} = -7$

    b. $\underline{\hspace{1cm}} \div -8 = -4$

    c. $\underline{\hspace{1cm}} \div -9 = 8$

    d. $-11 \div \underline{\hspace{1cm}} = 1$

    e. $\underline{\hspace{1cm}} \div -7 = 0$

    f. $-150 \div -25 = \underline{\hspace{1cm}}$

12. Evaluate the following expressions.

    a. $-4 + (-4)$

    b. 2×-4

    c. $-2 + (-2) + (-2)$

    d. 3×-2

    e. $-5 + (-5) + (-5) + (-5)$

    f. 4×-5

13. **MC** Select which of the following could be the missing numbers in the number sentence $16 \div \underline{\hspace{0.5cm}} = \underline{\hspace{0.5cm}}$.

    A. $2, -8$ **B.** $-2, -8$ **C.** $-4, 4$ **D.** $-2, 8$

14. **MC** Select which of the following could be the missing numbers in the number sentence $\underline{\hspace{0.5cm}} \div \underline{\hspace{0.5cm}} = -5$.

    A. $-15, 3$ **B.** $15, 5$ **C.** $25, 5$ **D.** $-30, -6$

15. **MC** Six people each owe the bank $50. Calculate the combined total of their six bank accounts.

    A. $300 **B.** $-$50 **C.** $-\$\dfrac{6}{5}$ **D.** $-$300

Communicating, reasoning and problem solving

16. Evaluate the following multiplications.

    a. i. $(-1)(-2)$

        ii. $(-3)(-2)(-1)$

        iii. $(-2)(-3)(-1)(-4)$

        iv. $(-4)(-1)(-5)(-2)(-1)$

    b. Write a rule for multiplying more than two negative numbers.

    c. Does this rule also apply to the division of negative numbers? Explain your answer.

17. When you multiply two integers, is the result an integer? What about when you divide two integers? Explain your answer.

18. A spider is running down the stairs from the first floor of an old lady's house to the basement below. It stops every 5 steps to catch a fly. If there are 26 steps above the ground floor and 14 steps below the ground floor, determine how many flies the spider catches.

19. Blake is jumping down the stairs, stepping on every second step only.
   He jumps a total of ten times.

   a. Determine the number of steps he has skipped.
   b. Determine the number of steps there are altogether.
   c. If each step is 25 cm high, determine how far he has travelled in total.
   d. If going up the stairs is considered going in a positive direction and going down the stairs is considered
     going in a negative direction, rewrite your answer to part c.

20. Manu is working on the expression $8 \div 2 \times (2 + 2)$ and says the answer is 1. Her friend Kareen says the
   answer should be 16. Show, with full working, who is correct.

21. On a test, each correct answer scores 5 points, each incorrect answer scores -2 points, and each question left
   unanswered scores 0 points. A student answers 16 questions on the test correctly, 3 incorrectly, and does not
   answer 1 question. Write an expression for the student's score and calculate the score.

LESSON
2.7 Order of operations with integers

LEARNING INTENTION

At the end of this lesson you should be able to:
• understand how to apply the order of operations with integers when more than one operation is
  involved
• apply the order of operations with positive and negative integers to solve expressions that involve more
  than one operation.

▶ 2.7.1 Applying BIDMAS to positive and negative integers

eles-3704

   • Lesson 1.8 in Topic 1 explored how the order of operations is applied with positive integers when more
     than one mathematical operation is involved.
   • The same order of operations is also applied for negative integers.

- The order of operations involves solving the brackets first, then solving powers or indices, then multiplication and division (working from left to right) and finally addition and subtraction (working from left to right).
- The order of operations can be written as BIDMAS (Brackets, Indices and roots, Division and Multiplication, Addition and Subtraction).

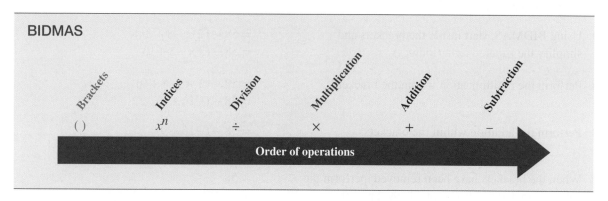

BIDMAS

- Examples of these operations are shown in this table.

B	Brackets	$8 \times (3 + 5) = 8 \times (8) = 64$
I	Indices	$5 + 3^2 = 5 + 9 = 14$
D	Division	$17 - 8 \div 4 = 17 - 2 = 15$
M	Multiplication	$7 + 6 \times 3 = 7 + 18 = 25$
A	Addition	$6 \times 2 + 5 = 12 + 5 = 17$
S	Subtraction	$21 \div 3 - 4 = 7 - 4 = 3$

WORKED EXAMPLE 19 Evaluating using the order of operations

a. $5 \times (2 - 8)$
b. $12 - 9 \div 3$
c. $\dfrac{15 + 9}{6}$

THINK

a. 1. Write the question.
  2. Using BIDMAS, evaluate the brackets.
  3. Perform the multiplication.
  4. Write the answer.

b. 1. Write the question.
  2. Using BIDMAS, perform the division first.
  3. Perform the subtraction.
  4. Write the answer.

c. 1. Write the question.
    Note: It can also be written as $\dfrac{(15 + 9)}{6}$ or $(15 + 9) \div 6$, as the 15 and 9 are both being divided by 6.

  2. Using BIDMAS, evaluate the addition in the brackets.

  3. Perform the division.

  4. Write the answer.

WRITE

a. $5 \times (2 - 8)$
   $= 5 \times -6$
   $= 30$

b. $12 - 9 \div 3$
   $= 12 - 3$
   $= 9$

c. $\dfrac{15 + 9}{6}$

   $= \dfrac{(15 + 9)}{6}$

   $= \dfrac{24}{6}$

   $= 4$

WORKED EXAMPLE 20 Calculating an expression using order of operations

Calculate $58 - (2 \times -8 + 3^2)$ **using the correct order of operations.**

THINK	WRITE
1. Write the question.	$58 - (2 \times -8 + 3^2)$
2. Using BIDMAS, start inside the brackets and simplify the squared term (indices).	$= 58 - (2 \times -8 + 3^2)$ $= 58 - (2 \times -8 + 9)$
3. Perform the multiplication within the brackets.	$= 58 - (2 \times -8 + 9)$ $= 58 - (-16 + 9)$
4. Perform the addition within the brackets.	$= 58 - (-16 + 9)$ $= 58 - (-7)$
5. When the brackets have been removed, perform the subtraction that had been outside the brackets.	$= 58 - -7$ $= 58 + 7$
6. Write the answer.	$= 65$

Digital technology

Scientific calculators can calculate expressions using order of operations.

$$(-4)^2 - 5 \times -3$$
$$31$$

Note: It is important to use brackets around indices. This will be explored more in Topic 3.

---

on Resources

Interactivities Combined operations (int-4040)

learn on

| 2.7 Quick quiz on | 2.7 Exercise |

Individual pathways

| ■ PRACTISE
1, 3, 7, 9, 12, 15 | ■ CONSOLIDATE
2, 4, 8, 10, 13, 16 | ■ MASTER
5, 6, 11, 14, 17, 18 |

Fluency

1. **WE19** Calculate the following using the correct order of operations.

   a. $6 + 3 \times -4$
   b. $18 - \dfrac{12}{-3}$
   c. $8 + -4 - 10$
   d. $17 - 3 + -8$

2. Evaluate each of the following expressions using the correct order of operations.

   a. $6 \times -3 \div 9$
   b. $72 \div 8 \times -3$
   c. $7 + (-3 - 4)$
   d. $\dfrac{(6 + 3)}{-9}$

3. Calculate each of the following expressions using the correct order of operations.

   a. $-3 \times -2 + 3 \times -1$
   b. $-6 \times 5 - 2 \times -6$
   c. $-4 \times (6 - -4)$
   d. $(-8 + 3) \times -7$

4. **WE20** Evaluate each of the expressions using the correct order of operations.

   a. $4 + 7 \times -3 - 2$
   b. $-6 - 4 + (-3)^2$
   c. $(-2)^3 - 3 \times -2$
   d. $3 + (2 - 8) + -6$

5. Calculate each of the expressions using the correct order of operations.

   a. $\dfrac{-8}{2} + (-2)^2$
   b. $-4 \times -8 - [2 + (-3)^2]$
   c. $(-7 + 5) - -24 \div 6$
   d. $\dfrac{-15}{(2 - 5)} - 10$

6. Evaluate each of the expressions using the correct order of operations.

   a. $54 \div -6 + 8 \times -9 \div -4$
   b. $\dfrac{(9 - -6)}{-5} + -8 \times 1$
   c. $-7 + -7 \div -7 \times -7 - -7$
   d. $-9 \times -5 - (3 - -2) + -48 \div 6$

7. **MC** The expression $6 + 2 \times -5 - -10 \div 2$ is equal to:

   A. -15
   B. -35
   C. -60
   D. 1

8. **MC** The expression $(9 - -6) \div -5 + -8 \times 0$ is equal to:

   A. -15
   B. -3
   C. -60
   D. 0

Understanding

9. A submarine dives 100 m below sea level, rises 60 m, then dives 25 m. Calculate its final position.

10. Jini has $2075 in the bank. She makes 2 withdrawals of $150 each, then 5 deposits of $60 each. Calculate the amount of money Jini now has in the bank.

11. A truck can hold 156 L of diesel. After driving for 4 days, there is 64 L left. Calculate the average fuel usage per day.

Communicating, reasoning and problem solving

12. Darcy says there is no difference between the expressions $8 - 2 \times (4 + 2)$ and $8 - 2 \times 4 + 2$. Do you agree with Darcy? Justify your answer with correct working.

13. Are the following expressions true or false? Justify your answer with correct working.
    a. $5 \times 6 - 3 + (5 - 2) < 5 \times (6 - 3) + 5 - 2$
    b. $2 + (15 \div 3) \times 1 = 2 + 15 \div (3 \times 1)$

14. Merlin is riding his bike east at a steady pace of $10 \, \text{km/h}$, while Morgan is riding her bike west at a steady $8 \, \text{km/h}$. They pass each other on Backpedal Bridge at 12 noon. Assume that east is the positive direction and west is negative, and that time before noon is negative and after noon is positive.

    a. Determine the location of each person with respect to the bridge at 9 am.
    b. Determine their locations with respect to the bridge at 2 pm.
    c. Determine how far apart they were at 10 am.
    d. Determine how far apart they will be at 4 pm.

15. Ryan buys 5 vanilla cupcakes at \$5.75 each, 2 cinnamon donuts at \$3.25 each, and 3 chocolate chip cookies at \$2.50 each. Determine the total amount Ryan has to pay.

16. A snail begins to climb up the side of a bucket. It climbs 3 cm and slips back 2 cm, then climbs a further 4 cm and slips back 1 cm. Write a number sentence to help you determine how far the snail is from the bottom of the bucket.

17. At many locations around the world, minimum and maximum temperatures are recorded. On one particular day, the minimum temperature in the Arctic circle was recorded as $-24\,^{\circ}\text{C}$ and the maximum temperature was recorded as $-7\,^{\circ}\text{C}$. In London the maximum temperature was $10\,^{\circ}\text{C}$ and the minimum temperature was $-2\,^{\circ}\text{C}$. In Mexico City the maximum temperature was $38\,^{\circ}\text{C}$ and the minimum temperature was $16\,^{\circ}\text{C}$. Calculate the difference in maximum and minimum temperatures at each of these three places.

18. In a Maths competition marks are awarded as follows: 3 marks for a hard question (H), 2 marks for a medium question (M) and 1 mark for an easy question (E).
    If an answer is incorrect (I), 1 mark is deducted.
    The top three students had the following scores:
    Student 1: 10H, 3M, 5E, 2I
    Student 2: 7H, 7M, 3E, 3I
    Student 3: 9H, 6M, 5E, 0I

    a. Determine the number of marks each student received.
    b. Determine who won the competition.

LESSON
2.8 Review

2.8.1 Topic summary

Number line

Integers can be represented on a number line.

- Positive integers are written to the right of 0.
- Negative integers are written to the left of 0.

Number plane

- The number plane, or Cartesian plane, has two axes:
  - the horizontal axis (x-axis)
  - the vertical axis (y-axis).

- The origin (0, 0) is where the two axes intersect.
  e.g. A (3, 2) move 3 right and 2 up
  B (−3, 2) move 3 left and 2 up
  C (−3, −2) move 3 left and 2 down
  D (3, −2) move 3 right and 2 down

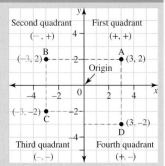

COMPUTATION WITH POSITIVE AND NEGATIVE INTEGERS

Integers

Integers are positive and negative whole numbers including 0.
- The integers −1, −2, −3… are called negative integers.
- The integers 1, 2, 3… are called positive integers.
- 0 is neither positive nor negative.

Addition of integers

A number line can be used to add integers.

- To add a positive integer, move to the right.
  e.g. $-4 + 1 = -3$
- To add a negative integer, move to the left.
  e.g. $5 + (-3) = 2$

Move right when adding a positive integer →

← Move left when adding a negative integer

Subtraction of integers

- Subtracting a number gives the same result as adding its opposite.
  e.g. $2 - (5) = 2 + (-5) = -3$ or
  $-6 - (-11) = -6 + (+11) = 5$

Division of integers

- When dividing integers with the same sign, the answer is positive.
  e.g. $-20 \div -4 = 5$ or $24 \div 6 = 4$
- When dividing integers with different signs, the answer is negative.
  e.g. $10 \div -5 = -2$ or $\dfrac{-35}{7} = -5$

Multiplication of integers

- When multiplying integers with the same sign, the answer is positive.
  e.g. $-4 \times -6 = 24$ or $7 \times 8 = 56$
- When multiplying integers with different signs, the answer is negative.
  e.g. $-5 \times 2 = -10$ or $9 \times -3 = -27$

Order of operations

- The order of operations is a set of rules that is followed so there is a common understanding of the order in which to solve mathematical operations.
- The set order in which we calculate problems is:
  1. **Brackets** — ( ) or [ ]
  2. **Indices or roots** — a^x or $\sqrt[n]{x}$
  3. **Division and Multiplication** (working left to right) — ÷ or ×
  4. **Addition and Subtraction** (working left to right) — + or −
- The acronym **BIDMAS** can be used to remember the correct order of operations.

2.8.2 Project

Directed numbers dice game

In this game, the winner is the first to complete a row of three numbers, horizontally, vertically or diagonally.

Equipment: 2 players, 2 standard dice, Connect-3 board (shown), dice placement sheet (shown), 9 blue counters, 9 red counters

Instructions:

- Roll the two dice and decide whether to add or subtract the numbers on the dice.
- Put the dice on one of the squares on the dice placement sheet to produce a total shown on the Connect-3 board shown.
- Once you have worked out the answer, place your counter on the appropriate number on the Connect-3 board. You should check each other's answers.
- You cannot cover a number that has already been covered. If you are unable to find a total that has not been covered, you must pass. The winner is the first to complete a row of three numbers either horizontally, vertically or diagonally.

Make your own board game using dice with negative numbers on each face and put different numbers on the board. The dice placement sheet could also be changed to include × or ÷ symbols.

Connect-3 board

	−5	−4	−3	−2
−1	0	1	2	3
4	5	6	7	8
9	10	11	12	

Dice placement sheet

 Resources

Interactivities Crossword (int-2604)
Sudoku puzzle (int-3172)

Exercise 2.8 Review questions

learnon

Fluency

1. Select which of the following are integers.
   a. $-2\dfrac{1}{2}$ b. 0.45   c. 0   d. -201

2. Complete each statement by inserting the correct symbol:>, < or =.
   a. $-6\;\square\;-2$ b. $-7\;\square\;7$ c. $0\;\square\;-5$ d. $-100\;\square\;9$

3. List the integers between -21 and -15.

4. Arrange the numbers $-3, 2, 0, -15$ in descending order.

5. State whether the following points are on the x-axis, the y-axis, both axes or in the first quadrant.
   a. $(0, 0)$ b. $(0, 5)$ c. $(3, 0)$ d. $(3, 2)$

6. Draw and appropriately label a set of axes. Plot the following points in the order given, joining each point to the next one. Name the shape that you have drawn.
   $$(-2, 3), (1, 3), (2, -2), (-1, -2), (-2, 3)$$

7. Determine in which quadrant or on which axes the following points lie.
   a. $(-2, 3)$ b. $(3, -1)$ c. $(-4, -1)$ d. $(0, 2)$ e. $(-1, 0)$ f. $(7, 9)$

▶

8. Calculate the following.

   a. $-12 + 7$ **b.** $-9 + -8$ **c.** $18 + -10$ **d.** $5 + 1$

9. Write the number that is 2 more than each of the following integers.

   a. -4 **b.** 5 **c.** -1 **d.** 0 **e.** -2

10. Calculate the following.

    a. $-5 - 3$ **b.** $17 - -9$ **c.** $-6 - -9$ **d.** $6 - 8$ **e.** $12 - 20$ **f.** $-10 - -12$

11. Evaluate the following.

    a. -6×7 **b.** 4×-8 **c.** -2×-5 **d.** $(-8)^2$ **e.** -8^2 **f.** $-2 \times (-8)^2$

12. Calculate the following.

    a. $-36 \div 3$ **b.** $-21 \div -7$ **c.** $45 \div -9$ **d.** $\dfrac{-18}{-2}$ **e.** $64 \div -4$ **f.** $\dfrac{-100}{25}$

Understanding

13. Evaluate the following expressions.

    a. $10 - 6 \times 2$

    b. $-7 - \dfrac{-8}{2}$

    c. $-3 \times -5 - -6 \times 2$

    d. $(-3 - 12) \div (-10 + 7)$

14. Replace the box with $=$, $<$ or $>$ to make each of the following statements correct.

    a. $-5 \,\square\, -3$ **b.** $-22 \times -2 \,\square\, 44$ **c.** $4 \,\square\, 2$

    d. $-5 \times -3 \,\square\, -15$ **e.** $0 \,\square\, -7$ **f.** $-2(-4 + 7) \,\square\, 6$

15. Replace the box with $=$, $<$ or $>$ to make each of the following statements correct.

    a. $-2 \times 5 \,\square\, -9$ **b.** $5 \times (-2 - 18) \div -4 \,\square\, 25$ **c.** $12 \div -4 \,\square\, 3$

    d. $5 \times (-2 - 18) \div 4 \,\square\, -25$ **e.** $-10 \times -5 \,\square\, 50$ **f.** $5 \times -2 - 18 \div 4 \,\square\, 2.5$

Communicating, reasoning and problem solving

16. Model the following situation with integers, then find the result. A scuba diver at 52 metres below sea level made her ascent in 3 stages of 15 metres each. Determine how far below sea level she was then.

17. Some historians believe that the Roman era began in 146 BCE and ended in 455 CE. According to this theory, calculate the length of the Roman era.

18. The lowest temperature recorded on Earth was $-90\,°C$ in Antarctica and the highest was $58\,°C$ in Africa. Determine the difference between these two temperatures.

19. Nazeem's bank account has a balance of $-\$43$. He pays a bill of $\$51$ using direct debit. Calculate his bank balance now.

20. For the following expressions, determine if the answer will be negative or positive. Explain your answers using mathematical reasoning.

    a. $-17\,489 - 25\,636$

    b. $-65\,234 + 123\,468$

    c. $-44\,322 + 31\,212$

21. Examine the graph shown.
    a. Determine the coordinates of a point, D, that can be added to the graph so that ABCD is a rectangle.
    b. Name the combination of points that make:
        i. a right-angled triangle
        ii. an isosceles triangle that doesn't include a right angle.
        (There may be more than one solution.)
    c. Determine the coordinates of a point, F, that can be added to the graph so that ACFE is a parallelogram. Mark point F on the grid and draw the parallelogram.
    d. State the coordinates of the midpoint of CA.
    e. State the coordinates of the midpoint of AE.
    f. Estimate the coordinates of G.

22. Answer the following questions.
    a. Add numbers to the empty sections in this diagram so that the numbers in each circle add up to zero.
    b. Add numbers to the empty sections in a different way so that the numbers in each circle add up to −2.

23. If $(X - Y > X)$, $(Y - X < Y)$ and X and Y are integers, determine whether X and Y are positive or negative values.

24. The negative integers −2 to −1000 can be arranged in 5 columns. A section of those 5 columns is shown.

Column 1	Column 2	Column 3	Column 4	Column 5
	−2	−3	−4	−5
−9	−8	−7	−6	
	−10	−11	−12	−13
−17	−16	−15	−14	

Identify which column −1000 would be in.

To test your understanding and knowledge of this topic, go to your learnON title at www.jacplus.com.au and complete the **post-test**.

Answers

Topic 2 Computation with positive and negative integers

2.1 Pre-test

1. a. -250 b. -15 c. $+20\,°C$
2. a. $-3 < 2$ b. $-5 > -7$ c. $-21 < -17$
   d. $0 > -8$
3. B
4. A $(3, -2)$
5. C
6. C
7. a. -2 b. 13   c. -19 d. -11
8. 7
9. 78 m
10. 48
11. a. 2   b. 8   c. 130
12. D
13. 6
14. $380
15. 108

2.2 Integers on the number line

1. a. -7 b. 10 or $+10$
   c. -422 d. 2228 or $+2228$
   e. -3 f. -30
2.
3. a. $-6,\ -2,\ 3,\ 5,\ 7$ b. $-6,\ -4,\ -1,\ 0,\ 3$
   c. $-5,\ -3,\ -2,\ 1,\ 7$ d. $-4, -3, 1, 2, 5$
4. a. 6   b. -7 c. -1 d. 8
5. a. >   b. <   c. >
   d. >   e. <   f. <
6. a. i. -5 ii. 7
   b. i. -4 ii. 5
   c. i. -10 ii. 7
   d. i. -4 ii. 3
7. $-2\,°C$
8. 14
9. Sample response: There were human settlements here 60 thousand years ago.
10. a.

Vanessa	School	Raj
-1.6	0	1.2

   b. 2.8 km

11. a.
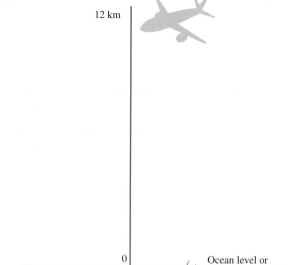

   12 km

   0   Ocean level or sea level

   -0.5 km

   b. 12.5 km
12. a. 2000 years in Common Era.
    b. i. 10   ii. -106
    c. i. 4   ii. 12   iii. 26   iv. 42   v. 57
    d.

   Neil Armstrong and Buzz Aldrin   *Discovery* mission
   Yuri Gagarin   Sally Ride   *Rosetta* mission

2.3 Integers on the number plane

1. a. $(3, 4)$ b. $(8, 4)$ c. $(4, 8)$
   d. $(0, 7)$ e. $(5, 0)$
2. a. G   b. F   c. H   d. J
3. C
4. D
5. $(-5, 2)$
6. a. $(3, 1)$; 1st quadrant
   b. $(-2, 4)$; 2nd quadrant
   c. $(-6, -5)$; 3rd quadrant
   d. $(0, -3)$; y-axis
   e. $(6, -6)$; 4th quadrant
7. a. C; 2nd quadrant   b. L; y-axis
   c. D; 4th quadrant   d. E; 1st quadrant
   e. G; x-axis
8. a. 3   b. 3   c. 7   d. 0
9. a. 2   b. -6 c. 0   d. -3
10. H
11. a. True   b. False   c. False
    d. True
12. D
13. B
14. B

15.

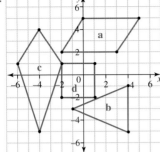

   a. Parallelogram   b. Isosceles triangle

   c. Kite   d. Square

16. According to the graph, both points F and D are 3 units down from the origin along the y-axis. Therefore they have the same y-coordinate. Similarly, both A and D are 3 units to the right along the x-axis. Therefore, they have the same x-coordinate.

17. a. (2, 1)

   b. (1, −2)

   c. Draw line DO

   d. (−3, 1), (−2, 1), (−1, 1) etc. Any two points on the line DE are correct.

   e. (−5, −2)

18. a. West   **b.** East

   c. West   **d.** East

   e. An odd number of negative signs turns you west, and an even number of negative signs turns you east.

   f. South

   g. An odd number of negative signs turns you south, and an even number of negative signs turns you north.

19. 9 units

20. a.

   b.

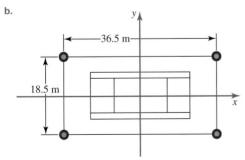

Top left post (−18.25, 9.25)
Top right post (18.25, 9.25)
Bottom right post (18.25, −9.25)
Bottom left post (−18.25, −9.25)

21. Points in the first quadrant are in the form (+, +). Swapping the x- and y-coordinates creates a point in the form (+, +), so the point will remain in the first quadrant. Points in the second quadrant are in the form (−, +). Swapping the x- and y-coordinates creates a point in the form (+, −), so the point will change to the fourth quadrant.
Points in the third quadrant are in the form (−, −). Swapping the x- and y-coordinates creates a point in the form (−, −), so the point will remain in the third quadrant. Points in the fourth quadrant are in the form (+, −). Swapping the x- and y-coordinates creates a point in the form (−, +), so the point will change to the second quadrant.

2.4 Adding integers

1. a. −15   **b.** −10   **c.** −1

   d. −13   **e.** 19   **f.** 0

2. a. $-4 + (+4) = 0$ **b.** $-1 + (+3) = +2$

   c. $-2 + (-4) = -6$ **d.** $+2 + (-4) = -2$

   e. $+6 + (-4) = +2$ **f.** $-1 + (-3) = -4$

3. a. −1   **b.** −1   **c.** 0

   d. +1   **e.** −2

4. a. $5 + (-2) = 3$ **b.** $-3 + (-4) = -7$

   c. $-2(+2) = 0$ **d.** $6 + (-5) = 1$

   e. $-5(+5) = 0$

5. a. $4 + (-6) = -2$ **b.** $-5 + 7 = 2$

   c. $6 + (-9) = -3$ **d.** $3 + (-3) = 0$

   e. $0 + (-6) = -6$

6. a. −7   **b.** −2   **c.** 0

   d. −4   **e.** 4   **f.** −10

7. a. 4   **b.** −3   **c.** −100

   d. 10   **e.** 2   **f.** −22

8.

+	−13	5
21	8	26
−18	−31	−13

9. a. −8   **b.** −6   **c.** 10

   d. −7   **e.** 3

10. a.

+	−2	−1	0	1	2
−2	−4	−3	−2	−1	0
−1	−3	−2	−1	0	1
0	−2	−1	0	1	2
1	−1	0	1	2	3
2	0	1	2	3	4

   b. **i.** Even numbers, increase by increments of 2

      ii. Zeros (addition of opposites)

      iii. Yes

11. a. $-2 + 3 = 1$ **b.** $-3 + 5 = 2$

   c. $-50 + 63 = 13$ **d.** $-50 + 26 = -24$

   e. $200 + (-350) = -150$

12. a. Sample response: A diver is 3 m deep and dives a further 2 m deeper until she is 5 m deep.

b. Sample response: A water pipe was drilled to 10 m deep, but a further 40 m was required to reach the water, which was 50 m deep.

c. Sample response: The lift was on the second floor and then travelled six floors down, stopping at the fourth-floor basement car park.

d. Sample response: Billy borrowed $20 from his mum. One week later he paid the loan back in full.

e. Sample response: Shalini gave eight lollies to her brother and later she got ten more lollies from her dad, leaving her with two extra lollies.

13. a. 'When adding two negative numbers, add the two numbers as if they were positive numbers and then write a negative sign in front of the answer.'

b. 'When adding a negative numbers and a positive number, ignore the signs, subtract the smaller number and then write the sign the large number has.'

14.

2	−12	7
4	−1	−6
−9	10	−4

15. $5575

16. 15 points

17. 2nd floor

18. If you add a number to its opposite, the answer will always be zero. For example: $-2 + 2 = 0$.

19. 4:00 pm

2.5 Subtracting integers

1. a. 5   b. −24   c. −5
   d. 26   e. −16   f. 11

2. a. 2   b. 10   c. −10
   d. 2   e. −9

3. a. −10   b. 0   c. −4
   d. 3   e. 10   f. 0

4. a. $2 - (+5) = -3$ b. $-3 - (+6) = -9$
   c. $-4 - (-9) = 5$ d. $5 - (-7) = 12$

5. a. −10   b. 13   c. 3
   d. 12

6. a. −16   b. −17   c. 23
   d. −12

7. a. 1   b. 9   c. −25   d. 5

8. a. $-2 - 6 = -8$ b. $-8 - 5 = -13$
   c. $-1 - 8 = -9$ d. $2 - 3 = -1$
   e. $4 - 3 = 1$ f. $-3 - 4 = -7$

9. D

10. a. $-2 - 3 = -5$ b. $3 - 5 = -2$
    c. $1 - 6 = -5$ d. $500 - $150 = 350

11. B

12. a. True   b. True
    c. True

13. −2 °C

14. $400 - 520 = -120$. Nihal has to cut 120 words.

15. 12 m

16. a. Balance at end of month: $10

b. During the second week her balance was negative.

17. Depth of 5 m below the surface $= -5$
    An ascent of 2 m $= +2$
    A descent of 12 m $= -12$
    Total depth $= -5 + 2 - 12 = -15$
    $-15 = 15$ m below the surface

18. a. 2nd floor

b. It travelled 22 floors in 10 minutes.

Subtract 5 from the number: $9 - 5 = 4$

20. If you subtract a number from its opposite, the answer will be double the original number value, but it can be positive or negative. For example: $2 - (-2) = 4$, or $-2 - 2 = -4$.

2.6 Multiplying and dividing integers

1. a.

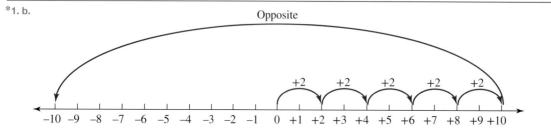

−8

b. See the bottom of the page*
   −10

---

*1. b.

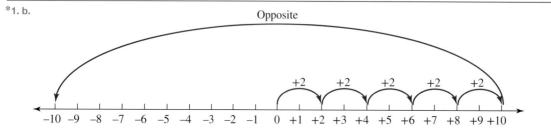

c. See the bottom of the page*

8

2. a. -10 b. -18 c. -63
   d. -30 e. 6

3. a. 18 b. 25 c. 0 d. 8

4. a. -10 b. -30 c. 200 d. 36

5. a. 4 b. -4 c. -4
   d. 2 e. 9

6. a. -8 b. -6 c. 0
   d. -21 e. -92

7. a. 25 b. 67 c. -43
   d. -71 e. -56

8. a. -3 b. 3 c. -1

9. a. -24 b. 36 c. 144

10. a. -3 b. -6 c. 1 d. -1 e. 0

11. a. 3 b. 32 c. -72
    d. -11 e. 0 f. 6

12. a. -8 b. -8 c. -6
    d. -6 e. -20 f. -20

13. B

14. A

15. D

16. a. i. 2 ii. -6 iii. 24 iv. -40

   b. When multiplying an even number of negative numbers, the answer is always positive. When multiplying an odd number of negative numbers, the answer is always negative.

   c. Yes, because multiplication and division have the same rules when working with negative numbers.

17. Agree. When you multiply two integers, the result is always an integer. When you divide two integers, the answer is not always an integer. If the integers do not divide equally, the answer will be a fraction.

18. 8 flies

19. a. 10 b. 20 c. 5 m   d. -5 m

20. According to Kareen: $8 \div 2 \times (2 + 2)$ applying BIDMAS to this expression
$$= 8 \div 2 \times (4)$$
$$= 4 \times 4$$
$$= 16$$
Kareen is correct, We do need to apply BIDMAS when more than one operation is used in an expression.

21. $16 \times 5 + 3 \times -2 + 0 = 80 - 6 = 74$ marks

2.7 Order of operations with integers

1. a. -6 b. 22 c. -6 d. 6

2. a. -2 b. -27 c. 0 d. -1

3. a. 3 b. -18 c. -40 d. 35

4. a. -19 b. -1 c. -2 d. -9

5. a. 0 b. 21 c. 2 d. -5

6. a. 9 b. -11 c. -7 d. 32

7. D

8. B

9. $-100 + 60 + -25 = -65$ m

10. $2075

11. 23 L per day

12. The two expressions are different due to the placement of the bracket in the first expression. See the working below.
First expression: $8 - 2 \times (4 + 2)$
$$= 8 - 2 \times (6)$$
$$= 8 - 12 = -4$$
Second expression: $8 - 2 \times 4 + 2 = 8 - 8 + 2$
$$= 0 + 2$$
$$= +2$$

13. a. False. See the working below.
$5 \times 6 - 3 + (5{-}2) < 5 \times (6 - 3) + 5{-}2$
Apply BIDMAS to both of the expressions.
$5 \times 6 - 3 + (5 - 2) = 5 \times 6 - 3 + (3)$
$= 30 - 3 + 3 = 33 - 3 = 30$ (larger)
$5 \times (6 - 3) + 5 - 2 = 5 \times (3) + 5 - 2$
$= 15 + 5 - 2 = 20 - 2 = 18$ (smaller)
$5 \times 6 - 3 + (5 - 2) >$
$5 \times (6 - 3) + 5 - 2.$
Therefore, the statement given is not true.

   b. True. See the working below.
$2 + (15 \div 3) \times 1 = 2 + 15 \div (3 \times 1)$
Apply BIDMAS to both of the expressions.
$2 + (15 \div 3) \times 1 = 2 + (5) \times 1 = 2 + 5 = 7$
$2 + 15 \div (3 \times 1) = 2 + 15 \div (3) = 2 + 5 = 7$
Because both the expressions are equal to 7, the statement is true.

14. a. Merlin: $-3 \times 10 = -30$, 30 km west of bridge
Morgan: $-3 \times -8 = 24$, 24 km east of bridge

   b. Merlin: $2 \times 10 = 20$, 20 km east of bridge
Morgan: $2 \times -8 = -16$, 16 km west of bridge

   c. 36 km

   d. 72 km

15. The total amount Ryan paid to the bakery is $42.75.

16. $3 + -2 + 4 + -1 = 4$
The snail is 4 cm from the bottom of the bucket.

*1. c.

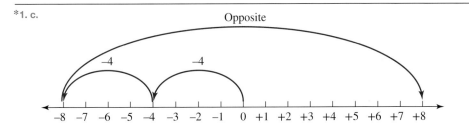

17. a. Arctic Circle 17 °C,
London 12 °C,
Mexico 22 °C

18. a. Student 1: 39 marks; Student 2: 35 marks; Student 3, 44 marks

b. Student 3 won the competition with 44 marks.

Project

Students should play the game to understand how it works. Then students should present a game they have created, which should include positive and negative numbers and any operation.

2.8 Review questions

1. a. No   **b.** No   **c.** Yes   **d.** Yes

2. a. <   **b.** <   **c.** >   **d.** <

3. $-20, -19, -18 - 17, -16$

4. $2, 0, -3, -15$

5. a. Both axes   **b.** y-axis
    c. x-axis   **d.** First quadrant

6.

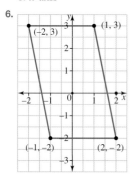

Parallelogram

7. a. Second quadrant   **b.** Fourth quadrant
    c. Third quadrant   **d.** y-axis
    e. x-axis   **f.** First quadrant

8. a. -5 **b.** -17 **c.** 8   **d.** 6

9. a. -2 **b.** 7   **c.** 1   **d.** 2
    e. 0

10. a. -8 **b.** 26   **c.** 3   **d.** -2
    e. -8 **f.** 2

11. a. -42 **b.** -32 **c.** 10   **d.** 64
    e. -64 **f.** -128

12. a. -12 **b.** 3   **c.** -5 **d.** 9
    e. -16 **f.** -4

13. a. -2 **b.** -3 **c.** 27   **d.** 5

14. a. $-5 < -3$ **b.** $-22 \times -2 = 44$
    c. $4 > 2$ **d.** $-5 \times 3 > -15$
    e. $0 > -7$ **f.** $-2(-4 + 7) < 6$

15. a. $-2 \times 5 < -9$ **b.** $5 \times (-2 - 18) \div -4 = 25$
    c. $12 \div -4 < 3$ **d.** $5 \times (-2 - 18) \div 4 = -25$
    e. $-10 \times -5 = 50$ **f.** $5 \times -2 - 18 \div 4 < 2.5$

16. $-52 + 3 \times 15 = -7$

17. 601 years

18. 148 °C

19. $-\$94$

20. a. Negative

b. Positive

c. Negative
Two negative numbers added together will always be negative. When a positive and a negative number are added together, if the larger number is positive the result will be positive. If the larger number is negative the result will be negative.

21. a. $D(-1, 0)$

b. **i.** CBA, BAD (others are possible).

     ii. ADE or BCE

c. $F(-3, -1)$

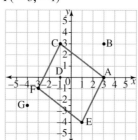

d. $(1, 1.5)$

e. $(2, -2)$

f. $G(-4, -2.5)$

22. a.

b.

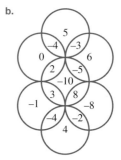

23. X is positive and Y is negative.

24. Column 2

3 Factors, multiples, indices and primes

LESSON
3.1 Overview

3.1.1 Why learn this?

Whole numbers can be written in different ways that can make calculations easier if you do not have your calculator handy. You have been using multiplication tables already without necessarily realising that the they give you numbers that are multiples of each other. Knowing how to write numbers in different ways helps in understanding how numbers are connected.

Indices are short ways of writing a repeated multiplication and are very useful in everyday life, because they allow us to write very large and very small numbers more easily. Indices can simplify calculations involving these very large or very small numbers. Astronomy is a branch of science in which very large distances are involved. The stars we see in the Southern Hemisphere night sky are different to the stars people see in the Northern Hemisphere. Alpha Crucis, the brightest and closest star to us in the Southern Cross constellation, is approximately 3000 million billion kilometres from Earth. That is a huge number — 3 followed by 15 zeros! Other scientists and engineers also need to be able to communicate and work with numbers of all sizes.

1. **MC** The first 3 multiples of 8 are:

   A. 2, 4, 8   B. 8, 16, 24   C. 4, 8, 16   D. 1, 2, 4

2. Kate earned $7 pocket money each week for 5 weeks for emptying the dishwasher. Given she saves all her money, calculate how much she would have saved at the end of the 5 weeks.

3. Determine the highest common factor of 42 and 66.

4. Fred takes 3 minutes to complete one lap of a running course, while Sam takes 4 minutes to complete one lap. After the start, determine when Fred will next be running beside Sam.

5. Write $5 \times 3 \times 5 \times 3 \times 3 \times 3$ in simplified index notation.

6. Select all the prime numbers from the following.

   A. 21   B. 109   C. 33   D. 81   E. 17   F. 31   G. 91

7. Select two pairs of prime numbers whose sum is 20.

   A. 1, 19   B. 2, 18   C. 3, 17   D. 5, 15   E. 7, 13   F. 9, 11

8. Select all the prime factors of 40.

   A. 1   B. 2   C. 3   D. 4   E. 5
   F. 8   G. 10   H. 20   I. 40

9. Determine the lowest common multiple (LCM) and highest common factor (HCF) of 54 and 96 by first expressing each number as a product of its prime factors.

10. A chessboard is a large square with sides that are 8 small squares in length. Calculate the total number of small squares on a chessboard.

11. Evaluate the following.
    $3^2 + \sqrt{16} \times 5^2$

12. Determine between which two integers $\sqrt{70}$ lies.

13. Evaluate the following.
    $\sqrt[3]{27}$

14. Sarah wants to organise a party with her friends at short notice. She decides to send a text to three of her friends, who contact another three friends each, and then they all contact a further three friends each.
    Determine the number of friends that are invited to the party.

15. Evaluate the following.
    $$\frac{11^2 - \sqrt[3]{216} \times \sqrt{144}}{\sqrt{49}}$$

LESSON
3.2 Factors and multiples

LEARNING INTENTION

At the end of this lesson you should be able to:
- list multiples and factors of a whole number
- determine multiples and factor pairs.

▶ 3.2.1 Multiples

eles-3725

- A **multiple** of a number is the result of **multiplying** that number by another whole number. For example, all numbers in the 5 times table are multiples of 5: so 5, 10, 15, 20, 25, ... are all multiples of 5.

×	1	2	3	4	5
1	1	2	3	4	5
2	2	4	6	8	10
3	3	6	9	12	15
4	4	8	12	16	20
5	5	10	15	20	25

WORKED EXAMPLE 1 Listing multiples of a whole number

List the first five multiples of 7.

THINK	WRITE
1. The first multiple is the number × 1, that is, 7 × 1.	7
2. The second multiple is the number × 2, that is, 7 × 2.	14
3. The third multiple is the number × 3, that is, 7 × 3.	21
4. The fourth multiple is the number × 4, that is, 7 × 4.	28
5. The fifth multiple is the number × 5, that is, 7 × 5.	35
6. Write the answer as a sentence.	The first five multiples of 7 are 7, 14, 21, 28 and 35.

WORKED EXAMPLE 2 Determining multiples

State which numbers in the following list are multiples of 8.
18, 8, 80, 100, 24, 60, 9, 40

THINK	WRITE
1. The biggest number in the list is 100. List multiples of 8 using the known multiplications of 8 just past 100; that is, 8 × 1 = 8, 8 × 2 = 16, 8 × 3 = 24, 8 × 4 = 32, 8 × 5 = 40, and so on.	8, 16, 24, 32, 40, 48, 56, 64, 72, 80, 88, 96, 104
2. Write any multiples that appear in the given list.	Numbers in the given list that are multiples of 8 are 8, 24, 40 and 80.

3.2.2 Factors

eles-3726

- A **factor** of a whole number **divides** into that whole number exactly. For example, the number 4 is a factor of 8 because 4 divides into 8 twice or $8 \div 4 = 2$.
- Factors of a number can be written as factor pairs. For example, since $4 \times 2 = 8$, it follows that 2 and 4 are a factor pair of 8.

> **Unique factors**
>
> **Except for the number 1, every whole number has at least two unique factors, 1 and itself.**
>
> **For example, the factors of 5 are 1 and 5.**

WORKED EXAMPLE 3 Listing factors of a whole number

List all the factors of 14.

THINK	WRITE
1. 1 is a factor of every number, and the number itself is a factor of itself, that is, $1 \times 14 = 14$.	1, 14
2. 14 is an even number, so 14 is divisible by 2; therefore, 2 is a factor. Divide the number by 2 to find the other factor ($14 \div 2 = 7$). There are no other whole numbers that divide evenly into 14.	2, 7
3. Write the answer as a sentence, placing the factors in order from smallest to largest.	The factors of 14 are 1, 2, 7 and 14.

WORKED EXAMPLE 4 Determining factor pairs

List the factor pairs of 30.

THINK	WRITE
1. 1 and the number itself are factors, that is, $1 \times 30 = 30$.	1, 30
2. 30 is an even number, so 2 and 15 are factors, that is, $2 \times 15 = 30$.	2, 15
3. Divide the next smallest number into 30. Therefore, 3 and 10 are factors, that is, $3 \times 10 = 30$.	3, 10
4. 30 ends in 0 so 5 divides evenly into 30, that is, $5 \times 6 = 30$.	5, 6
5. Write the answer as a sentence by listing the factor pairs.	The factor pairs of 30 are: 1, 30; 2, 15; 3, 10 and 5, 6.

COMMUNICATING — COLLABORATIVE TASK: Sorting multiples and factors

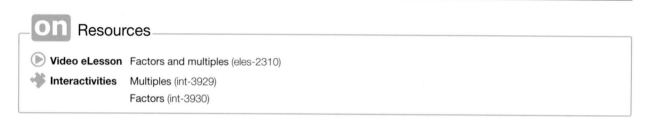

1. As a class, draw a large 10×5 number chart on the board and number the boxes from 1 to 50.
2. Your teacher will share the numbers 1 to 50 among the class. For each of your numbers, list the multiples up to and including 50.
3. Your teacher will call out the numbers between 1 and 50 one at a time. When one of your numbers is called, write it in each box on the chart that is a multiple of your number. For example, if your number is 10, write 10 in each box that corresponds to a multiple of 10.
4. At the end of the activity, the numbers written in each of the boxes are the factors of the box numbers. You can see that different numbers have different numbers of factors.
5. As a class, investigate:
   a. which numbers have one factor, two factors, three factors or four factors
   b. which numbers have more than four factors
   c. whether there are more numbers with an odd or even number of factors
   d. how factors and multiples are related.

Exercise 3.2 Factors and multiples   learn on

3.2 Quick quiz on	3.2 Exercise

Individual pathways

■ PRACTISE	■ CONSOLIDATE	■ MASTER
1, 4, 6, 8, 10, 12, 15, 18, 22, 23, 26	2, 5, 9, 13, 16, 19, 24, 27	3, 7, 11, 14, 17, 20, 21, 25, 28

Fluency

1. **WE1** List the first five multiples of the following numbers.
   a. 3   b. 6   c. 100   d. 11

2. List the first five multiples of the following numbers.
   a. 120   b. 45   c. 72   d. 33

3. **WE2** Select the numbers in the following list that are multiples of 10.
   10, 15, 20, 100, 38, 62, 70

4. Select the numbers in the following list that are multiples of 7.
   17, 21, 7, 70, 47, 27, 35

5. Write the numbers in the following list that are multiples of 16.
   16, 8, 24, 64, 160, 42, 4, 32, 1, 2, 80

6. List the multiples of 9 that are less than 100.

7. List the multiples of 6 between 100 and 160.

8. **MC** The first three multiples of 9 are:

   A. 1, 3, 9   B. 3, 6, 9
   C. 9, 18, 27   D. 18, 27, 36

9. **MC** The first three multiples of 15 are:

   A. 15, 30, 45   B. 30, 45, 60
   C. 1, 15, 30   D. 3, 5, 15

10. **WE3** List all the factors of the following numbers.

    a. 12   b. 40

11. A list of numbers is: 2, 7, 5, 20, 25, 15, 10, 3, 1.
    From this list, select all the numbers that are factors of each of the following values.

    a. 28   b. 60   c. 100

Understanding

12. List all factors of the following numbers in ascending order.

    a. 72   b. 250

13. A list of numbers is: 21, 5, 11, 9, 1, 33, 3, 17, 7.
    From this list, select all the numbers that are factors of each of the following values.

    a. 85   b. 99   c. 51

14. List the factor pairs of 20.

15. List the factor pairs of 132.

16. **MC** Select all the factor pairs of 18.

    A. 1, 18   B. 3, 6
    C. 2, 9   D. 6, 12

17. Three is a factor of 12. State the smallest number greater than 12 that has 3 as one of its factors.

18. **MC** A factor pair of 24 is:

    A. 2, 4   B. 4, 6
    C. 6, 2   D. 2, 8

19. **MC** A factor pair of 42 is:

    A. 6, 7   B. 20, 2
    C. 21, 1   D. 6, 8

20. Determine which of the numbers 3, 4, 5 and 11 are factors of 2004.

Communicating, reasoning and problem solving

21. Alex and Nadia were racing down a flight of stairs. Nadia took the stairs two at a time while Alex took the stairs three at a time. In each case, they reached the bottom with no steps left over.

    a. Determine how many steps there are in the flight of stairs. List three possible answers.
    b. Determine the smallest number of steps there could be.
    c. If Alex can also take the stairs five at a time with no steps left over, determine the smallest number of steps in the flight of stairs.

22. Place each of the first six multiples of 3 into the separate circles around the triangle so the numbers along each side of the triangle add up to 27. Use each number once only.

    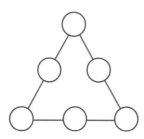

23. Ms Pythagoras is trying to organise their Year 4 class into rows for their class photograph.
    If Ms Pythagoras wishes to organise the 20 students into rows containing equal numbers of students, determine what possible arrangements they can have.
    Note: Ms Pythagoras will not be in the photograph.

24. My age is a multiple of 3 and a factor of 60. The sum of the digits of my age is 3. Determine my age. (There are three possible answers.)

25. A man and his grandson share the same birthday. For exactly six consecutive years the grandson's age is a factor of his grandfather's age. Determine how old each of them is at the sixth of these birthdays, given that the grandson is no more than 10.

26. I am a two-digit number that can be divided by 3 with no remainder. The sum of my digits is a multiple of 4 and 6. My first digit is double my second digit. Evaluate the number. Show your working.

27. Determine the following number. I am a multiple of 5 with factors of 6, 4 and 3. The sum of my digits is 6. Show your working.

28. Identify a two-digit number such that if you subtract 3 from it, the result is a multiple of 3; if you subtract 4 from it, the result is a multiple of 4; and if you subtract 5 from it, the result is a multiple of 5. Explain how you reached the answer.

LESSON
3.3 Lowest common multiple and highest common factor

LEARNING INTENTION

At the end of this lesson you should be able to:
- determine the lowest common multiple of two or more whole numbers
- determine the highest common factor of two or more whole numbers.

3.3.1 The LCM and HCF

eles-3727

- **Common multiples** are numbers that are multiples of more than one number.
  Multiples of 3: 3, 6, 9, **12**, 15, 18, 21, **24**, …
  Multiples of 4: 4, 8, **12**, 16, 20, **24**, 28, …
  The first two common multiples of 3 and 4 are 12 and 24.
- The **lowest common multiple (LCM)** is the smallest multiple that is common to two or more numbers. The LCM of 3 and 4 is 12.

- **Common factors** are numbers that are factors of more than one number.
  Factors of 16: **1**, **2**, **4**, **8** and 16
  Factors of 24: **1**, **2**, 3, **4**, 6, **8**, 12 and 24
  The common factors of 16 and 24 are 1, 2, 4 and 8.
- The common factors of 16 and 24 can be seen in the overlapping region of a Venn diagram, as shown.
- The **highest common factor (HCF)** is the largest of the common factors. The HCF of 16 and 24 is 8.

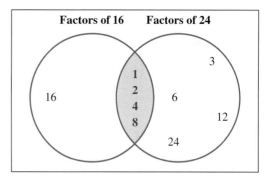

WORKED EXAMPLE 5 Determining the lowest common multiple of two numbers

a. **List the first nine multiples of 8.**
b. **List the first nine multiples of 12.**
c. **List the multiples that 8 and 12 have in common.**
d. **State the lowest common multiple of 8 and 12.**

THINK	WRITE
a. Write out the first nine numbers of the 8 times table.	a. 8, 16, 24, 32, 40, 48, 56, 64, 72
b. Write out the first nine numbers of the 12 times table.	b. 12, 24, 36, 48, 60, 72, 84, 96, 108
c. Choose the numbers that appear in both lists.	c. 24, 48, 72
d. State the lowest of these numbers to give the lowest common multiple (LCM).	d. LCM is 24.

WORKED EXAMPLE 6 Determining the highest common factor

a. Determine the common factors of 8 and 24 by:
  i. listing the factors of 8
  ii. listing the factors of 24
  iii. listing the factors common to both 8 and 24.
b. State the highest common factor of 8 and 24.

THINK	WRITE
a. i. 1. List the pairs of factors of 8.	a. i. 1, 8, 2, 4
2. Write them in order.	Factors of 8 are 1, 2, 4, 8.
ii. 1. List the pairs of factors of 24.	ii. 1, 24; 2, 12; 3, 8; 4, 6
2. Write them in order.	Factors of 24 are 1, 2, 3, 4, 6, 8, 12, 24.
iii. Write the common factors.	iii. Common factors are 1, 2, 4, 8.
b. State the highest common factor, that is, the largest of the common factors.	b. HCF is 8.

 Resources

 Interactivities Lowest common multiple (int-3931)

Highest common factor (int-4545)

Exercise 3.3 Lowest common multiple and highest common factor

learn

3.3 Quick quiz on	3.3 Exercise

Individual pathways

■ PRACTISE	■ CONSOLIDATE	■ MASTER
1, 2, 7, 10, 11, 12, 16, 19	3, 4, 8, 13, 14, 17, 20	5, 6, 9, 15, 18, 21

Fluency

1. a. **WE5** List the first ten multiples of 4.
   b. List the first ten multiples of 6.
   c. In your lists, circle the multiples that 4 and 6 have in common (that is, circle the numbers that appear in both lists).
   d. State the lowest multiple that 4 and 6 have in common. This is the lowest common multiple (LCM) of 4 and 6.

2. State the LCM for each of the following pairs of numbers.
   a. 3 and 9   b. 6 and 15   c. 7 and 10   d. 12 and 16   e. 4 and 15

3. State whether the following statement is true or false.
   20 is a multiple of 10 and 2 only.

4. State whether the following statement is true or false.
   15 and 36 are both multiples of 3.

5. State whether the following statement is true or false.
   60 is a multiple of 2, 3, 6, 10 and 12.

6. State whether the following statement is true or false.
   100 is a multiple of 2, 4, 5, 10, 12 and 25.

7. **WE6** By listing the factors of each number, determine the highest common factor (HCF) for each of the following pairs of numbers.

   a. 21 and 56
   b. 7 and 28
   c. 48 and 30

8. Determine the HCF for each of the following pairs of numbers.

   a. 9 and 36
   b. 42 and 77

9. Determine the HCF of the following numbers.

   a. 36 and 64
   b. 45, 72 and 108

Understanding

10. Kate goes to the gym every second evening, while Ian goes every third evening.

    a. On Monday evening, Kate and Ian are at the gym together. Determine how many days it will be before both attend the gym again on the same evening.
    b. Explain how this answer relates to the multiples of 2 and 3.

11. Vinod and Elena are riding around a mountain bike trail. Vinod completes one lap in the time shown on the stopwatch on the left, and Elena completes one lap in the time shown on the stopwatch on the right.

    a. If they both begin cycling from the starting point at the same time, determine how long it will be before they pass the starting point again at exactly the same time.
    b. Relate your answer to the multiples of 5 and 7.

12. A warehouse owner employs Bob and Charlotte as security guards. Each security guard checks the building at midnight. Bob then checks the building every 4 hours, and Charlotte checks every 6 hours.

    a. Determine how long it will be until both Bob and Charlotte are next at the warehouse at the same time.
    b. Relate your answer to the multiples of 4 and 6.

13. Two smugglers, Bill Bogus and Sally Seadog, have set up signal lights that flash continuously across the ocean. Bill's light flashes every 5 seconds and Sally's light flashes every 4 seconds. If they start together, determine how long it will take for both lights to flash again at the same time.

14. Twenty students in Year 7 were each given a different number from 1 to 20 and then asked to sit in numerical order in a circle. Three older girls — Milly, Molly and Mandy — came to distribute jelly beans to the class. Milly gave red jelly beans to every second student, Molly gave green jelly beans to every third student and Mandy gave yellow jelly beans to every fourth student.
    a. Determine which student had jelly beans of all 3 colours.
    b. Calculate how many students received exactly 2 jelly beans.
    c. Calculate how many students did not receive any jelly beans.

15. Wah needs to cut tubing into the largest pieces of equal length that he can, without having any offcuts left over. He has three sections of tubing: one 6 metres long, another 9 metres long and the third 15 metres long.
    a. Calculate how long each piece of tubing should be.
    b. Determine how many pieces of tubing Wah will end up with.

Communicating, reasoning and problem solving

16. Two candles of equal length are lit at the same time. The first candle takes 6 hours to burn out and the second takes 3 hours to burn out. Determine how many hours it will be until the slower-burning candle is twice as long as the faster-burning candle.

17. Mario, Luigi, Zoe and Daniella are playing a video game. Mario takes 2 minutes to play one level of the game, Luigi takes 3 minutes, Zoe takes 4 minutes and Daniella takes 5 minutes. They have 12 minutes to play.

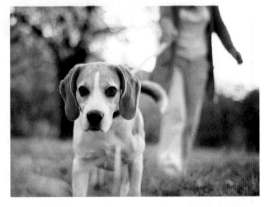

    a. If they play continuously, determine which player would be in the middle of a game as time ran out.
    b. Determine after how many minutes this player began the last game.

18. Six church bells ring repeatedly at intervals of 2, 4, 6, 8, 10 and 12 minutes respectively. Determine how many times three or more bells will ring together in 30 minutes.

19. Carmen types 30 words per minute on her laptop while Evan types 40 words per minute on his laptop. They have to type the same number of words and they both end up typing for a whole number of minutes.
    a. Determine how long Carmen has been typing for.
    b. Determine how long Evan has been typing for.

20. Alex enjoys running around the park after school. It takes him 10 minutes to complete one lap. His mother Claire walks the family dog at the same time. It takes her 15 minutes to complete one lap. They decide to go home when they meet next. Determine how long it will take them to meet again.

21. The runners in a 100 m race are lined up next to each other while those in a 400 m race are staggered around the track. Look at the following diagram of a standard 400 m athletic track and take note of both the starting position and finish line for a 100 m race.

a. State the shapes of the two curved ends of the track.
b. Explain why the starting line for a 100 m race has been positioned in the straight section.
c. Explain why the finish line for a 100 m race has been positioned where it is.

Track events such as the 200 m, 400 m and 800 m races have staggered starting positions. The following diagram illustrates the position of the starting blocks S_1–S_8 for each lane in the 400 m race.

d. Explain why there is a need to stagger the starting blocks in the 200 m, 400 m and 800 m track events.
e. A runner completing one lap of the inside lane (lane 1) runs a distance of 400 m, while if there is no staggered start a runner completing one lap in the next lane (lane 2) runs a distance of 408 m. If this pattern continues, determine how far runners in lanes 3–8 run. Enter these results in the appropriate column in the table.

Lane number	Distance travelled (m)	Difference
1	400 m	
2	408 m	
3		
4		
5		
6		
7		
8	456 m	

f. Calculate the difference between the distances travelled by the runners in each of the lanes compared to the distance travelled by the runner in lane 1. Enter these results in the appropriate column in the table.
g. Comment on what you notice about the values obtained in part f.

LESSON
3.4 Index notation

▶ 3.4.1 Introduction to indices

eles-3728

- An **index** (or **exponent** or **power**) is a short way of writing a repeated multiplication.
- The **base** is the number that is being repeatedly multiplied and the index (plural *indices*) is the number of times it is multiplied.

$$8^6 = 8 \times 8 \times 8 \times 8 \times 8 \times 8 = 262\ 144$$

Index form — Expanded form — Basic numeral

- Numbers in index form are read using the value of both the base and the power.

8^6 is read as '8 to the power of 6'.

12^4 is read as '12 to the power of 4'.

WORKED EXAMPLE 7 Identifying the base and index

For the following expressions, state:
i. the number or pronumeral that is the base
ii. the number or pronumeral that is the index.

   a. 6^7 **b.** x^9 **c.** 3^a **d.** y^b

THINK	WRITE

The base is the number or pronumeral that is repeatedly multiplied, and the index is the number of times that it is multiplied.	**a.** 6^7
	   i. Base is 6
	   ii. Index is 7

	b. x^9
	   i. Base is x
	   ii. Index is 9

	c. 3^a
	   i. Base is 3
	   ii. Index is a

	d. y^b
	   i. Base is y
	   ii. Index is b

WORKED EXAMPLE 8 Writing an expression using index notation

Write the following expressions using index notation.

a. $5 \times 5 \times 5 \times 5 \times 5 \times 5 \times 5$

b. $3 \times 3 \times 3 \times 3 \times 7 \times 7$

THINK	WRITE
a. 1. Write the multiplication.	**a.** $5 \times 5 \times 5 \times 5 \times 5 \times 5 \times 5$
2. Write the number being repeatedly multiplied as the base, and the number of times it is written as the index (the number 5 is written 7 times).	$= 5^7$
b. 1. Write the multiplication.	**b.** $3 \times 3 \times 3 \times 3 \times 7 \times 7$
2. Write the number being multiplied as the base, and the number of times it is written as the index (the number 3 is written 4 times, and 7 is written 2 times).	$= 3^4 \times 7^2$

WORKED EXAMPLE 9 Simplifying expressions using index notation

Simplify each of the following expressions by first writing each expression as a repeated multiplication and then in index notation.

a. $3^4 \times 3^6$

b. $\left(4^3\right)^3$

c. $(3 \times 5)^2$

THINK	WRITE
a. 1. Write the expression.	**a.** $3^4 \times 3^6$
2. Write the expression using repeated multiplication (that is, in expanded form).	$= (3 \times 3 \times 3 \times 3) \times (3 \times 3 \times 3 \times 3 \times 3 \times 3)$ $= 3 \times 3 \times 3 \times 3 \times 3 \times 3 \times 3 \times 3 \times 3 \times 3$
3. Write the repeated multiplication using index notation. The number being repeatedly multiplied is 3 (base), and the number of times it is written is 10 (index).	$= 3^{10}$
b. 1. Write the expression.	**b.** $\left(4^3\right)^3$
2. Write the expression using repeated multiplication (that is, in expanded form).	$= 4^3 \times 4^3 \times 4^3$ $= (4 \times 4 \times 4) \times (4 \times 4 \times 4) \times (4 \times 4 \times 4)$ $= 4 \times 4 \times 4 \times 4 \times 4 \times 4 \times 4 \times 4 \times 4$
3. Write the repeated multiplication using index notation. The number being repeatedly multiplied is 4 (base) and the number of times it is written is 9 (index).	$= 4^9$
c. 1. Write the expression.	**c.** $(3 \times 5)^2$
2. Write the expression using repeated multiplication (that is, in expanded form).	$= (3 \times 5) \times (3 \times 5)$ $= 3 \times 5 \times 3 \times 5$
3. Write the repeated multiplication using index notation. The numbers being repeatedly multiplied are 3 and 5 (base) and the number of times they are written is 2 (index) in each case.	$= 3 \times 3 \times 5 \times 5$ $= 3^2 \times 5^2$

3.4.2 Indices and place values

eles-3729

• By using place value, you can write numbers in expanded form with index notation. For example:

$$2700 = 2000 + 700$$
$$= 2 \times 10 \times 10 \times 10 + 7 \times 10 \times 10$$
$$= 2 \times 10^3 + 7 \times 10^2$$

WORKED EXAMPLE 10 Expressing numbers in expanded form using index notation

Write the following numbers in expanded form using index notation.
a. 59 176   **b. 108 009**

THINK	WRITE
a. 1. Write the number as the sum of each place value.	a. $59\,176 = 50\,000 + 9000 + 100 + 70 + 6$
2. Write each place value in multiples of 10.	$59\,176 = 5 \times 10 \times 10 \times 10 \times 10 + 9 \times 10 \times 10 \times 10$ $+ 1 \times 10 \times 10 + 7 \times 10 + 6$
3. Write each place value in index notation.	$59\,176 = 5 \times 10^4 + 9 \times 10^3 + 1 \times 10^2 + 7 \times 10^1 + 6$
b. 1. Write the number as the sum of each place value.	b. $108\,009 = 100\,000 + 8000 + 9$
2. Write each place value in multiples of 10.	$108\,009 = 1 \times 10 \times 10 \times 10 \times 10 \times 10$ $+ 8 \times 10 \times 10 \times 10 + 9$
3. Write each place value in index notation.	$108\,009 = 1 \times 10^5 + 8 \times 10^3 + 9$

WORKED EXAMPLE 11 Evaluating indices

Evaluate each of the following.
a. 2^5   **b. $3^2 \times 2^3$**   **c. $3^2 + 5^2 - 2^4$**

THINK	WRITE
a. 1. Write in expanded form.	a. $2^5 = 2 \times 2 \times 2 \times 2 \times 2$
2. Multiply the terms.	$= 32$
3. Write the answer.	$2^5 = 32$
b. 1. Write both terms in expanded form.	b. $3^2 \times 2^3 = (3 \times 3) \times (2 \times 2 \times 2)$
2. Calculate the product of the numbers in the brackets.	$= 9 \times 8$
3. Multiply the terms.	$= 72$
4. Write the answer.	$3^2 \times 2^3 = 72$
c. 1. Write all terms in expanded form.	c. $3^2 + 5^2 - 2^4 = (3 \times 3) + (5 \times 5) - (2 \times 2 \times 2 \times 2)$
2. Calculate the product of the numbers in the brackets.	$= 9 + 25 - 16$

3. Remember the order of operations. Since the operations are addition and subtraction, work left to right.

$= 34 - 16$

$= 18$

4. Write the answer.

$3^2 + 5^2 - 2^4 = 18$

Digital technology

Scientific calculators can evaluate expressions in index form where the base is a number.

Exercise 3.4 Index notation

learn on

3.4 Quick quiz on	3.4 Exercise

Individual pathways

■ PRACTISE	■ CONSOLIDATE	■ MASTER
1, 2, 5, 7, 10, 12, 13, 16	3, 6, 8, 14, 17, 18	4, 9, 11, 15, 19, 20, 21

Fluency

1. **WE 7** In each of the following expressions, enter the missing number into the box and state what the base and index are.

   a. $7^{\square} = 49$ 　　　 b. $4^{\square} = 4096$ 　　　 c. $\square^4 = 81$ 　　　 d. $\square^3 = 1000$

2. **WE8a** Write each of the following expressions in index notation.

   a. $7 \times 7 \times 7 \times 7$ 　　　　　　　　　　 b. $8 \times 8 \times 8 \times 8 \times 8 \times 8$

   c. $3 \times 3 \times 3 \times 3 \times 3 \times 3 \times 3 \times 3 \times 3$ 　　　 d. $13 \times 13 \times 13$

3. **WE8b** Write the following expressions in index notation.

   a. $4 \times 4 \times 4 \times 4 \times 4 \times 6 \times 6 \times 6$ 　　　 b. $2 \times 2 \times 3$

   c. $5 \times 5 \times 2 \times 2 \times 2 \times 2$ 　　　　　　 d. $3 \times 3 \times 2 \times 2 \times 5 \times 5 \times 5$

4. Write the following using repeated multiplication (that is, in expanded form).

   a. 6^5 　　　　　　　　　 b. 11^3

5. **WE9** Simplify each of the following expressions by first writing each expression as a repeated multiplication and then using index notation.

a. $5^8 \times 5^3$ b. $4^8 \times 4^5$ c. $\left(2^7\right)^3$

6. **WE10** Write the following numbers in expanded form using index notation.

a. 300   b. 4500   c. 6705   d. 10 000

Understanding

7. Simplify each of the following expressions by first writing each expression as a repeated multiplication and then using index notation.

a. $\left(9^5\right)^3$ b. $(3 \times 13)^6$

8. **MC** The value of 4^4 is:

A. 8   B. 16   C. 64   D. 256

9. **WE11** Evaluate each of the following.

a. 7^3 b. $6^2 \times 2^3$ c. $5^3 - 4^3 + 2^4$

10. Evaluate each of the following.

a. $2^3 \times 3^2$ b. $3^4 \times 4^3$ c. $3^5 \times 9^3$

11. Evaluate each of following.

a. $6^4 - 9^3$ b. $5^3 + 2^5 \times 9^2$ c. $2^7 - 4^5 \div 2^6$

Communicating, reasoning and problem solving

12. a. **MC** Which of the following expressions has the greatest value?

A. 2^8 B. 8^2 C. 3^4 D. 4^3

b. Justify your answer using mathematical reasoning.

13. We know that $12^2 = 144$ and $21^2 = 441$. It is also true that $13^2 = 169$ and $31^2 = 961$. If $14^2 = 196$, is $41^2 = 691$? Try to justify your answer without calculating 41^2.

14. Write < (less than) or > (greater than) in the boxes below to make each statement true.

a. $1^3 \ \square \ 3^1$ b. $2^3 \ \square \ 3^2$ c. $5^3 \ \square \ 3^5$ d. $4^4 \ \square \ 3^5$

15. You have a choice of how your weekly allowance is increased for 15 weeks.
Option 1: Start a 1 cent and double your allowance every week.
Option 2: Start at $1 and increase your allowance by $1 every week.

a. Explain which option you would choose and why.
b. Explain whether powers can be used to help with these calculations.

16. Jessica has a 'clean-up' button on her mobile phone that will clear half of the read messages in her inbox. She pressed the button six times, causing just one message to be left in her inbox. Calculate how many messages were in her inbox before the clean-up.

17. A student sent an email to eight of his cousins, who then each forwarded the email to eight different friends, who in turn forwarded the email to another eight friends each.

a. Calculate how many emails in total were sent during this process.
b. Assuming that everyone received the email only once, determine how many people in total received the email by the end of this process.

18. You received a text message from your friend. After 5 minutes, you forward the text message to two of your other friends. After 5 more minutes, those two friends forward it to two more friends. If the text message is sent every 5 minutes in this way, calculate how many people have received it in 30 minutes.

19. Bacteria such as those shown in the image grow at an alarming rate. Each bacterium can split itself in half, forming two new cells each hour.

    a. If you start with one bacterium cell, calculate how many you will have after 12 hours. Show your working to include the number of bacteria after each hour.
    b. If a bacterium could successfully divide into three new cells each hour, calculate how many bacteria you would have after 12 hours.
    c. Show how powers can be applied to the calculations in parts **a** and **b**.

20. A knock-out tennis competition ends with 2 players in the grand final. If the grand final is the 8th round of the competition, and half the players are knocked out in each round, determine how many players there were at the start of the competition.

21. A particular rule applies to each of these three sets of numbers.
    Set A: 1, 2, 5   Set B: 2, 5, 26   Set C: 3, 10, ?
    Explain how you can determine the rule and the missing number.

LESSON
3.5 Prime and composite numbers

LEARNING INTENTION

At the end of this lesson you should be able to:
- understand the difference between prime and composite numbers
- determine prime factors by drawing a factor tree
- write a composite number in index form
- determine the highest common factor (HCF) and lowest common multiple (LCM) of large numbers as products of their prime factors.

▶ 3.5.1 Prime numbers

eles-3730

- A **prime number** is a counting number that has exactly 2 unique factors: itself and 1.
  (Counting numbers are $1, 2, 3, 4, \ldots$)
  The number 3 is a prime number; its only factors are 3 and 1.
  The number 2 is the only even prime number; its only factors are 2 and 1.
- A **composite number** is one that has more than two factors.
  The number 4 is a composite number; its factors are 1, 2 and 4.
- The number 1 is a special number. It is neither a prime number nor a composite number because it has only one factor, 1.

COMMUNICATING — COLLABORATIVE TASK: The sieve of Eratosthenes

Eratosthenes (pronounced 'e-rah-toss-thee-knees') was a Greek mathematician who lived around 200 BC. He worked out a way to sort prime numbers from composite numbers.

1. As a class, draw a 10×10 grid on the board and number the boxes from 1 to 100.
2. Shade the number 1. It is not a prime or composite number.
3. Circle the number 2. Cross out all other multiples of 2.
4. Circle the next smallest uncrossed number, 3. Cross out all other multiples of 3.
5. Continue this process until all the numbers in the grid are either circled or crossed out.
6. Describe the circled numbers.
7. Describe the crossed out numbers.
8. a. Are there more prime numbers or composite numbers?
   b. What connections can you make between composite numbers and multiples?

1	2	3	4	5	6	7	8	9	10
11	12	13	14	15	16	17	18	19	20
21	22	23	24	25	26	27	28	29	30
31	32	33	34	35	36	37	38	39	40
41	42	43	44	45	46	47	48	49	50
51	52	53	54	55	56	57	58	59	60
61	62	63	64	65	66	67	68	69	70
71	72	73	74	75	76	77	78	79	80
81	82	83	84	85	86	87	88	89	90
91	92	93	94	95	96	97	98	99	100

WORKED EXAMPLE 12 Listing prime numbers

List the prime numbers between 50 and 70.

THINK

1. The only even prime number is 2. The prime numbers between 50 and 70 will be odd. Numbers ending in 5 are divisible by 5 so 55 and 65 are not primes.

2. Check the remaining odd numbers between 50 and 70:
   $51 = 3 \times 17$ and 1×51. It has four factors so it is composite.
   $53 = 1 \times 53$. It has two unique factors only so it is prime.
   $57 = 3 \times 19$ and 1×57. It has four factors so it is composite.
   $59 = 1 \times 59$. It has two unique factors only so it is prime.
   $61 = 1 \times 61$. It has two unique factors only so it is prime.
   $63 = 7 \times 9$, 3×21 and 1×63. It has six factors so it is composite.
   $67 = 1 \times 67$. It has two unique factors only so it is prime.
   $69 = 3 \times 23$ and 1×69. It has four factors so it is composite.

WRITE

50	51	52	(53)	54	55	56
57	58	(59)	60	(61)	62	63
64	65	66	(67)	68	69	70

The prime numbers are: $53, 59, 61, 67$.

WORKED EXAMPLE 13 Checking for composite and prime numbers

State whether the following numbers are prime or composite.

a. 45   b. 37   c. 86

THINK

a. Factors of 45 are 1, 3, 5, 9, 15 and 45.

b. The only factors of 37 are 1 and 37.

c. Factors of 86 are 1, 2, 43 and 86. All even numbers except 2 are composite.

WRITE

a. 45 is composite.

b. 37 is prime.

c. 86 is composite.

DISCUSSION

Explain your method of working out whether a number is a prime number.

3.5.2 Composite numbers and factor trees

eles-3731

- Every composite number can be written as the product of its prime factors.
- Writing any number as a product of its prime factors simply means to rewrite a number using **multiplication** of **prime numbers** only. For example:

$$12 = 2 \times 2 \times 3 = 2^2 \times 3$$

↑

Prime numbers that multiply to give 12

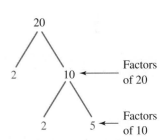

$$20 = 2 \times 2 \times 5 = 2^2 \times 5$$

- A **factor tree** shows the **prime factors** of a composite number.
- Each branch shows a factor of the number above it.
- Branches stop at factors that are prime numbers, as shown at right in pink.
- In the factor tree shown, 20 can be written as $2 \times 2 \times 5$; this is known as writing a number as a product of its prime factors.

WORKED EXAMPLE 14 Determining the prime factors by drawing a factor tree

a. **Determine the prime factors of 50 by drawing a factor tree.**
b. **Write 50 as a product of its prime factors in index form.**

THINK

a. 1. Write a factor pair of the given number and draw the factor tree ($50 = 5 \times 10$).

2. If a branch is prime, no other factors can be found (5 is prime). If a branch is composite, write factors of that number: 10 is composite, so $10 = 5 \times 2$.

WRITE

a.

3. Continue until all branches end in a prime number, then stop.

4. Write the prime factors.

The prime factors of 50 are 2 and 5.

b. Write 50 as a product of prime factors in index form found in part **a**.

b. $50 = 5 \times 5 \times 2$

$\quad = 5^2 \times 2$

WORKED EXAMPLE 15 Writing composite numbers in index form

Write 72 as a product of its prime factors in index form.

THINK

1. Draw a factor tree. When all factors are prime numbers you have found the prime factors.

WRITE

2. Write 72 as a product of its prime factors in index form.

$72 = 2 \times 2 \times 2 \times 3 \times 3$

$\quad = 2^3 \times 3^2$

WORKED EXAMPLE 16 Writing composite numbers in index form

Write 360 as a product of prime factors using index notation.

THINK

1. Draw a factor tree. If the number on the branch is a prime number, stop. If not, continue until a prime number is reached.

WRITE

2. Write the number as a product of prime factors.

$360 = 2 \times 2 \times 2 \times 3 \times 3 \times 5$

3. Write your answer using index notation.

$360 = 2^3 \times 3^2 \times 5$

There can often be more than one way of drawing a factor tree for a number and getting the same result, as shown below.

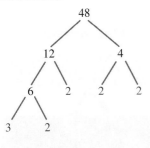

$48 = 12 \times 4$
$ = 3 \times 2 \times 2 \times 2 \times 2$
$ = 2^4 \times 3$

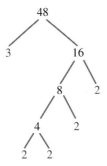

$48 = 3 \times 16$
$ = 3 \times 2 \times 2 \times 2 \times 2$
$ = 2^4 \times 3$

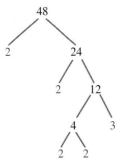

$48 = 2 \times 24$
$ = 2 \times 2 \times 2 \times 2 \times 3$
$ = 2^4 \times 3$

⏵ 3.5.3 Lowest common multiple and highest common factor using prime factors

eles-3732

- Expressing large numbers as products of their prime factors makes it easier to determine their lowest common multiple (LCM) or highest common factor (HCF).

Lowest common multiple (LCM)

The LCM of two numbers expressed as the products of their prime factors is the product of the prime factors that are factors of either number.

Highest common factor (HCF)

The HCF of two numbers expressed as the products of their prime factors is the product of the prime factors common to both numbers.

WORKED EXAMPLE 17 Determining HCF and LCM using prime factors

Determine the highest common factor and lowest common multiple of 270 and 900.

THINK	WRITE
1. Write 270 and 900 as products of their prime factors. Alternatively, a factor tree may be used to determine the prime factors.	$270 = 2 \times 3 \times 3 \times 3 \times 5$ $900 = 2 \times 2 \times 3 \times 3 \times 5 \times 5$
2. Circle the prime factors common to both numbers.	$270 = ②\times③\times③\times 3 \times⑤$ $900 = 2 \times②\times③\times③\times 5 \times⑤$
3. For the HCF, multiply the prime factors common to both numbers.	$2 \times 3 \times 3 \times 5 = 90.$ The HCF of 270 and 900 is 90.
4. For the LCM, multiply the prime factors that are factors of either number. *Note:* Only include one in the circled pair.	$2 \times 2 \times 3 \times 3 \times 3 \times 5 \times 5 = 2700$ The LCM of 270 and 900 is 2700.

COMMUNICATING — COLLABORATIVE TASK: Goldbach's conjecture

In 1742, mathematician Christian Goldbach suggested that every even number greater than 2 could be written as the sum of two prime numbers.

In pairs, test this for the even numbers greater than 2 and less than 40.

In pairs, choose 5 even numbers between 100 and 300. Test these numbers to see if Goldbach's conjecture is true.

Do you agree with Christian Goldbach?

PRIME NUMBERS

2,3,5,7,11,13,17,19,23, 29,31,37,41,43,47,53,59, 61,67,71,73,79,83,89,97

 Resources

 Interactivities Prime numbers (int-3933)

Composite numbers and factor trees (int-3934)

Exercise 3.5 Prime and composite numbers

learn on

3.5 Quick quiz **on**	3.5 Exercise

Individual pathways

■ PRACTISE	■ CONSOLIDATE	■ MASTER
1, 2, 4, 8, 10, 13, 15, 16, 21, 24, 25, 28	5, 7, 9, 12, 17, 19, 20, 22, 26, 29	3, 6, 11, 14, 18, 23, 27, 30

Fluency

1. **WE12** List four prime numbers that are between 20 and 40.

2. **WE13** State whether each of the following numbers is prime or composite.
   a. 9   b. 13   c. 55   d. 41

3. State whether each of the following numbers is prime or composite.
   a. 64   b. 79   c. 98   d. 101

4. i. **WE14** Determine the prime factors of each of the following numbers by drawing a factor tree.
   ii. Write each one as a product of its prime factors in index form.
   a. 15   b. 30   c. 100   d. 49

5. i. Determine the prime factors of each of the following numbers by drawing a factor tree.
   ii. Write each one as a product of its prime factors in index form.
   a. 72   b. 56   c. 45

6. i. Determine the prime factors of each of the following numbers by drawing a factor tree.
   ii. Write each one as a product of its prime factors in index form.
   a. 84   b. 112   c. 40

7. i. **WE15** Determine the prime factors of the following numbers by drawing a factor tree.
   ii. Express the number as a product of its prime factors in index form.
   a. 96   b. 32   c. 3000

8. Determine the prime factors of each of the following numbers.
   a. 48   b. 200   c. 81   d. 18

9. Determine the prime factors of each of the following numbers.
   a. 300   b. 60   c. 120   d. 80

Understanding

10. **WE16** Write the following as a product of prime factors using index notation.
    a. 60   b. 75   c. 220

11. Write the following as a product of prime factors using index notation.
    a. 192   b. 124

12. **WE17** By expressing the following pairs of numbers as products of their prime factors, determine their lowest common multiple and their highest common factor.
    a. 36 and 84   b. 48 and 60   c. 120 and 400   d. 220 and 800

13. State whether each of the following is true or false.
    a. All odd numbers are prime numbers.
    b. No even numbers are prime numbers.
    c. 1, 2, 3 and 5 are the first four prime numbers.
    d. A prime number has two factors only.
    e. 2 is the only even prime number.

14. State whether each of the following is true or false.
    a. The sum of two prime numbers is always even.
    b. The product of two prime numbers is always odd.
    c. There are no consecutive prime numbers.

15. **MC** The number of primes less than 10 is:
    A. 4   B. 3   C. 5   D. 2

16. **MC** The first three prime numbers are:
    A. 1, 3, 5   B. 2, 3, 4   C. 2, 3, 5   D. 3, 5, 7

17. **MC** The number 15 can be written as the sum of two prime numbers. These are:
    A. 3 + 12   B. 1 + 14   C. 13 + 2   D. 7 + 8

18. **MC** Factors of 12 that are prime numbers are:
    A. 1, 2, 3, 4   B. 2, 3, 6   C. 2, 3   D. 2, 4, 6, 12

19. The following numbers are not primes. Each of them is the product of two primes. Determine the two primes in each case.
    a. 365   b. 187

20. **MC** A factor tree for 21 is:

A.

C.

B.

D.

21. **MC** A factor tree for 36 is:

A.

C.

B.

D.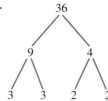

22. **MC** The prime factors of 16 are:

A. 1, 2   B. 1, 2, 4   C. 2   D. 1, 2, 4, 8, 16

23. **MC** The prime factors of 28 are:

A. 1, 28   B. 2, 7   C. 1, 2, 14   D. 1, 2, 7

Reasoning

24. Determine the largest three-digit prime number in which each digit is a prime number. Prove that this number is a prime number.

25. Is the sum of two prime numbers always a prime number? Explain your answer.

26. Determine a prime number greater than 10 with a sum of digits that equals 11. Show your working.

27. My age is a prime number. I am older than 50. The sum of the digits in my age is also a prime number. If you add a multiple of 13 to my age the result is 100. Determine my age.

28. Determine two prime numbers with a product of:

a. 21   b. 26   c. 323.

29. Determine how many even integers between 2 and 20 can be the sum of two different prime numbers.

30. Twin primes are pairs of prime numbers that differ by 2. Except for the pair of primes 2 and 3, this is the smallest difference between two prime numbers. The first twin primes are 3 and 5, followed by 5 and 7, then 11 and 13. Identify other twin primes below 100.

LESSON
3.6 Squares and square roots

LEARNING INTENTION

At the end of this lesson you should be able to:
- determine values of squares and square roots
- estimate the value of the square root of other numbers by using the perfect squares that lie on either side of the number
- evaluate squares or square roots using BIDMAS.

3.6.1 Square numbers

eles-3733

- The process of multiplying a number by itself is known as **squaring a number**.
- **Square numbers** or **perfect squares** are numbers that can be arranged in a square, with the same length and width. The amount of squares represents the square number.

$1^2 = 1 \times 1$
$= 1$

$2^2 = 2 \times 2$
$= 4$

$3^2 = 3 \times 3$
$= 9$

$4^2 = 4 \times 4$
$= 16$

- A square number is the number we get after multiplying a whole number by itself.
- The first four square numbers are: 1, 4, 9 and 16.
- All square numbers can be written in index notation, for example, $4^2 = 16$.

WORKED EXAMPLE 18 Determining square numbers

Determine the square numbers between 90 and 150.

THINK	WRITE
1. Use your knowledge of multiplication tables to determine the first square number after 90.	$10^2 = 10 \times 10 = 100$
2. Determine the square numbers that come after that one, but before 150.	$11^2 = 11 \times 11 = 121$
	$12^2 = 12 \times 12 = 144$
	$13^2 = 13 \times 13 = 169$ (too big)
3. Write the answer in a sentence.	The square numbers between 90 and 150 are 100, 121 and 144.

Write the two whole square numbers between which 5.7^2 will lie.

THINK	WRITE
1. Write the whole numbers either side of 5.7.	5.7 is between 5 and 6.
2. Consider the square of each whole number.	5.7^2 is between 5^2 and 6^2.
3. Simplify 5^2 and 6^2, then write the answer in a sentence.	So 5.7^2 is between 25 and 36.
4. Verify your answer with a calculator.	$5.7^2 = 32.49$, which lies between 25 and 36.
5. To determine the value of 5.7^2 press ⑤ ⊙ ⑦ x^2 ⓪.	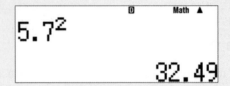

We have confirmed that 5.7^2 lies between 25 and 36.

DISCUSSION

Can you predict whether the perfect square of a number will be odd or even? Give some examples.

- When a composite number is squared, the result is equal to the product of the squares of its factors.

$$\text{For example: } 10^2 = (2 \times 5)^2$$
$$= 2^2 \times 5^2$$
$$= 4 \times 25$$
$$= 100$$

▶ 3.6.2 Square roots

eles-3734

- Evaluating the **square root** of a number is the opposite of squaring the number; for example, since $4^2 = 16$, then $\sqrt{16} = 4$.
- The symbol for square root is called the radical symbol. It is written as $\sqrt{}$.
- Using the illustrations of the squares shown earlier, a square of area 16 square units must have a side length of 4 units. This may help understand why $\sqrt{16} = 4$.

Square roots

When determining the square root of a number, you are determining the number that, when multiplied by itself, equals the number underneath the radical symbol.

For example, to determine $\sqrt{9}$, identify the number that when multiplied by itself gives 9.

Since $3 \times 3 = 9$, we can conclude that $\sqrt{9} = 3$.

- To determine the square roots of larger numbers, it helps to break the number up as a product of two smaller square roots with which we are more familiar. For example:

$$\sqrt{900} = \sqrt{9} \times \sqrt{100}$$
$$= 3 \times 10$$
$$= 30$$

WORKED EXAMPLE 20 Determining square roots

Evaluate the following square roots.

a. $\sqrt{49}$ **b.** $\sqrt{3600}$

THINK	WRITE
a. Determine a number that when multiplied by itself gives 49.	**a.** $\sqrt{49} = 7$ $(7 \times 7 = 49)$
b. 1. Write 3600 as the product of two smaller numbers for which we can calculate the square root.	**b.** $\sqrt{3600} = \sqrt{36} \times \sqrt{100}$
2. Take the square root of each of these numbers.	$= 6 \times 10$
3. Determine the product and write the answer.	$\sqrt{3600} = 60$

- Only perfect squares have square roots that are whole numbers.
- The value of the square root of other numbers can be estimated by using the perfect squares that lie on either side of the number.

WORKED EXAMPLE 21 Estimating square roots

Between which two numbers do the following numbers lie?

a. $\sqrt{74}$ **b.** $\sqrt{342}$

THINK	WRITE
a. 1. Write the square numbers on either side of 74.	**a.** 74 is between 64 and 81.
2. Consider the square root of each number.	$\sqrt{74}$ is between $\sqrt{64}$ and $\sqrt{81}$.
3. Simplify $\sqrt{64}$ and $\sqrt{81}$.	So $\sqrt{74}$ is between 8 and 9.
4. Verify your answer with a calculator. To determine the value of $\sqrt{74}$ press [√☐] [7] [4] [=].	$\sqrt{74} \approx 8.6023$

```
                        D      Math ▲
 √74
              8.602325267
```

We have confirmed that $\sqrt{74}$ lies between 8 and 9.

THINK	WRITE
b. 1. Write the square numbers on either side of 342.	**b.** 342 is between 324 $\left(18^2\right)$ and 361 $\left(19^2\right)$.
2. Consider the square root of each number.	$\sqrt{342}$ is between $\sqrt{324}$ and $\sqrt{361}$.

3. Simplify $\sqrt{324}$ and $\sqrt{361}$.

So $\sqrt{342}$ is between 18 and 19.

4. Verify your answer with a calculator.
To determine the value of $\sqrt{342}$ press

 3 4 2 = .

$\sqrt{342} \approx 18.4932$

```
√342
            18.49324201
```

We have confirmed that $\sqrt{342}$ lies between 18 and 19.

Order of operations

- The square or square root is an index represented by the letter I in BIDMAS, which determines the order in which operations are evaluated in a mathematical expression (as seen in previous topics).

WORKED EXAMPLE 22 Evaluating square roots using the order of operations

Evaluate:

a. $\sqrt{16 + 9}$

b. $\sqrt{16} + 9$

THINK

a. 1. Simplify the expression inside the square root and write the answer.

2. Complete the calculation by evaluating the square root and write the answer.

b. 1. Evaluate the square root.
Note: There is no expression inside the square root to evaluate first.

2. Perform the addition to complete the calculation and write the answer.

WRITE

a. $\sqrt{16 + 9} = \sqrt{25}$

$= 5$

b. $\sqrt{16} + 9 = 4 + 9$

$= 13$

COMMUNICATING — COLLABORATIVE TASK: How many squares?

Work in pairs to answer the following question.

How many squares of any size can be found in the following diagram?

Hint: Count the number of single squares, count the number of 2×2 squares, then count the number of 3×3 squares. Look for a pattern in these numbers, and use the pattern to help you.

Digital technology

Scientific calculators can evaluate squares of decimal numbers using the x^2 button.

	D	Math ▲
54.25^2		
		2943.0625

Scientific calculators can evaluate square roots of decimal numbers using the $\sqrt{}$ button.

	D	Math ▲
$\sqrt{4.6656}$		
		2.16

on Resources

 Interactivities Square numbers (int-3936)
Square roots (int-3937)

Exercise 3.6 Squares and square roots

learn on

3.6 Quick quiz **on**	3.6 Exercise

Individual pathways

■ PRACTISE	■ CONSOLIDATE	■ MASTER
1, 4, 6, 11, 13, 16, 19, 22	2, 5, 7, 9, 12, 17, 20, 23	3, 8, 10, 14, 15, 18, 21, 24, 25

Fluency

1. Evaluate the following and verify your answers with a calculator.

   a. 8^2 b. 11^2 c. 15^2 d. 25^2

2. **WE18** State the square numbers between 50 and 100.

3. State the square numbers between 160 and 200.

4. **WE19** Write two whole square numbers between which each of the following will lie.

   a. 6.4^2 b. 7.8^2 c. 9.2^2 d. 12.5^2

5. **WE20** Evaluate:

   a. $\sqrt{25}$ b. $\sqrt{81}$ c. $\sqrt{144}$ d. $\sqrt{400}$

6. Evaluate the following and verify your answers with a calculator.

   a. $\sqrt{4900}$ b. $\sqrt{14\,400}$ c. $\sqrt{360\,000}$ d. $\sqrt{160\,000}$

7. **WE21** Write the two whole numbers between which each of the following will lie.

   a. $\sqrt{60}$ b. $\sqrt{14}$ c. $\sqrt{200}$ d. $\sqrt{2}$

8. **WE22** Evaluate:

   a. $\sqrt{144 + 25}$ b. $\sqrt{144} + 25$

9. State the even square numbers between 10 and 70.

10. State the odd square numbers between 50 and 120.

Understanding

11. **MC** For which of the following square roots can we calculate an exact answer?

A. $\sqrt{10}$ **B.** $\sqrt{25}$ **C.** $\sqrt{50}$ **D.** $\sqrt{75}$

12. **MC** For which of the following square roots can we not calculate the exact value?

A. $\sqrt{160}$ **B.** $\sqrt{400}$ **C.** $\sqrt{900}$ **D.** $\sqrt{2500}$

13. Evaluate the following. Verify your answers with a calculator.

a. $2^2 + \sqrt{25}$ b. $9^2 - \sqrt{36}$ c. $5^2 \times 2^2 \times \sqrt{49}$ d. $3^2 + 2^2 \times \sqrt{16}$

14. Evaluate the following. Verify your answers with a calculator.

a. $3^2 - 2^2 \div \sqrt{4} + \sqrt{49}$ b. $\sqrt{9} \times 4^2 - \sqrt{144} \div 2^2$

15. a. Determine between which two whole numbers $\sqrt{150}$ will lie.
b. Use your answer to part **a** to estimate the value of $\sqrt{150}$ and check your answer with a calculator.

Communicating, reasoning and problem solving

16. a. Evaluate 15^2. b. Evaluate $3^2 \times 5^2$.
c. Are your answers to parts **a** and **b** equal? Explain why or why not.

17. a. Evaluate $\sqrt{225}$. b. Evaluate $\sqrt{25} \times \sqrt{9}$.
c. Are your answers to parts **a** and **b** equal? Explain why or why not.

18. a. Express 196 as a product of its prime factors.
b. Express 200 as a product of its prime factors.
c. Use the answers to parts **a** and **b** to explain how you can use prime factors to determine whether the square root of a number is an integer.

19. a. Evaluate $\sqrt{36} + \sqrt{64}$. b. Evaluate $\sqrt{36 + 64}$.
c. Are your answers to parts **a** and **b** equal? Explain why or why not.

20. a. Evaluate $\sqrt{25} + \sqrt{144}$. b. Evaluate $\sqrt{169}$.
c. $169 = 25 + 144$. Does $\sqrt{169} = \sqrt{25} + \sqrt{144}$? Explain your answer.

21. Megan has 3 different game scores that happen to be square numbers. The first 2 scores have the same three digits. The total of the 3 scores is 590.
Determine the 3 scores. Explain how you solved this question.

22. a. Evaluate each of the following using a calculator.
  i. $25^2 - 24^2$ ii. $24^2 - 23^2$ iii. $23^2 - 22^2$ iv. $22^2 - 21^2$ v. $21^2 - 20^2$
b. Comment on any pattern in your answers.
c. Discuss whether this pattern occurs with other numbers that are squared. Try some examples.

23. Guess the number that matches the following:
It is odd and it has an odd number of factors.
The sum of the digits is a two-digit prime.
The number is less than $\sqrt{10\,000}$ but greater than $\sqrt{100}$.

24. Rewrite the following in ascending order.
a. $8^2, \sqrt{56}, 62, \sqrt{72}, 8, 7, 3^2, \sqrt{100}$ b. $100, \sqrt{121}, \sqrt{10}, 3 \times 4, 9^2, \sqrt{169}, 2 \times 7$

25. The sum of two numbers is 20 and the difference between their squares is 40. Determine the two numbers.

LESSON
3.7 Divisibility tests

> **LEARNING INTENTION**
>
> At the end of this lesson you should be able to:
> - test for factors of a given number using divisibility tests.

▶ 3.7.1 Tests of divisibility

eles-3736

- Divisible means that one number divided by another gives a result that is a whole number.
- **Divisibility tests** are a quick way of testing for factors of a given number.
- These tests save time compared with dividing by each possible factor to determine whether the number is divisible by the chosen divisor.
- All positive integers are divisible by 1.

Divisibility by 2	**Divisibility by 3**	**Divisibility by 4**	**Divisibility by 5**
The last digit is 0, 2, 4, 6 or 8.	The sum of the digits can be divided by 3.	The last two digits together (as a 2-digit number) can be divided by 4.	The last digit is either 0 or 5.
e.g. 324 ÷ 2, since the last digit is 4.	e.g. 615 ÷ 3, since the sum of the digits is 12 (6 + 1 + 5) and 12 is divisible by 3.	e.g. 532 ÷ 4, since the last two digits are 32, and 32 is divisible by 4.	e.g. 135 ÷ 5, since the last digit is 5.
24, 74, 108, 324	12, 72, 234, 615	84, 184, 324, 532	25, 135, 640, 1870

Divisibility by 6	**Divisibility by 8**	**Divisibility by 9**	**Divisibility by 10**
The number can be divided by 2 and also divided by 3.	The last three digits can be divided by 8.	The sum of the digits can be divided by 9.	The last digit is 0.
e.g. 420 ÷ 6, since the last digit is 0 (0 ÷ 2), and the sum of the digits is 6 (4 + 2 + 0) and 6 is divisible by 3.	e.g. 1816 ÷ 8, since the last three digits are 816, and 816 is divisible by 8.	e.g. 756 ÷ 9, since the sum of the digits is 18 (7 + 5 + 6), and 18 is divisible by 9.	e.g. 770 ÷ 10, since the last digit is 0.
36, 144, 258, 420	96, 208, 1816, 5232	81, 198, 414, 756	80, 350, 600, 770

WORKED EXAMPLE 23 Applying divisibility tests

Determine whether the results of the following divisions are whole numbers.
a. **3454 ÷ 2**   b. **6381 ÷ 9**

THINK	WRITE
a. 1. To be divisible by 2, the last digit must be 0, 2, 4, 6 or 8.	a. In the number 3454, the last digit is 4.
2. Write the answer in a sentence.	Therefore, 3454 is divisible by 2.
b. 1. To be divisible by 9, the sum of all the digits must by divisible by 9.	b. The sum of all the digits: $6 + 3 + 8 + 1 = 18$ 18 is divisible by 9.
2. Write the answer in a sentence.	Therefore, 6381 is divisible by 9.

WORKED EXAMPLE 24 Checking divisibility

Check the divisibility of the following numbers.
a. $816 \div 4$ b. $7634 \div 6$

THINK	WRITE
a. 1. To be divisible by 4, the last two digits together must be divisible by 4.	**a.** In the number 816, the last two digits are 16. 16 is divisible by 4.
2. Write the answer in a sentence.	Therefore, 816 is divisible by 4.
b. 1. To be divisible by 6, the number must be divisible by 2 and also by 3.	**b.** In the number 7634, the last digit is 4. Therefore, 7634 is divisible by 2.
2. To be divisible by 2, the last digit must be 0, 2, 4, 6 or 8.	In the number 7634, the sum of all the digits: $7 + 6 + 3 + 4 = 20$
3. To be divisible by 3, the sum of all the digits must be divisible by 3.	20 is **not** divisible by 3.
4. Write the answer in a sentence.	The number 7634 is divisible by 2 but not divisible by 3. Therefore, 7634 is not divisible by 6.

 Resources

 Interactivities Crossword (int-2588)

Sudoko puzzle (int-3164)

Exercise 3.7 Divisibility tests

learn

3.7 Quick quiz **on**	3.7 Exercise

Individual pathways

■ PRACTISE	■ CONSOLIDATE	■ MASTER
1, 4, 6, 8, 11, 14	2, 5, 9, 12, 15	3, 7, 10, 13, 16

Fluency

1. State the divisibility tests for 2, 4 and 6 with an example for each.

2. State the divisibility tests for 3, 5 and 9 with an example for each.

3. State how the divisibility tests for 3 and 9 are similar.

4. **WE23** Determine whether or not the results of the following divisions are whole numbers.
   a. $54 \div 2$ b. $150 \div 9$ c. $295 \div 5$ d. $1293 \div 3$

5. **WE24** Check the divisibility of the following.
   a. $1816 \div 4$ b. $2638 \div 6$ c. $62\,873 \div 9$ d. $876 \div 8$

Understanding

6. State the factors (from 1 to 10) of the following numbers.
   a. 144   b. 264   c. 1456   d. 567 844

7. Use the divisibility rules to determine the factors (from 1 to 10) of the following numbers.
   a. 760   b. 535   c. 3457   d. 234 568

8. Apply the divisibility rules to determine the factors (from 1 to 10) of the following numbers.
   a. 76   b. 30   c. 458   d. 227

9. Without using a calculator, use divisibility tests to determine whether the following are True or False.
   a. 2 is a factor of 65   b. 3 is a factor of 96
   c. 4 is a factor of 22   d. 5 is a factor of 44
   e. 6 is a factor of 72   f. 10 is a factor of 85

10. a. Use the tests of divisibility to determine whether the following numbers are divisible by 2, 3, 4, 5, 6 or 10.
    i.   415   ii. 600   iii. 935
    iv. 1010   v.   4321   vi.   55 112

    b. Use a calculator to check your answers to part a.

Communicating, reasoning and problem solving

11. **WE24b** Jack says that 34 281 is divisible by 6, but Julie says that Jack is incorrect. Explain whether you agree with Jack or Julie, and why.

12. If a number is divisible by 9, determine another factor (2, 3, 4, 5, 6, 7, 8, or 10) for that number. Justify your answer.

13. 6 is a factor of 4572 and 2868. Is 6 a factor of $4572 + 2868$ and $4572 - 2868$? Show full working to justify your answer.

14. Determine the missing digit to create the number.
    a. Divisible by 3 : _____316   b. Divisible by 6 : 9_____32
    c. Divisible by 8 : 59_____16   d. Divisible by 9 : 1543_____

15. Determine 2 two-digit numbers that are divisible by 4, 5 and 8.

16. A number between 550 and 560 is divisible by 6 and 4. Determine the number.

LESSON
3.8 Cubes and cube roots

LEARNING INTENTION

At the end of this lesson you should be able to:
- determine values of cubes and cube roots
- evaluate cubes or cube roots using BIDMAS.

▶ 3.8.1 Cube numbers

eles-3737

- **Cube numbers** are numbers that can be arranged in a cube, with the same length, width and height, as shown.

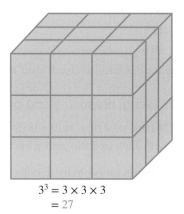

$$1^3 = 1 \times 1 \times 1$$
$$= 1$$

$$2^3 = 2 \times 2 \times 2$$
$$= 8$$

$$3^3 = 3 \times 3 \times 3$$
$$= 27$$

- The first cube number, 1, equals $1 \times 1 \times 1$.
- The second cube number, 8, equals $2 \times 2 \times 2$.
- The third cube number, 27, equals $3 \times 3 \times 3$.
- Each new perfect cube is found by multiplying a number by itself three times.
  This is known as *cubing a number* and is written using an index (or power) of 3. For example,
  $4^3 = 4 \times 4 \times 4 = 64$.

WORKED EXAMPLE 25 Calculating the value of a cube

Calculate the value of 5^3.

THINK	WRITE
1. Write 5^3 as the product of three lots of 5.	$5^3 = 5 \times 5 \times 5$
2. Evaluate and write the answer.	$= 125$

Order of operations

- When working with squares and cubes, remember to evaluate the index as part of the I in BIDMAS.

WORKED EXAMPLE 26 Calculating cubes using the order of operations

Calculate the value of $3^3 + 4 \times 5^2$.

THINK	WRITE
1. According to the order of operations (BIDMAS), first simplify both the cubed and the square terms.	$3^3 + 4 \times 5^2 = (3 \times 3 \times 3) + 4 \times (5 \times 5)$ $= 27 + 4 \times 25$
2. Then complete the multiplication.	$= 27 + 100$
3. Write the answer.	$= 127$

3.8.2 Cube roots

eles-3738

- Evaluating the **cube root** of a number is the opposite of cubing a number.
- The cube root symbol is similar to the square root symbol but with a small 3 written in front, and is written as $\sqrt[3]{}$.

  For example, since $2^3 = 8$, then $\sqrt[3]{8} = 2$.
- Using the illustrations of the cubes in section 3.8.1, a cube of 8 cubic units has a side length of 2 units.

WORKED EXAMPLE 27 Calculating cube roots

Calculate $\sqrt[3]{27}$.

THINK	WRITE
Look for a number that when multiplied by itself three times gives 27, and write the answer.	$27 = 3 \times 3 \times 3$ $\sqrt[3]{27} = 3$

WORKED EXAMPLE 28 Calculating cube roots using the order of operations

Calculate:
a. $\sqrt[3]{64 + 61}$ **b. $\sqrt[3]{64} + 61$.**

THINK	WRITE
a. 1. First simplify the expression inside the cube root.	a. $\sqrt[3]{64 + 61} = \sqrt[3]{125}$
2. To complete the calculation, evaluate the cube root by identifying the number that when multiplied by itself three times gives 125, and write the answer.	$= 5$
b. 1. Evaluate the cube root first. (Identify the number that when multiplied by itself three times gives 64.)	b. $\sqrt[3]{64} + 61 = 4 + 61$
2. Perform the addition to complete the calculation, and write the answer.	$= 65$

a. **Determine between which two whole numbers $\sqrt[3]{20}$ will lie.**
b. **Use your answer to part a to estimate the value of $\sqrt[3]{20}$ to 1 decimal place and check your answer with a calculator.**

THINK	WRITE
a. 1. Write the first few cubed numbers.	a. $1^3 = 1$ $2^3 = 8$ $3^3 = 27$ $\sqrt[3]{8} = 2$ and $\sqrt[3]{27} = 3$
2. Consider the cube root of each number.	Therefore $\sqrt[3]{20}$ lies between 2 and 3.
b. 1. The value of $\sqrt[3]{20}$ lies between 2 and 3. The value under the radical symbol is closer to 27 than 8. Estimate the answer.	b. Estimate: The value is more than halfway between the two numbers, so it is roughly: $\sqrt[3]{20} \approx 2.7$ or 2.8 Using a calculator: $\sqrt[3]{20} = 2.714\,417\,6...$
2. State your answer.	$\sqrt[3]{20} \approx 2.7$

Digital technology

Scientific calculators can evaluate cubes of decimal numbers.

> □   Math ▲
> 34.84^3
>
>   42289.6839

Scientific calculators can evaluate cube roots of decimal numbers.

> □   Math ▲
> $\sqrt[3]{17.576}$
>
>   2.6

Note: Scientific calculators will round any cube roots that are not exact.

 Resources

Interactivities   Cube numbers (int-3959)
            Cube roots (int-3960)

Exercise 3.8 Cubes and cube roots

3.8 Quick quiz on	3.8 Exercise

Individual pathways

■ PRACTISE	■ CONSOLIDATE	■ MASTER
1, 4, 7, 10, 12, 15	2, 5, 8, 13, 16	3, 6, 9, 11, 14, 17

Fluency

1. **WE25** Evaluate:

   a. 4^3 b. 2^3 c. 3^3

2. Evaluate:

   a. 6^3 b. 10^3

3. Write the first eight cube numbers.

4. **WE26** Evaluate $3^2 + 4^3 \times 2$.

5. Evaluate the following.

   a. $10^3 - 5^2 \times 2^3$ b. $4^3 - 3^3 - 2^3$

6. **WE27** Evaluate:

   a. $\sqrt[3]{8}$ b. $\sqrt[3]{64}$

Understanding

7. Evaluate each of the following. Verify your answers with a calculator.

   a. $\sqrt[3]{216}$ b. $\sqrt[3]{343}$

   (*Hint:* Use your answer to question **3**.)

8. **WE28** Determine the value of:

   a. $\sqrt[3]{125 + 91}$ b. $\sqrt[3]{125} + 91$

9. Evaluate each of the following:

   a. $\sqrt[3]{343 + 169}$ b. $\sqrt[3]{343} + \sqrt{169}$

10. **WE29** a. Determine between which two whole numbers $\sqrt[3]{50}$ will lie.

    b. Use your answer to part **a** to estimate the value of $\sqrt[3]{50}$ and check your answer with a calculator.

11. a. Determine between which two whole numbers $\sqrt[3]{90}$ will lie.

    b. Use your answer to part **a** to estimate the value of $\sqrt[3]{90}$ and check your answer with a calculator.

Communicating, reasoning and problem solving

12. Determine the first four numbers that could be arranged as a triangle-based pyramid (where all triangles are equilateral).

13. a. Express 216 as a product of its prime factors.
    b. Express 135 as a product of its prime factors.
    c. Use the answers to parts **a** and **b** to explain how you can use prime factors to determine whether the cube root of a number is an integer.

14. The first five square numbers are $1, 4, 9, 16, 25$. If we find the difference between these numbers, we get $4 - 1 = 3$, $9 - 4 = 5$, $16 - 9 = 7$ and $25 - 16 = 9$.

These numbers all differ by 2. Representing this in a table, we get:

Square numbers	1	4	9	16	25
First difference		3	5	7	9	
Second difference			2	2	2	

Repeat this process for the first six cube numbers. Explain how many times you needed to determine the difference until they were equal.

If you look at $1^4, 2^4, 3^4, 4^4, \ldots$, explain how many differences you would need to find until they are equal.

15. Dave said that $a^3 = \sqrt[a]{a}$ is a true statement for all numbers. Show your working to prove or disprove this statement.

16. Jack was organising a get-together of all his football mates from the last three years. He decided to phone 4 footballer mates and ask them to phone 4 footballers each, who in turn would phone 4 more footballers.

Each team member was given a list of 4 names so that no-one received more than one call.

a. Calculate how many footballers would receive a phone call about the party.

b. Jack's friend Karin liked the idea, so she decided to contact 258 people about a 10-year reunion using Jack's method.

Assuming there were three rounds of calls, calculate how many people each person should call.

17. It's an interesting fact that the sum of the digits of a cube of an integer, when reduced to a single digit, can only be $1, 8$ or 9.

For example:
$$12^3 = 1728$$
$$1 + 7 + 2 + 8 = 18$$
$$1 + 8 = 9$$

There is a pattern in the sequence of these numbers.

a. Use this rule with the cubes of the first 10 integers.

b. Determine the single-digit sum of 35^3.

c. Explain your answer.

LESSON
3.9 Review

3.9.1 Topic summary

Multiples

- A multiple of a number is the result of multiplying that number by another whole number.
- The times table gives lists of multiples.

e.g.
Multiples of 4 are: 4, 8, 12, 16, 20, …
Multiples of 6 are: 6, 12, 18, 24, 30, …

LCM and HCF

- The lowest common multiple (LCM) is the smallest multiple that is common to two or more numbers.
  e.g.
  Consider the numbers 8 and 12.
  Multiples of 8: 8, 16, **24**, 32, …
  Multiples of 12: 12, **24**, 36, 48, …
  LCM(8, 12) = 24
- The highest common factor (HCF) is the largest number that is a factor of both numbers.
  e.g.
  Factors of 8: 1, 2, **4**, 8
  Factors of 12: 1, 2, 3, **4**, 6, 12
  HCF(8, 12) = 4

Factors

- A factor of a whole number divides that number exactly.
- Factors are also numbers that multiply together to get another number.

e.g.
Since 20 = 4 × 5, or 2 × 10, or 1 × 20, then the factors of 20 are 1, 2, 4, 5, 10 and 20.

FACTORS, MULTIPLES, INDICES AND PRIMES

Index notation

- Index notation is a short way of writing a repeated multiplication.
  e.g. $2 \times 2 \times 2 \times 2 \times 2 \times 2$ can be written as 2^6, which is read as '2 to the power of 6'.
- The base is the number that is being repeatedly multiplied, and the index is the number of times it is multiplied.
  $2^6 = 2 \times 2 \times 2 \times 2 \times 2 \times 2 = 64$

Divisibility tests

Divisibility tests determine whether or not the number is divisible by the chosen divisor.
2: last digit is 0, 2, 4, 6 or 8
3: sum of the digits is divisible by 3
4: last 2 digits together are divisible by 4
5: last digit is either 0 or 5
6: divisible by 2 and divisible by 3
8: last 3 digits together are divisible by 8
9: sum of the digits is divisible by 9
10: last digit is 0

Squares

- The process of multiplying a number by itself is known as squaring a number.
  e.g.
  $1 \times 1 = 1^2 = 1$
  $2 \times 2 = 2^2 = 4$
- A square number is the number we get after multiplying a whole number by itself.
- The first four square numbers are 1, 4, 9 and 16.

Cubes

- A cube number is found by multiplying a number by itself three times.
  e.g.
  $1 \times 1 \times 1 = 1^3 = 1$
  $2 \times 2 \times 2 = 2^3 = 8$
  $3 \times 3 \times 3 = 3^3 = 27$
- The first four cube numbers are 1, 8, 27 and 64.

Prime and composite numbers

- A *prime* number has exactly two unique factors: 1 and itself.
- The first six prime numbers are 2, 3, 5, 7, 11 and 13.
- A *composite* number has more than two unique factors.
- Some examples of composite numbers are 4, 6, 8, 9 and 10.

Square roots

- The radical or square root symbol is $\sqrt{}$.
- The square root of a number is the opposite of squaring the number.
  e.g. If $4^2 = 16$ then $\sqrt{16} = 4$.

Cube roots

- The cube root symbol is $\sqrt[3]{}$.
- The cube root of a number is the opposite of cubing the number.
  e.g. If $5^3 = 125$ then $\sqrt[3]{125} = 5$.

One — a special number

The number 1 has only one unique factor: itself. It is neither prime nor composite, but a special number.

3.9.2 Project

Alphabet sizes

Bob's business, Reading Resources, produces individual letters of the alphabet printed on cardboard. He caters for the market of visually impaired students who are learning to read.

Teachers use the cards to form words for students to read. Because the students have varying difficulties with their sight, the letters are printed in heights of 1 cm, 2 cm, 3 cm, …, 10 cm. Bob has a set of all the letters at 1 cm and uses these as the basis for forming the larger letters.

Bob's factory has 10 machines. Each machine performs only one specific job, as indicated:
- Machine 1 makes letters 1 cm high.
- Machine 2 enlarges the letters so they are twice as high.
- Machine 3 enlarges the letters so they are 3 times as high.
- Machine 4 enlarges the letters so they are 4 times as high.
- Machine 10 enlarges the letters so they are 10 times as high.

Bob's business is thriving and he relies heavily on all his machines working. This morning he arrived to process a large order of letters that need to be 6 cm high, only to find that machine 6 was not working. He gathered his staff together to discuss the problem.

'No problem!' said Ken. 'As long as the other machines are working we can still get this order done.'
1. Suggest a solution that Ken might have proposed.
2. If Bob can do without machine 6, are there others he can also do without?
3. What is the minimum number of machines that Bob needs to make letters up to 10 cm high?

As Bob's business has become more successful, there has been a demand for letters of a greater height.
4. For Bob to make letters of every whole number up to a height of 20 cm, would he need twice as many machines as he currently has? Explain.
5. Having become very excited about needing fewer machines than he first thought, Bob considered expanding his business to make advertising signs using letters of all whole number heights up to 100 cm. To succeed in this new market, what is the minimum number of machines he would need?
6. Explain why it is necessary to have some specific machine numbers, while some others are not necessary.

 Resources

Interactivities	Crossword (int-2588)
	Sudoku puzzle (int-3164)

Fluency

1. List the first 5 multiples of each number.
   a. 11   b. 100   c. 5   d. 20   e. 35

2. Calculate the lowest common multiple (LCM) of the following pairs of numbers.
   a. 3 and 12   b. 6 and 15   c. 4 and 7   d. 5 and 8

3. List all the factors of each of the following numbers.
   a. 27   b. 50

4. List all the factors of each of the following numbers.
   a. 16   b. 42   c. 72

5. List the factor pairs of the following numbers.
   a. 24   b. 40   c. 48   d. 21   e. 99   f. 100

6. Evaluate:
   a. 6^2 b. 14^2 c. 19^2 d. 80^2

7. Write the two whole square numbers between which each of the following will lie.
   a. 3.8^2 b. 5.1^2 c. 10.6^2 d. 15.2^2

8. Evaluate:
   a. $\sqrt{49}$ b. $\sqrt{256}$ c. $\sqrt{900}$ d. $\sqrt{1369}$

9. Calculate:
   a. 4^3 b. 7^3 c. 15^3 d. 30^3

10. Evaluate each of the following.
    a. $\sqrt[3]{27}$ b. $\sqrt[3]{125}$ c. $\sqrt[3]{1000}$ d. $\sqrt[3]{8000}$

11. Write the following using index notation.
    a. $2 \times 3 \times 3 \times 3$
    b. $5 \times 5 \times 6 \times 6 \times 6 \times 6$
    c. $2 \times 5 \times 5 \times 5 \times 9 \times 9 \times 9$

12. Use your calculator to evaluate:
    a. 3^5 b. 7^3 c. 8^4 d. 11^5

Understanding

13. List all of the prime numbers less than 30.

14. List all of the single-digit prime numbers.

15. State the prime number that comes next after 50.

16. State the prime factors of:
    a. 99   b. 63   c. 125   d. 124.

17. Express the following numbers as products of their prime factors, in index form.
    a. 280   b. 144

18. Write the following numbers in expanded form using index notation.
    a. 1344   b. 30 601

19. Determine the highest common factor and lowest common multiple of 120 and 384 by first expressing each number as a product of its prime numbers.

20. Determine the highest common factor and lowest common multiple of 72 and 224 by first expressing each number as a product of its prime numbers.

Communicating, reasoning and problem solving

21. Hung and Frank are cyclists who are riding around a track. They ride past the finish line together, and from that point Hung takes 25 seconds to complete a lap and Frank takes 40 seconds.
    a. Calculate how long it will be until they next pass the finish line together.
    b. Determine how many laps each will have ridden when this occurs.

22. Joe, Claire and Daniela were racing up and down a flight of stairs. Joe took the stairs three at a time, Daniela took the stairs two at a time, while Claire took the stairs four at a time. In each case, they reached the bottom with no steps left over.
    a. Calculate how many steps are in the flight of stairs. List three possible answers.
    b. Determine the smallest number of steps there could be.
    c. If Joe can also take the stairs five at a time with no steps left over, determine the smallest number of steps in the flight of stairs.

23. A perfect number is one whose factors (all except the number itself) add up to the number. For example, 6 is a perfect number because $1 + 2 + 3$ (the sum of its factors, excluding 6) is equal to 6. Show why 496 is a perfect number.

24. In a race, one dirt bike rider completes each lap in 40 seconds while another completes a lap in 60 seconds. Determine how long after the start of the race the two bikes will pass the starting point together.

25. A three-digit number is divisible by 6. The middle digit is a prime number. The sum of the digits is 9. The number is between 400 and 500. The digits are in descending order. Determine the number.

26. Explain why we cannot determine the exact value of $\sqrt{20}$.

27. Write the following numbers in the order of size, starting with the smallest.
$$5, \sqrt{64}, 2, \sqrt[3]{27}$$

28. Calculate the smallest possible number which, when multiplied by 120, will give a result that is a square number. Now calculate the smallest possible number that will give a result that is a cube number.

29. List the prime numbers up to 100. Some prime numbers are separated by a difference of 2 or 6. (For example, $5 - 3 = 2$ and $19 - 17 = 2$.)
   a. Calculate the prime number pairs up to 100 that have a difference of 2.
   b. Calculate the prime number pairs up to 100 that have a difference of 6.
   c. Explain why there are no prime number pairs with a difference of 7.

30. Determine the smallest number that is both a square number and a cube number. Identify the next smallest number.

31. Complete the following sequence: $2, 3, 5, 7, 11, 13, \_\_\_, \_\_\_, \_\_\_$.

32. Write each of the numbers $250, 375$ and 625 as a product of its prime factors and:
   a. determine the highest common *prime* factor
   b. determine the HCF.

33. A motor boat requires an engine service every 5000 nautical miles, refuelling every 300 nautical miles and an oil change after 2250 nautical miles. Determine how many nautical miles the motor boat will have travelled before all three services are required simultaneously.

34. By folding a sheet of paper in half, you divide it into 2 regions.

   a. If you fold it in half again, determine how many regions there are.

   b. Set up a table similar to the one shown and fill it in.

   c. Explain how the number of regions increases with each new fold.

   d. Write a formula for the number of regions (R) for n folds.

n (number of folds)	R (number of regions made)
0	1
1	2
2	4
3	
4	
5	
6	

35. Look at the following table. As you read down each column, you'll notice that the equations follow a pattern. Continue the patterns to complete the table.

$2^5 = 32$	$3^5 = 243$
$2^4 = 16$	$3^4 = 81$
$2^3 = 8$	$3^? = 27$
$2^2 =$	$3^2 =$
$2^1 =$	$3^1 =$
$2^0 =$	$3^0 =$
$2^{-1} =$	$3^{-1} =$

   a. State what 5^0 equals.

   b. State what 2^{-2} is equal to.

To test your understanding and knowledge of this topic, go to your learnON title at www.jacplus.com.au and complete the **post-test**.

Answers

Topic 3 Factors, multiples, indices and primes

3.1 Pre-test

1. B
2. $35
3. 6
4. 12 minutes
5. $5^2 \times 3^4$
6. 17, 31, 109
7. C, E
8. B, E
9. LCM = 864; HCF = 6
10. 64
11. 109
12. 8 and 9
13. 3
14. 39
15. 7

3.2 Factors and multiples

1. a. 3, 6, 9, 12, 15   b. 6, 12, 18, 24, 30
   c. 100, 200, 300, 400, 500   d. 11, 22, 33, 44, 55

2. a. 120, 240, 360, 480, 600   b. 45, 90, 135, 180, 225
   c. 72, 144, 216, 288, 360   d. 33, 66, 99, 132, 165

3. 10, 20, 100, 70
4. 21, 7, 70, 35
5. 16, 64, 160, 32, 80
6. 9, 18, 27, 36, 45, 54, 63, 72, 81, 90, 99
7. 102, 108, 114, 120, 126, 132, 138, 144, 150, 156
8. C
9. A
10. a. 1, 2, 3, 4, 6, 12   b. 1, 2, 4, 5, 8, 10, 20, 40
11. a. 1, 2, 7
    b. 1, 2, 3, 5, 10, 15, 20
    c. 1, 2, 5, 10, 20, 25
12. a. 1, 2, 3, 4, 6, 8, 9, 12, 18, 24, 36, 72
    b. 1, 2, 5, 10, 25, 50, 125, 250
13. a. 1, 5, 17   b. 1, 3, 9, 11, 33   c. 1, 3, 17
14. 1, 20; 2, 10; 4, 5
15. 1, 132; 2, 66; 3, 44; 4, 33; 6, 22; 11, 12
16. A, B, C
17. 15
18. B
19. A
20. 3, 4
21. a. 6, 12, 18 (or any other multiple of 6)
    b. 6
    c. 30

22.

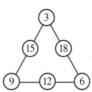

23. 4 rows of 5 students or 5 rows of 4 students; 2 rows of 10 students or 10 rows of 2 students; 1 row of 20 students or 20 rows of 1 student
24. 3, 12 or 30
25. Boy 6, grandfather 66
26. 84
27. 60
28. 60

3.3 Lowest common multiple and highest common factor

1. a. 4, 8, 12, 16, 20, 24, 28, 32, 36, 40
   b. 6, 12, 18, 24, 30, 36, 42, 48, 54, 60
   c. 12, 24, 36
   d. 12

2. a. 9   b. 30   c. 70   d. 48   e. 60
3. False
4. True
5. True
6. False
7. a. HCF = 7   b. HCF = 7   c. HCF = 6
8. a. 9   b. 7
9. a. 4   b. 9
10. a. 6 days
    b. 6 is the lowest common multiple of 2 and 3.
11. a. 35 minutes
    b. 35 is the lowest common multiple of 5 and 7.
12. a. 12 hours
    b. 12 is the lowest common multiple of 4 and 6.
13. 20 seconds
14. a. 12   b. 6   c. 7
15. a. 3 metres   b. 10
16. 2 hours
17. a. Daniella   b. 10 minutes
18. 6 times
19. a. 4 minutes and any multiple of 4 thereafter
    b. 3 minutes and any multiple of 3 thereafter
20. 30 minutes
21. a. Two semicircles
    b. The starting line for a 100 m race has been positioned in the extended section so the athletes can all start from the same starting line.

c. The finish line for a 100 m race has been positioned so that all athletes can run in a direct line without having to have staggered starting positions.

d. There is a need to stagger the starting blocks in the 200 m, 400 m and 800 m track events so that the outside runners will not be running further than the inside runners.

e. and f.

Lane number	Distance travelled	Difference
1	400 m	0
2	408 m	8
3	416 m	16
4	424 m	24
5	432 m	32
6	440 m	40
7	448 m	48
8	456 m	56

g. The runner in lane 8 is running 56 metres further than the runner in lane 1.

3.4 Index notation

1. a. $7^2 = 49$ Base is 7 and index is 2.
   b. $4^5 = 4096$ Base is 4 and index is 5.
   c. $3^4 = 81$ Base is 3 and index is 4.
   d. $10^3 = 1000$ Base is 10 and index is 3.

2. a. 7^4 b. 8^6 c. 3^9 d. 13^3

3. a. $4^5 \times 6^3$ b. $2^2 \times 3$
   c. $2^4 \times 5^2$ d. $2^2 \times 3^2 \times 5^3$

4. a. $6 \times 6 \times 6 \times 6 \times 6$ b. $11 \times 11 \times 11$

5. a. 5^{11} b. 4^{13} c. 2^{21}

6. a. $300 = 3 \times 10^2$
   b. $4500 = 4 \times 10^3 + 5 \times 10^2$
   c. $6705 = 6 \times 10^3 + 7 \times 10^2 + 0 \times 10^1 + 5$
   d. $10\,000 = 1 \times 10^4$

7. a. 9^{15} b. $3^6 \times 13^6$

8. D

9. a. 343   b. 288   c. 77

10. a. 72   b. 5184   c. 177 147

11. a. 567   b. 2717   c. 112

12. a. A
    b. At first the base number 2 would appear to be the smallest, but it has the highest power. Although the other numbers have a higher base, they have relatively small powers — therefore 2^8 will be the largest number. This can easily be verified by evaluating each number.

13. $41^2 \neq 691$. Notice that $31^2 = 961$, therefore 41^2 will need to be larger than 961.

14. a. <   b. <   c. <   d. >

15. a. Option 1: $327.68; Option 2 : $120 after the 15 weeks
    b. Option 1: 2^{15} cents after week 15

16. 64

17. a. $8^3 = 584$ b. 584

18. 127

19. a. $2^{12} = 4096$
    b. $3^{12} = 531\,441$
    c. Using powers is a quick way of getting answers that are really large.

20. 256

21. Square the first number and add one to calculate the second number. Square the second number and add one to calculate the third number. The missing number is 101.

3.5 Prime numbers and composite numbers

1. 23, 29, 31, 37

2. a. Composite   b. Prime
   c. Composite   d. Prime

3. a. Composite   b. Prime
   c. Composite   d. Prime

4. a. i. 15   3, 5   ii. $15 = 3 \times 5$

   b. i. 30   2, 3, 5   ii. $30 = 2 \times 3 \times 5$

   c. i. 100   2, 5   ii. $100 = 2^2 \times 5^2$

   d. i. 49   7   ii. $49 = 7^2$

5. a. i. 72   2, 3   ii. $72 = 2^3 \times 3^2$

   b. i. 56   2, 7   ii. $56 = 2^3 \times 7$

c. i. 3, 5   **ii.** $45 = 3^2 \times 5$

6. a. i. 2, 3, 7   **ii.** $84 = 2^2 \times 3 \times 7$

b. i. 112   2, 7   **ii.** $112 = 2^4 \times 7$

c. i. 40   2, 5   **ii.** $40 = 2^3 \times 5$

7. a. i. 96   2, 3   **ii.** $96 = 2^5 \times 3$

b. i. 32   2   **ii.** $32 = 2^5$

c. i. 3000   2, 3, 5   **ii.** $3000 = 2^3 \times 3 \times 5^3$

8. a. 2, 3   **b.** 2, 5   **c.** 3   **d.** 2, 3

9. a. 2, 3, 5   **b.** 2, 3, 5   **c.** 2, 3, 5   **d.** 2, 5

10. a. $2^2 \times 3 \times 5$ **b.** 3×5^2 **c.** $2^2 \times 5 \times 11$

11. a. $2^6 \times 3$ **b.** $2^2 \times 31$

12. a. LCM is 252; HCF is 12.
    b. LCM is 240; HCF is 12.
    c. LCM is 1200; HCF is 40.
    d. LCM is 8800; HCF is 20.

13. a. False   **b.** False   **c.** False
    d. True   **e.** True

14. a. False   **b.** False   **c.** False

15. A

16. C

17. C

18. C

19. a. $365 = 5 \times 73$ **b.** $187 = 11 \times 17$

20. A

21. D

22. C

23. B

24. 773

25. No. $3 + 7 = 10$, 3 and 7 are prime numbers, but their sum 10 is not a prime number.

26. 29 or 47 or 83 or 137. Others may be possible.

27. 61

28. a. 3, 7   **b.** 2, 13   **c.** 17, 19

29. 6 integers

30. 17 and 19, 29 and 31, 41 and 43, 59 and 61, 71 and 73

3.6 Squares and square roots

1. a. 64   **b.** 121   **c.** 225   **d.** 625

2. 64, 81

3. 169, 196

4. a. 36 and 49   **b.** 49 and 64
    c. 81 and 100   **d.** 144 and 169

5. a. 5   **b.** 9   **c.** 12   **d.** 20

6. a. 70   **b.** 120   **c.** 600   **d.** 400

7. a. 7 and 8   **b.** 3 and 4
    c. 14 and 15   **d.** 1 and 2

8. a. 13   **b.** 37

9. 16, 36, 64

10. 81

11. B

12. A

13. a. 9   **b.** 75   **c.** 700   **d.** 25

14. a. 14   **b.** 45

15. a. 12 and 13   **b.** 12.2

16. a. 225
    b. 225
    c. Yes, if two squares are multiplied together, then the result is equal to the square of the base numbers multiplied together.

17. a. 15
    b. 15
    c. Yes, if two square roots are multiplied together, then the result is equal to the square root of the base numbers multiplied together.

18. a. $2^2 \times 7^2$
    b. $2^3 \times 5^2$
    c. If all of the prime factors of a number are squares, then the square root of that number will be an integer.

19. a. 14
    b. 10

c. No, the square root of two numbers that have been added together is not equal to adding the square roots of these two numbers together.

20. a. 17

b. 13

c. No. You have to evaluate the individual square roots first before adding.

21. $169, 196, 225$

22. a. i. 49   ii. 47   iii. 45   iv. 43   v. 41

b. Answer is the sum of the two numbers.

c. No

23. 49

24. a. $7, \sqrt{56}, 8, \sqrt{74}, 3^2, \sqrt{100}, 62, 8^2$

b. $\sqrt{10}, \sqrt{121}, 3 \times 4, \sqrt{169}, 2 \times 7, 9^2, 100$

25. 9 and 11

3.7 Divisibility tests

1. A number is divisible by 2 if the last digit is 0, 2, 4, 6 or 8. For example, 56 is divisible by 2 because the last digit is 6. A number is divisible by 4 if the last two digits together (as a two-digit number) are divisible by 4. For example, 3416 is divisible by 4 because the last two digits are 16, which is divisible by 4.
A number is divisible by 6 if it is divisible by 2 and divisible by 3. For example, 354 is divisible by 2 because the last digit is 6; and the sum of the digits $(3 + 5 + 4)$ is 12, which is divisible by 3. Therefore, 354 is divisible by 6.

2. A number is divisible by 3 if the sum of the digits is divisible by 3. For example, 156 is divisible by 3 because the sum of the digits $(1 + 5 + 6)$ is 12, which is divisible by 3.
A number is divisible by 5 if the last digit is either 0 or 5. For example, 845 is divisible by 5 because the last digit is 5. A number is divisible by 9 if the sum of the digits is divisible by 9. For example, 54 is divisible by 9 because the sum of the digits $(5 + 4)$ is 9, which is divisible by 9.

3. The sum of the digits is calculated to test for divisibility by 3 and by 9.

4. a. 54 is divisible by 2 as the last digit is 4.

b. 150 is not divisible by 9 as the sum of the digits is $6 (1 + 5 + 0)$, which is not divisible by 9.

c. 295 is divisible by 5 as the last digit is 5.

d. 1293 is divisible by 3 as the sum of the digits is $15 (1 + 2 + 9 + 3)$, which is divisible by 3.

5. a. Divisible by 4 as the last two digits are divisible by 4

b. Not divisible by 6 as the number is divisible by 2 but not by 3

c. Not divisible by 9 as the sum of digits is not divisible by 9

d. Not divisible by 8; incorrect rule

6. a. 1, 2, 3, 4, 6, 8, 9   b. 1, 2, 3, 4, 6, 8
c. 1, 2, 4, 7, 8   d. 1, 2, 4

7. a. 1, 2, 4, 5, 8, 10   b. 1, 5
c. 1   d. 1, 2, 4, 8

8. a. 1, 2, 4   b. 1, 2, 3, 5, 6
c. 1, 2   d. 1

9. a. False   b. True
c. False   d. False
e. True   f. False

10. a. i. 5 only   ii. 2, 3, 4, 5, 6 and 10
iii. 5 only   iv. 2, 5 and 10
v. None   vi. 2 and 4

11. Julie is correct. 34 281 is not divisible by 6.

12. 3

13. Yes, $4572 + 2868$ and $4572 - 2868$ are divisible by 6.

14. a. 2, 5 or 8   b. 1, 4 or 7
c. 0, 2, 4, 6 or 8   d. 5

15. The two 2-digit numbers that are divisible by 4, 5 and 8 are 40 and 80.

16. 552

3.8 Cubes and cube roots

1. a. 64   b. 8   c. 27

2. a. 216   b. 1000

3. $1, 8, 27, 64, 125, 216, 343, 512$

4. 137

5. a. 800   b. 29

6. a. 2   b. 4

7. a. 6   b. 7

8. a. 6   b. 96

9. a. 8   b. 20

10. a. 3 and 4   b. 3.7

11. a. 4 and 5   b. 4.5

12. 3, 6, 10, 15 (not including 1, which would not really be a triangle at all).

13. a. $2^3 \times 3^3$

b. $3^3 \times 5$

c. If all of the prime factors of a number are cubes, then the cube root of that number will be an integer.

14. For cube numbers, it was necessary to find the third difference. If we look at the power 4, it will be necessary to find the fourth difference.

15. Dave is incorrect.

16. a. 84 footballers would receive a call.
b. Each person would call 6 people.

17. a. $1, 8, 9, 1, 8, 9, 1, 8, 9, 1$

b. 8

c. The single-digit sums of cube numbers repeat in the order 1, 8, 9.

Project

1. Enlarge the letters using machine 2, then machine 3.

2. 4, 6, 8, 9 and 10

3. 5

4. No. Numbers not required can be made using their prime factors.

5. Machine 1 and the prime numbers less than 100. Total of 26 machines.

6. Machine 1 is needed to make letters one centimetre high. Machines whose numbers are prime numbers can make all the sizes up to 100 centimetres.

3.9 Review questions

1. a. $11, 22, 33, 44, 55$
   b. $100, 200, 300, 400, 500$
   c. $5, 10, 15, 20, 25$
   d. $20, 40, 60, 80, 100$
   e. $35, 70, 105, 140, 175$

2. a. 12 b. 30 c. 28 d. 40

3. a. $1, 3, 9, 27$ b. $1, 2, 5, 10, 25, 50$

4. a. $1, 2, 4, 8, 16$
   b. $1, 2, 3, 6, 7, 14, 21, 42$
   c. $1, 2, 3, 4, 6, 8, 9, 12, 18, 24, 36, 72$

5. a. $1, 24; 2, 12; 3, 8; 4, 6$
   b. $1, 40; 2, 20; 4, 10; 5, 8$
   c. $1, 48; 2, 24; 3, 16; 4, 12; 6, 8$
   d. $1, 21; 3, 7$
   e. $1, 99; 3, 33; 9, 11$
   f. $1, 100; 2, 50; 4, 25; 5, 20; 10, 10$

6. a. 36 b. 196 c. 361 d. 6400

7. a. 9 and 16 b. 25 and 36
   c. 100 and 121 d. 225 and 256

8. a. 7 b. 16 c. 30 d. 37

9. a. 64 b. 343 c. 3375 d. $27\,000$

10. a. 3 b. 5 c. 10 d. 20

11. a. 2×3^3 b. $5^2 \times 6^4$ c. $2 \times 5^3 \times 9^3$

12. a. 243 b. 343 c. 4096 d. $161\,051$

13. $2, 3, 5, 7, 11, 13, 17, 19, 23, 29$

14. 4

15. 53

16. a. $3, 11$ b. $3, 7$ c. 5 d. $2, 31$

17. a. $280 = 2 \times 2 \times 2 \times 5 \times 7 = 2^3 \times 5 \times 7$
    b. $144 = 2^4 \times 3^2$

18. a. $1344 = 1 \times 10^3 + 3 \times 10^2 + 4 \times 10^1 + 4$
    b. $30\,601 = 3 \times 10^4 + 0 \times 10^3 + 6 \times 10^2 + 0 \times 10^1 + 1$

19. HCF is 24; LCM is 1920.

20. HCF is 8; LCM is 2016.

21. a. 200s

b. Hung 8 laps
   Frank 5 laps

22. a. $12, 24, 36$ b. 12 c. 60

23. The sum of its factors (except for 496) add up to give 496.

24. 120 seconds or 2 minutes

25. 432

26. 20 is not a perfect square, so we can only find an approximate answer. It will be between 4 and 5.

27. $2, \sqrt[3]{27}, 5, \sqrt{64}$

28. $30, 225$

29. a. $3, 5; 5, 7; 11, 13; 17, 19; 29, 31; 41, 43; 59, 61; 71, 73$
    b. $5, 11; 7, 13; 11, 17; 13, 19; 17, 23; 23, 29; 31, 37; 37, 43;$
       $41, 47; 47, 53; 53, 59; 61, 67; 67, 73; 73, 79; 83, 89$
    c. For 2, the number that's 7 higher is 9, which is not prime. For odd primes, the number that's 7 higher is even and therefore not prime.

30. $1, 64$

31. $2, 3, 5, 7, 11, 13, \underline{17}, \underline{19}, \underline{23}$

32. a. 5 b. 125

33. $45\,000$

34. a. 4 regions
    b.

n (number of folds)	0	1	2	3	4	5	6
R (number of regions made)	1	2	4	8	16	32	64

    c. Each time the page is folded the number of regions is multiplied by 2 (doubled).
    d. $R = 2^n$

35.

$2^5 = 32$	$3^5 = 243$
$2^4 = 16$	$3^4 = 81$
$2^3 = 8$	$3^3 = 27$
$2^2 = 4$	$3^2 = 9$
$2^1 = 2$	$3^1 = 3$
$2^0 = 1$	$3^0 = 1$
$2^{-1} = \dfrac{1}{2}$	$3^{-1} = \dfrac{1}{3}$

a. $5^0 = 1$ b. $2^{-2} = \dfrac{1}{4}$

4 Fractions and percentages

LESSON SEQUENCE

LESSON
4.1 Overview

Why learn this?

There are many everyday situations in which we do not work with whole numbers. You may share a pizza with a friend, so you would eat part of the pizza. Fractions are a way of expressing parts of whole things, amounts or quantities. They are important in many situations, such as cooking, shopping and telling the time. Did you know that approximately one third $\left(\dfrac{1}{3}\right)$ of the population of Australia lives in New South Wales?

Percentages, like fractions, are a way of expressing parts of a whole. They are commonly used in advertising, statistics and shopping. Did you know that the area of New South Wales is approximately 10%, or one tenth $\left(\dfrac{1}{10}\right)$, of the size of Australia? Fractions and percentages can be used interchangeably to represent the same part of a whole. Many different professions use fractions and percentages extensively in their work, including hospitality, finance, statistics and journalism. When you are next in a shopping centre, check out how many percentage (%) signs you see. You might be surprised.

Hey students! Bring these pages to life online

Watch videos

Engage with interactivities

Answer questions and check solutions

Find all this and MORE in jacPLUS

Reading content and rich media, including interactivities and videos for every concept

Extra learning resources

Differentiated question sets

Questions with immediate feedback, and fully worked solutions to help students get unstuck

1. **MC** Select the equivalent fraction to $\frac{2}{3}$:

   A. $\frac{3}{2}$　　　　**B.** $\frac{30}{45}$　　　　**C.** $\frac{30}{40}$　　　　**D.** $\frac{12}{15}$

2. Luke's football team scored 16 goals. Luke scored 6 of the team's goals. Answer the following questions, giving your answers in simplified form.
   a. Calculate the fraction of the team's goals that Luke scored.
   b. Calculate the fraction of the team's goals that the rest of the team scored.

3. A bowl of apples was handed out to kindergarten students. If 23 students ate $\frac{1}{4}$ of an apple each, calculate how many apples the students ate in total. Give your answer as a mixed number in simplified form.

4. Evaluate the following, giving your answer as a simplified fraction.

$$\frac{2}{3} - \frac{1}{6} + \frac{3}{12}$$

5. A movie went for 120 minutes. If the main character is on screen for $\frac{5}{6}$ of the movie, calculate for how many minutes they were on screen.

6. **MC** Select the simplified answer to $\frac{18}{56} \div \frac{27}{64}$.

   A. $\frac{16}{21}$　　　　**B.** $\frac{18}{21}$　　　　**C.** $\frac{2}{3}$　　　　**D.** $\frac{15}{24}$

7. Yuto gets \$16 a week for cleaning his room, but if he also empties the dishwasher and feeds the dogs he can get $2\frac{1}{4}$ times his pay. If Yuto completes all his chores in the week, calculate how much he earned.

8. When Thomas started school he was the shortest in his class, at 85 cm tall. By the time he finished grade 12 he was the tallest in his class. His height had increased by $2\frac{1}{5}$ times. Calculate how tall he was at the end of grade 12.

9. Newcastle High School is having a cupcake sale to raise money for a local charity. They have 250 cupcakes to sell. Of these, $\frac{1}{5}$ are chocolate and the remainder are vanilla. At the end of the day, $\frac{24}{25}$ of the chocolate cupcakes were sold, and $\frac{7}{10}$ of all the cupcakes were sold. Calculate what fraction of the vanilla cupcakes were sold.

10. **MC** Max spends 78% of his weekly pocket money on his mobile phone plan. Calculate what fraction of his pocket money is spent on the phone plan.

    A. $\frac{38}{50}$　　　　**B.** $\frac{37}{50}$　　　　**C.** $\frac{39}{50}$　　　　**D.** $\frac{18}{25}$

11. The cost of the new model PlayStation has increased by 12% compared to the previous model. Write the percentage increase as a:

   a. simplified fraction   **b.** decimal.

12. At a particular soccer game, 62% of the crowd supported the Matildas. If 17 800 people attended the game, determine how many supported the Matildas.

13. Alexander got 24 marks out of a possible 32. Calculate his percentage score to the nearest whole number.

14. **MC** A survey of 20 students found that 9 like Coke, 6 like Solo, 3 like Fanta and 2 like lemonade. Select the percentage of students who like Fanta.

   A. 3%   **B.** 6%   **C.** 10%   **D.** 15%

15. Sarah purchased a dress that was on sale for a discount of 10%.
    If the dress originally cost $155.75, calculate how much she saved, to the nearest cent.

LESSON
4.2 What are fractions?

LEARNING INTENTION

At the end of this lesson you should be able to:
- understand the concept of a fraction
- convert between equivalent fractions
- compare fractions.

▶ 4.2.1 Understanding fractions

eles-3798

- **Fractions** are used to describe parts of a whole.

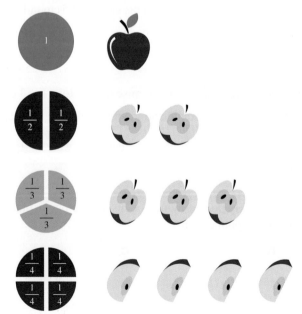

- The **numerator**, or top number of the fraction, shows how many parts are required. The numerator must be an integer.
- The **denominator**, or bottom number, shows the number of parts into which the whole can be divided. The denominator must be an integer, but cannot be zero.
- The horizontal bar separating the numerator from the denominator is called the **vinculum**.

Proper fractions	Improper fractions	Mixed numbers
Proper fractions have a numerator *less than* the denominator.	**Improper fractions** have a numerator *greater than* the denominator.	**Mixed numbers** consist of a whole number and a proper fraction.
$\dfrac{1}{4}, \dfrac{2}{3}, \dfrac{27}{68}$	$\dfrac{5}{3}, \dfrac{3}{2}, \dfrac{123}{17}$	$1\dfrac{1}{2}, 5\dfrac{7}{8}, 14\dfrac{152}{986}$

Understanding fractions

A fraction can be understood as numerator *out of* denominator.

For example: the fraction $\dfrac{6}{10}$ means 6 *out of* 10.

When the numerator is equal to the denominator, it is called a 'whole', which is equivalent to 1.

For example: $\dfrac{10}{10}$ means 10 *out of* 10 and is one 'whole'.

- To express one quantity as a fraction of another, write the first quantity as a fraction of the second quantity.

WORKED EXAMPLE 1 Identifying fractions of a quantity

a. **Identify the fraction of the rectangle that has been shaded.**
b. **Express the number of unshaded squares as a fraction of the number of squares in total.**

THINK

a. 1. Count how many equal parts the rectangle has been divided into.

 2. State how many parts are shaded.

 3. Write the number of shaded parts as a fraction of the total number of parts.

b. 1. Count the number of unshaded squares.

 2. State the number of squares in total.

 3. Write the number of unshaded squares as a fraction of the total number of squares.

WRITE

a. Total number of parts $= 8$

 3 parts are shaded.

 $\dfrac{3}{8}$ of the rectangle has been shaded.

b. There are 5 unshaded squares.

 There are 8 squares in total.

 $\dfrac{5}{8}$ of the squares are unshaded.

▶ 4.2.2 Equivalent fractions

eles-3799

- **Equivalent fractions** are equal fractions. For example, the fractions $\frac{1}{2}, \frac{2}{4}, \frac{5}{10}, \frac{8}{16}$ are all equivalent because they are all equal to a half.

- Equivalent fractions can be shown using diagrams. Equivalent fractions for $\frac{2}{3}$ are shown in the following diagrams.

Note that the same portion, $\frac{2}{3}$ of the rectangle, has been shaded in each case.

$$\frac{4}{6} = \frac{2}{3}$$

$$\frac{6}{9} = \frac{2}{3}$$

$$\frac{8}{12} = \frac{2}{3}$$

Equivalent fractions

Equivalent fractions are equal fractions. Equivalent fractions can be found by multiplying or dividing the numerator and denominator by the same number.

For example:

$$\overset{\times 2}{\underset{\times 2}{\frac{2}{3} = \frac{4}{6}}} \qquad \overset{\div 10}{\underset{\div 10}{\frac{10}{20} = \frac{1}{2}}}$$

WORKED EXAMPLE 2 Determining the multiplier

For the equivalent fractions $\frac{4}{5} = \frac{8}{10}$, determine the number (multiplier) that has been used to multiply both numerator and denominator of the first fraction in order to get the second.

THINK	WRITE
1. What number is 4 multiplied by to equal 8? $(4 \times \_ = 8)$ What number is 5 multiplied by to equal 10? $(5 \times \_ = 10)$	$\frac{4}{5} = \frac{4 \times 2}{5 \times 2} = \frac{8}{10}$
2. Write the answer.	The number used is 2.

WORKED EXAMPLE 3 Determining equivalent fractions

From the list, determine those fractions that are equivalent to $\dfrac{1}{2}$.

$$\dfrac{3}{9}, \ \dfrac{3}{6}, \ \dfrac{9}{18}, \ \dfrac{10}{15}, \ \dfrac{7}{14}, \ \dfrac{17}{34}$$

THINK	WRITE
1. Multiply the numerator and denominator of $\dfrac{1}{2}$ by the numerator of the first fraction in the list (3) to check whether the new fraction is in the list.	$\dfrac{1 \times 3}{2 \times 3} = \dfrac{3}{6}$
2. Multiply the numerator and denominator of $\dfrac{1}{2}$ by the next different numerator in the list (9) and check whether the new fraction is in the list.	$\dfrac{1 \times 9}{2 \times 9} = \dfrac{9}{18}$
3. Continue until all fractions have been considered.	$\dfrac{1 \times 10}{2 \times 10} = \dfrac{10}{20}$ $\dfrac{1 \times 7}{2 \times 7} = \dfrac{7}{14}$ $\dfrac{1 \times 17}{2 \times 17} = \dfrac{17}{34}$
4. Write the equivalent fractions.	From the given list, the equivalent fractions of $\dfrac{1}{2}$ are $\dfrac{3}{6}, \dfrac{9}{18}, \dfrac{7}{14}, \dfrac{17}{34}$.

WORKED EXAMPLE 4 Writing a sequence of equivalent fractions

Write the sequence of the first three equivalent fractions for $\dfrac{2}{3}$.

THINK	WRITE
1. Write the first three equivalent fractions in the sequence by multiplying both the numerator and denominator by 2, 3 and 4.	$\dfrac{2}{3} = \dfrac{2 \times 2}{3 \times 2} = \dfrac{4}{6}$ $= \dfrac{2 \times 3}{3 \times 3} = \dfrac{6}{9}$ $= \dfrac{2 \times 4}{3 \times 4} = \dfrac{8}{12}$
2. Write the equivalent fractions.	$\dfrac{2}{3} = \dfrac{4}{6} = \dfrac{6}{9} = \dfrac{8}{12}$
3. Write your answer.	The first three equivalent fractions for $\dfrac{2}{3}$ are $\dfrac{4}{6}, \dfrac{6}{9}$ and $\dfrac{8}{12}$.

▶ 4.2.3 Fractions on a number line

eles-3800

- The space between each whole number on a number line can be divided into equal parts. The number of equal parts is determined by the denominator of the fraction.

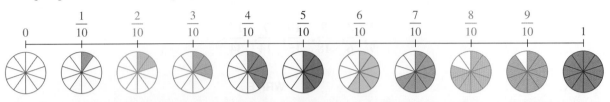

WORKED EXAMPLE 5 Placing fractions on a number line

Show the positions of $\dfrac{5}{6}$ and $1\dfrac{1}{6}$ on a number line.

THINK	WRITE
1. Draw a number line from 0 to 2.	
2. The denominator of both fractions is 6. Divide the sections between 0 and 1 and between 1 and 2 into 6 equal parts each.	
3. To place $\dfrac{5}{6}$, count 5 marks from 0. To place $1\dfrac{1}{6}$ count 1 mark from the whole number 1.	

- When choosing a scale for your number lines, take into account the number of equal parts each whole number has been divided into. For example, when plotting thirds it is easier to use a scale of 3 cm for each whole number (for each part) as opposed to a scale of 1 cm for each whole number.

Equivalent fractions on a number line

- Equivalent fractions can also been shown using a number line.
- A number line can be divided into smaller (or larger) equally sized parts.

COMMUNICATING — COLLABORATIVE TASK: Peg the fraction

Equipment: one long piece of string, one smaller piece of string for each group, laundry pegs, sticky notes

Your task is to place a peg on the string to show a fraction, following the steps below.
1. Two students hold the ends of a long piece of string across the front of the classroom.
2. The teacher calls out a fraction and, with the help of the class, another student places a peg on the string to show the position of that fraction.
3. The two students holding the ends of the string then fold the string into equal lengths to see how accurately the peg was placed.

4. Repeat steps **1** to **3** in small groups using the smaller lengths of string.
5. As a class, discuss the following.
   a. How do you visually estimate fractions when the numerator is 1?
   b. How can this strategy be used to estimate the position of a fraction in which the numerator is more than 1?
6. Two students divide the long piece of string in half and place a peg at the halfway mark. They hold the ends of the string so that it stretches across the front of the classroom. Another student writes the number 1 on a sticky note and attaches it to the peg.
7. The teacher calls out a fraction, which will be a mixed or an improper fraction, such as $1\frac{1}{2}$ or $\frac{7}{4}$. With the help of the class, another student places a peg on the string to show the position of that fraction on the string.
8. The two students holding the ends of the string then fold the string into equal lengths to see how accurately the peg was placed.
9. Repeat steps **6** to **8** in small groups using the smaller lengths of string.
10. As a class, discuss the effect the denominator has on determining the position of the peg.

▶ 4.2.4 Comparing fractions

eles-3801

- If we cut a block into 4 equal pieces, then 1 piece is less than 2 pieces. This means that $\frac{1}{4}$ of the block is smaller than $\frac{2}{4}$.

- Fraction walls, or number lines, are useful tools when comparing fractions. The following figure shows an example of a fraction wall.

- With a diagram such as a fraction wall or a number line, we can see that $\frac{1}{4}$ is less than $\frac{1}{3}$. That is, if we divided a block into 4 equal parts, each part would be smaller than if we divided it into 3 equal parts.

- If the numerators are the same, the smaller fraction is the one with the larger denominator.
  For example, $\frac{1}{7}$ is less than $\frac{1}{6}$, and $\frac{3}{10}$ is less than $\frac{3}{7}$.

Comparing fractions

If the fractions you wish to compare do not have the same denominator:
- **use equivalent fractions to convert both fractions so that they have the same denominator**
- **use the lowest common multiple of the two denominators.**

Once the fractions have the same denominator, compare the numerators. The fraction with the largest numerator is bigger.

WORKED EXAMPLE 6 Comparing fractions

a. **Identify which is the bigger fraction, $\dfrac{2}{3}$ or $\dfrac{3}{5}$.**

b. **Justify your answer using a fraction wall and a number line.**

THINK	WRITE
a. 1. Determine the lowest common multiple of the denominators. First, list the multiples of 3 and 5. Identify the lowest number that is common to both lists.	**a.** Multiples of 3 are $3, 6, 9, 12, \widehat{15}, 18, \dots$ Multiples of 5 are $5, 10, \widehat{15}, 20, \dots$ The lowest common multiple is 15.
2. Write each fraction as an equivalent fraction using the lowest common multiple (15) as the denominator.	$\dfrac{2}{3} = \dfrac{2 \times 5}{3 \times 5} = \dfrac{10}{15}$ and $\dfrac{3}{5} = \dfrac{3 \times 3}{5 \times 3} = \dfrac{9}{15}$
3. Decide which is bigger by comparing the numerators of the equivalent fractions.	$\dfrac{10}{15} = \dfrac{2}{3}$ is bigger than $\dfrac{9}{15} = \dfrac{3}{5}$.
4. Answer the question in words.	$\dfrac{2}{3}$ is bigger than $\dfrac{3}{5}$.
b. 1. Create a fraction wall showing thirds and fifths.	**b.**
2. Shade $\dfrac{2}{3}$ and $\dfrac{3}{5}$.	
3. Compare the lengths of the shaded areas to compare the fractions. Answer the question in words.	$\dfrac{2}{3}$ is bigger than $\dfrac{3}{5}$.

4. Draw a number line showing 0 to 1 in intervals of $\frac{1}{15}$ (found by using the lowest common multiple of the denominators).

5. Mark $\frac{9}{15}$ (the same as $\frac{3}{5}$) and $\frac{10}{15}$ (the same as $\frac{2}{3}$) and simplify any fractions on your number line.

6. Answer the question in words.

$\frac{2}{3}$ is bigger than $\frac{3}{5}$.

Comparison symbols

- The symbol '>' means *is bigger than* or *is greater than*.
- The symbol '<' means *is smaller than* or *is less than*.
- When we write $\frac{1}{2} < \frac{3}{4}$, it means that $\frac{1}{2}$ is less than $\frac{3}{4}$.

WORKED EXAMPLE 7 Comparing fractions

Insert the appropriate symbol, < or >, between each pair of fractions to make a true statement.

a. $\frac{6}{7}$ $\frac{7}{8}$

b. $\frac{7}{12}$ $\frac{5}{9}$

THINK

a. 1. Determine the lowest common multiple of the denominators.

2. Write each fraction as an equivalent fraction using the lowest common multiple (56) as the denominator.

3. Decide which fraction is bigger by comparing the numerators of the equivalent fractions.

4. Answer the question.

b. 1. Determine the lowest common multiple of the denominators.

2. Write each fraction as an equivalent fraction using the lowest common multiple (36) as the denominator.

3. Decide which fraction is bigger by comparing the numerators of the equivalent fractions.

4. Answer the question.

WRITE

a. Multiples of 7 are: 7, 14, 21, 28, 35, 42, 49, ⑤⑥, 63, …
Multiples of 8 are: 8, 16, 24, 32, 40, 48, ⑤⑥, 64, …
The lowest common multiple is 56.

$\frac{6}{7} = \frac{6 \times 8}{7 \times 8} = \frac{48}{56}$ and $\frac{7}{8} = \frac{7 \times 7}{8 \times 7} = \frac{49}{56}$

$\frac{48}{56} = \frac{6}{7}$ is less than $\frac{49}{56} = \frac{7}{8}$.

$\frac{6}{7} < \frac{7}{8}$

b. Multiples of 12 are: 12, 24, ㊱, 48, …
Multiples of 9 are: 9, 18, 27, ㊱, 45, …
The lowest common multiple is 36.

$\frac{7}{12} = \frac{7 \times 3}{12 \times 3} = \frac{21}{36}$ and $\frac{5}{9} = \frac{5 \times 4}{9 \times 4} = \frac{20}{36}$

$\frac{21}{36} = \frac{7}{12}$ is greater than $\frac{20}{36} = \frac{5}{9}$.

$\frac{7}{12} > \frac{5}{9}$

Resources

Exercise 4.2 What are fractions?

learn on

| **4.2 Quick quiz** | **4.2 Exercise** |

Individual pathways

■ PRACTISE	■ CONSOLIDATE	■ MASTER
1, 2, 4, 7, 9, 12, 14, 16, 19, 25, 28	3, 5, 8, 10, 13, 17, 20, 21, 23, 26, 29	6, 11, 15, 18, 22, 24, 27, 30

Fluency

1. **WE1** Identify the fraction of each of the following rectangles that has been shaded.

   a.   b.   c.

2. Identify the fraction of this flag that is coloured red.

3. Identify the fraction of this flag that is coloured red.

4. Calculate the fraction of the total number of pieces of fruit in the photograph that is made up of:
   a. bananas
   b. oranges.

5. Calculate the fraction of chocolate biscuits in this packet that has been eaten.

6. Express 17 minutes as a fraction of an hour.

7. **WE2** For the following equivalent fractions, determine the number that has been used to multiply both numerator and denominator.
   a. $\dfrac{3}{8} = \dfrac{9}{24}$
   b. $\dfrac{1}{4} = \dfrac{3}{12}$
   c. $\dfrac{3}{5} = \dfrac{12}{20}$
   d. $\dfrac{2}{3} = \dfrac{10}{15}$

8. For the following equivalent fractions, determine the number that has been used to multiply both numerator and denominator.
   a. $\dfrac{5}{6} = \dfrac{30}{36}$
   b. $\dfrac{9}{10} = \dfrac{81}{90}$
   c. $\dfrac{7}{8} = \dfrac{77}{88}$
   d. $\dfrac{7}{8} = \dfrac{84}{96}$

9. **WE3** From the list, identify the fractions that are equivalent to the fraction marked in pink.
   $$\dfrac{2}{3} \quad \dfrac{20}{30} \quad \dfrac{5}{8} \quad \dfrac{12}{16} \quad \dfrac{14}{21} \quad \dfrac{40}{60}$$

10. From the list, identify the fractions that are equivalent to the fraction marked in pink.
    $$\dfrac{4}{5} \quad \dfrac{12}{15} \quad \dfrac{15}{20} \quad \dfrac{36}{45} \quad \dfrac{16}{20} \quad \dfrac{28}{35} \quad \dfrac{80}{100}$$

11. From the list, identify the fractions that are equivalent to the fraction marked in pink.
    $$\dfrac{7}{10} \quad \dfrac{18}{25} \quad \dfrac{35}{50} \quad \dfrac{14}{21} \quad \dfrac{21}{30} \quad \dfrac{14}{20} \quad \dfrac{140}{200}$$

12. **WE4** Write a sequence of the first three equivalent fractions for $\dfrac{5}{6}$.

13. Write a sequence of the first three equivalent fractions for $\dfrac{7}{8}$.

14. **WE5** Draw a number line from 0 to 2. Show the position of each of the following fractions on the number line.
    a. $\dfrac{1}{5}$
    b. $\dfrac{4}{5}$
    c. $\dfrac{7}{5}$
    d. $1\dfrac{3}{5}$

15. Draw a number line from 0 to 3. Show the position of each of the following fractions on the number line.
    a. $\dfrac{1}{3}$
    b. $\dfrac{4}{3}$
    c. $\dfrac{8}{3}$
    d. $\dfrac{3}{3}$

Understanding

16. **WE6** Identify which is the bigger fraction. Justify your answer using a fraction wall or number line.

    a. $\dfrac{2}{5}$ or $\dfrac{3}{5}$

    b. $\dfrac{5}{8}$ or $\dfrac{7}{8}$

    c. $\dfrac{1}{5}$ or $\dfrac{1}{6}$

    d. $\dfrac{1}{8}$ or $\dfrac{1}{10}$

17. Identify which is the bigger fraction. Justify your answer using a fraction wall or number line.

    a. $\dfrac{3}{4}$ or $\dfrac{3}{5}$

    b. $\dfrac{2}{5}$ or $\dfrac{5}{8}$

    c. $\dfrac{3}{5}$ or $\dfrac{5}{8}$

    d. $\dfrac{4}{5}$ or $\dfrac{5}{8}$

18. **WE7** Insert the appropriate symbol, $<$ or $>$, between each pair of fractions to make a true statement.

    a. $\dfrac{5}{8}$ $\dfrac{3}{8}$

    b. $\dfrac{3}{8}$ $\dfrac{1}{4}$

    c. $\dfrac{3}{4}$ $\dfrac{3}{5}$

    d. $\dfrac{3}{10}$ $\dfrac{2}{5}$

19. **MC** Select which fraction is smaller than $\dfrac{5}{8}$.

    A. $\dfrac{7}{8}$

    B. $\dfrac{17}{24}$

    C. $\dfrac{11}{12}$

    D. $\dfrac{3}{5}$

20. **MC** Select which fraction is equivalent to $\dfrac{3}{4}$.

    A. $\dfrac{3}{8}$

    B. $\dfrac{6}{8}$

    C. $\dfrac{9}{16}$

    D. $\dfrac{12}{24}$

21. Write the following fractions with the same denominator and then write them in ascending order (smallest to largest).

$$\dfrac{3}{8}, \dfrac{1}{2}, \dfrac{1}{3}$$

22. Write the following fractions with the same denominator and then write them in descending order (largest to smallest).

$$\dfrac{3}{4}, \dfrac{2}{3}, \dfrac{7}{15}$$

23. Fill in the gaps.

$$\dfrac{3}{4} = \dfrac{}{40} = \dfrac{18}{} = \dfrac{}{36}$$

24. Fill in the gaps.

$$\dfrac{5}{6} = \dfrac{15}{} = \dfrac{}{42} = \dfrac{40}{}$$

Communicating, reasoning and problem solving

25. Explain why a scale of 2 cm per whole number would not be the most appropriate scale if you are displaying sevenths on a number line.

26. On the number line shown, the fractions have the same numerator but the denominator is always double the previous one. Explain, with reasons, the relationship between these denominators and the distance between the numbers.

27. Four hungry teenagers ordered individual pizzas, but none could eat the whole pizza. Sally ate $\frac{6}{8}$ of her seafood pizza. Hanna ate $\frac{1}{2}$ of her Hawaiian pizza. Sonya ate $\frac{3}{4}$ of her special pizza. Vanessa ate $\frac{5}{6}$ of her vegetarian pizza.

   a. Draw the four pizzas and shade the amount that each girl ate.
   b. Determine who ate the most pizza.
   c. State who ate more pizza, Sally or Sonya.

28. Five students came in at lunchtime to get help with mathematics. Macy used $\frac{1}{4}$ of the time asking questions, Lee Yin used $\frac{1}{3}$ of the time and Lachlan used $\frac{1}{6}$ of the time. Abbey also asked questions for $\frac{2}{12}$ of the time and Tao $\frac{1}{12}$ of the time. Determine which of the students used most of the lunchtime asking questions.

29. Meilin puts four equally spaced points on a number line. The first point is at the $\frac{3}{10}$ mark and the last point is at the $\frac{7}{20}$ mark. Determine where she placed the two points in between.

30. Shirly is sitting $2\frac{5}{7}$ m from the classroom door and Brent is $2\frac{2}{11}$ m from the classroom door. Place Shirly's and Brent's seats on the number line given.

LESSON
4.3 Simplifying fractions

LEARNING INTENTION

At the end of this lesson you should be able to:
- simplify fractions
- simplify mixed numbers.

▶ 4.3.1 Simplifying fractions

eles-3802

- Simplifying fractions means to write them in the simplest form possible. For example, $\frac{4}{8}$ can be simplified to $\frac{2}{4}$, which can be further simplified to $\frac{1}{2}$.

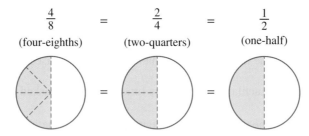

$$\frac{4}{8} \quad = \quad \frac{2}{4} \quad = \quad \frac{1}{2}$$
(four-eighths)   (two-quarters)   (one-half)

- Generally, fractions are written in their simplest form. In other words, we reduce the fraction to its lowest equivalent form.
- To simplify a fraction, divide the numerator and the denominator by the same number.

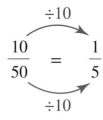

$$\frac{10}{50} = \frac{1}{5}$$

$\div 10$ (top) $\div 10$ (bottom)

- The highest common factor (HCF) of the numerator and denominator is used to simplify the fraction to its simplest form.
- A fraction is in simplest form when it cannot be simplified further. This occurs when the numerator and denominator share no common factor.

Simplifying fractions

There are two ways to convert a fraction into its simplest form:
1. **Divide the numerator and denominator by the highest common factor (HCF).**
2. **Simplify repeatedly, by whichever common factor you like, until there are no common factors between the numerator and denominator.**

WORKED EXAMPLE 8 Simplifying fractions

Write $\dfrac{3}{6}$ in simplest form.

THINK

1. Write the fraction and determine the highest common factor (HCF) or the largest number that is a factor of both the numerator and the denominator.

2. Divide the numerator and denominator by this factor $(3 \div 3 = 1, 6 \div 3 = 2)$.

3. Write the answer in simplest form.

WRITE

$\dfrac{3}{6}$, HCF = 3

$$\dfrac{3}{6} \overset{\div 3}{\underset{\div 3}{=}} \dfrac{1}{2}$$

$= \dfrac{1}{2}$

Simplifying mixed numbers

- To simplify a mixed number, leave the whole number part and simplify the fraction part.

WORKED EXAMPLE 9 Simplifying mixed numbers

Simplify $7\dfrac{40}{64}$.

THINK

1. Write the mixed number and determine the HCF of the numerator and denominator of the fraction part.

2. Divide both numerator and denominator by this factor $(40 \div 8 = 5, 64 \div 8 = 8)$.

3. Write your answer as a mixed number in simplest form.

WRITE

$7\dfrac{40}{64}$, HCF = 8

$$\dfrac{40}{64} \overset{\div 8}{\underset{\div 8}{=}} \dfrac{5}{8}$$

$= 7\dfrac{5}{8}$

DISCUSSION

When simplifying a fraction, how do you know when it has been fully simplified? What method do you use to simplify fractions?

 Resources

Interactivities Simplifying fractions (int-3943)

Exercise 4.3 Simplifying fractions

learn on

Individual pathways

■ PRACTISE	■ CONSOLIDATE	■ MASTER
1, 4, 6, 7, 11, 14	2, 5, 8, 10, 12, 15	3, 9, 13, 16

Fluency

1. **WE8** Write the following fractions in simplest form.

   a. $\dfrac{5}{10}$ b. $\dfrac{8}{12}$ c. $\dfrac{21}{24}$ d. $\dfrac{48}{60}$

2. Write the following fractions in simplest form.

   a. $\dfrac{100}{120}$ b. $\dfrac{48}{50}$ c. $\dfrac{63}{72}$ d. $\dfrac{49}{70}$

3. Write the following fractions in simplest form.

   a. $\dfrac{33}{36}$ b. $\dfrac{22}{50}$ c. $\dfrac{21}{56}$ d. $\dfrac{49}{56}$

4. **WE9** Simplify.

   a. $4\dfrac{21}{35}$ b. $7\dfrac{45}{54}$ c. $10\dfrac{10}{20}$ d. $1\dfrac{44}{48}$

5. Simplify.

   a. $6\dfrac{21}{28}$ b. $5\dfrac{16}{48}$ c. $3\dfrac{11}{55}$ d. $2\dfrac{16}{64}$

6. **MC** Select the equivalent fraction to $\dfrac{26}{28}$.

   A. $\dfrac{2}{3}$ B. $\dfrac{2}{6}$ C. $\dfrac{3}{8}$ D. $\dfrac{13}{14}$

Understanding

7. Kylie's netball team scored 28 goals. Kylie scored 21 of her team's goals. Calculate the fraction of the team's goals that Kylie scored. Simplify the answer.

8. Angelo's basketball team scored 36 points. Angelo scored 20 of the team's points.

   a. Calculate the fraction of the team's points that Angelo scored. Simplify the answer.
   b. Calculate the fraction of the team's points that the rest of the team scored. Simplify the answer.

9. Year 7 students at Springfield High School ran a car wash to raise money for the local hospital. They raised a total of $1000 and drew up a table showing how much money each class raised.

7A	7B	7C	7D	7E
$200	$150	$320	$80	$250

Express as a simple fraction how much of the total each class raised.

10. Below are the results of a book-reading competition.

Mark	5 books
Ken	10 books
Samantha	8 books
Jules	6 books
Ahmed	4 books
Darren	7 books

a. Determine how many books were read in total.
b. For each contestant, record the number of books read as a fraction of the total number of books read. Where possible, reduce the fractions to their simplest forms.

Communicating, reasoning and problem solving

11. The Newcastle to Sydney train runs on time 3 out of every 5 trains.
The Sydney to Newcastle train runs on time 4 out of every 7 trains.
Determine which train is more reliable.

12. a. Draw three lines to divide the triangle shown into 4 equal parts.
b. State what fraction of the original triangle is one of the new triangles.
c. Halve each smaller triangle.
d. State what fraction of the smaller triangle is one of these even smaller triangles.
e. Using all the triangles available, explain why $\dfrac{2}{8} = \dfrac{1}{4}$.

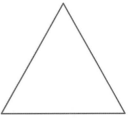

13. The fraction $\dfrac{13}{n}$ is not written in its simplest form. Explain what number the denominator could be.

14. Sarina orders pizza for 22 friends. Each friend must get $\dfrac{1}{4}$ of a pizza.

a. Determine the number of pizzas Sarina needs to order.
b. Calculate the amount of leftover pizza.

15. a. Arrange the digits $1, 2, 3, 4, 5, 6, 7, 8$ and 9 to form a fraction that is equivalent to $\dfrac{1}{2}$.

For example, one arrangement is $\dfrac{7329}{14\,658}$. Can you find two others?

Hint: Try starting with a numerator beginning with 72 for one of them and 67 for the other.

b. Arrange the nine digits so that they form a fraction equivalent to $\dfrac{1}{3}$.

16. One of the designs for a patchwork motif is made up of squares and right-angled isosceles triangles as shown. Determine what fraction of the whole has been shaded.

LESSON
4.4 Mixed numbers and improper fractions

LEARNING INTENTION

At the end of this lesson you should be able to:
- convert between mixed numbers and improper fractions.

4.4.1 Converting improper fractions to mixed numbers

eles-3803

- An improper fraction has a numerator larger than the denominator: for example, $\frac{3}{2}$.

- An improper fraction can be changed to a mixed number by dividing the denominator into the numerator and writing the remainder over the same denominator.

Mixed numbers and improper fractions

Mixed numbers and improper fractions are different ways of writing the same quantity.

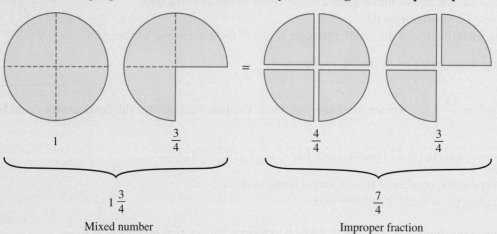

$$1 \qquad \frac{3}{4} \qquad = \qquad \frac{4}{4} \qquad \frac{3}{4}$$

$$1\frac{3}{4} \qquad\qquad\qquad \frac{7}{4}$$

Mixed number   Improper fraction

WORKED EXAMPLE 10 Converting improper fractions to mixed numbers

Draw a diagram to show $\dfrac{5}{4}$ as parts of a circle, then write the improper fraction as a mixed number.

THINK

1. Draw a whole circle and divide it into the number of parts shown by the denominator.

WRITE

This is 4 quarters or $\dfrac{4}{4}$.

2. Determine the number of parts left over and draw them.

Extra $\dfrac{1}{4}$ needed.

3. Write the improper fraction as a mixed number.

$\dfrac{5}{4} = 1\dfrac{1}{4}$

WORKED EXAMPLE 11 Converting improper fractions to mixed numbers

Express $\dfrac{11}{5}$ as a mixed number.

THINK

1. Write the improper fraction.

2. Determine how many times the denominator can be divided into the numerator and what the remainder is. The whole number is part of the answer and the remainder becomes the numerator of the fractional part with the same denominator as the original improper fraction.

3. Write the answer.

WRITE

$\dfrac{11}{5}$

$= 11 \div 5$

$= 2 \text{ remainder } 1$

$= 2\dfrac{1}{5}$

▶ 4.4.2 Converting mixed numbers to improper fractions

eles-3804

- A mixed number can be changed to an improper fraction by first multiplying the whole number by the denominator and then adding the numerator. The denominator stays the same.
- You may wish to draw a diagram to help visualise the conversion.

Mixed numbers to improper fractions

$$2\dfrac{1}{4} = \dfrac{(2 \times 4) + 1}{4} = \dfrac{8 + 1}{4} = \dfrac{9}{4}$$

Draw a diagram to illustrate $2\dfrac{1}{3}$ as pieces of a circle, then write $2\dfrac{1}{3}$ as an improper fraction.

THINK

1. Draw two whole circles and $\dfrac{1}{3}$ of a circle.

2. Divide the whole circles into thirds.

3. Count the number of thirds and write the mixed number as an improper fraction.

WRITE

$$2\dfrac{1}{3} = \dfrac{7}{3}$$

Express $2\dfrac{3}{4}$ as an improper fraction.

THINK

1. Write the mixed number.

2. Multiply the whole number by the denominator, then add the numerator. The result becomes the numerator, and the denominator stays the same.

3. Evaluate the top line of the fraction.

4. Write the answer.

WRITE

$$2\dfrac{3}{4}$$

$$= \dfrac{2 \times 4 + 3}{4}$$

$$= \dfrac{8 + 3}{4}$$

$$= \dfrac{11}{4}$$

Digital technology

Scientific calculators usually give answers in decimal form. Most calculators have a button that converts decimals and fractions. Depending on the brand of calculator, this may appear as S ⇔ d, f ◁▷ d or similar. The screenshot on the right shows how to convert a decimal to a fraction.

	Math ▲
85.25	
	$\dfrac{341}{4}$

Most scientific calculators have a button that can convert between improper fractions and mixed numbers. The button may appear as $a\dfrac{b}{c}$ ◁▷ $\dfrac{d}{c}$ or similar.

	Math ▲
85.25	
	$85\dfrac{1}{4}$

Resources

    Interactivities Mixed numbers to improper fractions (int-3944)

Exercise 4.4 Mixed numbers and improper fractions   **learn**on

4.4 Quick quiz on	**4.4 Exercise**

Individual pathways

■ PRACTISE	■ CONSOLIDATE	■ MASTER
1, 2, 4, 7, 10, 12, 14, 17, 20, 22, 25	3, 5, 9, 11, 15, 18, 21, 23, 26	6, 8, 13, 16, 19, 24, 27, 28

Fluency

1. **WE10** Draw a diagram to show the following improper fractions as pieces of a circle, then write each improper fraction as a mixed number.

   a. $\dfrac{5}{2}$ b. $\dfrac{4}{3}$ c. $\dfrac{13}{8}$

2. Aamira ate $\dfrac{3}{2}$ or 3 halves of an apple. Draw the amount of apple that Aamira ate and express $\dfrac{3}{2}$ as a mixed number.

3. Dean ate $\dfrac{9}{4}$ or 9 quarters of pizza. Draw the amount of pizza that Dean ate and express $\dfrac{9}{4}$ as a mixed number.

4. **WE11** Express these improper fractions as mixed numbers.

   a. $\dfrac{7}{5}$ b. $\dfrac{11}{4}$ c. $\dfrac{21}{2}$ d. $\dfrac{51}{12}$ e. $\dfrac{92}{11}$

5. Express these improper fractions as mixed numbers.

   a. $\dfrac{29}{13}$ b. $\dfrac{100}{3}$ c. $\dfrac{25}{2}$ d. $\dfrac{20}{3}$ e. $\dfrac{8}{5}$

6. Change these improper fractions to mixed numbers.

    a. $\dfrac{27}{8}$ **b.** $\dfrac{11}{2}$ **c.** $\dfrac{58}{7}$ **d.** $\dfrac{117}{10}$ **e.** $\dfrac{67}{8}$

7. **MC** Select which of the following is the same as $\dfrac{61}{8}$.

    A. $5\dfrac{5}{8}$ **B.** $6\dfrac{5}{8}$ **C.** $7\dfrac{5}{8}$ **D.** $8\dfrac{5}{8}$

8. **MC** Select which of the following is the same as $\dfrac{74}{10}$.

    A. $3\dfrac{7}{10}$ **B.** $7\dfrac{2}{5}$ **C.** $7\dfrac{7}{10}$ **D.** $7\dfrac{4}{5}$

9. **WE12** Draw a diagram to show the following mixed numbers as pieces of a circle and then write each one as an improper fraction.

    a. $1\dfrac{2}{3}$ **b.** $3\dfrac{1}{4}$ **c.** $2\dfrac{3}{5}$

10. **WE13** Express the following mixed numbers as improper fractions.

    a. $1\dfrac{1}{8}$ **b.** $3\dfrac{3}{4}$ **c.** $2\dfrac{1}{6}$ **d.** $1\dfrac{1}{7}$ **e.** $1\dfrac{4}{5}$

11. Express the following mixed numbers as improper fractions.

    a. $3\dfrac{1}{3}$ **b.** $6\dfrac{1}{4}$ **c.** $4\dfrac{1}{9}$ **d.** $11\dfrac{1}{2}$ **e.** $7\dfrac{4}{5}$

12. Express the following mixed numbers as improper fractions.

    a. $6\dfrac{3}{5}$ **b.** $9\dfrac{2}{5}$ **c.** $3\dfrac{1}{2}$ **d.** $4\dfrac{1}{5}$ **e.** $6\dfrac{5}{7}$

13. Express the following mixed numbers as improper fractions.

    a. $2\dfrac{9}{10}$ **b.** $3\dfrac{7}{12}$ **c.** $9\dfrac{4}{7}$ **d.** $1\dfrac{10}{11}$ **e.** $8\dfrac{5}{6}$

14. **MC** Select which of the following is the same as $10\dfrac{2}{7}$.

    A. $\dfrac{27}{10}$ **B.** $\dfrac{72}{10}$ **C.** $\dfrac{10}{27}$ **D.** $\dfrac{72}{7}$

15. **MC** Select which of the following is the same as $8\dfrac{5}{6}$.

    A. $\dfrac{85}{6}$ **B.** $\dfrac{58}{6}$ **C.** $\dfrac{56}{8}$ **D.** $\dfrac{53}{6}$

16. **MC** Select which of the following is the same as $7\dfrac{9}{11}$.

    A. $\dfrac{86}{11}$ **B.** $\dfrac{99}{7}$ **C.** $\dfrac{79}{11}$ **D.** $\dfrac{97}{11}$

Understanding

17. Kim and Carly arrived home from school quite hungry and cut up some fruit. Kim ate 9 quarters of apple and Carly ate 11 quarters. Calculate how many apples they each ate.

18. Thanh was selling pieces of quiche at the school fete. Each quiche was divided into 8 pieces. If Thanh sold 52 pieces, calculate how many quiches Thanh sold.

19. It was Cecilia's responsibility to supply oranges for her soccer team at half-time. If 19 players ate $\frac{1}{4}$ of an orange each, calculate how many oranges were eaten.

20. Insert the appropriate < or > sign between each pair of fractions to make a true statement.

    a. $\frac{8}{5}$ $1\frac{2}{5}$
    b. $\frac{19}{2}$ $8\frac{1}{2}$
    c. $\frac{33}{32}$ $1\frac{3}{32}$
    d. $\frac{15}{2}$ $5\frac{1}{2}$

21. Insert the appropriate < or > sign between each pair of fractions to make a true statement.

    a. $7\frac{1}{7}$ $\frac{65}{7}$
    b. $2\frac{1}{3}$ $\frac{8}{3}$
    c. $1\frac{16}{17}$ $\frac{35}{17}$
    d. $4\frac{3}{8}$ $\frac{36}{8}$

Communicating, reasoning and problem solving

22. A catering company provided 12 quiches that were each cut into 8 pieces for a luncheon. At the end of the luncheon, 12 pieces of quiche were left over. Calculate how many whole quiches were eaten. Show your full working.

23. a. Convert the following numbers to improper fractions.

    i. $3\frac{2}{7}$
    ii. $5\frac{3}{4}$

    b. Explain with the use of diagrams why $3\frac{2}{7} = \frac{23}{7}$.

24. A bakery makes birthday cakes to serve 24 people. Arabella takes one to school to celebrate her birthday, but there are 30 people in her class. Determine what fraction of the original slices each student will receive.

25. a. Nasira made apple pies for her Grandma's 80th birthday party. She divided each pie into 6 pieces, and after the party she noted that $4\frac{1}{6}$ pies had been eaten. Calculate how many pieces were eaten.

b. Nasira's cousin Ria provided cordial for the same party and calculated that she could make 20 drinks from each bottle. At the end of the party, $3\frac{17}{20}$ bottles had been used. Calculate how many drinks were consumed.

26. Baljinder walks to school four days a week and her father drives her to school on Fridays. If she lives $2\frac{1}{3}$ km from school, calculate:

a. the distance Baljinder walks to school and back home each day
b. the total distance walked by Baljinder over the four days
c. the total distance she travels to school either walking or by car over the whole week.
   State your answers in improper fractions and mixed numbers if possible.

27. The following mixed numbers are missing their denominators. The missing denominators are 3, 4, 6, 8 and 14. The mixed numbers are placed in order from smallest to largest. Determine the missing numbers.

$$1\frac{1}{a}, \qquad 2\frac{1}{b}, \qquad 2\frac{1}{c}, \qquad 3\frac{1}{d}, \qquad 3\frac{1}{e}$$

28. Explain the strategies you use to write an improper fraction as a mixed number.

LESSON
4.5 Adding and subtracting fractions

▶ 4.5.1 Adding and subtracting proper fractions

eles-3805

• Fractions can be added and subtracted if they have the same denominator.
• Addition and subtraction of fractions can be visualised using areas of shapes.

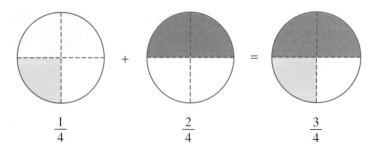

$$\frac{1}{4} \qquad \frac{2}{4} \qquad \frac{3}{4}$$

• When adding fractions, the model shows that the numerators (the number of pieces) are added, but the denominator (the size of the piece) stays the same.

Adding and subtracting fractions with equal denominators

To add or subtract fractions of the same denominator, simply add or subtract the numerators. Leave the denominator unchanged.

For example:

$$\frac{1}{5} + \frac{2}{5} = \frac{1+2}{5} = \frac{3}{5} \qquad \frac{5}{8} - \frac{2}{8} = \frac{5-2}{8} = \frac{3}{8}$$

WORKED EXAMPLE 14 Adding proper fractions

Evaluate $\frac{5}{9} + \frac{1}{9}$.

THINK

1. Write the question.

2. Since the denominators (piece sizes) are the same, add the numerators.

3. Simplify the answer if possible by cancelling. Divide the numerator and the denominator by the highest common factor ($6 \div 3 = 2, 9 \div 3 = 3$).

WRITE

$$\frac{5}{9} + \frac{1}{9}$$

  +

$$= \frac{6}{9}$$

$$= \frac{\overset{2}{\cancel{6}}}{\underset{3}{\cancel{9}}}$$

$$= \frac{2}{3}$$

Lowest common denominator

- The lowest common denominator (LCD) is the lowest number that is a multiple of the denominators. This is also known as the lowest common multiple (LCM) of the numbers in the denominators.

- Let us visualise $\frac{1}{2} + \frac{1}{3}$ using areas of circles. We can divide both circles into the same number of parts and then add the parts.

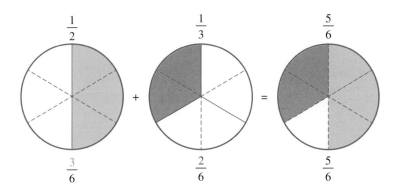

Adding and subtracting fractions with different denominators

To add or subtract fractions with different denominators, convert to equivalent fractions with the lowest common denominator (LCD), then add the numerators.

For example:
$$\frac{1}{2} + \frac{1}{3}$$

The LCD of 2 and 3 is 6.

Add the intermediate step to the equation and calculate the answer:

$$\frac{1}{2} + \frac{1}{3} = \frac{3}{6} + \frac{2}{6} = \frac{3+2}{6} = \frac{5}{6}$$

WORKED EXAMPLE 15 Subtracting proper fractions

Evaluate $\dfrac{5}{6} - \dfrac{1}{12}$, expressing the answer in simplest form.

THINK	WRITE
1. List multiples of each denominator.	Multiples of 6 are 6, 12, ... Multiples of 12 are 12, ...
2. Identify the lowest common denominator.	LCD is 12.
3. Rewrite the fractions as equivalent fractions with 12 as the LCD.	$\dfrac{5}{6} - \dfrac{1}{12} = \dfrac{5 \times 2}{6 \times 2} - \dfrac{1}{12} = \dfrac{10}{12} - \dfrac{1}{12}$

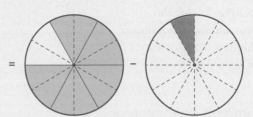

$$= \frac{9}{12}$$

4. Subtract the numerators.

5. Simplify the answer if possible by cancelling. Divide by the highest common factor ($9 \div 3 = 3, 12 \div 3 = 4$).

$$= \frac{\cancel{9}^{3}}{\cancel{12}^{4}}$$

$$= \frac{3}{4}$$

WORKED EXAMPLE 16 Subtracting a fraction from a whole number

Evaluate $4 - \dfrac{1}{3}$, expressing the answer as a mixed number.

THINK	WRITE
1. Express the whole number as a fraction over 1.	$4 - \dfrac{1}{3} = \dfrac{4}{1} - \dfrac{1}{3}$
2. Equate the denominators.	$= \dfrac{12}{3} - \dfrac{1}{3}$
3. Perform the subtraction.	$= \dfrac{11}{3}$
4. Convert to a mixed number.	$= 3\dfrac{2}{3}$

WORKED EXAMPLE 17 Application of adding and subtracting fractions

A farmer picks the fruit from $\dfrac{2}{3}$ of her orange trees and her partner collects fruit from another $\dfrac{1}{6}$ of the trees. Determine the fraction of the total number of trees that still need to have their fruit picked.

THINK	WRITE
1. Calculate the total fraction of trees that have had fruit picked, by adding the fractions.	$\dfrac{2}{3} + \dfrac{1}{6} = \dfrac{4}{6} + \dfrac{1}{6}$
	$= \dfrac{5}{6}$

2. Using the diagram, determine the remaining fraction of trees that have not yet had the fruit picked.

$\frac{1}{6}$

Or, alternatively

$$1 - \frac{5}{6} = \frac{6}{6} - \frac{5}{6}$$

$$= \frac{1}{6}$$

3. Write the answer.

$\frac{1}{6}$ of the orange trees still have fruit that needs to be picked.

COMMUNICATING — COLLABORATIVE TASK: Dice fractions

Equipment: dice

Work in pairs or in small groups to play this game.
1. The first player rolls both dice. One number rolled is the numerator of the fraction and the other number rolled is the denominator. The player makes the smallest possible fraction with the numbers rolled.
2. The first player rolls the dice again to make another fraction, which should be as close as possible to the first fraction.
3. The first player works out the difference between the two fractions rolled by subtracting one fraction from the other. This amount is recorded as their result for this round.
4. Repeat steps 1 to 3 until every player has a result for the first round.
5. Compare the results for the round. The player with the smallest difference between their two fractions receives 1 point. If two players have the smallest difference, they both receive 1 point. The game is over when a player has collected 5 points.
6. What strategy did you use to make the smallest difference between your fractions?

Digital technology

A calculator can be used to add and subtract proper, improper and mixed fractions using the fraction button $\frac{\square}{\square}$.

Exercise 4.5 Adding and subtracting fractions

4.5 Quick quiz on	4.5 Exercise

Individual pathways

■ PRACTISE	■ CONSOLIDATE	■ MASTER
1, 3, 5, 7, 11, 14, 16, 20, 23	2, 6, 8, 9, 12, 15, 17, 21, 24	4, 10, 13, 18, 19, 22, 25, 26

Fluency

1. **WE14** Evaluate:

   a. $\dfrac{1}{5} + \dfrac{3}{5}$ b. $\dfrac{9}{11} - \dfrac{3}{11}$ c. $\dfrac{3}{6} + \dfrac{2}{6}$ d. $\dfrac{20}{50} + \dfrac{11}{50}$

2. Evaluate the following, expressing the answers in simplest form.

   a. $\dfrac{5}{12} + \dfrac{1}{12}$ b. $\dfrac{7}{16} - \dfrac{3}{16}$ c. $\dfrac{9}{15} - \dfrac{4}{15}$ d. $\dfrac{37}{100} + \dfrac{13}{100}$

3. Evaluate:

   a. $\dfrac{15}{25} - \dfrac{6}{25}$ b. $\dfrac{12}{12} - \dfrac{1}{12}$ c. $\dfrac{5}{8} - \dfrac{1}{8} + \dfrac{3}{8}$ d. $\dfrac{4}{7} + \dfrac{5}{7} - \dfrac{3}{7}$

4. Evaluate the following, expressing the answers in simplest form.

   a. $\dfrac{4}{7} + \dfrac{5}{7} - \dfrac{2}{7}$ b. $\dfrac{13}{16} + \dfrac{9}{16} - \dfrac{10}{16}$ c. $\dfrac{11}{28} + \dfrac{10}{28}$ d. $\dfrac{21}{81} + \dfrac{21}{81} + \dfrac{30}{81}$

5. **WE15** Evaluate the following, expressing the answers in simplest form.

   a. $\dfrac{1}{2} + \dfrac{1}{4}$ b. $\dfrac{1}{2} - \dfrac{1}{4}$ c. $\dfrac{1}{3} + \dfrac{2}{21}$ d. $\dfrac{1}{8} + \dfrac{20}{32}$

6. Evaluate the following, expressing the answers in simplest form.

   a. $\dfrac{9}{18} + \dfrac{1}{6}$ b. $\dfrac{7}{9} - \dfrac{8}{27}$ c. $\dfrac{31}{35} - \dfrac{3}{7}$ d. $\dfrac{8}{10} - \dfrac{28}{70}$

7. a. Determine the lowest common multiple of each of the following pairs of numbers.

      i. 2 and 3   ii. 3 and 5   iii. 5 and 10   iv. 4 and 6

   b. Use the answers that were found in part **a** as the lowest common denominators to add or subtract these fractions.

      i. $\dfrac{1}{2} - \dfrac{1}{3}$ ii. $\dfrac{2}{3} - \dfrac{2}{5}$ iii. $\dfrac{3}{5} + \dfrac{3}{10}$ iv. $\dfrac{3}{4} - \dfrac{1}{6}$

8. a. Determine the lowest common multiple of each of the following pairs of numbers.

      i. 9 and 6   ii. 4 and 8   iii. 6 and 8   iv. 7 and 5

   b. Use the answers from part **a** as the lowest common denominators to add or subtract these fractions.

      i. $\dfrac{1}{9} + \dfrac{5}{6}$ ii. $\dfrac{1}{4} + \dfrac{3}{8}$ iii. $\dfrac{5}{6} - \dfrac{3}{8}$ iv. $\dfrac{5}{7} - \dfrac{2}{5}$

9. Answer the following by first identifying the lowest common denominator. Simplify your answer if necessary.

   a. $\dfrac{7}{10} + \dfrac{2}{15}$ b. $\dfrac{5}{12} + \dfrac{2}{9}$ c. $\dfrac{7}{11} + \dfrac{1}{4}$ d. $\dfrac{3}{10} - \dfrac{1}{20}$

10. Answer the following by first identifying the lowest common denominator. Simplify your answer if necessary.

   a. $\dfrac{5}{9} - \dfrac{2}{7}$

   b. $\dfrac{2}{3} - \dfrac{1}{13}$

   c. $\dfrac{4}{5} - \dfrac{2}{3}$

   d. $\dfrac{6}{11} - \dfrac{1}{3}$

11. **MC** The lowest common denominator of $\dfrac{1}{3}, \dfrac{1}{4}$ and $\dfrac{1}{6}$ is:

   A. 16   **B.** 24   **C.** 18   **D.** 12

Understanding

12. Answer the following:

   a. $\dfrac{1}{2} + \dfrac{1}{3} - \dfrac{1}{4}$

   b. $\dfrac{3}{10} + \dfrac{4}{15} - \dfrac{1}{6}$

   c. $\dfrac{7}{8} - \dfrac{3}{4} + \dfrac{4}{6}$

   d. $\dfrac{11}{12} - \dfrac{8}{15} - \dfrac{3}{20}$

13. Answer the following:

   a. $\dfrac{1}{16} + \dfrac{1}{8} + \dfrac{1}{4}$

   b. $\dfrac{5}{6} - \dfrac{7}{18} - \dfrac{1}{9}$

   c. $\dfrac{1}{2} + \dfrac{2}{3} + \dfrac{3}{4}$

   d. $\dfrac{3}{4} + \dfrac{1}{3} + \dfrac{5}{6}$

14. **WE16** Evaluate the following, expressing the answers as proper fractions or mixed numbers.

   a. $1 - \dfrac{1}{4}$

   b. $1 - \dfrac{3}{7}$

   c. $1 - \dfrac{11}{71}$

   d. $2 - \dfrac{2}{5}$

15. Evaluate the following, expressing the answers as proper fractions or mixed numbers.

   a. $3 - \dfrac{7}{15}$

   b. $5 - \dfrac{5}{6}$

   c. $10 - \dfrac{1}{3}$

   d. $23 - \dfrac{9}{10}$

16. Melissa and Antonio went berry picking. Melissa picked $\dfrac{3}{10}$ of a bucket of berries and Antonio picked $\dfrac{1}{6}$ of a bucket.

   a. Calculate how much they picked altogether.
   b. If they needed 1 full bucket of berries, calculate what fraction more they still need to pick.

17. A loaf of bread needed to be cooked at $200\,°C$ for $\dfrac{3}{4}$ hour then a further $\dfrac{1}{2}$ hour at $150\,°C$. Calculate the total cooking time in hours.

18. Robert, Jason and Luke play football in the same team. Last Saturday, Robert kicked $\dfrac{1}{4}$ of the team's goals, Jason kicked $\dfrac{1}{3}$ and Luke kicked $\dfrac{5}{12}$. Calculate what fraction of the team's goals were kicked by:

   a. Jason and Luke
   b. Robert and Jason
   c. the three together.

19. Suraya and her friend decided to climb the tree in her front yard. It took Suraya 10 minutes to climb halfway up, 10 more minutes to climb another sixth of the way and yet another 10 minutes to reach a further tenth of the way. Here she sat admiring the view for 5 minutes before beginning her descent.

   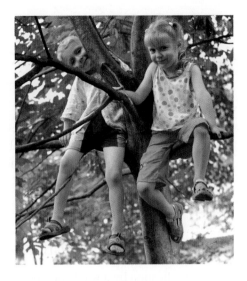

   a. Calculate what fraction of the height of the tree Suraya had climbed after 20 minutes.

   b. Calculate what fraction of the tree Suraya climbed in total.

Communicating, reasoning and problem solving

20. Yuna sold $\frac{2}{3}$ of the cupcakes she made. She had 45 cupcakes left. Calculate how many cupcakes she made.

21. Explain why $\frac{1}{3} + \frac{2}{5} \neq \frac{3}{8}$.

22. Chris lives on a farm and has a horse he rides every day. He fills up the horse's trough with $\frac{2}{3}$ hay, $\frac{1}{6}$ carrots, $\frac{1}{12}$ oats and the rest maize. Calculate the fraction of the trough that is filled with maize.

23. a. Without using a calculator, perform the following additions:

   i. $\frac{1}{2} + \frac{1}{3}$ ii. $\frac{1}{3} + \frac{1}{4}$ iii. $\frac{1}{4} + \frac{1}{5}$

   b. Explain any pattern you noticed.

24. a. Without using a calculator, perform the following subtractions:

   i. $\frac{1}{2} - \frac{1}{3}$ ii. $\frac{1}{3} - \frac{1}{4}$ iii. $\frac{1}{4} - \frac{1}{5}$

   b. Explain any pattern you noticed.

   c. Using your answer to part b, state the answer to the subtraction $\frac{1}{n} - \frac{1}{n+1}$.

25. Explain the difficulties you might experience when adding and subtracting fractions with different denominators and how you could overcome them.

26. Arrange the following fractions in the circles in the diagram so that each row adds up to the same value.

$$\frac{1}{12}, \frac{1}{4}, \frac{1}{2}, \frac{1}{6}, \frac{3}{4}, \frac{1}{3}, \frac{2}{3}$$

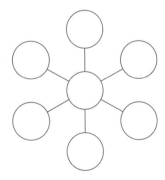

LESSON
4.6 Multiplying fractions

LEARNING INTENTION

At the end of this lesson you should be able to:
- multiply fractions by a whole number
- multiply fractions by another fraction
- calculate fractions of a fraction
- calculate fractions of an amount.

▶ 4.6.1 Multiplying fractions by a whole

eles-3806

- Multiplication by a whole number can be thought of as repeated addition.
  For example, $3 \times 4 = 4 + 4 + 4 = 12$.

  The same principle holds when multiplying a fraction by a whole number.
  For example, $3 \times \dfrac{1}{4} = \dfrac{1}{4} + \dfrac{1}{4} + \dfrac{1}{4} = \dfrac{3}{4}$.

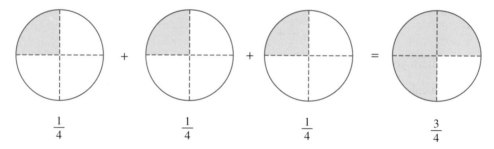

$$\frac{1}{4} \qquad + \qquad \frac{1}{4} \qquad + \qquad \frac{1}{4} \qquad = \qquad \frac{3}{4}$$

- A quicker way of multiplying a fraction by a whole number is to simply multiply the numerator by the whole number.

Multiplying a fraction by a whole number

To multiply a fraction by a whole number, simply multiply the numerator by the whole number and keep the denominator the same.

For example:

$$6 \times \frac{1}{5} = \frac{6 \times 1}{5} = \frac{6}{5}$$

This can be read as 'six groups of one-fifth', which gives 'six-fifths'.

WORKED EXAMPLE 18 Multiplying a fraction by a whole number

Calculate $6 \times \dfrac{1}{3}$.

THINK	WRITE
1. Write the question.	$6 \times \dfrac{1}{3}$

2. Multiply by a whole number using repeated addition. Add $\dfrac{1}{3}$ six times.

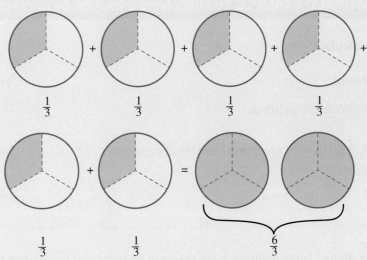

$$\dfrac{1}{3} \qquad \dfrac{1}{3} \qquad \dfrac{1}{3} \qquad \dfrac{1}{3}$$

$$\dfrac{1}{3} \qquad \dfrac{1}{3} \qquad \dfrac{6}{3}$$

3. Alternatively, multiply the numerator by the whole number.

$$= \dfrac{6 \times 1}{3}$$

$$= \dfrac{6}{3}$$

4. Simplify and write the answer.

$$= 2$$

4.6.2 Multiplying fractions by fractions

eles-####

- The multiplication $\dfrac{1}{3} \times \dfrac{1}{2}$ is the same as one-third of a half; this is the same as one-sixth of a whole unit, as shown here.

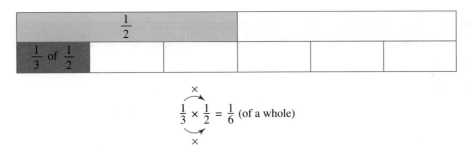

$$\dfrac{1}{3} \times \dfrac{1}{2} = \dfrac{1}{6} \text{ (of a whole)}$$

- Notice that when multiplying two fractions, the numerators are multiplied and the denominators are multiplied.
- When we multiply any number by a fraction between 0 and 1, the result will always be smaller.

Multiplying a fraction by another fraction

To multiply a fraction by another fraction, simply multiply the numerators and then multiply the denominators.

For example:

$$\dfrac{1}{2} \times \dfrac{3}{4} = \dfrac{1 \times 3}{2 \times 4} = \dfrac{3}{8}$$

Note: Fractions don't need to have the same denominator to be multiplied together.

Calculate $\dfrac{1}{5} \times \dfrac{3}{4}$.

THINK	WRITE
1. Write the question.	$\dfrac{1}{5} \times \dfrac{3}{4}$
2. Multiply the numerators and the denominators.	$= \dfrac{1 \times 3}{5 \times 4}$
3. Write the answer.	$= \dfrac{3}{20}$

- Before being multiplied, fractions can be simplified by dividing both the numerator and denominator by their highest common factor. This is often called *cancelling*.
- When multiplying fractions, cancelling can only occur:
  - vertically, for example, $\dfrac{3^1}{24^8} \times \dfrac{5}{7} = \dfrac{1 \times 5}{8 \times 7} = \dfrac{5}{56}$
  - diagonally, for example, $\dfrac{2^1}{3} \times \dfrac{5}{6^3} = \dfrac{1 \times 5}{3 \times 3} = \dfrac{5}{9}$

DISCUSSION

Why is cancelling a pair of numerators or a pair of denominators not possible?

Calculate $\dfrac{2}{3} \times \dfrac{9}{10}$.

THINK	WRITE
1. Write the question.	$\dfrac{2}{3} \times \dfrac{9}{10}$
2. Simplify by dividing the numbers in the numerator and denominator by their highest common factor. The HCF of 2 and 10 is 2, and the HCF of 3 and 9 is 3. *Note:* This may be done vertically or diagonally.	$= \dfrac{2^1 \times 9^3}{3^1 \times 10^5}$
3. Multiply the numerators, then the denominators, and simplify the answer if appropriate.	$\dfrac{1 \times 3}{1 \times 5} = \dfrac{3}{5}$

Equipment: string, pegs, counters

1. Form small groups. One student from each group should collect a piece of string and two pegs.
2. Your teacher will give each group a multiplication of two fractions.
3. As a group, place a peg on the length of string to show the position of one of your fractions.
4. Using this peg as a new marker for the end of the string, find the second fraction of this new string length and place a second peg in the correct position on the string.
5. Check the position of the second peg by calculating the product of the fractions and then folding the complete length of string into the appropriate number of parts to find the position of the product.
6. Repeat steps 2 to 5 with other fraction multiplications.
7. One student from each group should collect a handful of counters.
8. Your teacher will give each group a fraction of an amount.
9. As a group, divide up the counters to show your allocated fraction.
10. Write a description of the process that your group used to find your fraction of an amount.

4.6.3 Using the word *of*

eles-3807

- Sometimes the word *of* means 'multiply'. For example, $\frac{1}{2}$ of $\frac{1}{3}$ can be visualised as:

Half of $\frac{1}{3} = \frac{1}{2} \times \frac{1}{3} = \frac{1}{6}$

- A third of an hour, or $\frac{1}{3}$ of 60 minutes, is the same as $\frac{1}{3} \times \frac{60}{1} = 20$ minutes.

- If a customer wants to buy $\frac{1}{4}$ of the remaining 8 pizzas, it is the same as $\frac{1}{4} \times \frac{8}{1} = 2$ pizzas. The customer buys 2 pizzas.

- A diagram can also be used to display multiplication of a fraction by another fraction. The following area model shows how to calculate $\frac{2}{3}$ of $\frac{4}{5}$.

The rectangle is divided vertically into fifths; $\frac{4}{5}$ is shaded blue.

$\frac{4}{5}$

$\frac{2}{3}$

$\frac{4}{5}$

The rectangle is divided horizontally into thirds; $\frac{2}{3}$ of the blue rectangles are now shaded purple. The purple shading represents $\frac{2}{3}$ of $\frac{4}{5}$.

$\frac{2}{3} \times \frac{4}{5} = \frac{8}{15}$

8 out of 15 total rectangles are shaded.

WORKED EXAMPLE 21 Calculating a fraction of a fraction using a visual model

Calculate $\dfrac{2}{3}$ of $\dfrac{1}{4}$ using a visual model.

THINK	WRITE
1. Write the question.	$\dfrac{2}{3}$ of $\dfrac{1}{4}$

$\dfrac{1}{4}$			
$\dfrac{2}{3}$ of $\dfrac{1}{4}$			
$\dfrac{1}{6}$			

THINK	WRITE
2. Change 'of' to $\times$ and cancel if appropriate.	$= \dfrac{\overset{1}{2}}{3} \times \dfrac{1}{\underset{2}{4}}$
3. Perform the multiplication and write the answer.	$= \dfrac{1}{6}$

WORKED EXAMPLE 22 Calculating fractions of an amount

Mum put a 2-litre carton of flavoured milk in the refrigerator. When the children came home from school, Joanna drank one quarter of the milk, Kieren drank one half of it and Daisy drank one sixth of it. Calculate how much milk, in litres, each person drank.

THINK	WRITE
1. Write the fraction of the milk that Joanna drank and simplify the answer.	Joanna: $\dfrac{1}{4}$ of $2 = \dfrac{1}{4} \times 2$
2. Perform the multiplication.	$= \dfrac{1}{\underset{2}{4}} \times \dfrac{\overset{1}{2}}{1}$
	$= \dfrac{1}{2}$
3. Write the fraction of the milk that Kieren drank and simplify the answer.	Kieren: $\dfrac{1}{2}$ of $2 = \dfrac{1}{2} \times 2$
4. Perform the multiplication.	$= \dfrac{1}{\underset{1}{2}} \times \dfrac{\overset{1}{2}}{1}$
	$= 1$

5. Write the fraction of the milk that Daisy drank and simplify the answer.

$$\text{Daisy: } \frac{1}{6} \text{ of } 2 = \frac{1}{6} \times 2$$

6. Perform the multiplication.

$$= \frac{1}{\cancel{6}^3} \times \frac{\cancel{2}^1}{1}$$

$$= \frac{1}{3}$$

7. Write a sentence giving the answer in litres.

Joanna drank half a litre, Kieren drank 1 litre and Daisy drank a third of a litre of milk.

WORKED EXAMPLE 23 Calculating a fraction of a fraction of an amount

Before a road trip John completely fills up his car's fuel tank with 72 litres of petrol. On the first day of the trip, the car uses $\frac{3}{8}$ of the full tank. On the second day, the car uses $\frac{2}{5}$ of the remaining amount. Determine how much petrol was used on the second day.

THINK

1. Determine the fraction left after $\frac{3}{8}$ of the petrol is used on the first day.

2. Calculate the total fraction of the tank of petrol that has been used on day 2.

3. Calculate the amount of petrol used on day 2 as the answer. Remember to state the units.

WRITE

$1 - \frac{3}{8} = \frac{5}{8}$

$\frac{5}{8}$ remains after the first day.

$\frac{2}{5}$ of $\frac{5}{8}$ of a tank

$\frac{2 \times 5}{5 \times 8} = \frac{1}{4}$

$\frac{1}{4}$ of the full tank is used on day 2.

$\frac{1}{4} \times 72 = \frac{72}{4} = 18$

18 litres of petrol were used on the second day.

DISCUSSION

Is repeated addition an efficient method of multiplying fractions?

Digital technology

Scientific calculators can multiply a fraction by another fraction or a whole number.

$$\frac{12}{23} \times \frac{45}{84}$$

$$\frac{45}{161}$$

Exercise 4.6 Multiplying fractions

learn

| 4.6 Quick quiz **on** | 4.6 Exercise |

Individual pathways

| ■ PRACTISE | ■ CONSOLIDATE | ■ MASTER |
| 1, 2, 5, 8, 10, 13, 15, 18 | 3, 6, 9, 11, 16, 19 | 4, 7, 12, 14, 17, 20 |

Fluency

1. **WE18** Calculate the following.

   a. $\dfrac{2}{3} \times 3$
   b. $\dfrac{5}{14} \times 7$
   c. $\dfrac{15}{22} \times 11$
   d. $\dfrac{5}{16} \times 4$

2. **WE19** Calculate the following using written methods, then check your answers with a calculator.

   a. $\dfrac{1}{2} \times \dfrac{1}{4}$
   b. $\dfrac{1}{3} \times \dfrac{2}{3}$
   c. $\dfrac{5}{9} \times \dfrac{4}{3}$
   d. $\dfrac{11}{13} \times \dfrac{1}{2}$

3. Calculate the following using written methods, then check your answers with a calculator.

   a. $\dfrac{6}{7} \times \dfrac{9}{11}$
   b. $\dfrac{11}{12} \times \dfrac{11}{12}$
   c. $\dfrac{8}{9} \times \dfrac{5}{9}$
   d. $\dfrac{5}{8} \times \dfrac{1}{3}$

4. Calculate the following using written methods, then check your answers with a calculator.

   a. $\dfrac{5}{6} \times \dfrac{11}{12}$
   b. $\dfrac{5}{11} \times \dfrac{5}{12}$
   c. $\dfrac{2}{3} \times \dfrac{4}{13}$
   d. $\dfrac{4}{7} \times \dfrac{2}{3}$

5. **WE20** Calculate the following using written methods, then check your answers with a calculator.

   a. $\dfrac{4}{8} \times \dfrac{3}{9}$
   b. $\dfrac{2}{5} \times \dfrac{1}{2}$
   c. $\dfrac{3}{7} \times \dfrac{5}{6}$
   d. $\dfrac{3}{8} \times \dfrac{4}{15}$

6. Calculate the following using written methods, then check your answers with a calculator.

   a. $\dfrac{10}{11} \times \dfrac{22}{25}$
   b. $\dfrac{24}{27} \times \dfrac{9}{8}$
   c. $\dfrac{18}{32} \times \dfrac{64}{72}$
   d. $\dfrac{48}{56} \times \dfrac{24}{60}$

7. Calculate the following using written methods, then check your answers with a calculator.

   a. $\dfrac{15}{27} \times \dfrac{36}{45}$
   b. $\dfrac{8}{49} \times \dfrac{14}{16}$
   c. $\dfrac{4}{9} \times \dfrac{6}{8}$
   d. $\dfrac{5}{25} \times \dfrac{12}{48}$

8. **WE21** Calculate using a visual model:

   a. $\dfrac{1}{2}$ of $\dfrac{1}{4}$
   b. $\dfrac{3}{4}$ of $\dfrac{2}{3}$
   c. $\dfrac{5}{6}$ of $\dfrac{5}{6}$
   d. $\dfrac{8}{9}$ of $\dfrac{1}{4}$

9. Calculate:

   a. $\dfrac{7}{8}$ of 32
   b. $\dfrac{9}{10}$ of 50
   c. $\dfrac{3}{5}$ of 25
   d. $\dfrac{4}{5}$ of 20

Understanding

10. Joshua's dad made 6 litres of cordial for his birthday party. Emily drank $\frac{1}{3}$ of the cordial, Tracy drank $\frac{1}{12}$ and Jonathan drank $\frac{1}{18}$. Calculate how much cordial, in litres, each person drank.

11. Zoe and Jimin play basketball with the Sharp Shooters. Each game lasts 40 minutes. Zoe played $\frac{4}{5}$ of last week's game and Jimin played $\frac{7}{8}$. Calculate how many minutes of the game:

    a. Zoe played
    b. Jimin played.

12. Mark's uncle gave him a family-size block of chocolate. He divided it into thirds to share with his two sisters. His friends Tao and Nick then arrived to visit, and Mark shared his portion of the chocolate with them, so that they each had the same amount.

    a. Draw a block of chocolate with 9×6 pieces, and shade Mark's third of the block. Then draw lines to represent Tao's share.
    b. Identify the fraction of the block of chocolate that Nick received.

13. A car's petrol tank holds 48 litres of fuel. The tank was full at the start of a trip.

    a. By examining the gauge shown, determine what fraction of the tank of petrol has been used.
    b. Determine how many litres of petrol have been used.

14. Answer the following questions.

    a. State how many half pizzas are in 1 pizza.
    b. State how many half pizzas are in 2 whole pizzas.
    c. State how many half pizzas are in 5 whole pizzas.

Communicating, reasoning and problem solving

15. a. Draw a quarter of a pie.
    b. Shade half of this piece of pie.
    c. Determine what fraction of the whole pie is shaded.
    d. Complete the mathematical sentence $\frac{1}{4} \times \frac{1}{2} = \underline{\quad}$ to show how the answer was found.

16. To make a loaf of fruit bread in a breadmaker, the fruit needs to be added after the second rise stage and the top of the bread needs to be glazed two-thirds of the way through the baking stage. The recipe shows how long each stage of the total cycle takes.

Stage	Length of time
First knead	5 minutes
Second knead	20 minutes
First rise then punch down	40 minutes
Second rise then punch down	25 minutes
Third rise	50 minutes
Bake	40 minutes

a. State when the fruit should be added.

b. Express this length of time as a fraction of the total time in the breadmaker.

c. Calculate the length of time during which the bread should be glazed.

17. **a.** Without using a calculator, evaluate the following:

   i. $\dfrac{1}{2} \times \dfrac{2}{3}$

   ii. $\dfrac{1}{2} \times \dfrac{2}{3} \times \dfrac{3}{4}$

   iii. $\dfrac{1}{2} \times \dfrac{2}{3} \times \dfrac{3}{4} \times \dfrac{4}{5}$

   b. Describe any pattern you see in part **a.**

   c. Hence, evaluate the solution to:

   $$\dfrac{1}{2} \times \dfrac{2}{3} \times \dfrac{3}{4} \dots \times \dfrac{98}{99} \times \dfrac{99}{100}$$

18. Gustave's monthly take-home pay is $2400. From this he spends a quarter on his home loan payments, one half on food and drink and one sixth on clothing. One half of the remainder goes into his savings account. Determine how much money Gustave deposits into his savings account each month.

19. On a given map, one kilometre is represented by $\dfrac{2}{9}$ cm. If a distance on the map is 27 cm, determine how many kilometres this represents.

20. Andrea requires $5\dfrac{2}{7}$ m of fabric to make a pattern on a quilt. One third of this fabric has to be red and $\dfrac{1}{4}$ must be green. Determine how many metres of red and green fabric she needs.

LESSON
4.7 Dividing fractions

LEARNING INTENTION

At the end of this lesson you should be able to:
• divide fractions with or without a visual aid.

▶ 4.7.1 Dividing fractions

eles-3808

• To divide fractions you need to know what their reciprocal is.

• The **reciprocal** of a fraction is found by turning the fraction upside down. So $\dfrac{3}{5}$ is the reciprocal of $\dfrac{5}{3}$.

• A whole number can be written as a fraction by putting it over 1. Therefore, the reciprocal of a whole number is 1 over the number. For example, since $4 = \dfrac{4}{1}$, the reciprocal of 4 is $\dfrac{1}{4}$.

WORKED EXAMPLE 24 Reciprocal of a proper fraction

Identify the reciprocal of $\dfrac{2}{3}$.

THINK	WRITE
Turn the fraction upside down and write the answer in a sentence.	The reciprocal of $\dfrac{2}{3}$ is $\dfrac{3}{2}$.

- To determine the reciprocal of a mixed number, first express it as an improper fraction.

WORKED EXAMPLE 25 Reciprocal of a mixed number

Determine the reciprocal of $1\dfrac{2}{3}$.

THINK	WRITE
1. Write $1\dfrac{2}{3}$ as an improper fraction.	$1\dfrac{2}{3} = \dfrac{5}{3}$
2. Turn the improper fraction upside down and write the answer in a sentence.	The reciprocal of $1\dfrac{2}{3}$ is $\dfrac{3}{5}$.

- Try to think of division as a question. For example, $10 \div 4$ can be interpreted as 'How many groups of 4 are there in 10?'

$$10 \div 4 = 2\tfrac{1}{2} \text{ (groups of 4)}$$

- 10 ones
- Groups of 4 ones

- Division of fractions follows the same logic. For example, $3 \div \dfrac{1}{2}$ can be interpreted as 'How many halves are there in 3?'

$$3 \div \tfrac{1}{2} = 6\left(\text{groups of } \tfrac{1}{2}\right)$$

- 3 ones
- Groups of 1 half

- Division of fractions by fractions can also be explained with the same logic. For example, $\dfrac{5}{6} \div \dfrac{1}{3}$ can be interpreted as "How many thirds are there in five sixths?"

$$\tfrac{5}{6} \div \tfrac{1}{3} = 2\tfrac{1}{2}\left(\text{groups of } \tfrac{1}{3}\right)$$

- $\dfrac{5}{6}$
- Groups of $\dfrac{1}{3}$

1 group   1 group $\dfrac{1}{2}$ a group

- This method is great for understanding the concept of division of fractions, but is not so efficient for evaluating many divisions involving fractions.
- Notice that when 3 is divided into groups of $\frac{1}{2}$, the answer is the same as when 3 is multiplied by $\frac{2}{1}$ (its reciprocal). This also works when dividing fractions by fractions.

Dividing fractions

To divide fractions, multiply the first fraction by the reciprocal of the second fraction.

For example: $\frac{1}{3} \div \frac{2}{5} = \frac{1}{3} \times \frac{5}{2} = \frac{5}{6}$

This process can be remembered easily by the saying KEEP, CHANGE, FLIP.

$$\frac{3}{4} \qquad \div \qquad \frac{2}{7}$$
$$\text{KEEP} \quad \text{CHANGE} \quad \text{FLIP}$$
$$\frac{3}{4} \qquad \times \qquad \frac{7}{2}$$

WORKED EXAMPLE 26 Dividing fractions

Evaluate $\frac{2}{3} \div \frac{4}{9}$.

THINK	WRITE
1. Write the question.	$\frac{2}{3} \div \frac{4}{9}$
2. Keep the first fraction the same. Change the division sign to a multiplication sign and flip the second fraction.	$= \frac{2}{3} \times \frac{9}{4}$
3. Cancel diagonally.	$= \frac{2^1}{3^1} \times \frac{9^3}{4^2}$
4. Perform the multiplication.	$= \frac{3}{2}$

DISCUSSION

Explore what happens when you multiply and divide positive numbers by a fraction between 0 and 1. Discuss the results.

Resources

Interactivities Division of rational numbers (int-3948)

Exercise 4.7 Dividing fractions

learn on

4.7 Quick quiz on	4.7 Exercise

Individual pathways

■ PRACTISE 1, 4, 7, 10, 13	■ CONSOLIDATE 2, 6, 8, 11, 14	■ MASTER 3, 5, 9, 12, 15

Fluency

1. **WE24** Identify the reciprocals of each of the following.

   a. $\dfrac{3}{4}$ b. $\dfrac{2}{7}$ c. $\dfrac{6}{5}$ d. $\dfrac{5}{3}$ e. 5

2. Identify the reciprocals of each of the following.

   a. $\dfrac{5}{12}$ b. $\dfrac{9}{2}$ c. $\dfrac{5}{3}$ d. $\dfrac{10}{3}$ e. 20

3. Identify the reciprocals of each of the following.

   a. $\dfrac{1}{12}$ b. $\dfrac{3}{15}$ c. $\dfrac{2}{22}$ d. $\dfrac{1}{6}$ e. 1

4. **WE25** Determine the reciprocals of these mixed numbers.

   a. $1\dfrac{2}{3}$ b. $4\dfrac{3}{7}$ c. $9\dfrac{2}{7}$ d. $3\dfrac{3}{8}$ e. $6\dfrac{9}{10}$

5. Determine the reciprocals of these mixed numbers.

   a. $3\dfrac{6}{7}$ b. $5\dfrac{1}{3}$ c. $7\dfrac{1}{2}$ d. $10\dfrac{2}{9}$ e. $2\dfrac{5}{9}$

6. Multiply each of these numbers by its reciprocal.

   a. $\dfrac{5}{7}$ b. $\dfrac{3}{5}$ c. $2\dfrac{1}{8}$

Understanding

7. **WE26** Evaluate the following, then check your answers with a calculator.

   a. $\dfrac{1}{4} \div \dfrac{2}{3}$ b. $\dfrac{8}{9} \div \dfrac{7}{6}$ c. $\dfrac{1}{12} \div \dfrac{2}{3}$ d. $\dfrac{4}{11} \div \dfrac{5}{6}$ e. $\dfrac{10}{3} \div \dfrac{15}{21}$

8. Evaluate the following, then check your answers with a calculator.

   a. $\dfrac{2}{7} \div \dfrac{20}{21}$

   b. $\dfrac{4}{9} \div \dfrac{4}{9}$

   c. $\dfrac{4}{7} \div 10$

   d. $\dfrac{4}{3} \div 12$

   e. $\dfrac{15}{11} \div 5$

9. Evaluate the following, then check your answers with a calculator.

   a. $\dfrac{5}{6} \div \dfrac{3}{2}$

   b. $\dfrac{3}{11} \div 6$

   c. $\dfrac{9}{10} \div \dfrac{5}{7}$

   d. $\dfrac{20}{17} \div \dfrac{5}{3}$

   e. $\dfrac{50}{7} \div 5$

Communicating, reasoning and problem solving

10. Lucy says that the following statement is correct.

$$3\dfrac{1}{2} \div \dfrac{4}{6} > \dfrac{4}{3} \div \dfrac{2}{15}$$

However, Lucy's friend Lily does not agree. Explain who is correct.

11. Mai is making sushi. The recipe requires $\dfrac{1}{4}$ cup of rice. She has a 2 kg bag of rice. If 1 cup of rice weighs 250 g, calculate the number of batches of the recipe she can make.

12. Maytreyi bought $1\dfrac{1}{2}$ kg of chocolates for her birthday party. She is going to give each friend attending her birthday $\dfrac{1}{10}$ kg. Calculate the number of friends attending her birthday party.

13. A plumber has a piece of pipe that is $\dfrac{26}{4}$ cm long. He divides the pipe into pieces that are $\dfrac{3}{2}$ cm long. Determine the number of pieces he has now.

14. Calculate the average of $\dfrac{1}{4}, \dfrac{1}{5}, \dfrac{1}{10}$ and $\dfrac{1}{20}$.

15. Determine the value of the following expression.

$$\dfrac{1}{2 + \dfrac{1}{2\frac{1}{2}}}$$

LESSON
4.8 Working with mixed numbers

LEARNING INTENTION

At the end of this lesson you should be able to:
- add, subtract, multiply and divide mixed numbers.

▶ 4.8.1 Adding and subtracting mixed numbers

eles-3809

- Adding and subtracting mixed numbers follows the same principles as those for adding and subtracting proper and improper fractions.

- To add or subtract mixed numbers, first convert to improper fractions, and then add and subtract as usual.

- For example, $1\frac{1}{3} + 1\frac{1}{2}$ can be explained visually, similarly to in lesson 4.5.

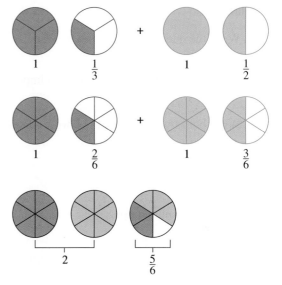

Working with mixed numbers

To add, subtract, multiply or divide mixed numbers, convert them to improper fractions and then perform the operation as usual.

WORKED EXAMPLE 27 Adding mixed numbers

Evaluate $2\frac{3}{4} + 1\frac{1}{2}$.

THINK	WRITE
1. Write the question.	$2\frac{3}{4} + 1\frac{1}{2}$
2. Convert each mixed number into an improper fraction.	$= \frac{11}{4} + \frac{3}{2}$
3. Write both fractions with the lowest common denominator (LCD).	$= \frac{11}{4} + \frac{3 \times 2}{2 \times 2}$
	$= \frac{11}{4} + \frac{6}{4}$

▶

4. Add the numerators.	$= \dfrac{17}{4}$
5. Convert to a mixed number if appropriate and write the answer.	$= 4\dfrac{1}{4}$

WORKED EXAMPLE 28 Subtracting mixed numbers

Evaluate $3\dfrac{1}{5} - 1\dfrac{3}{4}$.

THINK	WRITE
1. Write the question.	$3\dfrac{1}{5} - 1\dfrac{3}{4}$
2. Convert each mixed number into an improper fraction.	$= \dfrac{16}{5} - \dfrac{7}{4}$
3. Write both fractions with the lowest common denominator (LCD).	$= \dfrac{16 \times 4}{5 \times 4} - \dfrac{7 \times 5}{4 \times 5}$ $= \dfrac{64}{20} - \dfrac{35}{20}$
4. Subtract the numerators.	$= \dfrac{29}{20}$
5. Convert to a mixed number if appropriate and write the answer.	$= 1\dfrac{9}{20}$

Estimation

- Estimation can be used to determine approximate answers to addition and subtraction of mixed numbers.
- This allows you to check easily whether your answer to a calculation is reasonable.
- To estimate, add or subtract the whole number parts. The answer should be close to the result obtained by this method.

WORKED EXAMPLE 29 Checking answers by estimation

Evaluate $3\dfrac{1}{3} + 2\dfrac{1}{2}$ and check the answer by estimation.

THINK	WRITE
1. Write the question.	$3\dfrac{1}{3} + 2\dfrac{1}{2}$
2. Convert each mixed number into an improper fraction.	$= \dfrac{10}{3} + \dfrac{5}{2}$
3. Write both fractions with the lowest common denominator (LCD).	$= \dfrac{10 \times 2}{3 \times 2} + \dfrac{5 \times 3}{2 \times 3}$ $= \dfrac{20}{6} + \dfrac{15}{6}$

4. Add the numerators.

$$= \frac{35}{6}$$

5. Convert to a mixed number if appropriate.

$$= 5\frac{5}{6}$$

6. Check the answer by adding the whole numbers to determine an approximation.

$$3\frac{1}{3} + 2\frac{1}{2} \simeq 3 + 2$$

$$3\frac{1}{3} + 2\frac{1}{2} \simeq 5$$

The approximation shows that the answer we obtain should be close to 5.

⊙ 4.8.2 Multiplying and dividing mixed numbers

eles-3810

• Similar to proper and improper fractions, how to multiply and divide mixed numbers can be explained visually. For example, $1\frac{1}{2} \times 1\frac{2}{3}$ is shown below in an area model:

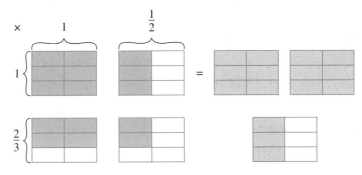

• The area model shows $1\frac{1}{2}$ groups of $1\frac{2}{3}$ meaning one whole group and $\frac{1}{2}$ of a group with only half shaded. Since all the pieces are the same size, they can be added together to give two wholes and one half.

• We can also see this model as $\frac{3}{2} \times \frac{5}{3} = \frac{15}{6}$

$$= \frac{5}{2}$$

$$= 2\frac{1}{2}$$

• We can then say that when multiplying mixed numbers, you should first convert to improper fractions, and then multiply as usual.

WORKED EXAMPLE 30 Multiplying mixed numbers

Evaluate $1\frac{1}{2} \times 3\frac{3}{5}$.

THINK

1. Write the question.

WRITE

$$1\frac{1}{2} \times 3\frac{3}{5}$$

▶

2. Change mixed numbers to improper fractions and cancel if possible.

$$= \frac{3}{\cancel{2}^{1}} \times \frac{\cancel{18}^{9}}{5}$$

3. Multiply the numerators and then the denominators.

$$= \frac{27}{5}$$

4. Convert to a mixed number if appropriate and write the answer.

$$= 5\frac{2}{5}$$

- Similarly to how we divided proper and improper fractions, mixed numbers can be divided by first converting them into improper fractions, and then dividing as usual.

WORKED EXAMPLE 31 Dividing mixed numbers

Evaluate $1\frac{1}{2} \div 3\frac{2}{5}$.

THINK	WRITE
1. Write the question.	$1\frac{1}{2} \div 3\frac{2}{5}$
2. Convert the mixed numbers to improper fractions.	$= \frac{3}{2} \div \frac{17}{5}$
3. Change the $\div$ to $\times$ and flip the second fraction.	$= \frac{3}{2} \times \frac{5}{17}$
4. Multiply the numerators, then the denominators and simplify the answer if necessary.	$= \frac{15}{34}$

Digital technology

Mixed fractions can be multiplied and divided using a scientific calculator.

To input a mixed fraction press the SHIFT button then $\frac{\square}{\square}$.

Resources

 Interactivities Addition and subtraction of mixed numbers (int-3951)

Exercise 4.8 Working with mixed numbers

4.8 Quick quiz on	4.8 Exercise

Individual pathways

■ PRACTISE	■ CONSOLIDATE	■ MASTER
1, 3, 5, 6, 8, 10, 14, 17, 20, 23	2, 7, 11, 12, 16, 18, 21, 24, 25	4, 9, 13, 15, 19, 22, 26

Fluency

1. **WE27&28** Evaluate the following, giving answers as mixed numbers in simplest form.

   a. $1\frac{1}{5} + 3\frac{1}{2}$
   b. $3\frac{2}{3} - 2\frac{1}{3}$
   c. $5\frac{2}{3} - 2\frac{1}{4}$
   d. $9\frac{3}{8} + 4$

2. Evaluate the following, giving answers as mixed numbers in simplest form.

   a. $2\frac{3}{5} - 1\frac{1}{2}$
   b. $3\frac{1}{2} + 2\frac{3}{4}$
   c. $7 - 5\frac{2}{3}$
   d. $2\frac{1}{3} - \frac{5}{6}$

3. Evaluate the following, giving answers as mixed numbers in simplest form.

   a. $3\frac{1}{2} - 1\frac{7}{8}$
   b. $4\frac{3}{5} - 2\frac{9}{10}$
   c. $5\frac{1}{4} - 1\frac{7}{12}$
   d. $5\frac{3}{4} - 3\frac{5}{6}$

4. Evaluate the following, giving answers as mixed numbers in simplest form.

   a. $\frac{1}{5} + 2\frac{2}{7}$
   b. $\frac{5}{8} + 3\frac{3}{4} - \frac{1}{2}$
   c. $2\frac{2}{9} + 1\frac{1}{3} + \frac{1}{6}$
   d. $4\frac{2}{3} + 1\frac{3}{8} - 3$

5. **WE29** Evaluate the following and check the answers by estimation.

   a. $5\frac{1}{3} + 2\frac{1}{4}$
   b. $1\frac{1}{6} + 2\frac{1}{2}$
   c. $3\frac{1}{3} + 1\frac{1}{6}$
   d. $6\frac{1}{2} - 2\frac{1}{4}$

6. **WE30** Evaluate:

   a. $1\frac{1}{4} \times \frac{2}{3}$
   b. $1\frac{1}{2} \times 1\frac{1}{2}$
   c. $2\frac{5}{8} \times 1\frac{3}{4}$
   d. $2\frac{1}{5} \times 3\frac{1}{2}$

7. Evaluate:

   a. $3\frac{5}{6} \times 9\frac{5}{7}$
   b. $5\frac{1}{5} \times 6\frac{2}{3}$
   c. $3\frac{3}{4} \times 2\frac{2}{3}$
   d. $\frac{7}{8} \times 3\frac{1}{3}$

8. **WE31** Evaluate:

   a. $1\frac{1}{4} \div \frac{3}{4}$
   b. $\frac{9}{4} \div 3\frac{1}{2}$
   c. $\frac{10}{3} \div 3\frac{1}{3}$
   d. $7\frac{4}{9} \div \frac{1}{3}$

9. Evaluate:

   a. $1\frac{1}{2} \div \frac{1}{2}$
   b. $10\frac{9}{10} \div 4$
   c. $3\frac{2}{7} \div 2\frac{1}{7}$
   d. $3\frac{5}{8} \div 1\frac{3}{4}$

Understanding

10. One-third of a litre of cordial is mixed with $1\frac{1}{2}$ litres of water. Calculate how many litres of drink have been made.

11. The lengths of bolts and nails are often measured in inches. For the bolt shown, calculate the length of the threaded section if the total length of the bolt is $3\frac{1}{4}$ inches.

12. Julia is planning a holiday to the US. She has 5 months and has worked out the following itinerary. She will be in California for $1\frac{1}{2}$ months, in Colorado for $1\frac{2}{3}$ months and in Florida for $\frac{3}{4}$ of a month. The other state she will be visiting is New York. Determine how long she will spend in New York.

13. Theo is building a house. Before the frame of the house can be assembled, the footings need to be dug. The digging will take $3\frac{1}{2}$ days. The concrete slab that is then needed will take $2\frac{3}{4}$ days to pour and set. Determine how long it will be before Theo can start assembling the frame.

14. Felicity is dividing $1\frac{1}{2}$ kilograms of minced steak into 8 hamburger portions. Calculate how many kilograms of mince are in each portion.

15. Ned spends his school holidays helping his father with shearing sheep.

    a. It takes Ned $\frac{1}{4}$ hour to shear one sheep. Calculate how many sheep Ned will shear in $5\frac{1}{2}$ hours.

    b. Ned's father, Wesley, shears a sheep every $\frac{1}{12}$ of an hour. If Wesley worked continuously for 8 hours, calculate how many sheep he would shear.

16. Lachlan was looking for something to do one Saturday afternoon and his dad suggested he cook something for afternoon tea. Lachlan found a recipe for peanut butter muffins.

    Ingredients:

    $\frac{1}{4}$ cup sugar $\qquad$ $\frac{1}{4}$ cup margarine

    $\frac{1}{2}$ cup peanut butter $\qquad$ 2 eggs

    $1\frac{1}{2}$ cups milk $\qquad$ $2\frac{1}{2}$ cups self-raising flour

    $\frac{1}{4}$ teaspoon baking soda

    Method:
    Blend the sugar, margarine and peanut butter. Beat in the eggs and milk. Add the self-raising flour and baking soda. Place the mixture in greased muffin pans and sprinkle with cinnamon sugar. Bake for 15–20 minutes in a 200 °C oven.

    This recipe makes 10 muffins, but Lachlan wants enough mixture to make 15. He needs to multiply the quantities by $1\frac{1}{2}$. Write the quantities of each ingredient Lachlan needs to make 15 muffins.

17. Anh bought $7\frac{3}{4}$ kg of prawns for a party and $\frac{3}{5}$ of them were eaten. Calculate how many kilograms of prawns were left over.

18. Teagan buys $8\frac{3}{4}$ kg of sugar. She wants to store it in containers that can hold $1\frac{1}{4}$ kg. Calculate how many containers she needs.

19. Max works as a train driver and is paid $21 per hour. If Max works on a public holiday, he is paid double time and a half, which means he earns $2\frac{1}{2}$ times his normal hourly rate. Calculate what Max earns per hour working on a public holiday.

Communicating, reasoning and problem solving

20. To make the casing for a simple collapsible telescope, you need two cylinders, one of which fits inside the other.

    a. Calculate the width of the cylinder on the right that would slide over the one shown to the left.

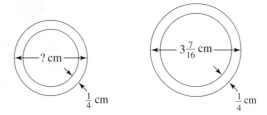

    b. Calculate the width of the cylinder in the left diagram that would slide into the one shown at right.

21. Worldwide figures indicate that $\frac{1}{10}$ of men are left-handed, while $\frac{2}{25}$ of women are left-handed. If there are 400 men in a room, calculate how many women would need to be there so that there are the same number of left-handed men as left-handed women.

22. Explain how you could you convince someone that $\frac{1}{2}$ of $3\frac{1}{3}$ is $1\frac{2}{3}$.

23. Five-eighths of a class were girls and $\frac{2}{3}$ of the girls were brunette.

    a. Calculate what fraction of the class were brunette girls.
    b. If there were 24 people in the class, calculate how many were brunette girls.

24. Michelle can run one lap of a cross-country course in $\frac{1}{5}$ hour.

a. Calculate how many laps she completed in $1\frac{1}{3}$ hours. (Assume she can keep up the same pace.)

b. Calculate the number of minutes it takes for Michelle to run one lap.

25. A skydiver opens his parachute at 846 m above ground level. He has already fallen five-sevenths of the distance to the ground. Calculate the height of the plane when he jumped.

26. A new vending machine stocking water and sports drinks was installed at the local gymnasium. The number of bottles sold each day over the first 5 weeks after its installation is shown.

4, 39, 31, 31, 50, 43, 60, 45, 57, 61, 18, 26, 3, 52, 51, 59, 33,
51, 27, 62, 30, 40, 3, 30, 37, 9, 33, 44, 53, 16, 22, 6, 42, 33, 19

The gymnasium realises that the number of daily sales varies, depending particularly on the weather. The machine is stocked with 60 bottles at the start of each day. The supplier advises that, because of electricity costs and other running expenses, the machine is not worthwhile having unless at least three-quarters of the bottles are sold on average 2 days per week.
Round these sales to the nearest 5 and analyse the figures to determine whether the gymnasium should keep the vending machine.

LESSON
4.9 Percentages as fractions

LEARNING INTENTION

At the end of this lesson you should be able to:
• convert percentages to fractions.

⏵ 4.9.1 Definition of a percentage

eles-3812

• The term **per cent** means 'per hundred' or 'out of a hundred'.
• The symbol for per cent is %. For example, 7% means 7 out of 100.

WORKED EXAMPLE 32 Writing a percentage as a fraction

Write 47% as a fraction.

THINK	WRITE
Recall the definition of a percentage. 47% means 47 out of 100.	$47\% = \dfrac{47}{100}$

Percentages as fractions

- All percentages can be expressed as fractions.
  - To convert a whole number percentage into a fraction, write it over 100 and simplify.
    For example, $10\% = \dfrac{10}{100} = \dfrac{1}{10}$

WORKED EXAMPLE 33 Converting a percentage into a fraction

Write 20% as a fraction in simplest form.

THINK	WRITE
1. Write the percentage as a fraction with a denominator of 100.	$20\% = \dfrac{20}{100}$
2. Simplify the fraction.	$= \dfrac{1}{5}$

- To convert a percentage containing a decimal number into a fraction:
  - write it as a fraction over 100
  - change the numerator to a whole number by multiplying both the numerator and denominator by the appropriate power of 10
  - simplify if possible.

  For example: $28.2\% = \dfrac{28.2}{100} = \dfrac{282}{1000} = \dfrac{141}{500}$

WORKED EXAMPLE 34 Converting percentages to fractions

Write 36.4% as a fraction in simplest form.

THINK	WRITE
1. Write the percentage as a fraction out of 100.	$36.4\% = \dfrac{36.4}{100}$
2. Change the numerator to a whole number by multiplying it by an appropriate power of 10, and multiply the denominator by the same power of 10.	$= \dfrac{36.4 \times 10}{100 \times 10}$ $= \dfrac{364 \div 4}{1000 \div 4}$
3. Simplify the fraction.	$= \dfrac{91}{250}$

- To convert a percentage that is a proper fraction into a fraction, multiply the denominator by 100 and simplify if possible.
- For example: $\dfrac{5}{8}\% = \dfrac{5}{800} = \dfrac{1}{160}$

- To convert a mixed number percentage into a fraction:
  - change the mixed number into an improper fraction
  - multiply the denominator by 100
  - simplify if possible.
  For example: $12\dfrac{1}{2}\% = \dfrac{25}{2}\% = \dfrac{25}{200} = \dfrac{1}{8}$

Converting percentages into fractions

In simple terms, to convert a percentage into a fraction, place the percentage as a fraction over 100, then simplify so that both the numerator and denominator are whole numbers with no common factors.

$$x\% = \dfrac{x}{100}$$

WORKED EXAMPLE 35 Converting percentages to fractions

Write the following percentages as fractions in simplest form.
a. $\dfrac{1}{4}\%$
b. $15\dfrac{1}{3}\%$

THINK	WRITE
a. 1. Write the percentage and then multiply the denominator by 100.	a. $\dfrac{1}{4}\% = \dfrac{1}{4 \times 100}$
2. Simplify.	$= \dfrac{1}{400}$
b. 1. Convert the mixed number to an improper fraction.	b. $15\dfrac{1}{3}\% = \dfrac{46}{3}\%$
2. Multiply the denominator by 100.	$= \dfrac{46}{3 \times 100}$
3. Simplify the fraction.	$= \dfrac{46}{300}$
	$= \dfrac{23}{150}$

 Resources

Interactivities Percentages as fractions (int-3992)

Exercise 4.9 Percentages as fractions

learnon

| 4.9 Quick quiz on | 4.9 Exercise |

Individual pathways

■ PRACTISE	■ CONSOLIDATE	■ MASTER
1, 3, 6, 9, 12, 15, 16, 17, 19, 22	2, 4, 7, 10, 13, 20, 23, 24	5, 8, 11, 14, 18, 21, 25

Fluency

1. **WE32** Write the following percentages as fractions.
   a. 17%   b. 29%   c. 79%   d. 99%   e. 3%

2. Write the following percentages as fractions.
   a. 19%   b. 33%   c. 9%   d. 243%   e. 127%

3. **WE33** Write the following percentages as fractions in simplest form.
   a. 50%   b. 80%   c. 35%   d. 60%   e. 10%

4. Write the following percentages as fractions in simplest form.
   a. 45%   b. 12%   c. 5%   d. 74%   e. 2%

5. Write the following percentages as fractions in simplest form. Change the answer to a mixed number where appropriate.
   a. 150%   b. 180%   c. 200%   d. 500%   e. 112%

6. **WE34** Write the following percentages as fractions in simplest form.
   a. 3.5%   b. 7.2%   c. 19.7%   d. 32.4%

7. Write the following percentages as fractions in simplest form.
   a. 15.5%   b. 16.2%   c. 28.3%   d. 41.38%

8. Write the following percentages as fractions in simplest form.
   a. 18.14%   b. 57.99%   c. 12.15%   d. 0.05%

9. **WE35** Write the following percentages as fractions in simplest form.
   a. $\frac{1}{2}\%$ b. $\frac{1}{5}\%$ c. $\frac{2}{3}\%$ d. $\frac{1}{10}\%$ e. $\frac{6}{11}\%$

10. Write the following percentages as fractions in simplest form.
    a. $8\frac{1}{4}\%$ b. $20\frac{2}{3}\%$ c. $9\frac{2}{3}\%$ d. $15\frac{1}{2}\%$ e. $11\frac{1}{5}\%$

11. Write the following percentages as fractions in simplest form. Leave the answer as a mixed number where appropriate.
    a. $10\frac{3}{8}\%$ b. $11\frac{2}{3}\%$ c. $150\frac{1}{2}\%$ d. $120\frac{1}{2}\%$ e. $33\frac{1}{3}\%$

Understanding

12. **MC** 40% as a fraction in simplest form is:
    A. $\frac{1}{4}$ B. $\frac{2}{5}$ C. $\frac{4}{100}$ D. $\frac{4000}{1}$

13. **MC** $10\frac{1}{2}\%$ as a fraction is:
    A. $\frac{21}{2}$ B. $\frac{10.5}{100}$ C. $\frac{2100}{2}$ D. $\frac{21}{200}$

14. **MC** 1.5% as a fraction in simplest form is:

A. $\dfrac{1.5}{100}$ **B.** $\dfrac{3}{100}$ **C.** $\dfrac{3}{200}$ **D.** $\dfrac{15}{100}$

15. Imran saves 20% of his pocket money each week. Calculate the fraction that he saves.

16. In one game, 35% of a football team is injured. Calculate what fraction of the team is injured.

17. Each week Jodie spends 45% of her wages at the supermarket. Calculate what fraction of Jodie's wages is spent at the supermarket.

18. Calculate what fraction of a class were born overseas if 68% were born in Australia.

Communicating, reasoning and problem solving

19. Of the students in a Year 7 class, 80% have pets and 20% have blue eyes.

   a. Calculate what fraction of students have pets.
   b. Calculate what fraction of students have blue eyes.
   c. Explain whether this means that none of the students with blue eyes have pets.

20. **a.** In a molecule of methane (natural gas), 80% of the atoms are hydrogen. Calculate the fraction of the methane molecule that is composed of hydrogen atoms.

Hydrogen atoms

Carbon atom

Methane molecule

b. The diagram represents a molecule of glycolaldehyde, a simple form of sugar found in the gas surrounding a young star called IRAS 16293-2422, located around 400 light-years away from Earth. Hydrogen atoms make up 50% of the atoms of this molecule. Calculate the fraction of the glycolaldehyde molecule that is composed of hydrogen atoms.

Carbon atoms

Hydrogen atoms

Oxygen atoms

Glycolaldehyde molecule

c. Although both molecules have the same number of hydrogen atoms, explain why they represent different fractions.

21. Shirley plays basketball in her school's team. In the last game she scored 50 baskets from 75 shots but in the previous game she scored 45 baskets from 60 shots.

a. Compare the percentage of baskets scored relative to the number of shots for each game to determine which game she performed best in.
b. Explain how the percentage is affected by:
   i. increasing the number of baskets when the number of shots stays the same
   ii. increasing the number of shots when the number of baskets stays the same
   iii. increasing both the number of baskets and the number of shots at the same percentage rate.

22. Less than 50% of the estimated 200 000 Australian invertebrate species have been described. Determine what fraction this is.

23. If the unemployment rate in Australia is 8%, calculate:
   a. what fraction of the population is unemployed
   b. what fraction of the population is employed
   c. out of 100 people, the number you would expect to be unemployed.

24. Determine what fraction remains if:
   a. 65% of winnings have been spent
   b. 19% of an audience hated a movie
   c. all stock was discounted by 15%
   d. 93.5% of the school population supports the uniform policy.

25. Seventeen per cent of visitors to Australia are from New Zealand.
   a. State the fraction of visitors to Australia from New Zealand.
   b. Calculate the fraction of visitors to Australia not from New Zealand.
   c. Out of 100 visitors to Australia, state how many you would expect to be from New Zealand.

LESSON
4.10 Calculating percentages of an amount

LEARNING INTENTION

At the end of this lesson you should be able to:
- calculate percentages of amounts using fractions.

▶ 4.10.1 Calculating percentages of an amount using fractions

eles-3814

- To calculate a percentage of an amount using fractions, follow these steps.
  - Step 1: Convert the percentage into a fraction.
  - Step 2: Multiply by the amount.
- When calculating a percentage of an amount, the unit of the answer will always be the unit of the initial quantity. For example, the percentage of an amount of kilograms will be a certain number of kilograms.

WORKED EXAMPLE 36 Calculating a percentage of an amount using fractions

Calculate 40% of $135.

THINK	WRITE
1. Write the question.	40% of $135
2. Write the percentage as a fraction with a denominator of 100, change *of* to × and write the amount as a fraction over 1.	$= \dfrac{40}{100} \times \dfrac{135}{1}$
3. Simplify as appropriate.	$= \dfrac{40}{20} \times \dfrac{27}{1}$ $= \dfrac{2}{1} \times \dfrac{27}{1}$
4. Perform the multiplication.	$= \dfrac{54}{1}$
5. State the answer.	$= \$54$

WORKED EXAMPLE 37 Calculating a percentage of an amount using fractions

Of the 250 students selected at random to complete a survey, 18% were in Year 11. Calculate how many students were in Year 11.

THINK	WRITE
1. Decide what percentage of the total is required and write an expression to determine the percentage of the total.	18% of 250
2. Write the percentage as a fraction, change *of* to × and write the total as a fraction.	$= \dfrac{18}{100} \times \dfrac{250}{1}$

---

3. Simplify as appropriate.

$$= \frac{18}{2} \times \frac{5}{1}$$

$$= \frac{9}{1} \times \frac{5}{1}$$

4. Perform the multiplication.

$$= \frac{45}{1}$$

$$= 45$$

5. Answer the question by writing a sentence.

45 of the 250 students were in Year 11.

COMMUNICATING — COLLABORATIVE TASK: Elastic percentages

Equipment: scissors, wide rubber bands, rulers, A4 paper

1. Working in pairs, cut a rubber band in half to make one long rubber strip.
2. Use your ruler to draw an even scale on the rubber strip from 0 to 100, using intervals of 10. Make sure the rubber strip is not stretched in any way during this process.
3. Select five different objects around the classroom for which you will identify a percentage of their length, width or height; an object may be a book, a pencil or even someone's arm.
4. On an A4 sheet of paper, write the percentage that you will measure for each object. Your teacher may allocate your pair a percentage, or you may like to choose your own. Measure the same percentage for each of your objects; for example, you may choose to calculate 60% of the heights of a chair, a table, a computer screen, a poster and a bookshelf.
5. For each of your five objects, work with your partner to record the measurement of your given percentage using the following steps.
   • Stretch the rubber strip along the edge of the object that you wish to measure. The 0 mark on the strip must be at one end of the object you are measuring, and the 100 mark must be at the other end of the object. Hold onto the rubber strip tightly.
   • Place a pencil point on the object next to the position of your percentage on the rubber strip.
   • Use a ruler to measure the distance from the end of the object to the pencil point.
   • Draw a line of this length on your sheet of paper, and then label the line with a description of the object that you measured.
6. Each line on your paper represents the same percentage of different lengths. With your partner, discuss why the lines are different lengths.
7. As a class, discuss the relationship between the size of a percentage and the size of the whole object being measured. For example, is 60% of the length of a book more or less than 60% of the height of a door? Why?

DISCUSSION

What method would you use to calculate 60% of an amount? Are there other ways? Which is the most efficient?

on Resources

Video eLesson Percentages of an amount (eles-1882)

Interactivities Percentages of an amount using fractions (int-3996)

Exercise 4.10 Calculating percentages of an amount

4.10 Quick quiz on | **4.10 Exercise**

Individual pathways

■ PRACTISE	■ CONSOLIDATE	■ MASTER
1, 3, 6, 9, 12, 16, 18, 23, 25, 28, 31	2, 4, 7, 10, 13, 15, 19, 21, 24, 27, 29, 32	5, 8, 11, 14, 17, 20, 22, 26, 30, 33

Fluency

1. Copy each of the following problems and then calculate the answers by completing the working.

   a. 90% of $200 = \dfrac{90}{100} \times \dfrac{200}{1}$
   b. 8% of $50 = \dfrac{8}{100} \times \dfrac{50}{1}$
   c. 50% of $120 = \dfrac{50}{100} \times \dfrac{120}{1}$

2. Copy each of the following problems and then calculate the answers by completing the working.

   a. 5% of $30 = \dfrac{5}{100} \times \dfrac{30}{1}$
   b. 15% of $70 = \dfrac{15}{100} \times \dfrac{70}{1}$
   c. 65% of $120 = \dfrac{65}{100} \times \dfrac{120}{1}$

3. **WE36** Calculate the following by converting the percentage to a fraction.

   a. 50% of 20 b. 20% of 80 c. 10% of 30 d. 9% of 200 e. 40% of 15

4. Calculate the following by converting the percentage to a fraction.

   a. 12% of 50 b. 70% of 110 c. 80% of 5000 d. 44% of 150 e. 68% of 25

5. Calculate the following by converting the percentage to a fraction.

   a. 52% of 75 b. 38% of 250 c. 110% of 50 d. 150% of 8 e. 125% of 20

6. Calculate the following by converting the percentage to a fraction. Write the answer as a mixed number.

   a. 18% of 20 b. 16% of 30 c. 11% of 70 d. 74% of 25 e. 66% of 20

7. Calculate the following by converting the percentage to a fraction. Write the answer as a mixed number.

   a. 55% of 45 b. 15% of 74 c. 95% of 62 d. 32% of 65 e. 82% of 120

8. **MC** 45% written as a fraction is:

   A. $\dfrac{45}{100}$ B. $\dfrac{45}{1}$ C. $\dfrac{450}{1}$ D. $\dfrac{1}{45}$

9. **MC** To evaluate 17% of 22, *of* should be changed to:

   A. $\div$ B. $+$ C. $\times$ D. $-$

10. **MC** Select which of the following would determine 15% of 33.

    A. 15 of 33 B. $\dfrac{15}{1} \times 33$ C. $\dfrac{15}{100} \times \dfrac{33}{100}$ D. $\dfrac{15}{100} \times \dfrac{33}{1}$

11. **MC** 60% of 30 is:

    A. $19\dfrac{4}{5}$ B. $\dfrac{31}{5}$ C. 19 D. 18

12. **MC** 8% of 50 equals:

    A. 40 B. 4 C. 140 D. 14

13. **MC** 40% of 160 equals:

    A. 400 B. 6400 C. 640 D. 64

14. **MC** 30% of 60 equals:

    A. 180 B. 1.8 C. 400 D. 18

224 Jacaranda Maths Quest 7 Stage 4 NSW Syllabus Third Edition

Understanding

15. **WE37** Of the 300 Year 7 students selected to complete a questionnaire on ice cream, 70% said that their favourite flavour was chocolate chip. Calculate how many students favoured chocolate chip ice cream.

16. A thunder day at a given location is a calendar day on which thunder is heard at least once. About 20% of days near Darwin are thunder days. Calculate the number of thunder days you would expect in Darwin in one year.

17. Two per cent of Australians play lawn bowls. In a group of 50 people, state how many you would expect to play lawn bowls.

18. Two thousand people entered a marathon. Some walked and the rest jogged. If 20% walked, calculate:
    a. what percentage jogged
    b. how many people jogged.

19. William earns $550 per week. He has just received a pay rise of 4%.
    a. Calculate how much more William will earn per week.
    b. Calculate William's total pay after the pay rise.

20. The lotto jackpot for Saturday night is $20 million. If you win 8% of the jackpot, calculate how much money you would win.

21. The Australian cricket team was fined 15% of their match payments for a slow over rate. If the team was paid $80 000 for the match, calculate how much the team was fined.

22. Australia has 315 species of mammals. Of these, 15% are threatened. Calculate how many threatened species of mammals Australia has. Round your answer to the nearest whole number.

23. Two million people attended at least one AFL game last year. Of these people, 30% attended 10 games or more.
    Calculate how many people attended 10 or more AFL games.

24. The water content of a particular brand of shampoo is 35%. After performing a calculation, Peter claims that an 800 mL bottle of this particular shampoo contains 28 mL of water.
    a. Explain why Peter's calculation is incorrect.
    b. Calculate how much of the shampoo in the 800 mL bottle is actually water.

25. In Sydney, 14% of adults use a telephone to pay bills and 38% use the internet. If 50 adults need to pay bills, calculate how many will use the:
    a. telephone
    b. internet.

26. In Weburbia, 85% of all households have access to the internet. If there were 34 houses in Website Street, calculate how many you would expect to have access to the internet. Write the answer to the nearest whole number.

27. Terry runs a factory that makes parts for trucks. He has increased his staff by 12%. If Terry originally had 34 workers, calculate how many workers he has after the increase. Write the answer to the nearest whole number.

28. In the year 2020, 60% of households in Geelong had a smoke alarm. Of these smoke alarms, 94% worked. A street in Geelong has 20 houses.

   a. Calculate how many houses you would expect to have a smoke alarm.
   b. Calculate how many would have a smoke alarm that works. Write the answer to the nearest whole number.

29. In some parts of Africa, 6% of women die during pregnancy or childbirth. Of 50 women living in these parts who are pregnant, calculate how many are likely to die during pregnancy or childbirth.

30. For a radioactive substance, a half-life is the time it takes for half of the radioactive atoms in the substance to decay (lose their radioactivity). For 400 g of a given radioactive substance, the decay rate is 50% every 20 days.

   a. Calculate the quantity of radioactive substance left after 20 days, 40 days, 60 days, 80 days and 100 days.
   b. Calculate the percentage of the initial quantity remaining after 20 days, 40 days, 60 days, 80 days and 100 days.
   c. The quantity of the radioactive substance halves every 20 days. Explain why the percentage decrease is smaller after every 20 days.

31. Squeezy fruit drink is 36% pure orange juice and 64% water. If 1 litre of Squeezy fruit drink is mixed with 600 mL of water, determine the concentration of juice in the new mixture.

32. The students in Year 7A were voting for class captain. There were only two candidates. Rachel received 84% of the votes and Susi received 28% of the votes. It was then discovered that some students had voted twice. If the class consisted of 25 students, determine how many of them voted twice.

33. A department store stocked a selection of denim jeans. At the start of December, the store had 320 of these jeans. During December, it sold 50% of the jeans. In January, it sold 25% of the remaining denim jeans. Determine how many were left at the end of January.

LESSON
4.11 One amount as a percentage of another

LEARNING INTENTION

At the end of this lesson you should be able to:
• express an amount as a percentage of another amount.

▶ 4.11.1 Fractions as percentages

eles-6125

• A percentage can be represented as a fraction out of 100.
• There are two ways to change a fraction into a percentage:
  • Convert the fraction to an equivalent fraction with a denominator of 100. The numerator will read the percentage (because it is out of 100).

$$\overset{\times 25}{\underset{\times 25}{\frac{3}{4} = \frac{75}{100}}} = 75\%$$

- Multiply the fraction by 100, simplify and add the % sign.

$$\frac{3}{4} \times \frac{100}{1} = \frac{300}{4} = 75\%$$

Converting fractions into percentages

To convert a fraction into a percentage:
- Convert to an equivalent fraction with a denominator of 100. The numerator will be the percentage; or:
- multiply the fraction by 100, simplify and add the % sign.

WORKED EXAMPLE 38 Converting fractions to percentages

Write the following fractions as percentages:

a. $\dfrac{3}{5}$ b. $\dfrac{4}{25}$ c. $\dfrac{6}{40}$

THINK

WRITE

a. 1. The denominator can be multiplied by 20 to get an equivalent fraction out of 100. Therefore, multiply the numerator and denominator by 20.

    2. Write the numerator as a percentage.

a. $\dfrac{3}{5} \times \dfrac{20}{20} = \dfrac{60}{100}$

$= 60\%$

b. 1. The denominator can be multiplied by 4 to get an equivalent fraction out of 100. Therefore, multiply the numerator and denominator by 4.

    2. Write the numerator as a percentage.

b. $\dfrac{4}{25} \times \dfrac{4}{4} = \dfrac{16}{100}$

$= 16\%$

c. 1. The denominator does not multiply by a whole number to get 100. Multiply the fraction by 100.

    2. Simplify by dividing the numerator and denominator by 15%.

c. $\dfrac{6}{40} \times \dfrac{100}{1} = \dfrac{600}{40}$

$= 15\%$

⊙ ## 4.11.2 One amount as a percentage of another

eles-3815

- To express one amount as a percentage of another, write the amounts as a fraction $\dfrac{\text{amount}}{\text{total amount}}$, then convert the fraction to a percentage.
- When expressing one amount as a percentage of another, make sure that both amounts are in the same unit.

WORKED EXAMPLE 39 Expressing one amount as a percentage of another

Express:

a. **15 as a percentage of 20**

b. **9 as a percentage of 33. Write the answer as a mixed number.**

THINK	WRITE
a. **1.** Write the amount as a fraction of the total: $\dfrac{\text{amount}}{\text{total amount}}$.	**a.** $\dfrac{15}{20}$
2. Multiply by $\dfrac{100}{1}$ and include the % sign.	$= \dfrac{15}{20} \times \dfrac{100}{1}\%$
3. Simplify and then multiply.	$= \dfrac{15}{1} \times \dfrac{5}{1}\%$
	$= \dfrac{75}{1}\%$
4. State the answer.	$= 75\%$
b. **1.** Write the amount as a fraction of the total: $\dfrac{\text{amount}}{\text{total amount}}$.	**b.** $\dfrac{9}{33}$
2. Multiply by $\dfrac{100}{1}$ and include the % sign.	$= \dfrac{9}{33} \times \dfrac{100}{1}\%$
3. Simplify and then multiply.	$= \dfrac{3}{11} \times \dfrac{100}{1}\%$
	$= \dfrac{300}{11}\%$
4. Write as a mixed number by dividing the denominator into the numerator.	$= 27\dfrac{3}{11}\%$

WORKED EXAMPLE 40 Calculating percentages with different units

Express 45c as a percentage of $2.

THINK	WRITE
1. Write the larger amount using the smaller unit.	$\$2 = 200\,\text{cents}$
2. Write the first amount as a fraction of the second amount: $\dfrac{\text{amount}}{\text{total amount}}$	$\dfrac{45}{200}$
3. Multiply by $\dfrac{100}{1}$ and include the % sign.	$= \dfrac{45}{{}_2 200} \times \dfrac{\overset{1}{100}}{1}\%$
4. Simplify and then multiply.	$= \dfrac{45}{2} \times \dfrac{1}{1}\%$
	$= \dfrac{45}{2}\%$

5. Convert to a mixed number.

$$= 22\frac{1}{2}\%$$

6. Write the answer.

45c is $22\frac{1}{2}\%$ of $2.

WORKED EXAMPLE 41 Calculating percentages of test marks

Kye obtained 17 out of 30 on their Science test. Calculate their percentage score. Round your answer to the nearest whole number.

THINK	WRITE
1. Write the amount as a fraction of the total: $\dfrac{\text{amount}}{\text{total amount}}$.	$\dfrac{17}{30}$
2. Multiply by $\dfrac{100}{1}$ and include the % sign.	$= \dfrac{17}{30} \times \dfrac{100}{1}\%$
3. Simplify and then multiply.	$= \dfrac{17}{3} \times \dfrac{10}{1}\%$
	$= \dfrac{170}{3}\%$
4. Convert to a mixed number.	$= 56\frac{2}{3}\%$
5. Round to the nearest whole number. (If the fraction is a half or more, add one to the number.)	$\approx 57\%$
6. Write the answer as a sentence.	Kye obtained 57% for their Science test.

Digital technology

Scientific calculators can calculate one amount as a percentage of another. The answer could be either in fraction or mixed number format.

 Resources

 Interactivities   One amount as a percentage of another (int-3998)

4.11 Quick quiz on	**4.11 Exercise**

Individual pathways

■ PRACTISE	■ CONSOLIDATE	■ MASTER
1, 2, 5, 8, 12, 14, 21, 23, 26	3, 6, 9, 11, 15, 17, 19, 22, 24, 27, 28	4, 7, 10, 13, 16, 18, 20, 25, 29, 30

Fluency

1. **WE38** Write the following fractions as percentages:

   a. $\dfrac{1}{2}$ b. $\dfrac{3}{10}$ c. $\dfrac{8}{20}$ d. $\dfrac{7}{8}$ e. $\dfrac{6}{15}$

2. **WE39a** Express:

   a. 2 as a percentage of 10   b. 13 as a percentage of 52   c. 8 as a percentage of 80
   d. 8 as a percentage of 20   e. 15 as a percentage of 60.

3. Express:

   a. 120 as a percentage of 150   b. 8 as a percentage of 25   c. 12 as a percentage of 30
   d. 40 as a percentage of 200   e. 10 as a percentage of 200.

4. **WE39b** Express (writing the answers as mixed numbers):

   a. 180 as a percentage of 720   b. 63 as a percentage of 200   c. 25 as a percentage of 400
   d. 32 as a percentage of 500   e. 620 as a percentage of 3000.

5. Express (writing the answers as mixed numbers):

   a. 70 as a percentage of 80   b. 12 as a percentage of 42   c. 52 as a percentage of 78
   d. 11 as a percentage of 30   e. 4 as a percentage of 15.

6. Express (writing the answers as mixed numbers):

   a. 5 as a percentage of 70   b. 4 as a percentage of 18   c. 3 as a percentage of 7
   d. 64 as a percentage of 72   e. 15 as a percentage of 21.

7. Express (rounding all answers to the nearest whole number):

   a. 7 as a percentage of 42   b. 18 as a percentage of 54
   c. 95 as a percentage of 150   d. 7 as a percentage of 65.

8. Express (rounding all answers to the nearest whole number):

   a. 30 as a percentage of 36   b. 14 as a percentage of 18
   c. 80 as a percentage of 450   d. 8 as a percentage of 68.

9. **WE40** Express the first amount as a percentage of the second amount.

   a. 30c and $3   b. 200 m and 6 km   c. 5 mm and 4 cm

10. Express the first amount as a percentage of the second amount.

    a. 6 days and 3 weeks   b. 15 minutes and 1 hour   c. 25 mL and 2 L

11. **MC** 9 out of 36 as a percentage is:

    A. 25%   B. 9%   C. 400%   D. 27%

12. **MC** 14 as a percentage of 35 is:

    A. 14%   B. 40%   C. 20%   D. 35%

13. **MC** Alice got 42 out of 60 for her Mathematics test. Her mark as a percentage is:

    A. 58%   B. 60%   C. 30%   D. 70%

Understanding

14. **WE41** Ash Barty has won 7 out of her last 10 tennis matches. Calculate the number of wins in her last 10 matches as a percentage.

15. Three out of five people prefer chocolate ice cream to vanilla. Calculate what percentage of people prefer chocolate ice cream to vanilla.

16. Caillan's pocket money increased from $15 per week to $20 per week.

    a. Calculate how much Caillan's pocket money has increased by.
    b. Calculate this as a percentage increase.

17. Daniel earns $500 per week. He spends $50 on petrol, $70 on rent and $60 on food.

    a. Calculate the percentage Daniel spends on:

       i. petrol   ii. rent   iii. food.

    b. Calculate what percentage of his total wage Daniel spends on petrol, rent and food combined.
    c. Calculate the percentage Daniel has left.

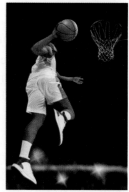

18. David shot 40 baskets from the free-throw line. Twenty-five of the 40 shots went in. Calculate the percentage of scored shots.

19. Jordan had a mass of 60 kg. After one month of not eating fast food, they lost 4 kg. Calculate their weight loss as a percentage.

20. After practising for one month, Tayla increased the speed of her tennis serve from 120 km/h to 140 km/h. Give exact answers for the following.

    a. Calculate how much her serve speed increased by.
    b. Calculate this as a percentage increase.

21. There are 28 students in a class. A survey showed that 12 walk home, 8 catch a bus and 8 ride in a car. Write the answers as mixed numbers.

    a. Calculate the percentage of students who walk.
    b. Calculate the percentage of students who catch a bus.
    c. Calculate the total percentage who walk or catch a bus.
    d. Calculate the percentage of students who do not walk.

22. During one AFL season, the GWS Giants won 16 out of 22 games. Calculate the percentage of games the Giants lost.

Communicating, reasoning and problem solving

23. A baker uses 200 g of flour for 500 g of dough in one recipe, and 250 g of flour for 500 g of dough in another recipe.

    a. Identify which recipe contains the higher percentage of flour. Explain your choice.
    b. Explain what happens when the quantity of flour is increased but the quantity of dough remains constant.

24. a. If you increase $100 by 25% and then decrease the new amount by 25%, will you end up with more than $100, less than $100 or exactly $100? Give reasons for your answer.
    b. Choose another amount and another percentage to test your reasoning.

25. My mother is four times older than I am. My sister is 75% of my age. She is also 10% of my grandfather's age. My father is 50, which is 2 years older than my mother. Calculate how old my sister and my grandfather are.

26. The teacher's mark book shows Karina has achieved the following test results in Maths.

Topic	Probability	Measurement	Algebra	Equations	Geometry
Score	$\dfrac{15}{20}$	$\dfrac{13}{15}$	$\dfrac{27}{30}$	$\dfrac{70}{80}$	$\dfrac{95}{100}$
Percentage					

a. Copy and complete the table to show Karina's percentage for each test. Round answers to the nearest whole number.

b. Compare the percentage results to determine which test Karina performed best in.

c. Calculate Karina's average percentage mark by adding up the percentages and dividing by 5, the number of tests.

27. Determine which of the following quantities represents:

a. the highest percentage

b. the lowest percentage

c. the same percentage.

200 g out of 500 g

300 g out of 450 g

420 g out of 840 g

350 g out of 700 g

28. Suggest why we need to be able to express one quantity as a percentage of another quantity.

29. A type of concrete uses 10 kg of cement for a quantity of 65 kg of concrete. Another type of concrete uses 10 kg of cement for 85 kg of concrete.

a. Explain which type of concrete contains the higher percentage of cement.

b. Explain what happens when the quantity of cement is constant but the quantity of concrete increases.

30. Roseanne is planning a tour of Australia. She calculates that 70% of the distance will be on sealed roads and the remainder on dirt roads. If there are 1356 km of dirt roads, determine how long Roseanne's journey will be.

LESSON
4.12 Common percentages and short cuts

4.12.1 Common percentages and short cuts

eles-3816

- Common percentages that are often used in everyday life include 10%, 20%, 25%, 50%, 75%.
- It is useful to remember the fraction equivalents of the commonly used percentages to make calculations involving them quicker and easier.
- The table below shows the fractional equivalents of some commonly used fractions and how to use these to speed up calculations.

Percentage	Fraction	How to calculate this percentage
1%	$\dfrac{1}{100}$	Divide by 100
5%	$\dfrac{1}{20}$	Divide by 20
10%	$\dfrac{1}{10}$	Divide by 10
$12\dfrac{1}{2}\%$	$\dfrac{1}{8}$	Divide by 8
20%	$\dfrac{1}{5}$	Divide by 5

Percentage	Fraction	How to calculate this percentage
25%	$\dfrac{1}{4}$	Divide by 4
$33\dfrac{1}{3}\%$	$\dfrac{1}{3}$	Divide by 3
50%	$\dfrac{1}{2}$	Divide by 2
$66\dfrac{2}{3}\%$	$\dfrac{2}{3}$	Divide by 3, then multiply by 2
75%	$\dfrac{3}{4}$	Divide by 4, then multiply by 3

- Calculating 10% without a calculator is easy. As it is equivalent to dividing by 10, the decimal place just needs to be moved 1 space to the left. Once you have found 10% of the amount, it is easy to calculate some other percentages. For example:
  - to calculate 5%, halve 10%
  - to calculate 20%, double 10%.

Short cut for calculating whole number percentages

Calculate 1% by dividing the amount by 100 (move the decimal point 2 spaces to the left), then multiply 1% by the whole number that you are looking for.

For example, to calculate 12%, calculate 1%, then multiply by 12.

WORKED EXAMPLE 42 Calculating 10% of an amount

Calculate 10% of each of the following.
a. $37   b. $120

THINK	WRITE
a. Write the question and move the position of the decimal point 1 place to the left for the answer. Remember that if there is no decimal point, put it at the end of the number. ($37 = $37.00)	a. 10% of $37 = $3.70
b. Write the question and move the position of the decimal point 1 place to the left for the answer.	b. 10% of $120 = $12

WORKED EXAMPLE 43 Calculating common percentages of an amount

Calculate:
a. 5% of $180   b. 20% of $7   c. 25% of $46

THINK	WRITE
a. 1. Calculate 10% of the amount.	a. 10% of $180 = $18
2. Calculate 5% by halving the amount obtained in step 1. Write the answer.	5% of $180 = $18 ÷ 2 = $9
b. 1. Calculate 10% of the amount.	b. 10% of $7 = $0.70
2. Calculate 20% by doubling the amount obtained in step 1. Write the answer.	20% of $7 = 2 × $0.70 = $1.40
c. 1. To calculate 25% of an amount divide it by 4.	c. 25% of $46 = $46 ÷ 4
2. Perform the division.	= $11.50
3. Write the answer.	25% of $46 = $11.50

WORKED EXAMPLE 44 Calculating percentages using short cuts

Calculate:
a. 12% of $53   b. 43% of $120
Round your answers to the nearest 5 cents.

THINK	WRITE
a. 1. Break up the percentage into lots of 10% and 1%.	a. 12% = 10% + 2 × 1%
2. Calculate 10% of the amount.	10% of $53 = $5.30
3. Calculate 1% of the amount and then double.	1% of $53 = $0.53 2% of $53 = 2 × $0.53 = $1.06
4. Determine 12% of the amount by adding 10% and 2 × 1%.	12% of 53% = $5.30 + $1.06 = $6.36
5. Round the answer to the nearest 5 cents.	= $6.35

Alternatively, you could use the short-cut method shown below.

1. First calculate 1% of $53 by moving the decimal point 2 places to the left.

$$1\% \text{ of } \$53 = \$0.53$$

2. Multiply by 12.

$$12\% \text{ of } \$53 = 12 \times 1\% \text{ of } \$53$$
$$= 12 \times \$0.53$$
$$= \$6.36$$

3. Round the answer to the nearest 5 cents.

$$= \$6.35$$

b. 1. Break up the percentage into lots of 10% and 1%.

b. $43\% = 4 \times 10\% + 3 \times 1\%$

2. Calculate 10% and 1% of the amount.

$$10\% \text{ of } \$120 = \$12$$
$$1\% \text{ of } \$120 = \$1.20$$

3. Use the values of 10% and 1% to calculate 43% of $120.

$$43\% \text{ of } \$120 = 4 \times \$12 + 3 \times \$1.20$$
$$= \$48 + \$3.60$$
$$= \$51.60$$

Alternatively, you could use the short-cut method shown below.

1. First calculate 1% of $120 by moving the decimal point 2 places to the left.

$$1\% \text{ of } \$120 = \$1.20$$

2. Multiply by 43 and write the answer.

$$43\% \text{ of } \$120 = 43 \times 1\% \text{ of } \$120$$
$$= 43 \times \$1.20$$
$$= \$51.60$$

 Resources

 Interactivities Common percentages and short cuts (int-3999)

Exercise 4.12 Common percentages and short cuts

learn on

| 4.12 Quick quiz on | 4.12 Exercise |

Individual pathways

■ PRACTISE	■ CONSOLIDATE	■ MASTER
1, 4, 7, 10, 13, 16, 19, 22, 24, 27, 30, 33	2, 5, 8, 11, 14, 17, 20, 23, 25, 28, 31, 34	3, 6, 9, 12, 15, 18, 21, 26, 29, 32, 35

Fluency

1. Calculate 10% of each of the following, rounding the answer to the nearest 5 cents.

   a. $10.00   b. $18.00   c. $81.00   d. $150.00   e. $93.00

2. Calculate 10% of each of the following, rounding the answer to the nearest 5 cents.

   a. $22.00   b. $17.20   c. $12.60   d. $1.50   e. $47.80

3. Calculate 10% of each of the following, rounding the answer to the nearest 5 cents.
   a. $192.40   b. $507.00   c. $1926.00   d. $3041.50   e. $1999.90

4. Calculate 10% of the following. Round your answers to the nearest 5 cents.
   a. $15   b. $51   c. $9   d. $137   e. $4.29

5. Calculate 10% of the following. Round your answers to the nearest 5 cents.
   a. $6.37   b. $39.17   c. $74.81   d. $102.75   e. $67.87

6. Calculate 10% of the following. Round your answers to the nearest 5 cents.
   a. $517.83   b. $304.77   c. $628.41   d. $100.37   e. $207.08

7. **WE43** Calculate 5% of each of the following. Round the answers to the nearest 5 cents.
   a. $8.20   b. $6.40   c. $2.20   d. $140.20   e. $42.40

8. Calculate 5% of each of the following. Round the answers to the nearest 5 cents.
   a. $10.60   b. $304.80   c. $1000   d. $103.27   e. $31.70

9. Calculate 20% of the following. Round the answers to the nearest 5 cents.
   a. $21.50   b. $42.30   c. $3.30   d. $74.10   e. $0.79

10. Calculate 20% of the following. Round the answers to the nearest 5 cents.
    a. $16.40   b. $135.80   c. $261.70   d. $1237   e. $5069

11. Calculate the following. Round the answers to the nearest 5 cents.
    a. 15% of $12   b. 15% of $8.00   c. 15% of $60.00

12. Calculate the following. Round the answers to the nearest 5 cents.
    a. 25% of $30.00   b. 25% of $90.00   c. 25% of $220.00

13. Calculate the following. Round the answers to the nearest 5 cents.
    a. 30% of $15.00   b. 30% of $25.00   c. 30% of $47.50

14. Calculate 1% of the following. Round the answers to the nearest 5 cents.
    a. $268   b. $713   c. $604   d. $5.60   e. $13

15. Calculate 1% of the following. Round the answers to the nearest 5 cents.
    a. $14.80   b. $81.75   c. $19.89   d. $4.25   e. $6.49

16. **WE44** Calculate the following. Round the answers to the nearest 5 cents.
    a. 12% of $11   b. 21% of $50   c. 3% of $22   d. 6% of $40   e. 13% of $14

17. Calculate the following. Round the answers to the nearest 5 cents.
    a. 35% of $210   b. 9% of $17   c. 2% of $53   d. 7% of $29   e. 45% of $71.50

18. Calculate the following. Round the answers to the nearest 5 cents.
    a. 33% of $14.50   b. 42% of $3.80   c. 31% of $1.45   d. 64% of $22.50   e. 41% of $1200

Understanding

19. **MC** 10% of $7.25 equals:
    A. $725   B. $7.30   C. $72.50   D. $0.73

20. **MC** 1% of $31.48 equals:
    A. $3.14   B. $0.31   C. $0.32   D. $31.50

21. **MC** 15% of $124 equals:

   A. $12.40   **B.** $1.24   **C.** $6.20   **D.** $18.60

22. **MC** 22% of $5050 equals:

   A. $60.60   **B.** $50.50   **C.** $1111   **D.** $43.56

23. Maria is buying a new set of golf clubs. The clubs are marked at $950, but if Maria pays cash, the shop will take 10% off the marked price. Determine how much the clubs will cost if Maria pays cash.

24. Thirty per cent of residents in a district are over the age of 65. If there are 180 000 residents, determine how many are over the age of 65.

25. Jay is buying a new lounge suite worth $2150. Jay has to leave a 15% deposit and then pay the balance in monthly instalments.

   a. Calculate the deposit Jay needs to pay.
   b. Determine how much Jay will have to pay each month if he plans to pay the balance off in one year.

26. Ninety per cent of students at a school were present for school photographs. If the school has 1100 students, calculate how many were absent on the day that the photographs were taken.

27. Jim can swim 50 m in 31 seconds. If he improves his time by 10%, calculate Jim's new 50 m time.

28. In a survey, 40 people were asked if they liked or disliked Vegemite. Of the people surveyed, 5% said they disliked Vegemite. Calculate how many people:

   a. disliked Vegemite   **b.** liked Vegemite.

29. 35 800 went to the SCG to watch a Swans versus Giants football match. Of the crowd, 30% went to the game by car and 55% caught public transport. Calculate how many people:

   a. arrived by car
   b. caught public transport
   c. did not arrive at the game by car or public transport.

Communicating, reasoning and problem solving

30. When I am 5% older than I am now, I will be 21 years old. How old am I now?

31. The price of bread is now 250% of what it cost 20 years ago, which is an increase of 150%. If a loaf of bread costs $2.00 now, determine how much it cost 20 years ago.

32. I am six months old. If I gain 10% of my current mass I will be three times my birth mass. If my birth mass was 3 kg, determine my current mass. Round your answer to 1 decimal place.

33. I am 33 years old. I have lived in England for 8 years. If I stay in England, determine how old I will be when the number of years I have lived here is 50% of my age. Show how you reached your answer.

34. Kate is a carpenter and has made a cutting board as shown.

15 cm

20 cm

a. One of her customers asked Kate to make her a similar cutting board but to double its width. Calculate the percentage increase of the width of the board.

b. A second customer asked Kate to make her a cutting board with dimensions 30 cm by 30 cm. Calculate:

   i. the percentage increase of the width of the board
   ii. the percentage increase of the length of the board
   iii. the percentage increase of the perimeter of the board.

c. Explain whether there is a connection between the percentage increase of a side and the percentage increase of the perimeter.

35. Ten per cent of the messages received by a business were from customers dissatisfied with the product bought. Of these customers, 32% were male.

a. Calculate the overall percentage of dissatisfied male customers.
b. Calculate the overall percentage of dissatisfied female customers.
c. Explain what happens when you apply a percentage to another percentage.

LESSON
4.13 Review

4.13.1 Topic summary

Working with fractions

- When working with fractions, mixed numbers provide the user with a good understanding of the value.
- However, when computing operations with fractions (addition, subtraction, multiplication and division), it is best to have all fractions expressed as proper fractions.

Types of fractions

- Fractions are used to represent parts of a whole.
- **Proper fractions** have a numerator that is less than the denominator, e.g. $\frac{2}{7}$.
- **Improper fractions** have a numerator that is greater than the denominator, e.g. $\frac{10}{3}$.
- A **mixed number** consists of an integer and a proper fraction, e.g. $2\frac{1}{4}$.

Finding a percentage of a quantity

- Finding a percentage of a number or a quantity is the same as finding an equivalent fraction of that number or quantity.
- Remember that in mathematics the word 'of' means multiply.

e.g. Determine 15% of $80.

$$15\% \text{ of } \$80 = \frac{15}{100} \times 80$$
$$= \frac{15}{100} \times \frac{80}{1}$$
$$= \frac{120}{10}$$
$$= 12$$

Simplifying fractions

- Fractions can be simplified by dividing the numerator and denominator by the same number.
- A fraction is in its simplest form when there are no factors (other than 1) that are common to both the numerator and denominator.

e.g. Simplify $\frac{80}{200}$.

Divide the numerator and denominator by 10.

$$\frac{80^{\div 10}}{200^{\div 10}} = \frac{8}{20}$$

There are still numbers that divide into both 8 and 20.

Divide the numerator and denominator by 4.

$$\frac{80^{\div 10}}{200^{\div 10}} = \frac{8^{\div 4}}{20^{\div 4}} = \frac{2}{5}$$

FRACTIONS AND PERCENTAGES

Multiplying fractions

To multiply fractions together, follow the three steps below:
1. Multiply the numerators.
2. Multiply the denominators.
3. Simplify the fraction if possible.

e.g. Evaluate $\frac{1}{2} \times \frac{4}{9}$.

$$\frac{1}{2} \times \frac{4}{9} = \frac{1 \times 4}{2 \times 9} = \frac{4}{18} = \frac{2}{9}$$

Dividing fractions

To divide two fractions, follow the KEEP, CHANGE, FLIP method outlined below and then multiply the fractions together.
1. **KEEP** the fraction to the left of the ÷ symbol the same.
2. **CHANGE** the division sign to a multiplication sign.
3. **FLIP** (invert) the fraction to the right of the symbol.

e.g. Evaluate $\frac{3}{4} \div \frac{2}{7}$.

KEEP CHANGE FLIP

$$\frac{3}{4} \qquad \times \qquad \frac{7}{2}$$

$$\frac{3}{4} \div \frac{2}{7} = \frac{3}{4} \times \frac{7}{2} = \frac{3 \times 7}{4 \times 2} = \frac{21}{8}$$

Addition and subtraction of fractions

To add or subtract fractions, follow the steps below:
1. Examine the denominators of all fractions and determine the lowest common multiple (LCM).
2. Use the method of equivalent fractions to change each fraction so that the denominators are all the same.
3. Depending on the symbol in the question, add or subtract the numerators leaving the denominator unchanged.

e.g. Evaluate $\frac{2}{3} + \frac{7}{4}$.

LCM = 12

$$\frac{2}{3} + \frac{7}{4} = \frac{2}{3} \times \frac{4}{4} + \frac{7}{4} \times \frac{3}{3} = \frac{8}{12} + \frac{21}{12}$$
$$= \frac{8 + 21}{12} = \frac{29}{12}$$

Expressing a percentage as a fraction

- To convert a percentage to a fraction, simply divide by 100 and remove the percentage symbol.

e.g. Express 80% as a common fraction.

$$80\% = 80 \div 100$$
$$= \frac{80^{\div 20}}{100^{\div 20}} = \frac{4}{5}$$
$$= 0.54$$

Equivalent fractions

- Equivalent fractions are fractions that have the same value but look different.

e.g. $\frac{1}{2}$, $\frac{2}{4}$ and $\frac{50}{100}$ are equivalent fractions. The value of each fraction is 0.5.

- Equivalent fractions can be created by multiplying both the numerator and denominator by the same number.

e.g.

$$\frac{1}{2} \times \frac{2}{2} = \frac{2}{4}$$
$$\frac{1}{2} \times \frac{50}{50} = \frac{50}{100}$$

Converting fractions to percentages

- To express a fraction as a percentage, simply multiply by 100 and include the percentage symbol.

e.g. Express $\frac{9}{20}$ as a percentage.

$$\frac{9}{20} \times 100\% = \frac{9}{20} \times \frac{100}{1}\%$$
$$= \frac{90}{2}\%$$
$$= 45\%$$

4.13.2 Project

The Memory game

The game Memory is played with a standard deck of playing cards. All cards are placed face-side down on a table and each player takes turns to reveal 2 cards. If the 2 cards contain the same number or suit, they are a pair and the player gets another turn. If they are not a pair, they are returned to their original spot and turned face-down. The object of the game is to remember where the same cards are and collect as many pairs as you can. The winner is the person with the most pairs.

Your task is to develop a version of Memory with a fractions flavour. Each of your cards will have a fraction written on it and its pair will be an equivalent fraction.

1. As an introduction to the game, match the pairs of equivalent fractions on the cards below.

$$\frac{6}{8} \quad \frac{3}{9} \quad \frac{1}{3} \quad \frac{3}{4} \quad \frac{4}{8} \quad \frac{1}{2}$$

    A   B   C   D   E   F

The requirements for your game are as follows.
- There are 15 pairs of equivalent fractions in this game — that makes 30 cards.
- One of the fractions of each pair must be in simplest form.
- For example, $\frac{1}{3}$ and $\frac{4}{12}$ are an acceptable pair, but $\frac{10}{30}$ and $\frac{4}{12}$ are not.
- Fractions don't have to be proper fractions. Equivalent mixed numbers and improper fractions are allowed.

2. Using the information in step 1, write down the 15 pairs of fractions that will be a part of your memory game.

When the game is packaged for sale, additional information accompanies the cards. This information includes a list of the fractions enclosed and instructions for the game. The list of fractions is given in order from smallest to largest.

3. List your fraction pairs in order from smallest to largest.

When items are placed for sale, presentation of the packaging is important. Eye-catching designs will attract attention, and mistakes must be avoided at all costs. You are now the designer for a company that is going to market the new game.

4. In your workbook or on a blank piece of paper, draw two rectangles the size of playing cards. In the first rectangle, design the back of your cards, and in the second rectangle, design how the fractions will appear on the face of the cards.

The playing instructions are very important pieces of information in games. They explain the number of players allowed, how the game is to be played and who wins.

5. Taking into consideration the information that has been provided for you throughout this task, write a set of instructions that show how your Memory game is played.

6. Now construct your own game cards and present them in a suitable package. Remember, presentation of the game is important. When you have finished, try the game out with your friends.

Exercise 4.13 Review questions

learn on

Fluency

1. State the fraction of the eggs in this carton that are white.

2. Write the sequence of the first three equivalent fractions for the following.
   a. $\dfrac{3}{8}$
   b. $\dfrac{4}{9}$
   c. $\dfrac{5}{11}$
   d. $\dfrac{6}{7}$

3. Fill the gaps to make equivalent fractions.
   a. $\dfrac{1}{2} = \dfrac{}{4} = \dfrac{3}{}$
   b. $\dfrac{1}{3} = \dfrac{3}{} = \dfrac{}{15}$
   c. $\dfrac{3}{4} = \dfrac{9}{} = \dfrac{}{28}$
   d. $\dfrac{2}{5} = \dfrac{}{10} = \dfrac{16}{}$

4. Ricky, Leo and Gabriel were the highest goal scorers for their local football team throughout the season. Ricky scored $\dfrac{2}{7}$ of the team's goals, Leo scored $\dfrac{1}{3}$ of the goals and Gabriel scored $\dfrac{1}{4}$ of the goals.

   a. Determine who scored most of the goals.
   b. State who scored more goals: Ricky or Gabriel.

5. Simplify these fractions.
   a. $\dfrac{42}{49}$
   b. $\dfrac{81}{108}$
   c. $\dfrac{18}{20}$
   d. $\dfrac{21}{28}$

6. Simplify these fractions.
   a. $\dfrac{50}{150}$
   b. $\dfrac{75}{125}$
   c. $\dfrac{2}{80}$
   d. $\dfrac{36}{72}$

▶

7. Simplify the following mixed numbers.

a. $6\dfrac{8}{10}$ b. $3\dfrac{6}{9}$ c. $4\dfrac{15}{20}$ d. $1\dfrac{20}{100}$

8. Determine what fraction of months of the year begin with the letter J. Express your answer in simplest form.

9. Convert these improper fractions to mixed numbers.

a. $\dfrac{16}{3}$ b. $\dfrac{21}{5}$ c. $\dfrac{12}{7}$ d. $\dfrac{11}{2}$

10. Convert these improper fractions to mixed numbers.

a. $\dfrac{5}{4}$ b. $\dfrac{80}{7}$ c. $\dfrac{55}{9}$ d. $\dfrac{72}{10}$

11. Express these mixed numbers as improper fractions.

a. $2\dfrac{3}{4}$ b. $9\dfrac{7}{8}$ c. $3\dfrac{5}{7}$ d. $5\dfrac{5}{6}$

12. Express these mixed numbers as improper fractions.

a. $3\dfrac{2}{3}$ b. $9\dfrac{9}{10}$ c. $3\dfrac{11}{12}$ d. $8\dfrac{1}{2}$

13. Marcella was in charge of baking and selling lemon slice at the school fete. Each tray she baked contained 15 pieces of lemon slice. If Marcella sold 68 slices, determine how many empty trays there were.

14. Evaluate:

a. $\dfrac{5}{6}+\dfrac{1}{8}$ b. $\dfrac{7}{9}-\dfrac{2}{5}$ c. $\dfrac{12}{5}-\dfrac{4}{3}$

15. Evaluate:

a. $\dfrac{4}{7}\times\dfrac{5}{9}$ b. $\dfrac{2}{5}\times\dfrac{3}{5}$ c. $\dfrac{25}{36}\times\dfrac{18}{20}$ d. $\dfrac{56}{81}\times\dfrac{18}{64}$

16. Evaluate:

a. $\dfrac{4}{5}$ of 55   b. $\dfrac{2}{11}$ of 121   c. $\dfrac{7}{9}$ of $\dfrac{18}{84}$ d. $\dfrac{1}{12}$ of 200

17. Write the reciprocals of:

a. $\dfrac{5}{2}$ b. $\dfrac{1}{9}$ c. $\dfrac{1}{4}$ d. $5\dfrac{3}{4}$

18. Evaluate:

a. $\dfrac{4}{5}\div\dfrac{8}{15}$ b. $\dfrac{9}{10}\div\dfrac{27}{40}$ c. $\dfrac{7}{9}\div 4$ d. $\dfrac{3}{7}\div 10$

Understanding

19. Evaluate:

    a. $1\dfrac{7}{8} + 2\dfrac{5}{6}$

    b. $10\dfrac{3}{7} - 8\dfrac{1}{2}$

    c. $6\dfrac{1}{3} - 2\dfrac{5}{9}$

    d. $5\dfrac{1}{2} - 2\dfrac{5}{6}$

20. Evaluate:

    a. $8\dfrac{2}{9} - 4\dfrac{5}{8}$

    b. $12\dfrac{6}{11} - 11\dfrac{7}{9}$

    c. $1\dfrac{1}{5} + 3\dfrac{1}{3} - \dfrac{13}{15}$

    d. $4\dfrac{7}{9} - 1\dfrac{5}{6} + 3\dfrac{1}{3}$

21. Evaluate:

    a. $10\dfrac{2}{7} \times 6\dfrac{2}{9}$

    b. $\dfrac{4}{9} \times 2\dfrac{1}{4} \times \dfrac{3}{5}$

    c. $3\dfrac{5}{12} \times 1\dfrac{3}{5} \times 2\dfrac{1}{2}$

    d. $\dfrac{6}{17} \times 1\dfrac{1}{4} \times 3\dfrac{2}{5}$

22. Evaluate:

    a. $1\dfrac{1}{2} \div 4\dfrac{4}{5}$

    b. $6\dfrac{2}{3} \div 4\dfrac{1}{6}$

    c. $4\dfrac{1}{5} \div 2\dfrac{7}{8}$

    d. $10\dfrac{1}{3} \div 12\dfrac{5}{7}$

23. Write the following percentages as fractions in simplest form.

    a. 13%

    b. 70%

    c. 26%

    d. 132%

24. Change the following fractions to percentages.

    a. $\dfrac{15}{100}$

    b. $\dfrac{11}{20}$

    c. $\dfrac{3}{5}$

    d. $\dfrac{18}{36}$

25. Change the following fractions to percentages.

    a. $\dfrac{30}{80}$

    b. $\dfrac{8}{15}$

    c. $\dfrac{5}{6}$

    d. $\dfrac{4}{11}$

26. Calculate the following percentages using fractions.

    a. 5% of 120

    b. 60% of 20

    c. 75% of 52

    d. 130% of 30

27. Calculate the following percentages using fractions.

    a. 14% of 45

    b. 7% of 530

    c. 32% of 15

    d. 17% of 80

28. Express:

    a. 30 as a percentage of 50

    b. 18 as a percentage of 60

    c. 32 as a percentage of 80

    d. 28 as a percentage of 70.

29. Express:

    a. 4 days as a percentage of 8 weeks

    b. 20 minutes as a percentage of 3 hours.

30. Calculate 5% of the following by finding 10% and halving your answer. Round your answer to the nearest 5 cents.

    a. $8.00

    b. $21.00

    c. $64.00

    d. $104.00

31. Calculate the following using short cuts. Round your answer to the nearest 5 cents.

    a. 1% of $16.00

    b. 1% of $28.00

    c. 12% of $42.00

    d. 30% of $90.00

Communicating, reasoning and problem solving

32. Year 7 students at Camberwell Secondary College raised a total of $1600 for the Royal Children's Hospital Good Friday Appeal. The table below shows how much money each class raised.

Class	7.1	7.2	7.3	7.4	7.5	7.6
Amount raised ($)	$200	$400	$250	$250	$280	$220

Express as a simple fraction how much of the total each class raised.

33. Anthony's monthly take-home pay is $2800. He spends a quarter on his home loan payments, one-half on food and drink, and one-seventh on clothing. One-half of the remainder goes into his savings account. Calculate how much money Anthony saves each month.

34. On a balance scale, a brick exactly balances $\frac{2}{3}$ of a brick and a $\frac{2}{3}$ kg weight. How heavy is the brick?

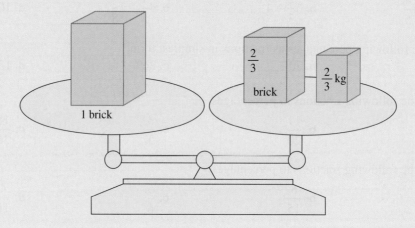

35. Complete the following:

a. $\dfrac{5}{12} = \dfrac{1}{4} + \,?$ b. $\dfrac{5}{12} = \dfrac{2}{3} - \,?$ c. $\dfrac{5}{12} = \dfrac{3}{4} \times \,?$ d. $\dfrac{5}{12} = \dfrac{4}{5} \div \,?$

36. Explain how you can determine a fraction whose numerator and denominator have a product of 252, and that is equivalent to $\frac{4}{7}$.

37. Tomas buys a 24-piece pizza and eats 8 pieces. Drew buys a 16-piece pizza of the same size, and eats 5 pieces. Identify who ate the most pizza and explain your choice.

38. An Egyptian fraction is one that can be written as the sum of a set of different unit fractions: for example, $\dfrac{3}{4} = \dfrac{1}{2} + \dfrac{1}{4}$. (*Note:* The numerator of each unit fraction is 1.) Write $\dfrac{4}{5}$ as the sum of a set of different unit fractions.

39. Mr Thompson earned $2400 per month. Recently, he had a pay increase. He now earns $1\frac{1}{4}$ times his old salary. Calculate his new salary.

Mr Thompson's colleague, Mr Goody, earned $3000 per month and had a pay cut. He now gets $\frac{1}{6}$ less of his old salary. Calculate Mr Goody's new salary.

40. A shopping centre has 250 parking spaces. Of these, $\dfrac{3}{10}$ are suitable for small cars only. The remaining

    spaces are for standard cars. Last Friday small cars took up $\dfrac{14}{15}$ of their allocated spots, but overall the

    car park was $\dfrac{7}{10}$ full.

    Determine the fraction of standard car spaces that were filled.

41. Byron needs $\dfrac{2}{5}$ kg of catalyst mixed with $4\dfrac{1}{2}$ kg of resin to create one surfboard.

    a. If he has $2\dfrac{1}{4}$ kg of catalyst and $20\dfrac{1}{4}$ kg of resin, how many whole surfboards can Byron build?

    b. Calculate how much more catalyst and resin Byron needs to build one more surfboard. Explain your
       reasoning.

42. Calculate the percentage that remains if:
    a. 38% of the winnings have been spent
    b. all stock was discounted 25%.

43. Eighty per cent of a Year 7 Maths class got question 1 correct on their percentages test. If there are
    30 students in the class, calculate how many got the question correct.

44. Forty per cent of primary-school children can sing the national anthem. In a group of 675
    primary-school children, calculate how many you would expect to be able to sing the national anthem.

45. Lake Eyre in South Australia has an annual rainfall of
    around 100 mm. If 8 mm fell in one week, give this as a
    percentage of the annual rainfall.

46. A tennis player agreed to pay her coach 8% of her winnings
    from tournaments. If she wins $145 000 in a tournament,
    calculate how much her coach receives.

47. At Queens High School, there are 900 students.
    45% walk to school.
    55% don't walk to school.
    12% of students wear glasses.

    Identify which of the following statements is definitely
    false. (There may be more than one answer.)
    a. There are 100 students who don't walk to school and
       who wear glasses.
    b. 10% of the students who don't walk to school and 2%
       of the students who do walk to school wear glasses.
    c. 20% of the students who walk to school wear
       glasses.
    d. More than 22% of the students who don't walk to
       school wear glasses.

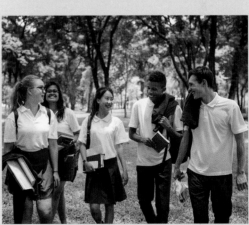

48. The label on a 500 g pack of butter states that it contains 81% fat. A milk bottle label says that the milk contains 3 grams of fat per 100 mL. Determine the number of litres of this milk that contains the same amount of fat as there is in the butter.

49. In a recent survey of house occupancy in a town, it was discovered that 40% of the houses contained 2 or more people. Of those houses that contain only 1 person, 25% of them contained a male. Of all the houses in a town, calculate the percentage that contains only 1 female and no males.

50. A television set has a ticketed price of $2500. A 10% discount is allowed on the set.
    a. Calculate the negotiated price of the television set.
    b. If a further 10% is negotiated off the price of the set, calculate the price paid.
    c. Calculate the total discount as a percentage of the original ticketed price.
    d. Explain why two successive 10% discounts do not equal a 20% discount.

51. Joseph wants to spend $2500 on a new barbecue. Since he works at a BBQ store he can get a staff discount of 10%. He can also get a discount of 30% in the stocktake sale.
    Joseph wants to know whether he should ask the salesman to apply his staff discount or the sale discount first to achieve the cheapest price.
    a. Calculate how much the barbecue will cost if the staff discount is taken off first and then the sale discount is taken off.
    b. Calculate how much it will cost if done the opposite way.
    c. Decide which method Joseph should select to get the best price.

52. Two triangles have the same area. Show that if one triangle has a 25% wider base than the other, then it must be 20% shorter (that is, its height must be 20% less).

To test your understanding and knowledge of this topic, go to your learnON title at www.jacplus.com.au and complete the **post-test**.

Answers

Topic 4 Fractions and percentages

4.1 Pre-test

1. B

2. a $\dfrac{3}{8}$ b $\dfrac{5}{8}$

3. $5\dfrac{3}{4}$

4. $\dfrac{3}{4}$

5. 100 minutes

6. A

7. 36

8. 187 cm

9. $\dfrac{127}{200}$

10. C

11. a. $\dfrac{3}{25}$ b. 0.12

12. 11 036

13. 75%

14. D

15. $15.58

4.2 What are fractions?

1. a. $\dfrac{3}{4}$ b. $\dfrac{15}{16}$ c. $\dfrac{7}{8}$

2. $\dfrac{1}{3}$

3. $\dfrac{3}{4}$

4. a. $\dfrac{3}{11}$ b. $\dfrac{2}{11}$

5. $\dfrac{3}{11}$

6. $\dfrac{17}{60}$ of an hour

7. a. 3   b. 3   c. 4   d. 5

8. a. 6   b. 9   c. 11   d. 12

9. $\dfrac{20}{30}, \dfrac{14}{21}, \dfrac{40}{60}$

10. $\dfrac{12}{15}, \dfrac{36}{45}, \dfrac{16}{20}, \dfrac{28}{35}, \dfrac{80}{100}$

11. $\dfrac{35}{50}, \dfrac{21}{30}, \dfrac{14}{20}, \dfrac{140}{200}$

12. $\dfrac{5}{6} = \dfrac{10}{12} = \dfrac{15}{18} = \dfrac{20}{24}$

13. $\dfrac{7}{8} = \dfrac{14}{16} = \dfrac{21}{24} = \dfrac{28}{32}$

14.

```
◄─┬──┬────────┬──┬──────┬───┬───┬─►
  0  1/5      4/5  1     7/5 1 3/5  2
```

15.

```
◄─┬──┬────┬──┬──────┬──────┬──┬─►
  0  1/3  3/3 4/3   2      8/3 3
```

16. a. $\dfrac{3}{5}$ b. $\dfrac{7}{8}$ c. $\dfrac{1}{5}$ d. $\dfrac{1}{8}$

17. a. $\dfrac{3}{4}$ b. $\dfrac{5}{8}$ c. $\dfrac{5}{8}$ d. $\dfrac{4}{5}$

18. a. >   b. >   c. >   d. <

19. D

20. B

21. $\dfrac{9}{24}, \dfrac{12}{24}, \dfrac{8}{24}; \dfrac{1}{3}, \dfrac{3}{8}, \dfrac{1}{2}$

22. $\dfrac{45}{60}, \dfrac{40}{60}, \dfrac{28}{60}; \dfrac{7}{15}, \dfrac{2}{3}, \dfrac{3}{4}$

23. $\dfrac{30}{40} = \dfrac{18}{24} = \dfrac{27}{36}$

24. $\dfrac{15}{18} = \dfrac{35}{42} = \dfrac{40}{48}$

25. 7 does not divide into 2 evenly, so it would be difficult to mark the equal parts. It would be more appropriate to use a scale of 3.5 cm or 7 cm per whole number.

26. As the denominator doubles, the distance between fractions halves, because doubling the denominator is the same as dividing the fraction by 2.

27. a.

Sally   Hanna   Sonya   Vanessa

b. Vanessa

c. They ate the same amount.

28. Lee Yin

29. $\dfrac{19}{60}, \dfrac{1}{3}$

30.

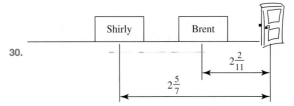

4.3 Simplifying fractions

1. a. $\dfrac{1}{2}$ b. $\dfrac{2}{3}$ c. $\dfrac{7}{8}$ d. $\dfrac{4}{5}$

2. a. $\dfrac{5}{6}$ b. $\dfrac{24}{25}$ c. $\dfrac{7}{8}$ d. $\dfrac{7}{10}$

3. a. $\dfrac{11}{12}$ b. $\dfrac{11}{25}$ c. $\dfrac{3}{8}$ d. $\dfrac{7}{8}$

4. a. $4\dfrac{3}{5}$ b. $7\dfrac{5}{6}$ c. $10\dfrac{1}{2}$ d. $1\dfrac{11}{12}$

5. a. $6\dfrac{3}{4}$ b. $5\dfrac{1}{3}$ c. $3\dfrac{1}{5}$ d. $2\dfrac{1}{4}$

6. D

7. $\dfrac{21}{28} = \dfrac{3}{4}$

8. a. $\dfrac{20}{36} = \dfrac{5}{9}$ b. $\dfrac{16}{36} = \dfrac{4}{9}$

9. 7A: $\dfrac{1}{5}$, 7B: $\dfrac{3}{20}$, 7C: $\dfrac{8}{25}$, 7D: $\dfrac{2}{25}$, 7E: $\dfrac{1}{4}$

10. a. 40

   b. Mark: $\dfrac{5}{40} = \dfrac{1}{8}$, Ken: $\dfrac{10}{40} = \dfrac{1}{4}$, Samantha: $\dfrac{8}{40} = \dfrac{1}{5}$,

   Jules: $\dfrac{6}{40} = \dfrac{3}{20}$, Ahmed: $\dfrac{4}{40} = \dfrac{1}{10}$, Darren: $\dfrac{7}{40}$

11. Newcastle to Sydney is more reliable.

12. a.

   b. $\dfrac{1}{4}$

   c.

   d. $\dfrac{1}{8}$

   e. Because two smallest triangles form one middle size triangle

13. The denominator could be any multiple of 13 because the fraction is not written in its simplest form and 13 is a prime number.

14. a. Six pizzas
   b. Two quarters (one half) of a pizza are left over.

15. a. $\dfrac{7293}{14\,586}$, $\dfrac{6792}{13\,584}$ b. $\dfrac{5823}{17\,469}$

16. $\dfrac{2}{5}$

4.4 Mixed numbers and improper fractions

1. a. $2\dfrac{1}{2}$

   b. $1\dfrac{1}{3}$

   c. $1\dfrac{5}{8}$

2. $\dfrac{3}{2} = 1\dfrac{1}{2}$

3. $\dfrac{9}{4} = 2\dfrac{1}{4}$

4. a. $1\dfrac{2}{5}$ b. $2\dfrac{3}{4}$ c. $10\dfrac{1}{2}$ d. $4\dfrac{1}{4}$ e. $8\dfrac{4}{11}$

5. a. $2\dfrac{3}{13}$ b. $33\dfrac{1}{3}$ c. $12\dfrac{1}{2}$ d. $6\dfrac{2}{3}$ e. $1\dfrac{3}{5}$

6. a. $3\dfrac{3}{8}$ b. $5\dfrac{1}{2}$ c. $8\dfrac{2}{7}$ d. $11\dfrac{7}{10}$ e. $8\dfrac{3}{8}$

7. C

8. B

9. a. $\dfrac{5}{3}$

   b. $\dfrac{13}{4}$

   c. $\dfrac{13}{5}$

10. a. $\dfrac{9}{8}$ b. $\dfrac{15}{4}$ c. $\dfrac{13}{6}$ d. $\dfrac{8}{7}$ e. $\dfrac{9}{5}$

11. a. $\dfrac{10}{3}$ b. $\dfrac{25}{4}$ c. $\dfrac{37}{9}$ d. $\dfrac{23}{2}$ e. $\dfrac{39}{5}$

12. a. $\dfrac{33}{5}$ b. $\dfrac{47}{5}$ c. $\dfrac{7}{2}$ d. $\dfrac{21}{5}$ e. $\dfrac{47}{7}$

13. a. $\dfrac{29}{10}$ b. $\dfrac{43}{12}$ c. $\dfrac{67}{7}$ d. $\dfrac{21}{11}$ e. $\dfrac{53}{6}$

14. D

15. D

16. A

17. Kim: $2\dfrac{1}{4}$, Carly: $2\dfrac{3}{4}$

18. $6\dfrac{1}{2}$

19. $4\dfrac{3}{4}$

20. a. >   b. >   c. <   d. >

21. a. <   b. <   c. <   d. <

22. $10\dfrac{1}{2}$

23. a. i. $\dfrac{23}{7}$ ii. $\dfrac{23}{4}$

b.

7 parts in a whole

7 parts in a whole

7 parts in a whole

} 21 parts

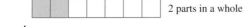
2 parts in a whole

24. $\dfrac{4}{5}$

25. a. 25   b. 77

26. a. $\dfrac{14}{3}$ or $4\dfrac{2}{3}$km

b. $\dfrac{56}{3}$ or $18\dfrac{2}{3}$km

c. $\dfrac{70}{3}$ or $23\dfrac{1}{3}$km

27. $1\dfrac{1}{4},2\dfrac{1}{12},2\dfrac{1}{6},3\dfrac{1}{8},3\dfrac{1}{3}$

28. To write an improper fraction as a mixed number we divide the denominator into the numerator and write the remainder over the same denominator. Alternatively, we can draw circles to determine how many whole numbers we have, and the parts left over will be the fraction part of the mixed number.

4.5 Adding and subtracting fractions

1. a. $\dfrac{4}{5}$ b. $\dfrac{6}{11}$ c. $\dfrac{5}{6}$ d. $\dfrac{31}{50}$

2. a. $\dfrac{1}{2}$ b. $\dfrac{1}{4}$ c. $\dfrac{1}{3}$ d. $\dfrac{1}{2}$

3. a. $\dfrac{9}{25}$ b. $\dfrac{11}{12}$ c. $\dfrac{7}{8}$ d. $\dfrac{6}{7}$

4. a. 1   b. $\dfrac{3}{4}$ c. $\dfrac{3}{4}$ d. $\dfrac{8}{9}$

5. a. $\dfrac{3}{4}$ b. $\dfrac{1}{4}$ c. $\dfrac{3}{7}$ d. $\dfrac{3}{4}$

6. a. $\dfrac{2}{3}$ b. $\dfrac{13}{27}$ c. $\dfrac{16}{35}$ d. $\dfrac{2}{5}$

7. a.   i. 6   ii. 15   iii. 10   iv. 12

b.   i. $\dfrac{1}{6}$ ii. $\dfrac{4}{15}$ iii. $\dfrac{9}{10}$ iv. $\dfrac{7}{12}$

8. a. i. 18   ii. 8   iii. 24   iv. 35

b. i. $\dfrac{17}{18}$ ii. $\dfrac{5}{8}$ iii. $\dfrac{11}{24}$ iv. $\dfrac{11}{35}$

9. a. $\dfrac{5}{6}$ b. $\dfrac{23}{36}$ c. $\dfrac{39}{44}$ d. $\dfrac{1}{4}$

10. a. $\dfrac{17}{63}$ b. $\dfrac{23}{39}$ c. $\dfrac{2}{15}$ d. $\dfrac{7}{33}$

11. D

12. a. $\dfrac{7}{12}$ b. $\dfrac{2}{5}$ c. $\dfrac{19}{24}$ d. $\dfrac{7}{30}$

13. a. $\dfrac{7}{16}$ b. $\dfrac{1}{3}$ c. $1\dfrac{11}{12}$ d. $1\dfrac{11}{12}$

14. a. $\dfrac{3}{4}$ b. $\dfrac{4}{7}$ c. $\dfrac{60}{71}$ d. $1\dfrac{3}{5}$

15. a. $2\dfrac{8}{15}$ b. $4\dfrac{1}{6}$ c. $9\dfrac{2}{3}$ d. $22\dfrac{1}{10}$

16. a. $\dfrac{7}{15}$ b. $\dfrac{8}{15}$

17. $1\dfrac{1}{4}$ hours

18. a. $\dfrac{3}{4}$ b. $\dfrac{7}{12}$ c. 1

19. a. $\dfrac{2}{3}$ b. $\dfrac{23}{30}$

20. 135

21. To add fractions, you must convert to equivalent fractions with the same denominators. You cannot simply add the numerators and add the denominators.

22. $\dfrac{1}{12}$

23. a. i. $\dfrac{5}{6}$ ii. $\dfrac{7}{12}$ iii. $\dfrac{9}{20}$

b. When adding two fractions, both with the numerator of 1 and denominators that are consecutive numbers, the answer is $\dfrac{\text{sum of denominators}}{\text{product of denominators}}$.

24. a. i. $\dfrac{1}{6}$ ii. $\dfrac{1}{12}$ iii. $\dfrac{1}{20}$

b. When subtracting two fractions that both have a numerator of 1 and denominators that are consecutive numbers, the answer is $\dfrac{1}{\text{product of denominators}}$.

c. $\dfrac{1}{n(n+1)}$

25. Fractions cannot be added or subtracted unless they have the same denominator. This is because we need to be adding or subtracting like fractions. If the denominators are different, we can make the two fractions into equivalent fractions by finding the lowest common denominator (LCD). The LCD is the smallest multiple of all the denominators in a set of fractions.

26.

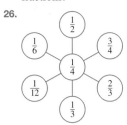

4.6 Multiplying fractions

1. a. 2   **b.** $2\dfrac{1}{2}$ **c.** $7\dfrac{1}{2}$ **d.** $1\dfrac{1}{4}$

2. a. $\dfrac{1}{8}$ **b.** $\dfrac{2}{9}$ **c.** $\dfrac{20}{27}$ **d.** $\dfrac{11}{26}$

3. a. $\dfrac{54}{77}$ **b.** $\dfrac{121}{144}$ **c.** $\dfrac{40}{81}$ **d.** $\dfrac{5}{24}$

4. a. $\dfrac{55}{72}$ **b.** $\dfrac{25}{132}$ **c.** $\dfrac{8}{39}$ **d.** $\dfrac{8}{21}$

5. a. $\dfrac{1}{6}$ **b.** $\dfrac{1}{5}$ **c.** $\dfrac{5}{14}$ **d.** $\dfrac{1}{10}$

6. a. $\dfrac{4}{5}$ **b.** 1   **c.** $\dfrac{1}{2}$ **d.** $\dfrac{12}{35}$

7. a. $\dfrac{4}{9}$ **b.** $\dfrac{1}{7}$ **c.** $\dfrac{1}{3}$ **d.** $\dfrac{1}{20}$

8. a. $\dfrac{1}{8}$ **b.** $\dfrac{1}{2}$ **c.** $\dfrac{25}{36}$ **d.** $\dfrac{2}{9}$

9. a. 28   **b.** 45   **c.** 15   **d.** 16

10. Emily drank 2 L. Tracy drank $\dfrac{1}{2}$ L. Jonathan drank $\dfrac{1}{3}$ L.

11. a. 32 minutes   **b.** 35 minutes

12. a.

Here is one example answer.

b. $\dfrac{1}{9}$

13. a $\dfrac{3}{4}$ **b** 36 litres

14. a. 2   **b.** 4   **c.** 10

15. a.   **b.**

c. $\dfrac{1}{8}$ **d.** $\dfrac{1}{2} \times \dfrac{1}{4} = \dfrac{1}{8}$

16. a. 90 minutes

b. $\dfrac{90}{180} = \dfrac{1}{2}$

c. $140 + 26\dfrac{2}{3} = 166\dfrac{2}{3}$ minutes

17. a. i. $\dfrac{1}{3}$ **ii.** $\dfrac{1}{4}$ **iii.** $\dfrac{1}{5}$

b. The result from the numerator of the first fraction and the denominator of the last fraction

c. $\dfrac{1}{100}$

18. $\dfrac{1}{24} \times 2400 = \$100$

19. $121\dfrac{1}{2}$ km

20. Red $= 1\dfrac{16}{21}$ m, green $= 1\dfrac{9}{28}$ m

4.7 Dividing fractions

1. a. $\dfrac{4}{3}$ **b.** $\dfrac{7}{2}$ **c.** $\dfrac{5}{6}$ **d.** $\dfrac{3}{5}$ **e.** $\dfrac{1}{5}$

2. a $\dfrac{12}{5}$ **b** $\dfrac{2}{9}$ **c** $\dfrac{3}{5}$ **d** $\dfrac{3}{10}$ **e** $\dfrac{1}{20}$

3. a. 12   **b.** 5   **c.** 11   **d.** 6   **e.** 1

4. a. $\dfrac{3}{5}$ **b.** $\dfrac{7}{31}$ **c.** $\dfrac{7}{65}$ **d.** $\dfrac{8}{27}$ **e.** $\dfrac{10}{69}$

5. a. $\dfrac{7}{27}$ **b.** $\dfrac{3}{16}$ **c.** $\dfrac{2}{15}$ **d.** $\dfrac{9}{92}$ **e.** $\dfrac{9}{23}$

6. a. 1   **b.** 1   **c.** 1

7. a. $\dfrac{3}{8}$ **b.** $\dfrac{16}{21}$ **c.** $\dfrac{1}{8}$ **d.** $\dfrac{24}{55}$ **e.** $4\dfrac{2}{3}$

8. a. $\dfrac{3}{10}$ **b.** 1   **c.** $\dfrac{2}{35}$ **d.** $\dfrac{1}{9}$ **e.** $\dfrac{3}{11}$

9. a. $\dfrac{5}{9}$ **b.** $\dfrac{1}{22}$ **c.** $\dfrac{63}{50}$ **d.** $\dfrac{12}{17}$ **e.** $\dfrac{10}{7}$

10. Lily is correct.

11. 32

12. 15 people

13. $\dfrac{13}{3}$

14. $\dfrac{3}{20}$

15. $\dfrac{5}{12}$

4.8 Working with mixed numbers

1. a. $4\dfrac{7}{10}$ b. $1\dfrac{1}{6}$ c. $3\dfrac{5}{12}$ d. $13\dfrac{3}{8}$

2. a. $1\dfrac{1}{10}$ b. $6\dfrac{1}{4}$ c. $1\dfrac{1}{3}$ d. $1\dfrac{1}{2}$

3. a. $1\dfrac{5}{8}$ b. $1\dfrac{7}{10}$ c. $3\dfrac{2}{3}$ d. $1\dfrac{11}{12}$

4. a. $2\dfrac{17}{35}$ b. $3\dfrac{7}{8}$ c. $3\dfrac{13}{18}$ d. $3\dfrac{1}{24}$

5. a. $7\dfrac{7}{12};7$ b. $3\dfrac{2}{3};3$ c. $4\dfrac{1}{2};4$ d. $4\dfrac{1}{4};4$

6. a. $\dfrac{5}{6}$ b. $2\dfrac{1}{4}$ c. $4\dfrac{19}{32}$ d. $7\dfrac{7}{10}$

7. a. $37\dfrac{5}{21}$ b. $34\dfrac{2}{3}$ c. 10 d. $2\dfrac{11}{12}$

8. a. $1\dfrac{2}{3}$ b. $\dfrac{9}{14}$ c. 1 d. $22\dfrac{1}{3}$

9. a. 3 b. $2\dfrac{29}{40}$ c. $1\dfrac{8}{15}$ d. $2\dfrac{1}{14}$

10. $1\dfrac{5}{6}$ L

11. $1\dfrac{5}{16}$ inches

12. $1\dfrac{1}{12}$ months

13. $6\dfrac{1}{4}$ days

14. $\dfrac{3}{16}$ kg

15. a. 22   b. 96

16. $\dfrac{3}{8}$ cup sugar

$\dfrac{3}{8}$ cup margarine

$\dfrac{3}{4}$ cup peanut butter

3 eggs

$2\dfrac{1}{4}$ cups milk

$3\dfrac{3}{4}$ cups self-raising flour

$\dfrac{3}{8}$ teaspoon baking soda

17. $3\dfrac{1}{10}$ kg

18. 7 containers

19. $52.50

20. a. $3\dfrac{15}{16}$ cm   b. $2\dfrac{15}{16}$ cm

21. 500 women

22. We could convert $3\dfrac{1}{3}$ to an improper fraction, $\dfrac{10}{3}$. We would then multiply it by $\dfrac{1}{2}$.

$$\dfrac{10}{3} \times \dfrac{1}{2} = \dfrac{10}{6} = \dfrac{5}{3} = 1\dfrac{2}{3}$$

Alternatively, we could draw a picture of $3\dfrac{1}{3}$ circles and divide each by $\dfrac{1}{2}$, and then add these fractions together, giving us a result of $1\dfrac{2}{3}$.

23. $\dfrac{5}{12}$, 10

24. a. $6\dfrac{2}{3}$ laps   b. 12 minutes

25. 2961 m

26. The gymnasium should keep the drink machine.

4.9 Percentages as fractions

1. a. $\dfrac{17}{100}$ b. $\dfrac{29}{100}$ c. $\dfrac{79}{100}$ d. $\dfrac{99}{100}$ e. $\dfrac{3}{100}$

2. a. $\dfrac{19}{100}$ b. $\dfrac{33}{100}$ c. $\dfrac{9}{100}$ d. $\dfrac{243}{100}$ e. $\dfrac{127}{100}$

3. a. $\dfrac{1}{2}$ b. $\dfrac{4}{5}$ c. $\dfrac{7}{20}$ d. $\dfrac{3}{5}$ e. $\dfrac{1}{10}$

4. a. $\dfrac{9}{20}$ b. $\dfrac{3}{25}$ c. $\dfrac{1}{20}$ d. $\dfrac{37}{50}$ e. $\dfrac{1}{50}$

5. a. $1\dfrac{1}{2}$ b. $1\dfrac{4}{5}$ c. $\dfrac{2}{1} = 2$ d. $\dfrac{5}{1} = 5$ e. $1\dfrac{3}{25}$

6. a. $\dfrac{7}{200}$ b. $\dfrac{9}{125}$ c. $\dfrac{197}{1000}$ d. $\dfrac{81}{250}$

7. a. $\dfrac{31}{200}$ b. $\dfrac{81}{500}$ c. $\dfrac{283}{1000}$ d. $\dfrac{2069}{5000}$

8. a. $\dfrac{907}{5000}$ b. $\dfrac{5799}{10\,000}$ c. $\dfrac{243}{2000}$ d. $\dfrac{1}{2000}$

9. a. $\dfrac{1}{200}$ b. $\dfrac{1}{500}$ c. $\dfrac{1}{150}$ d. $\dfrac{1}{1000}$ e. $\dfrac{3}{550}$

10. a. $\dfrac{33}{400}$ b. $\dfrac{31}{150}$ c. $\dfrac{29}{300}$ d. $\dfrac{31}{200}$ e. $\dfrac{14}{125}$

11. a. $\dfrac{83}{800}$ b. $\dfrac{7}{60}$ c. $1\dfrac{101}{200}$ d. $1\dfrac{41}{200}$ e. $\dfrac{1}{3}$

12. B

13. D

14. C

15. $\dfrac{20}{100} = \dfrac{1}{5}$

16. $\dfrac{35}{100} = \dfrac{7}{20}$

17. $\dfrac{45}{100} = \dfrac{9}{20}$

18. 32% were born overseas $\dfrac{8}{25}$

19. a. $\dfrac{4}{5}$

    b. $\dfrac{1}{5}$

    c. Not necessarily. There could be an overlap of students with blue eyes who have pets.

20. a. $\dfrac{4}{5}$

    b. $\dfrac{1}{2}$

    c. Although the two molecules contain the same number of hydrogen atoms, the total number of atoms differs between the two molecules.

21. a. She scored a higher percentage in the second game (75%).

    b.   i. Increased percentage

        ii. Unchanged

        iii. Decreased percentage

22. Less than $\dfrac{1}{2}$

23. a. $\dfrac{2}{25}$ b. $\dfrac{23}{25}$ c. 8 people

24. a. $\dfrac{7}{20}$ b. $\dfrac{81}{100}$ c. $\dfrac{17}{20}$ d. $\dfrac{13}{200}$

25. a. $\dfrac{17}{100}$ b. $\dfrac{83}{100}$ c. 17

4.10 Calculating percentages of an amount

1. a. 180   b. 4   c. 60

2. a. $1\dfrac{1}{2}$ b. $10\dfrac{1}{2}$ c. 78

3. a. 10   b. 16   c. 3   d. 18

4. a. 6   b. 77   c. 4000   d. 66   e. 17

5. a. 39   b. 95   c. 55   d. 12   e. 25

6. a. $3\dfrac{3}{5}$ b. $4\dfrac{4}{5}$ c. $7\dfrac{7}{10}$ d. $18\dfrac{1}{2}$ e. $13\dfrac{1}{5}$

7. a. $24\dfrac{3}{4}$ b. $11\dfrac{1}{10}$ c. $58\dfrac{9}{10}$ d. $20\dfrac{4}{5}$ e. $98\dfrac{2}{5}$

8. A

9. C

10. C

11. D

12. B

13. D

14. D

15. 210 students

16. 73 days

17. 1 person

18. a. 80%   b. 1600 people

19. a. $22   b. $572

20. $1 600 000

21. $12 000

22. 47 species

23. 600 000 people

24. a. Peter calculated 35% of 80 mL rather than 800 mL.

    b. 280 mL

25. a. 7 adults   b. 19 adults

26. 29 houses

27. 38 workers

28. a. 12 houses   b. 11 houses

29. 3

30. a. 200 g, 100 g, 50 g, 25 g and 12.5g

    b. 50%, 25%, 12.5%, 6.25% and 3.125%

    c. Every 20 days the quantity of the radioactive substance is smaller so there is a 50% reduction of an already reduced quantity.

31. 22.5%

32. 3 students voted twice.

33. 120

4.11 One amount as a percentage of another

1. a. 50%   b. 30%   c. 40%   d. 87.5%   e. 40%

2. a. 20%   b. 25%   c. 10%   d. 40%   e. 25%

3. a. 80%   b. 32%   c. 40%   d. 20%   e. 5%

4. a. 25%   b. $31\dfrac{1}{2}$%   c. $6\dfrac{1}{4}$%   d. $6\dfrac{2}{5}$%   e. $20\dfrac{2}{3}$%

5. a. $87\dfrac{1}{2}$%   b. $28\dfrac{4}{7}$%   c. $66\dfrac{2}{3}$%   d. $36\dfrac{2}{3}$%   e. $26\dfrac{2}{3}$%

6. a. $7\dfrac{1}{7}$%   b. $22\dfrac{2}{9}$%   c. $42\dfrac{6}{7}$%   d. $88\dfrac{8}{9}$%   e. $71\dfrac{3}{7}$%

7. a. 17%   b. 33%   c. 63%   d. 11%

8. a. 83%   b. 78%   c. 18%   d. 12%

9. a. 10%   b. $3\dfrac{1}{3}$%   c. $12\dfrac{1}{2}$%

10. a. $28\dfrac{4}{7}$%   b. 25%   c. $1\dfrac{1}{4}$%

11. A

12. B

13. D

14. 70%

15. 60%

16. a. $5 b. $33\dfrac{1}{3}\%$

17. a. i. 10% ii. 14% iii. 12%
    b. 36%
    c. 64%

18. $62\dfrac{1}{2}\%$

19. $6\dfrac{2}{3}\%$

20. a. 20 km/h b. $16\dfrac{2}{3}\%$

21. a. $42\dfrac{6}{7}\%$ b. $28\dfrac{4}{7}\%$ c. $71\dfrac{3}{7}\%$ d. $57\dfrac{1}{7}\%$

22. $27\dfrac{3}{11}\%$

23. a. Second recipe. The second recipe contains 50% flour compared to 40% in the first recipe.
    b. The percentage of flour increases

24. a. Less than $100 because 25% of the new amount is larger than the amount added. So, a higher amount is subtracted.
    b. Sample responses can be found in the worked solutions in the online resources.

25. Sister is 9; grandfather is 90.

26. a.

Topic	Prob.	M'ment	Algebra	Eqns	Geo.
Score	$\dfrac{15}{20}$	$\dfrac{13}{15}$	$\dfrac{27}{30}$	$\dfrac{70}{80}$	$\dfrac{95}{100}$
Percentage	75	87	90	88	95

b. Geometry
c. 87%

27. a. 300 g out of 450 g (300 g out of 450 g represents 67%.)
    b. 200 g out of 500 g (200 g out of 500 g represents 40%.)
    c. 420 g out of 840 g and 350 g out of 700 g (420 g out of 840 g and 350 g out of 700 g both represent 50%.)

28. Expressing one quantity as a percentage of another quantity allows us to compare the two quantities.

29. a. First recipe. The first type of concrete contains 15.4% cement, and the second type of concrete contains 11.8% cement.
    b. The percentage of cement decreases.

30. 4520 km

4.12 Common percentages and short cuts

1. a. $1.00 b. $1.80 c. $8.10 d. $15.00 e. $9.30
2. a. $2.20 b. $1.70 c. $1.25 d. $0.15 e. $4.80
3. a. $19.25 b. $50.70 c. $192.60
   d. $304.15 e. $200.00
4. a. $1.50 b. $5.10 c. $0.90 d. $13.70 e. $0.45
5. a. $0.65 b. $3.90 c. $7.50 d. $10.30 e. $6.80
6. a. $51.80 b. $30.50 c. $62.85 d. $10.05 e. $20.70
7. a. $0.40 b. $0.30 c. $0.10 d. $7.00 e. $2.10
8. a. $0.55 b. $15.25 c. $50 d. $5.15 e. $1.60
9. a. $4.30 b. $8.45 c. $0.65 d. $14.80 e. $0.15
10. a. $3.30 b. $27.15 c. $52.35
    d. $247.40 e. $1013.80
11. a. $1.80 b. $1.20 c. $9.00
12. a. $7.50 b. $22.50 c. $55.00
13. a. $4.50 b. $7.50 c. $14.25
14. a. $2.70 b. $7.15 c. $6.05 d. $0.05 e. $0.15
15. a. $0.15 b. $0.80 c. $0.20 d. $0.05 e. $0.05
16. a. $1.30 b. $10.50 c. $0.65 d. $2.40 e. $1.80
17. a. $73.50 b. $1.55 c. $1.05 d. $2.05 e. $32.20
18. a. $4.80 b. $1.60 c. $0.45 d. $14.40 e. $492
19. D
20. B
21. D
22. C
23. $855
24. 54 000 residents
25. a. $322.50 b. $152.30
26. 110 students
27. 27.9 seconds
28. a. 2 people b. 38 people
29. a. 10 740 people
    b. 19 690 people
    c. 5370 people
30. 20 years old
31. $0.80
32. 8.2 kg
33. 50 years old
34. a. 100%
    b. i. 100% ii. 50% iii. 71.4%
    c. No, because the two lengths have changed at different percentages
35. a. 3.2%
    b. 6.8%
    c. The overall percentage is a lot smaller.

Project

1. A and D, B and C, E and F
2–6. An individual response is required by the student.

4.13 Review questions

1. $\dfrac{3}{20}$

2. a. $\dfrac{6}{16}, \dfrac{9}{24}, \dfrac{12}{32}$ b. $\dfrac{8}{18}, \dfrac{12}{27}, \dfrac{16}{36}$
   c. $\dfrac{10}{22}, \dfrac{15}{33}, \dfrac{20}{44}$ d. $\dfrac{12}{14}, \dfrac{18}{21}, \dfrac{24}{28}$

3. a. $\dfrac{2}{4} = \dfrac{3}{6}$ b. $\dfrac{3}{9} = \dfrac{5}{15}$

   c. $\dfrac{9}{12} = \dfrac{21}{28}$ d. $\dfrac{4}{10} = \dfrac{16}{40}$

4. a. Leo   b. Ricky

5. a. $\dfrac{6}{7}$ b. $\dfrac{3}{4}$ c. $\dfrac{9}{10}$ d. $\dfrac{3}{4}$

6. a. $\dfrac{1}{3}$ b. $\dfrac{3}{5}$ c. $\dfrac{1}{40}$ d. $\dfrac{1}{2}$

7. a. $6\dfrac{4}{5}$ b. $3\dfrac{2}{3}$ c. $4\dfrac{3}{4}$ d. $1\dfrac{1}{5}$

8. $\dfrac{3}{12} = \dfrac{1}{4}$

9. a. $5\dfrac{1}{3}$ b. $4\dfrac{1}{5}$ c. $1\dfrac{5}{7}$ d. $5\dfrac{1}{2}$

10. a. $1\dfrac{1}{4}$ b. $11\dfrac{3}{7}$ c. $6\dfrac{1}{9}$ d. $7\dfrac{1}{5}$

11. a. $\dfrac{11}{4}$ b. $\dfrac{79}{8}$ c. $\dfrac{26}{7}$ d. $\dfrac{35}{6}$

12. a. $\dfrac{11}{3}$ b. $\dfrac{99}{10}$ c. $\dfrac{47}{12}$ d. $\dfrac{17}{2}$

13. 4 trays

14. a. $\dfrac{23}{24}$ b. $\dfrac{17}{45}$ c. $1\dfrac{1}{15}$

15. a. $\dfrac{20}{63}$ b. $\dfrac{6}{25}$ c. $\dfrac{5}{8}$ d. $\dfrac{7}{36}$

16. a. 44   b. 22

   c. $\dfrac{1}{6}$ d. $\dfrac{50}{3}$ or $16\dfrac{2}{3}$

17. a. $\dfrac{2}{5}$ b. 9   c. 4   d. $\dfrac{4}{23}$

18. a. $\dfrac{3}{2}$ or $1\dfrac{1}{2}$ b. $\dfrac{4}{3}$ or $1\dfrac{1}{3}$

   c. $\dfrac{7}{36}$ d. $\dfrac{3}{70}$

19. a. $4\dfrac{17}{24}$ b. $1\dfrac{13}{14}$ c. $3\dfrac{7}{9}$ d. $2\dfrac{2}{3}$

20. a. $3\dfrac{43}{72}$ b. $\dfrac{76}{99}$ c. $3\dfrac{2}{3}$ d. $6\dfrac{5}{18}$

21. a. 64   b. $\dfrac{3}{5}$ c. $13\dfrac{2}{3}$ d. $1\dfrac{1}{2}$

22. a. $\dfrac{5}{16}$ b. $1\dfrac{3}{5}$ c. $1\dfrac{53}{115}$ d. $\dfrac{217}{267}$

23. a. $\dfrac{13}{100}$ b. $\dfrac{7}{10}$

   c. $\dfrac{13}{50}$ d. $\dfrac{33}{25} = 1\dfrac{8}{25}$

24. a. 15%   b. 55%   c. 60%   d. 50%

25. a. $37\dfrac{1}{2}\%$ b. $53\dfrac{1}{3}\%$ c. $83\dfrac{1}{3}\%$ d. $36\dfrac{4}{11}\%$

26. a. 6   b. 12   c. 39   d. 39

27. a. $6\dfrac{3}{10}$ b. $37\dfrac{1}{10}$ c. $4\dfrac{4}{5}$ d. $13\dfrac{3}{5}$

28. a. 60%   b. 30%   c. 40%   d. 40%

29. a. $7\dfrac{1}{7}\%$ b. $11\dfrac{1}{9}\%$

30. a. $0.40   b. $1.05   c. $3.20   d. $5.20

31. a. $0.15   b. $0.30   c. $5.05   d. $27.00

32. $7.1 = \dfrac{1}{8}, 7.2 = \dfrac{1}{4}, 7.3 = \dfrac{5}{32}, 7.4 = \dfrac{5}{32}, 7.5 = \dfrac{7}{40}, 7.6 = \dfrac{11}{80}$

33. $150

34. 2kg

35. a. $\dfrac{1}{6}$ b. $\dfrac{1}{4}$

   c. $\dfrac{5}{9}$ d. $\dfrac{48}{25} = 1\dfrac{23}{25}$

36. All fractions that are equivalent to $\dfrac{4}{7}$ will be of the form $\dfrac{4n}{7n}$ where n is some number. The product of the numerator and denominator is therefore $4n \times 7n = 28n^2$. Equating this product to 252:

$252 = 28n^2$

$n^2 = 9$

$n = \pm 3$

Substituting $n = +3$ gives a fraction of $\dfrac{12}{21}$, which is equivalent to $\dfrac{4}{7}$ and for which the product of the numerator and denominator is 252.

$\dfrac{12}{21}$

37. Tomas

38. $\dfrac{1}{2} + \dfrac{1}{4} + \dfrac{1}{20}$ or $\dfrac{1}{2} + \dfrac{1}{5} + \dfrac{1}{10}$

39. Mr Thompson $3000, Mr Goody $2500

40. $\dfrac{3}{5}$ of the standard spots were filled.

41. a. 4 complete surfboards

   b. $2\dfrac{1}{4}$kg of resin, 0 catalyst

42. a. 62%   b. 75%

43. 24 students

44. 270 children

45. 8%

46. $11 600

47. b, d

48. 13.5 litres of milk would be needed to make the fat of 500 g of butter.

49. 45%

50. a. $2250

    b. $2025

    c. 19%

    d. The second 10% discount is not 10% of the original price but 10% of an amount, which is only 90% of the original.

51. a. $1575

    b. $1575

    c. It doesn't matter which discount is applied first; the final price will be the same.

52. Sample responses can be found in the worked solutions in the online resources.

5 Decimals

LESSON
5.1 Overview

Why learn this?

Whole numbers cannot be used to measure everything. How much did you pay for that burger and chips? $12.60? Our money system is in dollars and cents and uses decimals to represent parts of the dollar. Decimals play an important part in our daily lives. You could have been using decimals without realising how important they are. Since our lives are so dependent on money, knowing how to work with decimals is very important. We all need to be skilled in the basic operations of addition, subtraction, multiplication and division with decimals.

Decimals are used not only with money but also with time and measurement. They are also used as another way of expressing a fraction or percentage of something. Did you know that Cathy Freeman won the gold medal in the 400-metre event at the 2000 Sydney Olympics in a time of 49.11 seconds, a winning margin of just 0.02 seconds? Usain Bolt set the men's 100-metre world record in 2009 with a time of just 9.58 seconds. Every person will use decimals in some aspect of their lives, including when checking their pay or their bank balance. Understanding and using decimals is vital in retail, hospitality, design, commerce, construction, the health industry and all areas of business.

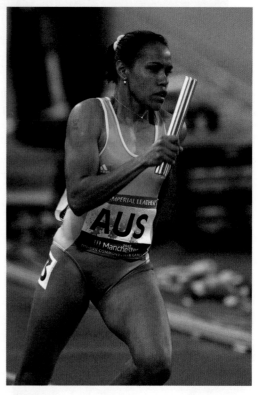

Hey students! Bring these pages to life online

- **Watch videos**
- **Engage with interactivities**
- **Answer questions and check solutions**

Find all this and MORE in jacPLUS

Reading content and rich media, including interactivities and videos for every concept

Extra learning resources

Differentiated question sets

Questions with immediate feedback, and fully worked solutions to help students get unstuck

1. **MC** Select which of the following is the largest number.

   A. 3.098   **B.** 3.089   **C.** 3.100   **D.** 2.999

2. The cost of the new PlayStation has increased by 12% compared to the previous model. Write the percentage increase as a decimal.

3. **MC** The decimal 0.56 in simplest fractional form is:

   A. $\dfrac{56}{100}$ **B.** $\dfrac{28}{50}$ **C.** $\dfrac{15}{24}$ **D.** $\dfrac{14}{25}$

4. Josh swam his race in 22.370 seconds and Ren swam his in $22\dfrac{3}{8}$ seconds. State who was the fastest.

5. Write the number 23.8951 rounded to 2 decimal places.

6. Walter wants to withdraw all the money from his bank account. He has a total of $378.93, but can only withdraw an amount to the nearest 5 cents. State how much he can withdraw.

7. **MC** The third largest number from the following choices is:

   A. $0.27\dot{6}$ **B.** $0.27\ddot{6}$ **C.** $0.26\dot{9}$ **D.** $0.27\dot{5}$

8. Calculate:

$$
\begin{array}{r}
2\ .\ 3\ 7\ 5 \\
+\ 3\ .\ 4\ 4\ 2 \\
\hline
\end{array}
$$

9. **MC** Michael works at McDonald's. If a customer had an order of $23.75 and he handed over a $50 note, calculate the change Michael gave to the customer.

   A. $25.25   **B.** $26.75   **C.** $26.25   **D.** $23.75

10. Calculate 8.2×400.

11. **MC** Callum coaches tennis on Sundays and he gets 1.5 times his usual hourly pay of $18.70. His Sunday hourly rate is:

    A. $37.40   **B.** $28.05   **C.** $25.70   **D.** $27.70

12. Kat bought 2.6 kg of sausages for her BBQ at a cost of $5.25 per kg. Calculate how much Kat paid for her sausages.

13. Calculate $34.792 \div 0.4$.

14. If Mei filled her car with petrol and it cost $53.68 at $1.22 per litre, calculate how many litres of petrol she put in her car.

15. Leah pours drinks for a group of people at a volume of 0.32 litres each. If she has 40 litres of the beverage to use, calculate how many drinks she can pour.

LESSON
5.2 Decimals and place value

LEARNING INTENTION

At the end of this lesson you should be able to:
- understand the place values used in decimal numbers
- determine the value of digits in decimal numbers
- compare the sizes of decimal numbers.

▶ 5.2.1 Whole numbers and decimal parts

eles-3940

- Each position in a number has its own place value.

Hundred thousands	Ten thousands	Thousands	Hundreds	Tens	Units
100 000	10 000	1000	100	10	1

- Each place to the left of another position has a value that is 10 times larger.
- Each place to the right of another position has a value that is $\frac{1}{10}$ of the previous position.
- The position of a digit in a number gives the value of the digit.
- The table shows the value of the digit 3 in some numbers.

Number	Value of 3 in number
132	3 tens or 30
3217	3 thousands or 3000
4103	3 units (ones) or 3

- In a decimal number, the whole number part and the fractional part are separated by a **decimal point**.

Whole number part — **73.**064 — Fractional part

Decimal point

- A place value table can be extended to include decimal place values. It can be continued to an infinite number of **decimal places**.

Thousands	Hundreds	Tens	Units	.	Tenths	Hundredths	Thousandths	Ten thousandths
1000	100	10	1	.	$\frac{1}{10}$	$\frac{1}{100}$	$\frac{1}{1000}$	$\frac{1}{10\,000}$

- The example of 73.064 can be placed in a place value table.

Thousands	Hundreds	Tens	Units	.	Tenths	Hundredths	Thousandths	Ten thousandths
		7	3	.	0	6	4	

- The table shows the value of the digit 3 in some decimal numbers.

Number	Value of 3 in the number
3.65	The 3 is 3 wholes or 3.
14.32	The 3 is 3 tenths or $\dfrac{3}{10}$.
106.013	The 3 is 3 thousandths or $\dfrac{3}{1000}$.
0.000 03	The 3 is 3 hundred thousandths or $\dfrac{3}{100\,000}$.

- The number of decimal places in a decimal is the number of digits after the decimal point. The number 73.064 has 3 decimal places.
- The zero (0) in 73.064 means that there are no tenths. The zero must be written to hold the place value; otherwise the number would be written as 73.64, which does not have the same value.

DISCUSSION

How do you know how many zeros the denominator should have when calculating the value of a digit to the right of the decimal point?

WORKED EXAMPLE 1 Stating the place value

State the value of 7 in each of the following.
a. **10.74**   b. **0.173**   c. **321.037**

THINK   **WRITE**

a. The value of the first place to the right of the decimal point is tenths, so the value of the 7 is seven tenths.

a. $\dfrac{7}{10}$

b. The second place after the decimal point is hundredths, so the value of the 7 is seven hundredths.

b. $\dfrac{7}{100}$

c. The third place after the decimal point is thousandths, so the value of the 7 is seven thousandths.

c. $\dfrac{7}{1000}$

WORKED EXAMPLE 2 Expressing place values in words

For the number 76.204, write the value of each digit in words and numbers.

THINK   **WRITE**

1. 7 is in the tens position.   Seventy, 70

2. 6 is in the units position.   Six, 6

3. 2 is in the first position after the decimal point, so it is tenths.   Two tenths, $\dfrac{2}{10}$

4. 0 is in the hundredths position.   Zero hundredths, $\dfrac{0}{100}$

5. 4 is in the thousandths position.   Four thousandths, $\dfrac{4}{1000}$

Expanded notation

A number can be written in expanded notation by adding the values of each digit.

The number 76.204 can be written in expanded notation as:

$$(7 \times 10) + (6 \times 1) + \left(2 \times \frac{1}{10}\right) + \left(4 \times \frac{1}{1000}\right)$$

WORKED EXAMPLE 3 Writing in expanded notation

Write 3.4501 in expanded notation.

THINK	WRITE
1. Write the decimal.	3.4501
2. Determine the place value of each digit. 3: 3 units 4: 4 tenths 5: 5 hundredths 0: 0 thousandths 1: 1 ten thousandth	$= 3$ units $+ 4$ tenths $+ 5$ hundredths $+ 0$ thousandths $+ 1$ ten thousandth
3. Write the number in expanded notation.	$= (3 \times 1) + \left(4 \times \frac{1}{10}\right) + \left(5 \times \frac{1}{100}\right) + \left(1 \times \frac{1}{10\,000}\right)$

5.2.2 Comparing decimals

eles-3941

- Decimals are compared using digits with the same place value.

Comparing decimals

- **The decimal with the largest number in the highest place-value column is the largest decimal,** *regardless of the number of decimal places.*
  For example, 15.71 is larger than 15.702 because the first place value with different digits (moving from left to right) is hundredths, and 1 is greater than 0; that is, 15.71 > 15.702.

Tens	Units (ones)	.	Tenths	Hundredths	Thousandths
1	5	.	7	1	
1	5	.	7	0	2

State the largest number in each of the following.

a. 0.126, 0.216, 0.122   b. 2.384, 2.388, 2.138   c. 0.506, 0.605, 0.612

THINK	WRITE
a. 1. As the units digit is 0 in each number, compare the tenths. The number 0.216 has 2 tenths, the others have 1 tenth so 0.216 is the largest number.	a. 0.216 is larger than 0.126 and 0.122.
2. Answer the question.	The largest number is 0.216.
b. 1. As the units digits are the same, compare the tenths. 2.138 has 1 tenth, the others have 3 tenths, so 2.138 is the smallest.	b. 2.384 and 2.388 are both larger than 2.138.
2. The hundredths are the same, so compare the thousandths. 2.388 has 8 thousandths, 2.384 has 4 thousandths, so 2.388 is the largest.	2.388 is larger than 2.384 and 2.138.
3. Answer the question.	The largest number is 2.388.
c. 1. As the units digits are the same, compare the tenths. 0.506 has 5 tenths, the others have 6 tenths, so 0.506 is the smallest.	c. 0.605 and 0.612 are larger than 0.506.
2. Compare the hundredths. 0.612 has 1 hundredth, 0.605 has 0 hundredths, so 0.612 is the largest.	0.612 is larger than 0.605 and 0.506.
3. Answer the question.	The largest number is 0.612.

- Decimals can also be compared using the *area model*. For example, we could use the area model to determine whether 0.67 or 0.7 is larger. From the following area model, it can be seen that 0.7 covers a greater area than 0.67 and is therefore larger.

0.67   0.7

- When comparing two numbers, it is easier to use symbols instead of words, as shown in topic 4 on fractions and percentages.
- In Worked example 4a, the result '0.216 is larger than 0.126' could be written as 0.216 > 0.126. We could also say that 0.126 is less than 0.216, or write it as 0.126 < 0.216.

WORKED EXAMPLE 5 Comparing numbers with decimals

Insert the appropriate < or > sign between the following pairs of numbers to make true statements.
a. 0.312   0.318   b. 0.0246   0.0168

THINK	WRITE
a. Compare the numbers. Both numbers have the same number of tenths and the same number of hundredths, so compare thousandths and insert the correct sign.	**a.** $0.312 < 0.318$
b. Compare the numbers. Both numbers have no tenths, so compare hundredths and insert the correct sign.	**b.** $0.0246 > 0.0168$

DISCUSSION

You may read statements in the media quoting decimals that are not achievable every day. Some examples are 'The average number of tries scored by the Roosters is 3.6' or 'The number of children in the average family is 2.3'. Do you think these statements are reasonable? How would you interpret these figures?

COMMUNICATING — COLLABORATIVE TASK: Some points about decimals

In English-speaking countries, the decimal point is a full stop and spaces are left between groups of three digits: for example, 12 345.67. In many other countries, a comma is used as a decimal point and full stops are used to separate groups of three digits; for example, 12.345, 67.

In groups, discuss the following:
- What are some of the problems associated with these different methods of writing decimal numbers?
- What could be done to fix these problems?
- Which method of displaying decimals do you think we should use and why?

Use the internet to investigate how the number 12 345.67 would be written in each of the following countries.

a. China	b. India	c. Japan
d. United States of America	e. United Kingdom	f. France
g. Turkey	h. Saudi Arabia	

 Resources

Interactivities   Whole numbers (int-3874)
                      Decimal parts (int-3975)
                      Comparing decimals (int-3976)
                      Comparing decimals using the area model (int-3977)

Exercise 5.2 Decimals and place value

| 5.2 Quick quiz on | 5.2 Exercise |

Individual pathways

■ PRACTISE	■ CONSOLIDATE	■ MASTER
1, 4, 7, 10, 13, 16, 19, 21, 24, 27	2, 5, 8, 11, 14, 17, 20, 23, 25, 28	3, 6, 9, 12, 15, 18, 22, 26, 29, 30

Fluency

1. **WE1** State the value of the 2 in each of the following.
   a. 5.2 b. 19.12 c. 100.29 d. 0.982 047

2. State the value of the 2 in each of the following.
   a. 6.1342 b. 90.0002 c. 12.14 d. 1.8902

3. State the value of the 9 in each of the following.
   a. 0.9 b. 14.98 c. 18.89 d. 12.090

4. State the value of the 9 in each of the following.
   a. 3.4629 b. 1.276 89 c. 900.76 d. 9.612

5. For each of the following numbers write the value of each digit in *numbers*.
   a. 0.4 b. 2.7 c. 5.23 d. 0.763

6. For each of the following numbers write the value of each digit in *numbers*.
   a. 0.1101 b. 7.2964 c. 0.330 24 d. 300.03

7. **WE2** For the following numbers write the value of each digit in *words* and *numbers*.
   a. 1.85 b. 0.271 c. 16.001 d. 3.402 07

8. **WE3** Write the following numbers in expanded notation.
   a. 1.25 b. 56.01 c. 39.01 d. 5.987

9. Write the following numbers in expanded notation.
   a. 13.482 b. 0.3062 c. 0.5002 d. 2.47

10. **MC** Five hundredths, 2 thousandths and 7 ten thousandths equals:
    A. 527 B. 52.7 C. 5.27 D. 0.0527

11. Copy and complete the table by putting only one digit in each box.

		Tens	Units		Tenths	Hundredths	Thousandths
Example	37.684	3	7	.	6	8	4
a	0.205			.			
b	1.06			.			
c	74.108			.			
d	0.108			.			
e	50.080			.			

12. **WE4** State the largest number in each of the following.
    a. 0.24, 0.32, 0.12 b. 0.57, 0.51, 0.59 c. 0.192, 0.191, 0.901

13. State the largest number in each of the following.

    a. 0.0392, 0.039 90, 0.0039   b. 2.506, 2.305, 2.559   c. 0.110 43, 0.110 49, 0.110 40

14. **WE5** Insert the appropriate < or > sign between each of the following pairs of numbers to make true statements.

    a. 3.2   2.9   b. 8.6   8.9   c. 0.64   0.67
    d. 0.29   0.39   e. 13.103   13.112

15. Insert the appropriate < or > sign between each of the following pairs of numbers to make true statements.

    a. 0.427   0.424   b. 0.580   0.508   c. 0.0101   0.0120
    d. 0.048 01 0.4801   e. 1.3830   1.3824

Understanding

16. a. State how many decimal places we use in our currency.
    b. Write the amount of money shown in the photograph:

      i. in words   ii. in numbers.

17. Write the following in order from smallest to largest (ascending order).

    a. 0.21, 0.39, 0.17, 0.45, 0.33
    b. 0.314, 0.413, 0.420, 0.391, 0.502
    c. 0.821, 0.803, 0.811, 0.807, 0.902

18. Write the following in order from smallest to largest (ascending order).

    a. 0.9864, 0.9812, 0.9943, 0.9087, 0.9189
    b. 4.6249, 4.5097, 4.802, 4.6031, 4.0292
    c. 0.004 65, 0.005 02, 0.003, 0.0056, 0.009

19. Write the following in order from largest to smallest (descending order).

    a. 0.36, 0.31, 0.39, 0.48, 0.19
    b. 0.91, 0.97, 0.90, 0.95, 0.99
    c. 1.264, 1.279, 1.273, 1.291, 1.288

20. Write the following in order from largest to smallest (descending order).

    a. 0.372, 0.318, 0.390, 0.309, 0.317
    b. 0.8207, 0.8889, 0.8823, 0.8217, 0.8448
    c. 1.349 54, 1.486 59, 1.702 96, 1.843 21, 1.486 13

21. **MC** The largest number in the list 0.4265, 0.4273, 0.4199, 0.3999 is:

    A. 0.4199   B. 0.4265   C. 0.3999   D. 0.4273

22. **MC** The smallest number in the list 0.4265, 0.4273, 0.4199, 0.3999 is:

    A. 0.4199   B. 0.4265   C. 0.3999   D. 0.4273

23. **MC** The list 0.4265, 0.4273, 0.4199, 0.3999 when arranged from smallest to largest is:

    A. 0.4273, 0.4265, 0.4199, 0.3999   B. 0.3999, 0.4199, 0.4265, 0.4273
    C. 0.3999, 0.4199, 0.4273, 0.4265   D. 0.3999, 0.4199, 0.4265, 0.4273

Communicating, reasoning and problem solving

24. Write True (T) or False (F) for each of the following and justify your response.

    a. 76.34 has 4 decimal places.

    b. $\dfrac{6}{10} + \dfrac{3}{100} + \dfrac{4}{10\,000}$ is the same as 0.6304.

    c. 4.03 has the same value as 4.3.

25. Write True or False for each of the following and justify your response.

    a. 29.60 has the same value as 29.6.

    b. 1.2804 could be written as $1 + \dfrac{2}{10} + \dfrac{8}{100} + \dfrac{4}{1000}$.

    c. 1090.264 51 has 5 decimal places.

26. a. Using a calculator, determine the decimal values of the following fractions.

    $$\dfrac{1}{7}, \dfrac{2}{7}, \dfrac{3}{7}, \dfrac{4}{7}, \dfrac{5}{7}, \dfrac{6}{7}$$

    Write as many decimals as possible.

    b. Is there a pattern? Explain.

27. For each of the numbers listed from a to e:

    i. state the place value of the zero
    ii. state whether the value of the number would change if the zero was not there. (Write yes or no.)
        a. 6.02   b. 7.360   c. 0.65   d. 20   e. 108.62

28. Year 7 students competing in their school swimming sports recorded the following times in the 50-metre freestyle, backstroke and breaststroke events.

Time (seconds) recorded by contestants							
Event	**Carolyn**	**Jessica**	**Mara**	**Jenika**	**Robyn**	**Shelley**	**Kyah**
Freestyle	37.23	39.04	40.90	38.91	37.45	37.02	37.89
Backstroke	40.23	43.87	44.08	42.65	41.98	40.29	41.05
Breaststroke	41.63	42.70	41.10	41.21	42.66	41.33	41.49

a. State who won the freestyle event and calculate how much they won by.
b. State who won the backstroke event and calculate how much they won by.
c. State who won the breaststroke event and calculate how much they won by.
d. List the first 3 placings of the freestyle event.
e. List the first 3 placings of the backstroke event.
f. List the first 3 placings of the breaststroke event.
g. Determine whether any students obtained a placing in all three events.

29. Determine my number.

My number contains four different even digits and has 3 decimal places. The digit in the thousandths position is half the value of the digit in the units position. The sum of the digits in the units and tenths positions is the same as the sum of the digits in the hundredths and thousandths positions. The digit in the thousandths position is bigger than the digit in the tenths position.

30. Your friend said, 'You don't have to worry about the zeros in decimals; you can just leave them out.' Explain why your friend is correct or incorrect.

LESSON
5.3 Converting between decimals and fractions

LEARNING INTENTION

At the end of this lesson you should be able to:
- convert decimals to fractions and fractions to decimals.

▶ 5.3.1 Converting decimals to fractions

- Decimal numbers can be written as fractions by applying knowledge of place values.

Number	Ones	·	Tenths	Hundredths
0.3	0	·	3	0
1.25	1	·	2	5

- The decimal number 0.3 can be written as $\dfrac{3}{10}$.

- The decimal number 1.25 can be thought of as $1 + \dfrac{2}{10} + \dfrac{5}{100} = 1 + \dfrac{20}{100} + \dfrac{5}{100}$

$$= 1\dfrac{25}{100}.$$

Converting decimals to fractions

- **If there is non-zero digit before the decimal point, it becomes the whole number part of the fraction.**
- **The digits after the decimal point become the numerator, and the place value of the last digit gives the denominator.**
- **Simplify fractions by dividing the denominator and numerator by a common factor,**
  e.g. $1\dfrac{25}{100} = 1\dfrac{1}{4}.$

WORKED EXAMPLE 6 Converting decimals to fractions

Write the following decimals as fractions, then simplify where appropriate.
a. **0.2**   b. **0.86**   c. **0.6021**

THINK	WRITE
a. 1. Write the decimal.	a. 0.2
2. The numerator is 2 and the last decimal place is tenths so the denominator is 10.	$= \dfrac{2}{10}$
3. Simplify the fraction and write the answer.	$= \dfrac{1}{5}$
b. 1. Write the decimal.	b. 0.86
2. The numerator is 86. The last decimal place is hundredths so the denominator is 100.	$= \dfrac{86}{100}$
3. Simplify the fraction and write the answer.	$= \dfrac{43}{50}$
c. 1. Write the decimal.	c. 0.6021
2. The numerator is 6021. The last place is tens of thousandths so the denominator is 10 000.	$= \dfrac{6021}{10\,000}$

WORKED EXAMPLE 7 Converting mixed numbers to fractions

Write each of the following as a mixed number in its simplest form.
a. **3.041**   b. **7.264**

THINK	WRITE
a. 1. Write the decimal.	a. 3.041
2. Write the whole number part and change the decimal part to a fraction. The numerator is 41. The last decimal place is thousandths so the denominator is 1000.	$= 3\dfrac{41}{1000}$
b. 1. Write the decimal.	b. 7.264
2. Write the whole number part and change the decimal part to a fraction. The numerator is 264 and the denominator is 1000.	$= 7\dfrac{264}{1000}$
3. Simplify the fraction.	$= 7\dfrac{264 \div 8}{1000 \div 8}$
4. Write the answer in simplest form.	$= 7\dfrac{33}{125}$

▶ 5.3.2 Converting fractions to decimals

eles-3944

- Fractions can be written as decimal numbers by dividing the numerator by the denominator.

  For example, to convert $\dfrac{1}{4}$ into a decimal, divide 1 by 4 : $\quad 4\overline{)1.00}^{\,0.25}$

- Add trailing zeros to the numerator where required.

WORKED EXAMPLE 8 Converting fractions to decimals

Change the following fractions into decimals.

a. $\dfrac{2}{5}$

b. $\dfrac{1}{8}$

THINK	WRITE
a. 1. Set out the question as for division of whole numbers, adding a decimal point and the required number of zeros.	a. $5\overline{)2.0}$
2. Divide, writing the answer with the decimal points aligned.	$5\overline{)2.0}^{\,0.4}$
3. Write the answer.	$\dfrac{2}{5} = 0.4$
b. 1. Set out the question as for division of whole numbers, adding a decimal point and the required number of zeros. *Note:* $\dfrac{1}{8} = 1 \div 8$.	b. $8\overline{)1.000}^{\,0.125}$
2. Divide, writing the answer with the decimal point exactly in line with the decimal point in the question, and write the answer.	$\dfrac{1}{8} = 0.125$

- By knowing the decimal equivalent of any fraction, it is possible to determine the equivalent of any multiple of that fraction. The following worked example illustrates this.

WORKED EXAMPLE 9 Converting multiples of a known fraction to decimals

Use the results of Worked example 8 to determine decimal equivalents for:

a. $\dfrac{3}{8}$

b. $4\dfrac{5}{8}$

THINK	WRITE
a. 1. Write the decimal equivalent for the fraction with 1 as the numerator.	a. $\dfrac{1}{8} = 0.125$
2. Multiply both sides of this equation by the appropriate multiple (3 in this case).	$\dfrac{1}{8} \times 3 = 0.125 \times 3$
3. Write the answer.	$\dfrac{3}{8} = 0.375$
b. 1. Consider only the fraction part of the mixed number. Write the decimal equivalent of this fraction with 1 as the numerator.	b. $\dfrac{1}{8} = 0.125$
2. Multiply both sides of this equation by the appropriate multiple (5 in this case).	$\dfrac{1}{8} \times 5 = 0.125 \times 5$

---

3. Simplify both sides.

$$\frac{5}{8} = 0.625$$

4. Combine with the whole number and write the answer.

$$4\frac{5}{8} = 4.625$$

 Resources

 Interactivities Conversion of decimals to fractions (int-3978)

Conversion of fractions to decimals (int-3979)

Exercise 5.3 Converting decimals to fractions and fractions to decimals

learn

5.3 Quick quiz on	5.3 Exercise

Individual pathways

■ PRACTISE	■ CONSOLIDATE	■ MASTER
1, 4, 7, 10, 12, 15, 18	2, 5, 8, 11, 13, 16, 19	3, 6, 9, 14, 17, 20

Fluency

1. **WE6** Write the following decimals as fractions, then simplify where appropriate.

   a. 0.3   b. 0.5   c. 0.21   d. 0.4   e. 0.44

2. Write the following decimals as fractions, then simplify where appropriate.

   a. 0.49   b. 0.502   c. 0.617   d. 0.30   e. 0.64

3. Write the following decimals as fractions, then simplify where appropriate.

   a. 0.28   b. 0.9456   c. 0.9209   d. 0.4621   e. 0.120

4. **WE7** Write the following decimals as mixed numbers in simplest form.

   a. 1.3   b. 2.7   c. 9.4   d. 1.2   e. 4.2

5. Write the following decimals as mixed numbers in simplest form.

   a. 8.5   b. 5.27   c. 19.182   d. 3.15   e. 6.25

6. Write the following decimals as mixed numbers in simplest form.

   a. 9.140   b. 16.682   c. 2.4917   d. 4.3386   e. 100.0048

Understanding

7. **MC** 0.13 as a fraction is:

   A. $\dfrac{13}{10}$ B. $\dfrac{13}{100}$ C. $\dfrac{13}{1000}$ D. $\dfrac{1.3}{100}$

8. **MC** 0.207 as a fraction is:

   A. $\dfrac{207}{1000}$
   B. $\dfrac{207}{100}$
   C. $2\dfrac{7}{10}$
   D. $20\dfrac{7}{10}$

9. **MC** 0.52 as a fraction in simplest form is:

   A. $\dfrac{52}{100}$
   B. $\dfrac{26}{50}$
   C. $\dfrac{13}{25}$
   D. $\dfrac{26}{100}$

10. **MC** 0.716 as a fraction in simplest form is:

    A. $\dfrac{716}{10\,000}$
    B. $\dfrac{368}{560}$
    C. $\dfrac{716}{1000}$
    D. $\dfrac{179}{250}$

11. **MC** 5.325 as a fraction in simplest form is:

    A. $\dfrac{5325}{1000}$
    B. $5\dfrac{325}{1000}$
    C. $5\dfrac{65}{200}$
    D. $5\dfrac{13}{40}$

12. **WE8** Change the following fractions to decimals.

    a. $\dfrac{3}{4}$
    b. $\dfrac{1}{2}$
    c. $\dfrac{4}{5}$
    d. $\dfrac{1}{20}$

13. Change the following fractions to decimals.

    a. $\dfrac{3}{12}$
    b. $\dfrac{1}{50}$
    c. $\dfrac{8}{25}$
    d. $\dfrac{3}{15}$

14. Write $\dfrac{1}{8}$ as a decimal. Using this value, write:

    a. $\dfrac{3}{8}$ as a decimal
    b. $\dfrac{7}{8}$ as a decimal
    c. $\dfrac{1}{16}$ as a decimal.

Communicating, reasoning and problem solving

15. The twin sisters in the photograph are keen tennis players. One of the twins has won 37 of her last 51 games and claims to be a much better player than her sister, who has won 24 out of her last 31 games.

    a. Compare the twins' results to determine who is the better player.
    b. Explain your answer to the twins.

Alyona
Games played = 51
Games won = 37

Darya
Games played = 31
Games won = 24

16. In a survey, 24 people were asked: 'What is the maximum distance you would walk to a train station?' A person who says '800 m' will walk up to 800 m, but no more. The survey responses are:

100 m	200 m	250 m	1 km	$\dfrac{1}{2}$ km	$\dfrac{3}{4}$ km
600 m	450 m	500 m	100 m	1.2 km	800 m
1.5 km	1.4 km	200 m	300 m	1.2 km	350 m
900 m	750 m	300 m	650 m	320 m	100 m

Determine the longest distance (as a multiple of 100 m) that at least $\dfrac{3}{4}$ of the people would walk to the train station. Justify your response.

17. You are competing in a long jump contest. Your first jump was 3.78 m, your second jump was $3\frac{8}{9}$ m and your third jump is $3\frac{4}{5}$ m. You are congratulated on your third jump, as being your best. Explain whether you agree.

18. A student received the following scores in his last three tests: History $\frac{12}{17}$, English $\frac{7}{10}$ and Science $\frac{71}{101}$. Place his test scores in order from highest to lowest score (by first converting each score to a decimal).

19. Place the following fractions in order from smallest to largest by first converting the fractions to decimals:

$$\frac{3}{7}, \frac{4}{15}, \frac{8}{25}, \frac{3}{11}, \frac{30}{100}, \frac{4}{12}, \frac{325}{1000}$$

20. The decimal equivalents of $\frac{1}{7}, \frac{2}{7}, \frac{3}{7} \cdots \frac{6}{7}$ can be found in the sequence 142 857 142 857 142 857 ...
If $\frac{1}{7} = 0.\dot{1}42\,85\dot{7}$ and $\frac{2}{7} = 0.\dot{2}85\,71\dot{4}$, without using a calculator, write $\frac{3}{7}, \frac{4}{7}, \frac{5}{7}$ and $\frac{6}{7}$ as decimals.

LESSON
5.4 Rounding decimals and terminating and recurring decimals

LEARNING INTENTION

At the end of this lesson, you should be able to:
- round terminating and recurring decimals to the required decimals of decimal places.

▶ 5.4.1 Rounding

eles-3945

- When rounding decimals, look at the first digit after the number of decimal places required.

Rules for rounding

- **If the first digit after the number of decimal places required is 0, 1, 2, 3 or 4, write the number without any change.**
- **If the first digit after the number of decimal places required is 5, 6, 7, 8 or 9, add one to the digit in the last required decimal place.**

- We can use the ≈ symbol to represent approximation when rounding has occurred, as shown in Worked example 10.

Round the following to 2 decimal places.

a. **3.641 883**
b. **18.965 402 0**

THINK	WRITE
a. 1. Write the number and underline the required decimal place.	a. 3.6<u>4</u>1 883
2. Circle the next digit and round according to the rule. *Note:* Since the circled digit is less than 5, we leave the number as it is.	= 3.64①883 ≈ 3.64
b. 1. Write the number and underline the required decimal place.	b. 18.9<u>6</u>5 402 0
2. Circle the next digit and round according to the rule. *Note:* Since the circled digit is greater than or equal to 5, add 1 to the last decimal place that is being kept.	= 18.96⑤4020 ≈ 18.97

- If you need to add 1 to the last decimal place and the digit in this position is a 9, the result is 10. The 0 is put in the last required place and the 1 is added to the digit in the next place to the left. For example, 0.298 rounded to 2 decimal places is 0.30.

Round the following to the number of decimal places shown in the brackets.

a. **27.462 973 (4)**
b. **0.009 94 (3)**

THINK	WRITE
a. 1. Write the number and underline the required decimal place.	a. 27.462 9<u>7</u>3
2. Circle the next digit and round according to the rule. *Note:* Since the circled digit is greater than 5, add 1 to the last decimal place that is being kept. As 1 is being added to 9, write 0 in the last place and add 1 to the previous digit.	= 27.462 9⑦3 ≈ 27.4630
b. 1. Write the number and underline the required decimal place.	b. 0.009 <u>9</u>4
2. Circle the next digit and round according to the rule. *Note:* Since the circled digit is greater than 5, add 1 to the last decimal place that is being kept. As 1 is being added to 9, write 0 in the last place and add 1 to the previous digit.	= 0.009 ⑨4 ≈ 0.0<u>1</u>0

Round 8.672 to the nearest unit.

THINK	WRITE
1. Write the decimal.	8.672
2. Look at the first digit after the decimal point. Since it is greater than 5, add 1 to the whole number.	≈ 9

- When trying to answer Worked example 12, you can think of the question this way: 'Is 8.672 closer to 8 or 9?'

5.4.2 Rounding to the nearest 5 cents

eles-3946

- The smallest denomination of physical money in Australia is the 5-cent coin, so prices must be rounded to the nearest 5 cents upon payment.
- Rounding to the nearest 5 cents works in the same way as rounding to the nearest 10: determine whether the number is closest to 0, 5 or 10, then round accordingly.

Rounding guidelines for cash transactions	
Final cash amount	**Round to:**
1 and 2 cents	nearest 10
3 and 4 cents	nearest 5
6 and 7 cents	nearest 5
8 and 9 cents	nearest 10

Source: www.accc.gov.au

WORKED EXAMPLE 13 Rounding to the nearest 5 cents

Melinda had $51.67 in her bank account. She wanted to withdraw all her money, so the bank rounded the amount to the nearest 5 cents. State how much money the teller gave Melinda.

THINK	WRITE
1. Write the actual amount she had in her account.	$51.67
2. Determine whether the last digit is closer to 5 or closer to 10, then rewrite the approximate value. *Note:* Alternatively it can be seen that 67 cents is closer to 65 cents than 70 cents.	$\approx$ $51.65
3. Write the answer in a sentence.	Melinda will receive $51.65 from the bank and leave 2c in the bank account.

5.4.3 Terminating and recurring (repeating) decimals

eles-3947

- A **terminating decimal** is a decimal number that ends (terminates) after any number of decimal places; for example, $\dfrac{9}{20} = 0.45$ is a terminating decimal.

- A **recurring decimal**, or a repeating decimal, occurs when we divide the denominator into the numerator and the answer keeps repeating.

  For example, $\dfrac{1}{3} = 0.333 \ldots$

  $\dfrac{9}{11} = 0.181\,81 \ldots$

WORKED EXAMPLE 14 Repeating decimals

Convert $\dfrac{1}{11}$ to a decimal. Continue dividing until a pattern emerges, then round the answer to 2 decimal places.

THINK	WRITE
1. Set out the question as for division of whole numbers, adding a decimal point and enough zeros to see a pattern emerging.	$11\overline{)1.0000}$

2. Divide, writing the answer with the decimal point exactly in line with the decimal point in the question. (The amount left over each time is 10 then 1, then 10 then 1 again. The decimal answer is also repeating.)

$$0.0\ \ 9\ \ 0\ \ 9\ \ 0\ \ 9\ \ 0\ \ 9...$$
$$11\overline{)1.^{1}0^{10}0^{1}0^{10}0^{1}0^{10}0^{1}0^{10}0}$$

3. Write the approximate answer rounded to 2 decimal places.

$$\frac{1}{11} \approx 0.09$$

- Recurring decimals can be written in one of the following shorter ways for an exact answer.
  - 4.6666 ... could be written as $4.\dot{6}$ (with a dot above the repeating part of the decimal).
  - 3.512 512 ... could be written as $3.\dot{5}1\dot{2}$ (with a dot above the first and last digits of the repeating part) or as $3.\overline{512}$ (with a line, called vinculum, above the repeating part of the decimal).
  - Like terminating decimals, the decimal equivalent of a fraction can be used to determine the decimal equivalent of any multiple of that fraction.

WORKED EXAMPLE 15 Equivalent repeating fractions

Use the result from Worked example 14 to calculate the decimal equivalent for $\frac{6}{11}$.

THINK

1. Write the decimal equivalent for the fraction with 1 as the numerator. In this case, it is an infinite recurring decimal. Therefore 0.090 909 ... can be written as $0.\overline{09}$.

2. Multiply both sides of this equation by the appropriate multiple (6 in this case).

3. Simplify and write the answer.

WRITE

$$\frac{1}{11} = 0.090\,909\,...$$

$$\frac{1}{11} \times 6 = 0.090\,909\,... \times 6$$
$$= 0.545\,454\,...$$

$$\frac{6}{11} = 0.\overline{54}$$

DISCUSSION

When using a calculator to perform a calculation, for example $2 \div 3$, the answer may show an approximation to a recurring decimal, for example 0.666 666 667. Why do you think this may be?

ᴰ   Math ▲

2÷3

0.6666666667

  Resources

Interactivities Rounding (int-3980)
Repeating decimals (int-3981)

Exercise 5.4 Rounding and repeating decimals

5.4 Quick quiz on	5.4 Exercise

Individual pathways

■ PRACTISE	■ CONSOLIDATE	■ MASTER
1, 4, 8, 12, 14, 17, 20, 23, 24, 28	2, 5, 7, 9, 10, 15, 16, 18, 21, 27, 29	3, 6, 11, 13, 19, 22, 25, 26, 30

Fluency

1. **WE10** Round the following to 2 decimal places.

    a. 0.3241   b. 0.863   c. 1.246 10   d. 13.049 92   e. 3.0409

2. Round the following to 2 decimal places.

    a. 7.128 63   b. 100.813 82   c. 71.260 39   d. 0.0092   e. 2.386 214

3. Round the following to 2 decimal places.

    a. 0.185 00   b. 19.6979   c. 0.3957   d. 0.999   e. 8.902

4. Round the following to 1 decimal place.

    a. 0.410   b. 0.87   c. 9.27   d. 25.25   e. 3.92

5. Round the following to 1 decimal place.

    a. 300.06   b. 12.82   c. 99.91   d. 8.88   e. 14.034 59

6. Round the following to 1 decimal place.

    a. 17.610 27   b. 0.8989   c. 93.994   d. 0.959 027   e. 96.280 49

7. **MC** 2.998 rounded to 1 decimal place is:

    A. 3.0   B. 2.9   C. 2.8   D. 3.1

8. **WE12** Round the following to the nearest unit.

    a. 10.7   b. 8.2   c. 3.6   d. 92.7   e. 112.1

9. Round the following to the nearest unit.

    a. 21.76   b. 42.0379   c. 2137.50   d. 0.12   e. 0.513

10. Write the following infinite recurring decimals using one of the short forms.

    a. 2.555 ...   b. 0.666 ...   c. 49.111 11 ...   d. 0.262 626 ...

11. Write the following infinite recurring decimals using one of the short forms.

    a. 0.913 913 ...   b. 8.641 864 18 ...   c. 133.946 246 2 ...   d. 0.127 7777 ...

Understanding

12. **WE11** Convert each of the following to a decimal. Continue dividing until a pattern emerges, then round the answer to the number of decimal places indicated in the brackets.

    a. $\dfrac{1}{3}$ (1 d.p.)   b. $\dfrac{1}{9}$ (1 d.p.)   c. $\dfrac{5}{12}$ (3 d.p.)

13. Convert each of the following to a decimal. Continue dividing until a pattern emerges, then round the answer to the number of decimal places indicated in the brackets.

    a. $\dfrac{1}{6}$ (2 d.p.)   b. $\dfrac{2}{11}$ (3 d.p.)   c. $\dfrac{7}{15}$ (2 d.p.)

14. **WE14** Convert $\frac{5}{9}$ to a decimal. Continue dividing until a pattern emerges, then round to 2 decimal places.

15. Convert $\frac{1}{6}$ to a decimal. Continue dividing until a pattern emerges, then round to 3 decimal places.

16. **WE15** Calculate the decimal equivalent of $\frac{5}{6}$.

17. **MC** 1.8888 ... written as an exact answer is:

    A. $1.\dot{8}$ **B.** 1.888   **C.** 1.88   **D.** 1.9   **E.** 1.889

18. **MC** $\frac{3}{7}$ as a decimal in exact form is:

    A. $0.\dot{4}285\,7\dot{1}$ **B.** $0.\overline{428}$ **C.** 0.4   **D.** 0.428 517 4   **E.** $2.\dot{3}$

19. Determine decimal equivalents for the following fractions.

    a. $\frac{7}{2}$ **b.** $4\frac{2}{3}$ **c.** $\frac{3}{4}$

20. Round the following to the nearest ten.

    a. 13   **b.** 76   **c.** 138   **d.** 262   **e.** 175

21. Round the following to the nearest hundred.

    a. 320   **b.** 190   **c.** 138   **d.** 6751   **e.** 9750.051

22. Round the following to the nearest thousand.

    a. 3426   **b.** 12 300   **c.** 18 640   **d.** 9462   **e.** 1098

Communicating, reasoning and problem solving

23. **WE13** In the supermarket, Christine's shopping bill came to $27.68. As there are no 1- or 2-cent pieces, this amount must be rounded to the nearest 5 cents. State how much Christine will pay in cash for her shopping.

24. Rank the following decimals from smallest to largest:
$0.295, 0.29\dot{5}, 0.2\dot{9}\dot{5}, 0.\dot{2}9\dot{5}$

25. Write a decimal that is greater than $0.\dot{5}432\dot{1}$ and less than 0.543 22.

26. Using a calculator, Greg worked out that the piece of timber required to finish making a support for a gate should be 3.567 82 metres. Realistically, the timber can be measured only to the nearest millimetre (nearest thousandth of a metre). What measurement should be used for the length of the timber? Explain why 3.567 82 m is unreasonable as a measurement for timber.

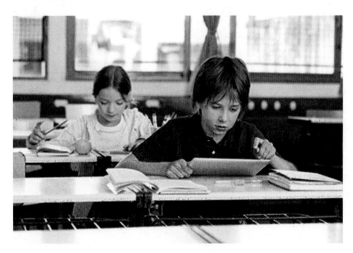

27. Barney rounds a decimal to 0.6 correct to 1 decimal place. Fred then says that the number before being rounded off could not have been bigger than 0.64. Explain whether Fred is correct.

28. The maximum temperature was recorded as 24.7 °C. In the news broadcast, the presenter quoted this to the nearest degree. Write the temperature quoted.

29. a. Round the decimal number 0.8375 to the nearest:
    i. tenth   ii. hundredth   iii. thousandth.
    b. Explain whether the rules for rounding a decimal up or down depend on the number of decimal places.

30. Divide 100 by the first 10 prime numbers. Write the answers correct to 4 decimal places.

LESSON
5.5 Adding and subtracting decimals

LEARNING INTENTION

At the end of this lesson you should be able to:
• add and subtract decimal numbers.

▶ 5.5.1 Adding decimals

eles-3948

• Decimals can be added using a method similar to that for whole numbers.
    Step 1: Set out the addition in vertical columns and line up the decimal points so that the digits with the same place value are underneath each other.
    Step 2: If the question is not written in columns, it is necessary to rewrite it with the decimal points lined up.
    Step 3: Fill the empty places with zeros.

Calculate:

a. $\begin{array}{r} 1.3 \\ +0.5 \\ \hline \end{array}$

b. $\begin{array}{r} 12.84 \\ +2.33 \\ \hline \end{array}$

c. $\begin{array}{r} 1.25 \\ 3.146 \\ +7.0 \\ \hline \end{array}$

THINK	WRITE
a. Copy the question exactly and add the digits as for whole numbers, working from right to left. Write the decimal point directly below the decimal points in the question.	a. $\begin{array}{r} 1.3 \\ +0.5 \\ \hline 1.8 \end{array}$
b. Copy the question exactly and add the digits as for whole numbers, working from right to left. Write the decimal point directly below the decimal points in the question.	b. $\begin{array}{r} {}^{1} \\ 12.48 \\ +2.33 \\ \hline 14.81 \end{array}$
c. Write the question, replacing the spaces with zeros. Add the digits as for whole numbers, working from right to left. Write the decimal point directly below the decimal points in the question.	c. $\begin{array}{r} 1.250 \\ 3.146 \\ +7.000 \\ \hline 11.396 \end{array}$

Rewrite in columns, then add $0.26 + 1.8 + 12.214$.

THINK	WRITE
1. Write the question in columns with the decimal points directly beneath each other with the zeros included.	$\begin{array}{r} {}^{1} \\ 0.260 \\ 1.800 \end{array}$
2. Add the digits as for whole numbers. Write the decimal point directly below the decimal points in the question.	$\begin{array}{r} +12.214 \\ \hline 14.274 \end{array}$

5.5.2 Subtracting decimals

eles-3949

- Decimals can be subtracted using a method similar to that for whole numbers.
- Set out the subtraction in vertical columns and line up the decimal points so that the digits with the same place value are underneath each other.
- If the question is not written in columns, it is necessary to rewrite it with the decimal points lined up.
- Fill the empty places with zeros.

WORKED EXAMPLE 18 Subtracting numbers with decimals

Calculate:
$$\begin{array}{r} 0.56 \\ -0.14 \\ \hline \end{array}$$

THINK	WRITE
Copy the question exactly and subtract the digits as for whole numbers, working from right to left. Write the decimal point directly below the decimal points in the question.	$\begin{array}{r} 0.56 \\ -0.14 \\ \hline 0.42 \\ \hline \end{array}$

WORKED EXAMPLE 19 Subtracting numbers with decimals

Rewrite the following in columns, then subtract.
a. $1.82 - 0.57$
b. $2.641 - 0.85$

THINK	WRITE
a. Write in columns with the decimal points directly under each other. Subtract, and insert the decimal point directly below the other decimal points in the question.	a. $\begin{array}{r} {}^{7}1 \\ 1.8\!\!\!/2 \\ -0.57 \\ \hline 1.25 \\ \hline \end{array}$
b. Write in columns with the decimal points directly under each other, adding zeros as appropriate. Subtract as for whole numbers and insert the decimal point directly below the other decimal points.	b. $\begin{array}{r} {}^{5}1 \\ 2.6\!\!\!/41 \\ -0.850 \\ \hline 1.791 \\ \hline \end{array}$

Checking by estimating

- Answers to decimal addition and subtraction can be checked by estimating. Round each decimal to the nearest whole number or to a similar number of decimal places and then add or subtract them.
  For example, in Worked example 17, you could check the answer by rounding to get an estimate by adding $0.3 + 2 + 12 = 14.3$, which is close to 14.274.
  In Worked example 18, you could check the answer by rounding to get an estimate $0.6 - 0.1 = 0.5$, which is close to 0.42.

DISCUSSION

What is your method for estimating the answer to a decimal addition or subtraction?

---

on Resources

Interactivities Addition of decimals (int-3982)
Checking by estimating (int-3983)
Subtraction of decimals (int-3984)

Exercise 5.5 Adding and subtracting decimals

5.5 Quick quiz on	5.5 Exercise

Individual pathways

■ PRACTISE	■ CONSOLIDATE	■ MASTER
1, 3, 6, 8, 11, 15, 18, 21, 24, 25, 28, 29	2, 4, 9, 12, 16, 19, 22, 26, 30, 31, 32	5, 7, 10, 13, 14, 17, 20, 23, 27, 33, 34, 35

Fluency

1. **WE16a,b** Calculate the following.

   a. $\begin{array}{r} 1.2 \\ +2.3 \\ \hline \end{array}$
   b. $\begin{array}{r} 1.67 \\ +1.02 \\ \hline \end{array}$
   c. $\begin{array}{r} 8.062 \\ +5.177 \\ \hline \end{array}$
   d. $\begin{array}{r} 10.0364 \\ +92.1494 \\ \hline \end{array}$

2. **WE16c** Calculate the following, after filling the blank spaces.

   a. $\begin{array}{r} 6.27 \\ +0.5 \\ \hline \end{array}$
   b. $\begin{array}{r} 3.26 \\ +18.6460 \\ \hline \end{array}$
   c. $\begin{array}{r} 4.2 \\ 62.013 \\ +1946.12 \\ \hline \end{array}$
   d. $\begin{array}{r} 48.129\,06 \\ 9 \\ +204.32 \\ \hline \end{array}$

3. **WE17** Rewrite the following in columns, then add. Check your answer by rounding to get an estimate.

   a. $1.4 + 3.2$
   b. $6.5 + 0.4$
   c. $3.261 + 0.21$
   d. $15.987 + 1.293$

4. Rewrite the following sums, then add. Check your answer by rounding to get an estimate.

   a. $10.8271 + 6.5$
   b. $1.8 + 18.6329$
   c. $0.24 + 3.16 + 8.29$
   d. $14.23 + 1.06 + 86.29 + 3.64$

5. Rewrite the following sums, then add. Check your answer by rounding to get an estimate.

   a. $5.27 + 1.381 + 12.3$
   b. $100 + 4.3 + 0.298 + 1.36$
   c. $82.3 + 100.6 + 0.9949 + 9$
   d. $126 + 372.8 + 100.0264 + 2020.13$

6. **WE18** Calculate the following, filling the spaces with zeros as required. Check your answer by rounding to get an estimate.

   a. $\begin{array}{r} 6.87 \\ -6.27 \\ \hline \end{array}$
   b. $\begin{array}{r} 12.231 \\ -8.026 \\ \hline \end{array}$
   c. $\begin{array}{r} 0.6301 \\ -0.5495 \\ \hline \end{array}$
   d. $\begin{array}{r} 3.0091 \\ -1.6723 \\ \hline \end{array}$

7. Calculate the following, filling the spaces with zeros as required. Check your answer by rounding to get an estimate.

   a. $\begin{array}{r} 31.02 \\ -26 \\ \hline \end{array}$
   b. $\begin{array}{r} 98.26 \\ 9.07 \\ \hline \end{array}$
   c. $\begin{array}{r} 146 \\ -58.91 \\ \hline \end{array}$
   d. $\begin{array}{r} 3.2 \\ -0.467 \\ \hline \end{array}$

8. **WE19** Rewrite the following in columns, then subtract. Check your answer by rounding to get an estimate.

   a. $5.64 - 2.3$
   b. $12.07 - 6.14$
   c. $0.687 - 0.36$
   d. $15.226 - 11.08$

9. Rewrite the following in columns, then subtract. Check your answer by rounding to get an estimate.

   a. $6.734 - 4.8$
   b. $12.2 - 8.911$
   c. $100.562 - 86.0294$
   d. $38 - 21.234$

10. Rewrite the following in columns, then subtract. Check your answer by rounding to get an estimate.

    a. $47 - 8.762$
    b. $5 - 0.8864$
    c. $0.2 - 0.0049$
    d. $3.279 - 2.506\,84$

Understanding

11. **MC** $1.6 + 4.8$ equals:

    A. 5.4   **B.** 6.4   **C.** 0.54   **D.** 0.64

12. **MC** $3.26 + 0.458$ equals:

    A. 3.718   **B.** 0.784   **C.** 0.037 18   **D.** 3.484

13. **MC** $1.84 + 0.61 + 4.07$ equals:

    A. 6.52   **B.** 6.42   **C.** 5.42   **D.** 5.52

14. **MC** $216 + 1.38 + 0.002\,64$ equals:

    A. 217.4064   **B.** 0.618   **C.** 217.644   **D.** 217.382 64

15. **MC** $0.39 - 0.15$ equals:

    A. 0.0024   **B.** 0.024   **C.** 0.24   **D.** 2.4

16. **MC** $1.4 - 0.147$ would be rewritten as:

    A. $\begin{array}{r} 1.4 \\ -0.417 \\ \hline \end{array}$ **B.** $\begin{array}{r} 1.400 \\ -0.417 \\ \hline \end{array}$ **C.** $\begin{array}{r} 1.40 \\ -1.47 \\ \hline \end{array}$ **D.** $\begin{array}{r} 1.004 \\ -0.147 \\ \hline \end{array}$

17. **MC** $0.3 - 0.024$ equals:

    A. 0.06   **B.** 0.276   **C.** 0.7   **D.** 0.76

18. **MC** $150.278 - 0.99$ equals:

    A. 150.728   **B.** 149.288   **C.** 1.49288   **D.** 159.388

19. Josh deposited $27.60 into his bank account. If his balance before the deposit was $139.40, determine Josh's new bank balance.

20. Jessica bought the following items at the school canteen: 1 can of Coke for $1.60, 1 sausage roll for $1.20, 1 packet of chips for $1.50 and 2 red frogs for $0.40 (red frogs cost 20 cents each). Calculate how much Jessica spent.

21. A triathlon consists of a 0.5-kilometre swim, a 15.35-kilometre bike ride and a 4.2-kilometre run. Calculate how far the competitors travel altogether.

22. Amy walked 3.6 kilometres to school, 0.8 kilometres from school to the shops, 1.2 kilometres from the shops to a friend's house and finally 2.5 kilometres from her friend's house to her their home. Calculate how far Amy walked.

23. For lunch Paula ordered 1 potato cake, 1 dim sim, minimum chips and a milkshake from the menu below. Calculate how much Paula spent on her lunch.

MENU			
Flake	$3.50	Coffee	$2.20
Whiting	$3.50	Tea	$2.20
Dim sims	$0.60	Soft drinks	$1.80
Potato cakes	$0.50	Milkshakes	$3.00
Minimum chips	$2.50	Water	$1.80

24. Ryan works in a newsagency. A customer buys $9.65 worth of goods and gives Ryan a $20 note. Calculate the change Ryan gives the customer.

Communicating, reasoning and problem solving

25. A jockey has a mass of 52.3 kilograms. After exercising and training for 2 days and spending time in a sauna, the jockey has lost 1.82 kilograms. Determine the jockey's mass now.

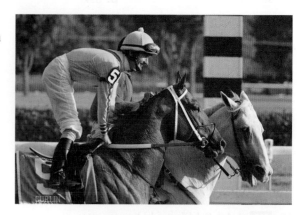

26. If 1.27 metres is cut from a piece of material that is 13 metres long, calculate how much material is left.

27. Gary and Liz are replacing the skirting boards in their lounge room. They know the perimeter of the room is 34.28 metres. If there is a door 0.82 metres wide and a fireplace 2.18 metres wide that do not require skirting boards, explain (with working) how much wood they will need to buy for their lounge room.

28. Cathy Freeman won a particular 400-m race in 51.35 seconds. In her next race, her time was 2.97 seconds faster than this. Determine Cathy's time for the later race.

29. Without using a calculator, determine the average of the numbers 0.1, 0.11 and 0.111.

30. Explain how rounding is used in estimation.

31. Jo and Anton made a certain number of telephone calls over a 2-month period. There is a charge of 25 cents for each call, and the monthly rental of the phone line is $13.60. The bill for the 2 months comes to $39.45. Determine how many calls they made in the 2-month period.

32. The following table shows the times recorded for each swimmer in the under-13, 50-metre freestyle relay for 6 teams.

	Times for each swimmer (seconds)			
Team	Swimmer 1	Swimmer 2	Swimmer 3	Swimmer 4
1	36.7	41.3	39.2	35.8
2	38.1	46.5	38.8	35.9
3	34.6	39.2	39.9	35.2
4	41.6	40.8	43.7	40.5
5	37.9	40.2	38.6	39.2
6	38.3	39.1	40.8	37.6

a. Determine the total time for each team. Put your results in a table.
b. Identify which team won the relay.
c. Determine the difference in time between the first and second placed teams.

33. Australian coins, which are minted by the Royal Australian Mint, have the following masses:

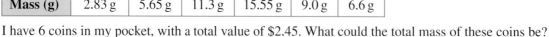

Coin	5c	10c	20c	50c	$1	$2
Mass (g)	2.83 g	5.65 g	11.3 g	15.55 g	9.0 g	6.6 g

I have 6 coins in my pocket, with a total value of $2.45. What could the total mass of these coins be?

34. You purchased some shares over a 4-month period. The price fell by $4.23 in the first month, rose by $6.67 in the second month, rose by $1.35 in the third month and fell by $3.28 in the fourth month. Explain whether the shares increased or decreased in value over the four months, and by how much.

35. Arrange the digits 1 to 9 into the boxes at right to make a true decimal addition.

$$
\begin{array}{r}
\square\,.\,\square\;\square \\
+\;\square\,.\,\square\;\square \\
\hline
\square\,.\,\square\;\square
\end{array}
$$

LESSON
5.6 Multiplying decimals

LEARNING INTENTION

At the end of this lesson you should be able to:
- multiply decimals, including by multiples of 10
- square decimal numbers.

▶ 5.6.1 Multiplying decimals

eles-3950

- The following calculation shows the multiplication 1.7×2.3. The diagram is a visual representation (area model) of each step in the calculation. There are 1.7 rows of 2.3, or 1.7 groups of 2.3.

$$
\begin{array}{r}
1.7 \\
\times 2.3 \\
\hline
0.21 \\
0.3 \\
1.4 \\
2.0 \\
\hline
3.91 \\
\end{array}
$$

$\leftarrow\;\;0.3 \times 0.7 = 0.21$
$\leftarrow\;\;0.3 \times 1.0 = 0.3$
$\leftarrow\;\;2.0 \times 0.7 = 1.4$
$\leftarrow\;\;2.0 \times 1.0 = 2.0$
$\leftarrow\;\;$**Total**

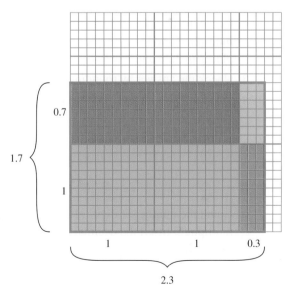

- The place value of the answer is determined by the place value of each of the decimals being multiplied. The multiplication above has decimal place values of 1 decimal place (tenths) multiplied by 1 decimal place (tenths) = 2 decimal places (hundredths) in the answer.

Multiplying decimals

The number of decimal places in the answer is equal to the total number of decimal places in the numbers being multiplied.

For example:

$$2.43 \times 1.3 = 3.159$$

2 decimals + 1 decimal = 3 decimals

WORKED EXAMPLE 20 Multiplying numbers with decimals

Calculate the following.

a. 12.6×7
b.
$$\begin{array}{r} 3.26 \\ \times\,0.4 \\ \hline \end{array}$$
c.
$$\begin{array}{r} 0.4629 \\ \times\,2.6 \\ \hline \end{array}$$

THINK	WRITE
a. 1. Rewrite and multiply the digits, ignoring the decimal places.	**a.** $\begin{array}{r} 126 \\ \times\,7 \\ \hline 882 \end{array}$
2. Count the number of decimal places altogether (1) and put in the decimal point.	12.6×7 $= 88.2$
3. Check the answer by rounding: $10 \times 7 = 70$, which is close to 88.2.	
b. 1. Multiply, ignoring the decimal places.	**b.** $\begin{array}{r} 326 \\ \times\,4 \\ \hline 1304 \end{array}$
2. Count the number of digits after the point in both the decimals being multiplied and insert the decimal point in the answer. There are 2 decimal places in 3.26 and 1 in 0.4 so there will be 3 decimal places in the answer.	3.26×0.4 $= 1.304$
3. Check the answer by rounding: $3 \times 0.4 = 1.2$, which is close to 1.304.	
c. 1. Multiply, ignoring the decimal places.	**c.** $\begin{array}{r} 4629 \\ \times\,26 \\ \hline 27\,774 \\ 92\,580 \\ \hline 120\,354 \end{array}$
2. Count the number of digits after the point in both the decimals being multiplied. Insert the decimal point in that position in the answer. There are 4 decimal places in 0.4629 and 1 decimal place in 2.6, so there will be 5 decimal places in the answer.	0.4629×2.6 $= 1.203\,54$

Digital technology

Scientific calculators can evaluate multiplication of numbers with decimals.

$$3.26 \times 0.4$$

D   Math ▲

$$1.304$$

DISCUSSION

Does multiplication always result in a larger number?

▶ 5.6.2 Squaring decimals

eles-3951

- To square a decimal, multiply the number by itself.

Squaring decimals

- **The number of decimal places in the square is twice the number of decimal places in the original number.**

- The following diagrams show how squaring decimal numbers can be represented visually.

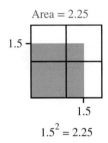
Area = 2.25
1.5
1.5
$1.5^2 = 2.25$

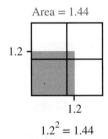
Area = 1.44
1.2
1.2
$1.2^2 = 1.44$

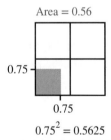
Area = 0.56
0.75
0.75
$0.75^2 = 0.5625$

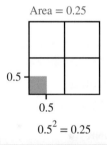
Area = 0.25
0.5
0.5
$0.5^2 = 0.25$

WORKED EXAMPLE 21 Squaring decimal numbers

Calculate the following.
a. 0.5^2
b. 1.2^2

THINK	WRITE
a. 1. Multiply the number by itself, ignoring the decimal places.	**a.** $5 \times 5 = 25$
2. Count the number of digits after the point in both the decimals being multiplied and insert the decimal point in the answer. There will be 2 decimal places in the answer.	$0.5 \times 0.5 = 0.25$
3. Write the answer.	$0.5^2 = 0.25$
b. 1. Multiply the number by itself, ignoring the decimal places.	**b.** $12 \times 12 = 144$

2. Count the number of digits after the point in both the decimals being multiplied and insert the decimal point in the answer. There will be 2 decimal places in the answer.

$1.2 \times 1.2 = 1.44$

3. Write the answer.

$1.2^2 = 1.44$

▶ 5.6.3 Multiplying by multiples of 10

eles-3952

- The multiples of 10 are $10, 20, 30, 40, \ldots 120, \ldots 1000, \ldots 1500, \ldots 20\,000$ and so on; they always end in zero.
- Multiplying by a multiple of 10 can be carried out by writing the multiple of 10 as the product of two factors.
- The factors used are the non-zero digits from the multiple of 10 and the corresponding power of 10.
  For example, $8700 = 87 \times 100$, so multiplying by 8700 is the same as multiplying by 87 and then multiplying by 100.
- To multiply a number with a decimal by a power of 10, move the decimal point to the right by the number of zeros.

Multiplying by 10	Multiplying by 100	Multiplying by 1000
$2.436 \times 10 = 24.36$	$2.436 \times 10 = 243.6$	$2.436 \times 10 = 2436$
Move the decimal point 1 place to the right.	Move the decimal point 2 places to the right.	Move the decimal point 3 places to the right.

WORKED EXAMPLE 22 Multiplying by multiples of 10

Calculate:

a. **5.1×600**

b. **$0.0364 \times 24\,000$**

THINK	WRITE
a. 1. Multiplying by 600 is the same as first multiplying by 6 and then multiplying by 100. Calculate 5.1×6.	a. $\begin{array}{r} 5.1 \\ \underline{\times\ 6} \\ \underline{30.6} \end{array}$
2. Multiply the result by 100. Move the position of the decimal point 2 places to the right.	$30.6 \times 100 = 3060$
3. Write the answer.	$5.1 \times 600 = 3060$
b. 1. Multiplying by 24 000 is the same as first multiplying by 24 and then multiplying by 1000. Calculate 0.0364×24.	b. $\begin{array}{r} 0.0364 \\ \underline{\times\ 24} \\ 1456 \\ \underline{7280} \\ \underline{0.8736} \end{array}$
2. Multiply the result by 1000. Move the position of the decimal point 3 places to the right.	$0.8736 \times 1000 = 873.6$
3. Write the answer.	$0.0364 \times 24\,000 = 873.6$

Recognising short cuts

- When multiplying by some common decimals, such as 0.1 or 0.25, you can use short cuts to help you complete the calculation more efficiently.

  For example, the fractional equivalent of 0.1 is $\frac{1}{10}$, so instead of multiplying by 0.1, you can multiply by $\frac{1}{10}$, which is the same as dividing by 10. Similarly, the fractional equivalent of 0.25 is $\frac{1}{4}$, so 0.25 of $60 is equivalent to $\frac{1}{4}$ of $60, which is equivalent to $60 ÷ 4, so $15.

Exercise 5.6 Multiplying decimals   **learn** on

5.6 Quick quiz on	**5.6 Exercise**

Individual pathways

■ PRACTISE	■ CONSOLIDATE	■ MASTER
1, 4, 7, 10, 13, 15, 20, 22, 27	2, 5, 8, 11, 14, 16, 19, 24, 25, 28	3, 6, 9, 12, 17, 18, 21, 23, 26, 29

Fluency

1. Calculate the following.
   - a. 3.5×4
   - b. 15.7×8
   - c. 10.2×6
   - d. 22.34×5

2. Calculate the following.
   - a. 27.18×7
   - b. 64.87×8
   - c. 1.064×6
   - d. $0.264\,81 \times 3$

3. **WE20a** Calculate the following.
   - a. 1.4×0.6
   - b. 4.2×0.7
   - c. 0.8×0.4
   - d. 9.7×0.8

4. Calculate the following.
   - a. 0.35×0.4
   - b. 0.64×0.3
   - c. 0.77×0.5
   - d. 0.49×0.9

5. Calculate the following.
   - a. 5.38×0.8
   - b. 0.347×0.6
   - c. 4.832×0.6
   - d. 12.2641×0.4

6. Calculate the following.
   - a. 0.002×0.05
   - b. 0.003×0.004
   - c. 0.7×0.09
   - d. 0.037×0.006

7. Calculate the following.
   - a. $0.000\,061 \times 0.04$
   - b. 0.004×0.09
   - c. 0.56×0.7
   - d. $0.030\,31 \times 0.02$

8. **WE20b** Calculate the following.
   - a. 0.25×1.2
   - b. 0.37×2.3
   - c. 0.47×5.4
   - d. 0.05×3.5

9. Calculate the following.
   a. 0.79×8.3 **b.** 4.68×3.6 **c.** 8.04×7.5 **d.** 5.06×6.2

10. **MC** When calculating 8.32×0.64, the number of decimal places in the answer is:
    A. 1   **B.** 2   **C.** 3   **D.** 4

11. **MC** 0.2×0.2 equals:
    A. 0.004   **B.** 0.04   **C.** 0.4   **D.** 4

12. **MC** 1.4×0.8 equals:
    A. 1.12   **B.** 8.2   **C.** 82   **D.** 11.2

13. **MC** 0.0312×0.51 equals:
    A. 0.001 591 2   **B.** 0.015 912   **C.** 0.156 312   **D.** 0.159 12

Understanding

14. **WE21** Calculate the following.
    a. 0.02^2 **b.** 1.3^2 **c.** 2.05^2

15. **WE22** Calculate the following.
    a. 0.127×8000 **b.** $0.543 \times 11\,000$ **c.** $8.1 \times 54\,000$

16. Calculate the following.
    a. 3.64×300 **b.** 7.58×600 **c.** 42.6×900

17. Calculate the following by moving the position of the decimal point.
    a. 6.48×10 **b.** 13.896×100 **c.** 0.2708×1000
    d. $217.148\,96 \times 1000$ **e.** $0.820\,496\,783 \times 100\,000$

18. Evaluate the following by moving the position of the decimal point.
    a. $32.689\,043\,267 \times 100\,000$ **b.** $0.984\,326\,641 \times 1\,000\,000$
    c. $0.000\,278\,498\,32 \times 1\,000\,000$ **d.** 0.46×1000

19. For the following calculations:
    i. Estimate the answer by rounding each number to the first digit.
    ii. Calculate the answers using written methods.
    iii. Check your answer by using a calculator.
    a. 5.6×4400 **b.** $10.2 \times 32\,510$ **c.** 33.3×1515

20. Change the following amounts of money to cents. (*Hint:* There are 100 cents in one dollar.)
    a. $35   **b.** $127   **c.** $11   **d.** $25.35   **e.** $58.20

21. One thousand Year 7 students contributed 75 cents each to the bushfire appeal. Calculate how many dollars they contributed altogether.

22. Benjamin and Robyn were providing ice cream for 600 children. Calculate how much ice cream would be needed if each child was expected to eat 0.18 litres.

23. Calculate the area of the tennis court shown, given that area = length × width.

10.97 m

23.77 m

Communicating, reasoning and problem solving

24. Judy bought 34.5 litres of petrol at 92.9 cents per litre. She paid 3200 cents at the counter.
    a. Convert 3200 cents into dollars.
    b. Do you think Judy paid the correct amount for her petrol?
       Explain your answer.

25. A sheet of paper is 0.012 cm thick. Explain whether you would notice the difference in the height between a stack of 500 sheets and a stack of 550 sheets.

26. A ball is dropped from a height of 1 metre. If it reaches 0.7 times the height of the previous bounce at each bounce, determine the number of times the ball bounces until the height is less than 1 centimetre.

27. Michael bought 0.65 kilograms of cubed steak at a butcher's shop. The meat costs $8.50 per kilogram. Calculate how much Michael paid for the steak. *Note:* Round to the nearest 5c.

28. James is using the recipe for chocolate chip muffins to make 1.5 times the given amount. If the recipe lists 0.25 litres of milk in the ingredients, determine the amount of milk James should use for his muffins.

29. Sophie, Hamish and Thomas shared a pizza and paid according to what they ate. Sophie ate 4 slices, Hamish ate 3 slices and Thomas ate 5 slices. If Thomas's share of the bill was $4.50, calculate the amount Sophie and Hamish had to pay.

LESSON
5.7 Dividing decimals

LEARNING INTENTION

At the end of this lesson you should be able to:
- divide a decimal number by a whole number or a decimal number
- divide a decimal number by a multiple of 10.

▶ 5.7.1 Dividing a decimal by a whole number

eles-####

- Dividing decimals by a whole number can be represented visually.

$$1.6 \div 8 = 0.2$$

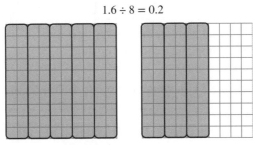

- $1.6 \div 8$ can be interpreted as '16 tenths divided into 8 groups $= 2$ tenths'.
- When dividing a decimal by a whole number, the amount of decimals will be the same in the answer.
- Knowing this, we can divide the digits as we would whole numbers and place the decimal point at the end.

$$8\overline{)16}^{\,0.2} \longrightarrow 8\overline{)1.6}^{\,0.2}$$

WORKED EXAMPLE 23 Dividing a decimal by a whole number

Calculate the value of 2.4896 ÷ 4.

THINK	WRITE
1. Set out the question as you would for whole numbers.	$4\overline{)2.4896}$
2. Divide 4 into the first digit of 2.4896 ($2 \div 4 = 0$ remainder 2). Write the 0 above the 2, and write the remainder beside the next digit, as shown in black.	$\begin{array}{r} 0 \\ 4\overline{)2.^24896} \end{array}$
3. The second digit in the number being divided is to the right of the decimal point, so write the decimal point in the answer directly above the decimal point in the question, as shown in pink.	$\begin{array}{r} 0. \\ 4\overline{)2.^24896} \end{array}$
4. Divide 4 into the second digit, which includes the carried 2 ($24 \div 4 = 6$ remainder 0). Write the 6 above the 4, as shown in plum.	$\begin{array}{r} 0.6 \\ 4\overline{)2.^24896} \end{array}$
5. Divide 4 into the third digit ($8 \div 4 = 2$). Write the 2 above the 8, as shown in pink.	$\begin{array}{r} 0.62 \\ 4\overline{)2.^24896} \end{array}$
6. Divide 4 into the fourth digit ($9 \div 4 = 2$ remainder 1). Write the 2 above the 9, and write the remainder beside the next digit, as shown in purple.	$\begin{array}{r} 0.622 \\ 4\overline{)2.^2489^16} \end{array}$
7. Divide 4 into the fifth digit, which includes the carried 1 ($16 \div 4 = 4$). Write the 4 above the 6, as shown in orange.	$\begin{array}{r} 0.6224 \\ 4\overline{)2.^2489^16} \end{array}$
8. Write the answer.	$2.4896 \div 4 = 0.6224$

- Sometimes, when you are dividing numbers, there will be a remainder. For example, $15.3 \div 4$:

$$\begin{array}{r} 3.8 \\ 4\overline{)15.^33} \end{array} \quad \text{remainder 1}$$

- Instead of leaving a remainder, add zeros to the end of the decimal and keep dividing until there is no remainder.

$$\begin{array}{r} 3.\ 8\ 2\ 5 \\ 4\overline{)15.^33^10^20} \end{array}$$

WORKED EXAMPLE 24 Dividing a decimal by a whole number

Calculate 21.76 ÷ 5. Add zeros and keep dividing until there is no remainder.

THINK	WRITE
1. Set up the division. Write the decimal point in the answer directly above the decimal point in the question and divide as for short division, adding zeros as required.	$\begin{array}{r} 4.\ 3\ 5\ 2 \\ 5\overline{)21.^17^26^10} \end{array}$
2. Check the answer by rounding: $20 \div 5 = 4$, which is close to 4.352.	

▶ 5.7.2 Dividing a decimal number by a multiple of 10

eles-3955

- When dividing by a multiple of 10, factorise the multiple to give a power of 10 and its other factor. Divide by the other factor first, and then by the power of 10.
  For example, to divide by 500, divide first by 5 and then by 100.

- To divide by a power of 10, move the decimal point to the left by the number of zeros.
  For example, to divide by 10, move the decimal point one place to the left;
  to divide by 100, move the decimal point two places to the left;
  to divide by 1000, move the decimal point three places to the left, and so on.

WORKED EXAMPLE 25 Dividing a decimal by a multiple of 10

Calculate:

a. $4.8 \div 40$ b. $19.2 \div 6000$

THINK	WRITE
a. 1. Dividing by 40 is the same as first dividing by 4 and then dividing by 10. First, divide by 4.	a. $\begin{array}{r} 1.2 \\ 4\overline{)4.8} \end{array}$
2. To divide by 10, move the position of the decimal point 1 place to the left.	$1.2 \div 10 = 0.12$
3. Write the answer.	$4.8 \div 40 = 0.12$
b. 1. Dividing by 6000 is the same as dividing by 6 and then dividing by 1000. First, divide by 6.	b. $\begin{array}{r} 3.2 \\ 6\overline{)19.2} \end{array}$
2. To divide by 1000, move the position of the decimal point 3 places to the left.	$3.2 \div 1000 = 0.0032$
3. Write the answer.	$19.2 \div 6000 = 0.0032$

COMMUNICATING — COLLABORATIVE TASK: Decimal division patterns

1. As a class, work out the following divisions using a calculator and write each part with its answer on the board.

   a. $10 \div 0.1$ b. $10 \div 0.2$ c. $10 \div 0.3$ d. $10 \div 0.4$ e. $10 \div 0.5$
   f. $10 \div 0.6$ g. $10 \div 0.7$ h. $10 \div 0.8$ i. $10 \div 0.9$ j. $10 \div 1.0$

2. Have you noticed any pattern in your answers? Discuss.
3. Now repeat the calculations, this time using $20 \div 0.1$, $20 \div 0.2$ and so on.
4. Discuss with your class what is happening when you divide by a decimal.
5. Make up some divisions of your own and check if your assumptions about dividing by a decimal hold true.

▶ 5.7.3 Dividing a decimal by a decimal

eles-####

- A visual representation of $0.6 \div 0.2$ is shown below.

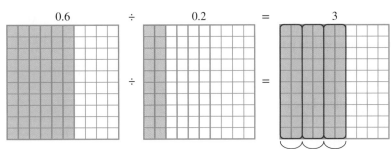

There are 3 groups of 0.2 in 0.6

- $0.6 \div 0.2$ can be interpreted as '6 tenths divided into groups of 2 tenths = 3 groups'.
- $0.6 \div 0.2$ has the same result as $6 \div 2$. Therefore, we can multiply both numbers by a multiple of 10 to make the divisor a whole number.

Dividing a decimal by a decimal

When dividing one decimal by another, multiply the decimal you are dividing (divisor) by a power of 10 to make it a whole number.

Multiply the other decimal by the same power of 10, as shown in the following example.

First multiply this number by 10 to make a whole number.

$6.4 \div 0.4$

Then multiply this number by 10 also.

This is the same as writing an equivalent fraction with a whole number as the denominator.

$$\frac{6.4}{0.4} \times \frac{10}{10} = \frac{64}{4}$$
$$= 16$$

WORKED EXAMPLE 26 Dividing a decimal by a decimal

Calculate:

a. $26.724 \div 0.4$

b. $3.0276 \div 0.12$

THINK

WRITE

a. 1. Rewrite the question as a fraction.

a. $26.724 \div 0.4 = \dfrac{26.724}{0.4}$

2. Multiply both the numerator and the denominator by the appropriate power of 10.

$= \dfrac{26.724}{0.4} \times \dfrac{10}{10}$

$= \dfrac{267.24}{4}$

3. Divide the decimal by the whole number.

$$\begin{array}{r} 6\ 6.\ 81 \\ 4\overline{)26^27.^324} \end{array}$$

4. Write the answer.

$26.724 \div 0.4 = 66.81$

b. 1. Rewrite the question as a fraction.

b. $3.0276 \div 0.12 = \dfrac{3.0276}{0.12}$

2. Multiply both the numerator and the denominator by the appropriate power of 10.

$= \dfrac{3.0276}{0.12} \times \dfrac{100}{100}$

$= \dfrac{302.76}{12}$

3. Divide the decimal by the whole number.

$$\begin{array}{r} 2\ 5.\ 2\ 3 \\ 12\overline{)30^62.^27^36} \end{array}$$

4. Write the answer.

$3.0276 \div 0.12 = 25.23$

Digital technology

Scientific calculators can evaluate division of numbers with decimals.

```
26.724÷0.4                    Math ▲

                              66.81
```

WORKED EXAMPLE 27 Dividing a decimal by a decimal

Calculate how many litres of petrol could be purchased for $50.88 if 1 litre costs $1.06.

THINK	WRITE
1. Write the question.	$50.88 \div 1.06$
2. Rewrite the question as a fraction.	$= \dfrac{50.88}{1.06}$
3. Multiply the numerator and denominator by the appropriate multiple of 10, in this case 100. Alternatively, the decimal point could be moved twice to the right in both numbers so that the divisor is a whole number $(50.88 \div 1.06 = 5088 \div 106)$.	$= \dfrac{50.88}{1.06} \times \dfrac{100}{100}$ $= \dfrac{5088}{106}$
4. Divide the decimal by the whole number. Alternatively, use a calculator.	$\quad\begin{array}{r} 4\ \ 8 \\ 106 \overline{)508\,^{84}8} \end{array}$ $50.88 \div 1.06 = 48$
5. Write the answer in a sentence.	Forty-eight litres of petrol could be purchased for $50.88.

COMMUNICATING — COLLABORATIVE TASK: Decimal divided by a decimal patterns

1. As a class, work out the following divisions and write each part with its answer on the board.
   a. $8 \div 4$
   b. $8 \div 0.4$
   c. $8 \div 0.04$
   d. $8 \div 0.004$
2. Have you noticed any pattern in your answers? Discuss.
3. As a class, work out the following divisions and write each part with its answer on the board.
   a. $0.8 \div 0.4$
   b. $0.08 \div 0.04$
   c. $0.008 \div 0.004$
4. Have you noticed any pattern in your answers? Discuss.

---

on Resources

▶ **Video eLesson** Division of decimals (eles-1877)

✦ **Interactivities** Division of decimals by a multiple of 10 (int-3990)
Division of a decimal by another decimal (int-3991)

Exercise 5.7 Dividing decimals

5.7 Quick quiz on	**5.7 Exercise**

Individual pathways

■ PRACTISE	■ CONSOLIDATE	■ MASTER
1, 4, 10, 13, 15, 17, 20, 23, 24, 25	2, 5, 6, 8, 11, 14, 18, 21, 26, 27	3, 7, 9, 12, 16, 19, 22, 28, 29, 30

Fluency

1. **WE23** Calculate:
   a. $3.6 \div 6$
   b. $21.7 \div 7$
   c. $4.86 \div 9$
   d. $8.05 \div 5$
   e. $9.68 \div 4$

2. Calculate:
   a. $1.576 \div 2$
   b. $8.029 \div 7$
   c. $32.5608 \div 8$
   d. $20.5782 \div 3$
   e. $126.4704 \div 4$

3. Calculate:
   a. $46.80 \div 15$
   b. $24.541 \div 11$
   c. $17.108 \div 14$
   d. $77.052 \div 12$
   e. $121.3421 \div 11$

4. **WE24** Calculate the following. In each case, add zeros and keep dividing until there is no remainder.
   a. $3.7 \div 2$
   b. $9.5 \div 2$
   c. $7.3 \div 5$

5. Calculate the following. In each case, add zeros and keep dividing until there is no remainder.
   a. $9.8 \div 4$
   b. $7.5 \div 6$
   c. $55.6 \div 8$

6. Calculate the following by changing the position of the decimal point.
   a. $14.07 \div 10$
   b. $968.13 \div 100$
   c. $620.8 \div 1000$
   d. $3592.87 \div 1000$
   e. $2349.78 \div 100\,000$

7. Calculate the following by changing the position of the decimal point.
   a. $9.0769 \div 100\,000$
   b. $802\,405.6 \div 1\,000\,000$
   c. $152.70 \div 1\,000\,000$
   d. $0.7205 \div 10\,000$
   e. $0.0032 \div 1\,000\,000$

8. **WE25** Calculate the following.
   a. $15.9 \div 60$
   b. $23.7 \div 30$
   c. $238 \div 400$
   d. $8.79 \div 6000$
   e. $5.22 \div 3000$

9. **WE26a** Calculate each of the following.
   a. $2.5 \div 0.5$
   b. $4.2 \div 0.6$
   c. $8.1 \div 0.9$
   d. $2.8 \div 0.7$
   e. $0.248 \div 0.8$

10. Calculate each of the following.
    a. $1.32 \div 0.6$
    b. $39.6 \div 0.6$
    c. $57.68 \div 0.8$
    d. $0.2556 \div 0.3$
    e. $0.5468 \div 0.4$

11. **WE26b** Calculate:
    a. $172.0488 \div 0.11$
    b. $0.510\,48 \div 0.12$
    c. $21.470\,10 \div 0.15$

12. Calculate:
    a. $142.888 \div 0.08$
    b. $0.028\,692 \div 0.06$
    c. $32.619 \div 0.02$

Understanding

13. **MC** To calculate $9.84 \div 0.8$, it can be rewritten as:
    A. $9.84 \div 8$
    B. $0.984 \div 0.8$
    C. $98.4 \div 0.8$
    D. $98.4 \div 8$

14. **MC** To calculate $151.368 \div 1.32$, it can be rewritten as:
    A. $151.368 \div 132$
    B. $151.368 \div 13.2$
    C. $1513.68 \div 132$
    D. $15\,136.8 \div 132$

15. **MC** 0.294 ÷ 0.7 equals:

   A. 0.042   B. 0.42   C. 4.2   D. 42

16. **MC** 21.195 ÷ 0.15 equals:

   A. 0.1413   B. 1.413   C. 14.13   D. 141.3

17. For the following calculations:

   i. Estimate the answer by rounding each number to the first digit.
   ii. Calculate the answers by using written methods.
   iii. Check your answer by using a calculator.

   a. 6600 ÷ 24   b. 1044 ÷ 4.5   c. 384 ÷ 0.16

18. Change the following to dollars ($) by dividing by 100.

   a. 365 cents   b. 170 cents   c. 5685 cents
   d. 75 cents   e. 90 cents   f. 6350 cents

19. Calculate $100\,987.5412 \div 10^4$.
   Round the answer to the nearest hundredth.

Communicating, reasoning and problem solving

20. The carat is a unit of measure used to weigh precious stones. A 2.9-carat diamond weighs 8.9494 grains. Calculate how many grains one carat is equal to.

21. You are helping to beautify the gardens at your school and have been given the responsibility for paving the path. The path is 6.715 m (or 671.5 cm) long and the width of only two pavers. The pavers you will use are shown in the photograph.

   a. Calculate how many pavers you should buy to minimise waste.
   b. If each paver costs $2.32, calculate how much the pavers cost in total.

22. A jug is $\frac{1}{5}$ full of water. Jack added more water to the jug until it was $\frac{3}{4}$ full. At that stage, there was 1.5 L of water in the jug. Show with working how much water was originally in the jug.

23. Stephanie spent $6.95 on the 12 chocolates in the photograph. Determine the cost of each chocolate. Assume that they all cost the same. Give your answer to the nearest 5 cents.

24. **WE27** Calculate how many 1.25-litre bottles of water could be poured into a 25-litre drink dispenser.

25. Calculate how many burgers could be bought for $562.80 if each burger costs $2.80.

26. If you have $22.50 for bus fares to school for the week, calculate how much you would spend on each of the 5 days.

27. The area of an office floor is 85.8 square metres. Calculate how many people could fit in the office if each person takes up 1.2 square metres.

28. Emily wants to make 10 cushions from 6.75 metres of material that she found on a table of remnants at a fabric shop. Determine how much material she would have for each cushion.

29. Calculate how many books can be stacked on a shelf that is 28.6 centimetres high if each book is 1.1 centimetres high.

30. Anna has a wooden rod that is 2.4 metres long. She needs to cut 3 lengths, each of 0.4 metres, and 2 lengths, each of 0.2 metres. Determine the fraction of the original rod left over after she has cut the required lengths.

LESSON
5.8 Decimals and percentages

LEARNING INTENTION

At the end of this lesson you should be able to:
- convert percentages to decimals
- convert decimals to percentages.

▶ 5.8.1 Percentages as decimals

eles-3957

- Percentages can be expressed as decimals.
- To convert a percentage to a decimal, divide the percentage by 100; that is, move the decimal point 2 spaces to the left.

Converting percentages to decimals

To express a percentage as a decimal, move the decimal point 2 spaces to the left.

$$45\% \to 0.45$$

WORKED EXAMPLE 28 Converting percentages to decimals

Write the following percentages as decimals.
a. **81%**
b. **16.8%**

THINK	WRITE
a. 1. To convert to a decimal, divide by 100.	a. $81\% = 81 \div 100$
2. Move the position of the decimal point 2 places to the left and place a zero in the units column.	$= \dfrac{81}{100}$ $= 0.81$

b. 1. To convert to a decimal divide by 100.

b. $16.8\% = 16.8 \div 100$

2. Move the position of the decimal point 2 places to the left and place a zero in the units column.

$$= \frac{16.8}{100}$$

$$= 0.168$$

5.8.2 Decimals as percentages

eles-6129

- To convert a decimal to a percentage, multiply the decimal by 100; that is, move the decimal point 2 spaces to the right and add a % sign.

Converting decimals to percentages

To express a decimal as a percentage, move the decimal point 2 spaces to the right and add a % sign.

$$0.45 \rightarrow 45\%$$

WORKED EXAMPLE 29 Converting decimals to percentages

Write the following decimals as percentages.

a. 0.94

b. 2.54

THINK	WRITE
a. 1. To convert to a percentage, multiply by 100.	**a.** $0.94 \times 100 = 94$
2. Add % on the end.	$= 94\%$
b. 1. To convert to a percentage, multiply by 100.	**b.** $2.54 \times 100 = 254$
2. Add % on the end.	$= 254\%$

 Resources

 Interactivity Percentages as decimals (int-3993)

Exercise 5.8 Decimals and percentages

learn**on**

5.8 Quick quiz on	**5.8 Exercise**

Individual pathways

■ PRACTISE	■ CONSOLIDATE	■ MASTER
1, 4, 8, 11, 13, 15, 18, 21, 24, 25	2, 5, 7, 9, 13, 16, 19, 22, 26	3, 6, 10, 12, 14, 17, 20, 23, 27, 28

Fluency

1. **WE28a** Write the following percentages as decimals.

   a. 36%   **b.** 14%   **c.** 28%   **d.** 73%   **e.** 66%

2. Write the following percentages as decimals.

    a. 59%   **b.** 99%   **c.** 9%   **d.** 4%   **e.** 1%

3. Write the following percentages as decimals.

    a. 25%   **b.** 200%   **c.** 150%   **d.** 360%   **e.** 100%

4. **WE28b** Write the following percentages as decimals.

    a. 12.3%   **b.** 31.6%   **c.** 59.2%   **d.** 84.9%   **e.** 42.1%

5. Write the following percentages as decimals.

    a. 37.6%   **b.** 21.9%   **c.** 16.9%   **d.** 11.1%   **e.** 3.1%

6. Write the following percentages as decimals.

    a. 9.2%   **b.** 5.9%   **c.** 8.8%   **d.** 14.25%   **e.** 31.75%

7. Write the following percentages as decimals.

    a. 23.55%   **b.** 45.75%   **c.** 0.05%   **d.** 4.01%   **e.** 0.02%

8. **WE29a** Write the following decimals as percentages.

    a. 0.34   **b.** 0.24   **c.** 0.13   **d.** 0.99

9. **WE29b** Write the following decimals as percentages.

    a. 2.35   **b.** 5.42   **c.** 1.2   **d.** 3.04

10. Write the following decimals as percentages.

    a. 0.045   **b.** 0.032   **c.** 0.004   **d.** 0.0091

Understanding

11. **MC** 41% as a decimal is:

    A. 41   **B.** 0.41   **C.** 4.100   **D.** 4.1

12. **MC** 8% as a decimal is:

    A. 8   **B.** 0.008   **C.** 0.08   **D.** 0.8

13. **MC** 43.64% as a decimal is:

    A. 0.4364   **B.** 4.364   **C.** 43.6400   **D.** 436.4

14. **MC** 110% as a decimal is:

    A. $\dfrac{110}{100}$ **B.** 1100   **C.** 0.11   **D.** 1.10

15. Car prices dropped by 17% during a recent 8-year period. Write this percentage as a:

    a. fraction   **b.** decimal.

16. The maximum legal blood alcohol concentration (BAC) for drivers in New South Wales is 0.05%. Write the BAC as a:

    a. fraction   **b.** decimal.

17. In 2020, the number of visitors to Australia was 80.7% lower than the previous year. Write this percentage as a decimal.

18. Over the past decade, insurance premiums have risen by 218%. Write this percentage as a decimal.

19. In the three months following the end of a drought, the price of vegetables fell by 13.8%. Write this percentage as a decimal.

20. Government funding to state schools has risen by 8.35%. Write this percentage as a decimal.

Communicating, reasoning and problem solving

21. a. Round the following percentages to the nearest whole number: 23.2%, 23.5%, 23.8%, 23.9%
    b. Round the following decimal numbers to 2 decimal places: 0.232, 0.235, 0.238, 0.239
    c. The percentages in part a are the same as the decimal numbers in part b. Explain whether it matters if you round the percentages or the decimal values up or down.

22. Consider the following percentages: 23.6%, 23.67%, 23.66%

    a. Which one of the three percentages is the largest? Justify your answer.
    b. Convert these percentages to decimals. State your answer to 4 decimal places.

23. Consider the following percentages: 37.8%, 39.6%, 30.9%, 34.5%, 32.8%

    a. Convert these numbers to decimals.
    b. Arrange the percentages in order from highest to lowest.
    c. Arrange the decimal numbers in order from highest to lowest.
    d. Explain whether it was easier to arrange the numbers in the order required using decimals or percentages.

24. Brent buys a used car for $7500. The dealer requires a 15% deposit. Calculate the value of the deposit he they must pay the dealer.

25. Alfie's car dashboard says he has used 0.35 of a full fuel tank. Determine the percentage of fuel remaining.

26. The price of entry into a theme park has increased by 10% every year since the theme park opened.
    If the latest price rise increased the ticket price to $8.80, determine the cost of a ticket 2 years ago.

27. Consider the following percentages:
    7.291%, 72.91%, 729.1%.

    a. Convert the following percentages to decimal numbers.

        i. 7.291%   ii. 72.91%   iii. 729.1%

    b. Explain the effect of the conversion on the numbers.

28. Suggest when it is more convenient to express a percentage as a decimal rather than as a fraction.

LESSON
5.9 Review

5.9.1 Topic summary

Decimal numbers and place value

- Decimal numbers can be identified easily because the whole number part and the fractional part are separated by a decimal point (.).
  e.g. 11.234 is a decimal number.
- Decimal numbers can also be written in expanded form.
  e.g. $11.234 = 11 + \frac{2}{10} + \frac{3}{100} + \frac{4}{1000}$

Rounding decimals

When rounding decimal numbers, follow the steps below.
1. Decide which digit is the last digit to keep (underline it).
2. Leave this underlined digit the same if the **next digit** is 0, 1, 2, 3 or 4 (this is called rounding down).
- Increase the underlined digit by 1 if the **next digit** is 5, 6, 7, 8 or 9 (this is called rounding up).
e.g. 2.3456 = 2.3 correct to 1 decimal place
2.3456 = 2.35 correct to 2 decimal places

Types of decimals

- **Terminating (finite) decimals** are decimal numbers that finish (terminate).
  e.g. $\frac{3}{8} = 0.375$
- **Recurring (repeating) decimals** are decimal numbers that keep on going forever and never finish.
  e.g. $\frac{1}{3} = 0.333\,333\,3... = 0.\dot{3}$
  (The 3 never stops.)
  $\frac{5}{11} = 0.454\,545\,45... = 0.\overline{45}$
  (The 45 pattern continues forever.)

DECIMALS

Addition and subtraction

- To add or subtract decimal numbers, set the problem out in columns according to place value and make sure the decimal points are aligned.
- Add and subtract using the same method as for whole numbers.
e.g. Calculate 1.23 + 7.894.

```
  1 . 2 3 0
+ 7 . 8 9 4
-----------
  9 . 1 2 4
```

Multiplying decimal numbers

- To multiply decimal numbers, use the method for multiplying whole numbers.
- To know where to place the decimal point in the answer, *count the total number of decimal places in the entire question*.
- The answer must have the same number of decimal places as your count.

Multiplying by powers of 10

- Count the number of zeros in the power of 10 being multiplied.
  e.g. 1000 has three zeros.
- Move the decimal point to the *right* by the same number.
e.g. $23.4567 \times 1000 = 23\,456.7$
(1000 has three zeros, so move the decimal point three places to the right.)

Dividing by powers of 10

- Count the number of zeros in the power of 10 being divided.
  e.g. 100 has two zeros.
- Move the decimal point to the *left* by the same number.
e.g. $23.4567 \div 100 = 0.234\,567$
(100 has two zeros, so move the decimal point two places to the left.)

Converting between percentages and decimals

- To convert a percentage to a decimal, simply divide by 100 and remove the percentage symbol.
  e.g. Express 54% as a decimal number.
  $54\% = 54 \div 100$
  $= 0.54$
- To convert a decimal to a percentage, simply multiply by 100 and add the percentage symbol.
  e.g. Express 0.23 as a percentage.
  $0.23 = 0.23 \times 100$
  $= 23\%$

Division by a whole number

- When dividing a decimal number by a whole number, use the same method as for division with whole numbers.
  e.g.
  $$5\overline{)10.375}\,^{2.075}$$

Dividing decimal numbers

- When dividing by a decimal number, convert the denominator to a whole number by multiplying by a power of 10.
  (Put simply, just move the decimal point to the right as many places as needed to make the denominator a whole number.)
- Multiply the numerator by the same power of 10 as the denominator to keep it balanced.
  (Move the decimal point in the numerator exactly the same number of places to the right as you did in the denominator.)
- Divide as normal.
e.g. Calculate $\frac{0.135}{0.5}$.
To make the denominator a whole number, multiply it by 10 (or move the decimal point one place to the right). Do the same to the numerator.
$$\frac{0.135}{0.5} = \frac{1.35}{5} = 0.27$$

Eating out

When we eat out or buy takeaway food, a number of factors are taken into consideration before we make our purchases. What types of foods are available? What is the size of each serving? What is the cost of the meal? These are examples of things we consider when we try to decide what and where we eat.

In a group of four, you and your friends have decided to eat at a local restaurant. Your parents have given each of you $30.00 to spend on your meal.

EAT EASY RESTAURANT MENU

SALADS

Tossed Green Salad	$5.50
Caesar Salad	$7.50
Chicken Caesar Salad	$8.95
Greens with Tomato	$6.50

STARTERS

Garlic or Herb Bread	$3.50
Bruschetta	$4.50
Dips & Bread	$3.50
Potato Skins	$6.50
– served with sour cream	
Basket of fries	$4.25
Nachos with cheese, salsa	
and sour cream	$6.50
Italian style baby potatoes	$6.50

SANDWICHES

All served with salad and fries on your choice of bread

Bacon, Lettuce & Tomato	$5.50
Ham & Cheese	$5.00
Ham, Cheese & Tomato	$5.50
Grilled Chicken & Cheese	$6.00
Egg & Lettuce	$5.50
Top Quality Steak & Onions	$9.95

BURGERS AND HOT DOGS

All served with salad and fries

Jumbo Hot Dog	$5.95
Gourmet Dog	$6.95
– includes smoked sausage	
Double Beef Burger	$5.95
Grilled Chicken Burger	$6.95
Fresh Fish Burger	$7.55
Vegetable Burger	$4.95

Additional items	
Bacon	$0.50
Grilled onion	$0.50
Olives	$0.75
Cheddar	$0.75

HOUSE SPECIAL

Today's Special is 250 gram Rib Eye Steak served with fries, vegetables with your choice of sauce $18.95

PASTA

Your choice of Spaghetti, Fettuccine, Gnocchi, Ravioli or Tagliatelle with the sauces below

Mama's Meaty Bolognaise	$9.95
Creamy Chicken	$9.95
Fresh Seafood Marinara	$10.95
Pesto	$9.95
Tomato and Hot Chilli	$8.95
Vegetable Lasagne	$9.95
Rich and Meaty Lasagne	$9.95

DESSERTS

Homemade Fruit Pie	$3.75
Brownie	$3.50
Cheesecake	$3.95
Frozen Yoghurt	$2.50
– Chocolate or Vanilla	
Jumbo Banana Split	$4.50

1. Using the menu provided, work out how much each of the following meals would cost. On a separate piece of paper, set out each order as it would appear on the final bill. An example is shown at right.
   a. Tossed green salad and Homemade fruit pies.
   b. Chicken Caesar salad, Jumbo hot dog and Vanilla-flavoured yoghurt.
   c. Garlic bread, Potato skins, House special and Jumbo banana split.
   d. Vegetable burger with cheddar, Basket of fries and Brownie.

EAT EASY RESTAURANT

Item	Cost
Tossed green salad	$5.50
Homemade fruit pies	$3.75
Total	$9.25

Your group has decided to order a three-course meal. The first course consists of items from the Salads and Starters sections and the third course is from the Desserts section. The second course is ordered from the remaining sections.

The group also decides to buy a large jug of orange juice to drink, and plans to share its cost of $4.60 evenly.

2. Calculate your share of the cost of the drinks.
3. How much does this leave you to spend on your food order?
4. Select your three-course meal, keeping in mind the amount of money you have to spend. Write down your order, and set it out as you did for question 1.

Obtain the orders from three of your classmates. This will represent your group of four.

5. Write down each group member's order on a separate sheet of paper. Present the information as you would like it to appear on your final bill. Include the total cost of each person's order.
6. Calculate how much change each group member received from their $30.
7. Imagine that the restaurant's final bill for your group showed only the total amount owing; that is, there was no information on the breakdown of charges. Comment on how you would feel about this and how you think bills should be presented for group bookings.

 Resources

Interactivities Crossword (int-2594)
                 Sudoku puzzle (int-3167)

Exercise 5.9 Review questions

Fluency

1. Give the value of the 7 in each of the following, in words.
   a. 1.719   b. 3.0726   c. 0.2078

2. Give the vaue of the 7 in each of the following, in words.
   a. 23.1487   b. 0.002 57   c. 17.592

3. Write the following numbers in expanded notation.
   a. 2.64   b. 18.406   c. 96.3428

4. Add 2 tenths to the following.
   a. 6.2   b. 0.743   c. 3.91

5. Add 3 thousandths to the following.
   a. 0.456   b. 1.6   c. 2.79

6. Put < or > between the following.
   a. 8.72   8.27   b. 0.35   0.37   c. 10.214   10.219

7. Put < or > between the following.
   a. 13.0496   13.149   b. 0.804 06   0.804 17   c. 0.000 879   0.000 876

8. Write the following decimals in order from smallest to largest.
   a. 0.13, 0.86, 0.34, 0.71, 0.22   b. 0.247, 0.274, 0.124, 0.258, 0.285

9. Write the following decimals in order from smallest to largest.
   a. 0.834, 0.826, 0.859, 0.888, 0.891   b. 0.356, 0.358, 0.365, 0.385, 0.217

10. Write the following decimals as fractions in simplest form.
    a. 0.8   b. 0.17   c. 0.36   d. 0.187

11. Write the following decimals as fractions in simplest form.
    a. 0.125   b. 0.568   c. 0.205   d. 0.950

12. Write the following decimals as mixed numerals in simplest form.
    a. 1.5   b. 4.60   c. 3.48   d. 5.25

13. Write the following decimals as mixed numerals in simplest form.
    a. 2.75   b. 2.625   c. 1.56   d. 8.32

14. Round the following to 2 decimal places.

    a. 2.047   **b.** 13.8649   **c.** 17.898 193

15. Round the following to the number of decimal places indicated in the brackets.

    a. 1.29 (1)   **b.** 0.0482 (3)   **c.** 1.925 96 (4)

16. Round the following to the nearest whole number.

    a. 13.6   **b.** 29.02   **c.** 86.99   **d.** 100.09

Understanding

17. Calculate the following.

    a. $1.8 + 7.3$ **b.** $4.21 + 5.88$ **c.** $6.75 + 0.243$

18. Calculate the following.

    a. $12.047 + 3.6$ **b.** $194 + 18.62 + 3.1$ **c.** $34.1 + 7.629 + 0.008\,45$

19. Calculate the following.

    a. $9.6 - 4.3$ **b.** $18.25 - 9.18$ **c.** $3.92 - 1.88$

20. Calculate the following.

    a. $100 - 9.341$ **b.** $4.876 - 3.927$ **c.** $1.6 - 0.025$

21. Calculate the following.

    a. 6.2×3 **b.** 4.67×9 **c.** 13.2036×5 **d.** 0.7642×7

22. Calculate the following.

    a. 0.23×11 **b.** 16.28×41 **c.** 182.94×28 **d.** $0.028\,94 \times 32$

23. Calculate the following.

    a. 0.26×10 **b.** 1.345×10 **c.** 0.0645×100 **d.** 1.8294×100

24. Calculate the following.

    a. 146.6281×100 **b.** $0.048\,064\,3 \times 1000$ **c.** $0.839\,204 \times 1000$ **d.** 0.368×1000

25. Calculate the following.

    a. 3.2×0.41 **b.** 1.72×0.3 **c.** 0.87×0.9 **d.** 0.03×0.006

26. Calculate the following.

    a. 0.58×1.5 **b.** 2.83×0.96 **c.** 11.468×1.3 **d.** 1.248×0.82

27. Calculate each of the following.

    a. 2.5^2 **b.** 0.03^2 **c.** 0.64^2 **d.** 0.0025^2

28. Calculate each of the following.

   a. $1.64 \div 4$ b. $12.48 \div 6$ c. $147.24 \div 2$ d. $1.76 \div 11$

29. Calculate each of the following.

   a. $14.623 \div 10$ b. $102.36 \div 10$ c. $9612.347 \div 1000$

30. Calculate each of the following.

   a. $20.032 \div 100$ b. $264\,983.0026 \div 1000$ c. $3462.94 \div 100$

31. Write each of these fractions as a terminating decimal or as an infinite recurring decimal.

   a. $\dfrac{1}{20}$ b. $\dfrac{5}{8}$ c. $\dfrac{5}{16}$ d. $\dfrac{2}{16}$

32. Write these infinite recurring decimals using a short form.

   a. $4.555\ldots$ b. $0.8282\ldots$ c. $19.278\,127\,81\ldots$ d. $83.016\,262\,62\ldots$

33. Calculate the following.

   a. $4.8 \div 0.6$ b. $35.7 \div 0.7$ c. $12.1 \div 1.1$ d. $13.72 \div 0.4$

34. Calculate the following.

   a. $17.8946 \div 0.02$ b. $372.045\,72 \div 0.06$ c. $0.289\,56 \div 0.12$ d. $3214.0170 \div 0.15$

35. Write the following percentages as decimals.

   a. 42% b. 5% c. 94% d. 139%

36. Write the following decimals as percentages.

   a. 0.06 b. 0.197 c. 0.5803 d. 0.008

37. Calculate the following percentages by converting the percentage to a decimal.

   a. 20% of 25 b. 12% of 31 c. 4.5% of 50 d. 9.2% of 75

38. Calculate the following percentages by converting the percentage to a decimal.

   a. 21.4% of 90 b. 32.3% of 120 c. 76.5% of 8 d. 42.3% of 96.2

Communicating, reasoning and problem solving

39. Jim saved the following amounts of pocket money to take away on holiday:
    $12.50, $15.00, $9.30, $5.70 and $10.80.
    Calculate how much money Jim had to spend while on holiday.

40. Mandie poured 0.375 litres from a 1.5-litre bottle of juice. Calculate how much juice was left in the bottle.

41. Tara bought 0.350 kilograms of shaved ham at $10.50 per kilogram. Calculate how much Tara paid for the ham.

42. The decimal equivalent of $\dfrac{3}{13}$ is $0.\overline{230\,769}$.

    a. Determine the decimal equivalent of $\dfrac{1}{13}$.

    b. Determine the decimal equivalent of $\dfrac{5}{13}$.

43. Lena saved \$32 when she purchased a pair of jeans at a sale. If the sale had 40% marked off any item, calculate the original price for the pair of jeans.

44. You are given two options at work:
Option 1: Your salary can be raised by 10% this month and then reduced by 10% next month.
Option 2: Your salary will be reduced by 10% this month and increased by 10% next month.
Compare the two options to determine which is the better option.

45. Leesa has 3 m of ribbon. It takes 0.15 m of ribbon to make a bow. Calculate how many bows they can make with the ribbon.

46. Steven lives $\dfrac{3}{4}$ km from school. He can walk at 6.25 km/h. Determine how long it takes him to walk to school.

on To test your understanding and knowledge of this topic, go to your learnON title at www.jacplus.com.au and complete the **post-test**.

Answers

Topic 5 Decimals

5.1 Pre-test

1. C
2. 0.12
3. D
4. Josh
5. 23.90
6. $378.95
7. D
8. 5.817
9. C
10. 3280
11. B
12. $13.65
13. 86.98
14. 44 litres
15. 125

5.2 Decimals and place value

1. a. $\dfrac{2}{10}$ b. $\dfrac{2}{100}$ c. $\dfrac{2}{10}$ d. $\dfrac{2}{1000}$

2. a. $\dfrac{2}{10\,000}$ b. $\dfrac{2}{10\,000}$ c. 2 d. $\dfrac{2}{10\,000}$

3. a. $\dfrac{9}{10}$ b. $\dfrac{9}{10}$ c. $\dfrac{9}{100}$ d. $\dfrac{9}{100}$

4. a. $\dfrac{9}{10\,000}$ b. $\dfrac{9}{100\,000}$ c. 900   d. 9

5. a. $\dfrac{4}{10}$ b. $2 + \dfrac{7}{10}$

   c. $5 + \dfrac{2}{10} + \dfrac{3}{100}$ d. $\dfrac{7}{10} + \dfrac{6}{100} + \dfrac{3}{1000}$

6. a. $\dfrac{1}{10} + \dfrac{1}{100} + \dfrac{1}{10\,000}$

   b. $7 + \dfrac{2}{10} + \dfrac{9}{100} + \dfrac{6}{1000} + \dfrac{4}{10\,000}$

   c. $\dfrac{3}{10} + \dfrac{3}{100} + \dfrac{2}{10\,000} + \dfrac{4}{100\,000}$

   d. $300 + \dfrac{3}{100}$

7. a. One unit, 1; eight tenths, $\dfrac{8}{10}$; five hundredths, $\dfrac{5}{100}$;

   $1 + \dfrac{8}{10} + \dfrac{5}{100}$

   b. Two tenths, $\dfrac{2}{10}$; seven hundredths, $\dfrac{7}{100}$; one thousandth,

   $\dfrac{1}{1000}$; $\dfrac{2}{10} + \dfrac{7}{100} + \dfrac{1}{1000}$

   c. One ten, 10; six units, 6; one thousandth, $\dfrac{1}{1000}$;

   $16 + \dfrac{1}{1000}$

   d. Three units, 3; four tenths, $\dfrac{4}{10}$; two thousandths, $\dfrac{2}{1000}$;

   seven hundred thousandths, $\dfrac{7}{100\,000}$; $3 + \dfrac{4}{10}$

   $+ \dfrac{2}{1000} + \dfrac{7}{100\,000}$

8. a. $(1 \times 1) + \left(2 \times \dfrac{1}{10}\right) + \left(5 \times \dfrac{1}{100}\right)$

   b. $(5 \times 10) + (6 \times 1) + \left(1 \times \dfrac{1}{100}\right)$

   c. $(3 \times 10) + (9 \times 1) + \left(1 \times \dfrac{1}{100}\right)$

   d. $(5 \times 1) + \left(9 \times \dfrac{1}{10}\right) + \left(8 \times \dfrac{1}{100}\right) + \left(7 \times \dfrac{1}{1000}\right)$

9. a. $(1 \times 10) + (3 \times 1) + \left(4 \times \dfrac{1}{10}\right) + \left(8 \times \dfrac{1}{100}\right)$

   $+ \left(2 \times \dfrac{1}{1000}\right)$

   b. $\left(3 \times \dfrac{1}{10}\right) + \left(6 \times \dfrac{1}{1000}\right) + \left(2 \times \dfrac{1}{10\,000}\right)$

   c. $\left(5 \times \dfrac{1}{10}\right) + \left(2 \times \dfrac{1}{10\,000}\right)$

   d. $(2 \times 1) + \left(4 \times \dfrac{1}{10}\right) + \left(7 \times \dfrac{1}{100}\right)$

10. D
11. See the table at the foot of the page.*
12. a. 0.32   b. 0.59   c. 0.901
13. a. 0.039 90   b. 2.559   c. 0.110 49
14. a. >   b. <   c. <   d. <   e. <
15. a. >   b. >   c. <   d. <   e. >

---

*11.

		Tens	Units		Tenths	Hundredths	Thousandths
a.	0.205	0	0	.	2	0	5
b.	1.06	0	1	.	0	6	0
c.	74.108	7	4	.	1	0	8
d.	0.108	0	0	.	1	0	8
e.	50.080	5	0	.	0	8	0

16. a. 2
    b. i. Seventy-eight dollars and fifty-five cents
       ii. $78.55

17. a. 0.17, 0.21, 0.33, 0.39, 0.45
    b. 0.314, 0.391, 0.413, 0.420, 0.502
    c. 0.803, 0.807, 0.811, 0.821, 0.902

18. a. 0.9087, 0.9189, 0.9812, 0.9864, 0.9943
    b. 4.0292, 4.5097, 4.6031, 4.6249, 4.802
    c. 0.003, 0.004 65, 0.005 02, 0.0056, 0.009

19. a. 0.48, 0.39, 0.36, 0.31, 0.19
    b. 0.99, 0.97, 0.95, 0.91, 0.90
    c. 1.291, 1.288, 1.279, 1.273, 1.264

20. a. 0.390, 0.372, 0.318, 0.317, 0.309
    b. 0.8889, 0.8823, 0.8448, 0.8217, 0.8207
    c. 1.843 21, 1.702 96, 1.486 59, 1.486 13, 1.349 54

21. D
22. C
23. D
24. a. False   b. True   c. False
25. a. True   b. False   c. True

26. a. $0.\dot{1}42\,857\,142\,\dot{9}$
       i. $0.\dot{2}85\,714\,285\,\dot{7}$
       ii. $0.\dot{4}28\,571\,428\,5\dot{7}$
       iii. $0.\dot{5}71\,428\,571\,\dot{4}$
       iv. $0.\dot{7}14\,285\,714\,\dot{3}$
       v. $0.\dot{8}57\,142\,857\,1\dot{4}$
    b. Each number has a repeated group of decimals in the same sequence: 142 857.

27. a. i. Tenths   ii. Yes
    b. i. Thousandths   ii. No
    c. i. Units   ii. No
    d. i. Units   ii. Yes
    e. i. Tens   ii. Yes

28. a. Shelley, 0.21 seconds
    b. Carolyn, 0.06 seconds
    c. Mara, 0.11 seconds
    d. Shelley, Carolyn, Robyn
    e. Carolyn, Shelley, Kyah
    f. Mara, Jenika, Shelley
    g. Yes, Shelley

29. 8.264

30. They are correct for trailing zeros (to the right) of a decimal. They are incorrect for zeros that are enclosed by non-zero digits, as they would indicate a place value of zero.

5.3 Converting between decimals and fractions

1. a. $\dfrac{3}{10}$ b. $\dfrac{1}{2}$ c. $\dfrac{21}{100}$ d. $\dfrac{2}{5}$ e. $\dfrac{11}{25}$

2. a. $\dfrac{49}{100}$ b. $\dfrac{251}{500}$ c. $\dfrac{617}{1000}$ d. $\dfrac{3}{10}$ e. $\dfrac{16}{25}$

3. a. $\dfrac{7}{25}$ b. $\dfrac{591}{625}$ c. $\dfrac{9209}{10\,000}$ d. $\dfrac{4621}{10\,000}$ e. $\dfrac{3}{25}$

4. a. $1\dfrac{3}{10}$ b. $2\dfrac{7}{10}$ c. $9\dfrac{2}{5}$ d. $1\dfrac{1}{5}$ e. $4\dfrac{1}{5}$

5. a. $8\dfrac{1}{2}$ b. $5\dfrac{27}{100}$ c. $19\dfrac{91}{500}$ d. $3\dfrac{3}{20}$ e. $6\dfrac{1}{4}$

6. a. $9\dfrac{7}{50}$ b. $16\dfrac{341}{500}$ c. $2\dfrac{4917}{10\,000}$
   d. $4\dfrac{1693}{5000}$ e. $100\dfrac{3}{625}$

7. B
8. A
9. C
10. D
11. D
12. a. 0.75   b. 0.5   c. 0.8   d. 0.05
13. a. 0.25   b. 0.02   c. 0.32   d. 0.2
14. a. 0.375   b. 0.875   c. 0.0625

15. a. Darya
    b. Compare the fraction of wins as decimals. Darya has a larger fraction of wins.

16. 300 m
17. No, the second jump was best.
18. History, Science, English

19. $\dfrac{4}{15}, \dfrac{3}{11}, \dfrac{30}{100}, \dfrac{8}{25}, \dfrac{325}{1000}, \dfrac{4}{12}, \dfrac{3}{7}$

20. $\dfrac{3}{7} = 0.\dot{4}28\,57\dot{1}, \dfrac{4}{7} = 0.\dot{5}71\,42\dot{8}, \dfrac{5}{7} = 0.\dot{7}14\,28\dot{5},$
    $\dfrac{6}{7} = 0.\dot{8}57\,14\dot{2}$

5.4 Rounding decimals and terminating and recurring decimals

1. a. 0.32   b. 0.86   c. 1.25   d. 13.05   e. 3.04
2. a. 7.13   b. 100.81   c. 71.26   d. 0.01   e. 2.39
3. a. 0.19   b. 19.70   c. 0.40   d. 1.00   e. 8.90
4. a. 0.4   b. 0.9   c. 9.3   d. 25.3   e. 3.9
5. a. 300.1   b. 12.8   c. 99.9   d. 8.9   e. 14.0
6. a. 17.6   b. 0.9   c. 94.0   d. 1.0   e. 96.3
7. A
8. a. 11   b. 8   c. 4   d. 93   e. 112
9. a. 22   b. 42   c. 2138   d. 0   e. 1
10. a. $2.\dot{5}$ b. $0.\dot{6}$ c. $49.\dot{1}$ d. $0.\dot{2}\dot{6}$
11. a. $0.\dot{9}1\dot{3}$ b. $8.\dot{6}41\dot{8}$ c. $133.\dot{9}46\overline{2}$ d. $0.12\dot{7}$
12. a. 0.3   b. 0.1   c. 0.417
13. a. 0.17   b. 0.182   c. 0.47
14. $0.\overline{5}$
15. $0.1\overline{6}$
16. $0.8\overline{3}$
17. A
18. A

19. a. 3.5　　　　b. 4.$\dot{6}$　　　　c. 0.75

20. a. 10　b. 80　c. 140　d. 260　e. 180

21. a. 300　b. 200　c. 100　d. 6800　e. 9800

22. a. 3000　b. 12 000　c. 19 000　d. 9000　e. 1000

23. $27.70

24. 0.295, 0.2$\dot{9}$5, 0.29$\dot{5}$, 0.2$\dot{9}\dot{5}$

25. An example answer is 0.543 216.

26. A more reasonable measurement would be 3.568 m. With the instruments we have available, it would not be possible to cut the timber more accurately.

27. Fred is incorrect; any number between 0.5 and 0.65 will round to 0.6.

28. 25 °C

29. a.　i. 0.8　　　　ii. 0.84　　　　iii. 0.838
    b. The rules are the same regardless of the number of decimals.

30. 50.0000, 33.3333, 20.0000, 14.2857, 9.0909, 7.6923, 5.8824, 5.2632, 4.3478, 3.4483

5.5 Adding and subtracting decimals

1. a. 3.5　　b. 2.69　　c. 13.239　　d. 102.1858

2. a. 6.77　　b. 21.906　　c. 2012.333　d. 261.449 06

3. a. 4.6　　b. 6.9　　c. 3.471　　d. 17.28

4. a. 17.3271　b. 20.4329　c. 11.69　　d. 105.22

5. a. 18.951　b. 105.958　c. 192.8949　d. 2618.9564

6. a. 0.60　　b. 4.205　　c. 0.0806　　d. 1.3368

7. a. 5.02　　b. 89.19　　c. 87.09　　d. 2.733

8. a. 3.34　　b. 5.93　　c. 0.327　　d. 4.146

9. a. 1.934　　b. 3.289　　c. 14.5326　d. 16.766

10. a. 38.238　b. 4.1136　　c. 0.1951　　d. 0.772 16

11. B

12. A

13. A

14. D

15. C

16. B

17. B

18. B

19. $167.00

20. $4.70

21. 20.05 kilometres

22. 8.1 kilometres

23. $6.60

24. $10.35

25. 50.48 kilograms

26. 11.73 metres

27. 31.28 metres

28. 48.38 seconds

29. 0.107

30. When we estimate an answer we use rounding to make the numbers simple to calculate; for example, 11 may be rounded to 10.

31. 49 calls

32. a. See the table at the foot of the page.*
    b. Team 3
    c. 4.1 seconds

33. 65.53 g, 43.43 g, 32.03 g or 32.04 g

34. Increase 51c

35. One possible solution is 5.83 + 1.46 = 7.29.

5.6 Multiplying decimals

1. a. 14.0　　b. 125.6　　c. 61.2　　d. 111.70

2. a. 190.26　b. 518.96　c. 6.384　　d. 0.794 43

3. a. 0.84　　b. 2.94　　c. 0.32　　d. 7.76

4. a. 0.140　　b. 0.192　　c. 0.385　　d. 0.441

5. a. 4.304　　b. 0.2082　c. 2.8992　d. 4.905 64

6. a. 0.0001　b. 0.000 012　c. 0.063　d. 0.000 222

7. a. 0.000 002 44　　　　b. 0.000 36
    c. 0.392　　　　　　　d. 0.000 606 2

8. a. 0.3　　b. 0.851　　c. 2.538　　d. 0.175

9. a. 6.557　　b. 16.848　c. 60.300　d. 31.372

10. E

11. B

12. A

13. B

14. a. 0.0004　　b. 1.69　　　c. 4.2025

15. a. 1016　　b. 5973　　c. 437 400

16. a. 1092　　b. 4548　　c. 38 340

17. a. 64.8　　b. 1389.6　　c. 270.8
    d. 217 148.96　e. 82 049.6783

18. a. 3 268 904.3267　　b. 984 326.641
    c. 278.498 32　　　　d. 460

19. a.　i. 24 000　　　　ii. 24 640
    b.　i. 300 000　　　　ii. 331 602
    c.　i. 60 000　　　　ii. 50 449.5

20. a. 3500　b. 12 700　c. 1100　d. 2535　e. 5820

21. $750

22. 108 L

23. 260.7569 m^2

24. a. $32.00　　　　　　b. $32.05

25. The 50 extra sheets are 6 mm thick. You might notice the difference.

---

*32.

Team	1	2	3	4	5	6
Time (seconds)	153	159.3	148.9	166.6	155.9	155.8

26. 13

27. $5.55

28. 0.375 L

29. Together they pay $6.30: Sophie pays $3.60 and Hamish pays $2.70.

5.7 Dividing decimals

1. a. 0.6   b. 3.1   c. 0.54   d. 1.61   e. 2.42

2. a. 0.788   b. 1.147   c. 4.0701   d. 6.8594   e. 31.6176

3. a. 3.12   b. 2.231   c. 1.222   d. 6.421   e. 11.0311

4. a. 1.85   b. 4.75   c. 1.46

5. a. 2.45   b. 1.25   c. 6.95

6. a. 1.407   b. 9.6813   c. 0.6208
   d. 3.592 87   e. 0.023 497 8

7. a. 0.000 090 769   b. 0.802 405 6   c. 0.000 152 7
   d. 0.000 072 05   e. 0.000 000 003 2

8. a. 0.265   b. 0.79   c. 0.595
   d. 0.001 465   e. 0.001 74

9. a. 5   b. 7   c. 9   d. 4   e. 0.31

10. a. 2.2   b. 66   c. 72.1   d. 0.852   e. 1.367

11. a. 1564.08   b. 4.254   c. 143.134

12. a. 1786.1   b. 0.4782   c. 1630.95

13. D

14. D

15. B

16. D

17. a. i. 350   ii. 275
    b. i. 200   ii. 232
    c. i. 2000   ii. 2400

18. a. $3.65   b. $1.70   c. $56.85
    d. $0.75   e. $0.90   f. $63.50

19. 10.10

20. 3.086 grains

21. a. 32   b. $74.24

22. 0.4 L

23. $0.60

24. 20

25. 201

26. $4.50

27. 71

28. 0.675 m

29. 26

30. $\dfrac{1}{3}$

5.8 Decimals and percentages

1. a. 0.36   b. 0.14   c. 0.28   d. 0.73   e. 0.66

2. a. 0.59   b. 0.99   c. 0.09   d. 0.04   e. 0.01

3. a. 0.25   b. 2.00   c. 1.5   d. 3.6   e. 1

4. a. 0.123   b. 0.316   c. 0.592   d. 0.849   e. 0.421

5. a. 0.376   b. 0.219   c. 0.169   d. 0.111   e. 0.031

6. a. 0.092   b. 0.059   c. 0.088   d. 0.1425   e. 0.3175

7. a. 0.2355   b. 0.4575   c. 0.0005   d. 0.0401   e. 0.0002

8. a. 34%   b. 24%   c. 13%   d. 99%

9. a. 235%   b. 542%   c. 120%   d. 304%

10. a. 4.5%   b. 3.2%   c. 0.4%   d. 0.91%

11. B

12. C

13. A

14. D

15. a. $\dfrac{17}{100}$ b. 0.17

16. a. $\dfrac{1}{2000}$ b. 0.0005

17. 0.807

18. 2.18

19. 0.138

20. 0.807

21. a. 23, 24, 24, 24
    b. 0.23, 0.24, 0.24, 0.24
    c. It does not matter as long as both values are rounded up or down at the same digit.

22. a. 23.67%
    b. 0.2360, 0.2367, 0.2366

23. a. 0.378, 0.396, 0.309, 0.345, 0.328
    b. 39.6%, 37.8%, 34.5%, 32.8%, 30.9%
    c. 0.396, 0.378, 0.345, 0.328, 0.309
    d. Sample response: It is easier to arrange the numbers in the order using percentages as you can actually see the numbers and order them without any calculation. Decimals are more difficult to determine order, because there are more place values to look at.

24. $1125

25. 65%

26. $7.27

27. a. i. 0.072 91   ii. 0.7291   iii. 7.291
    b. When converting a percentage to a decimal number, the number becomes 100 times smaller.

28. When dealing with measurements it is easier to use decimals rather than percentages or fractions.

Project

1. a. $9.25   b. $17.40   c. $33.45   d. $13.45

2. $1.15

3. $28.85

4. Sample response:

Item	Cost
Nachos with cheese, salsa and sour cream	$6.50
Vegetable burger	$4.95
Cheesecake	$3.95
Total	**$15.40**

5. Sample response:

Your order	
Item	**Cost**
Nachos with cheese, salsa and sour cream	$6.50
Vegetable burger	$4.95
Cheesecake	$3.95
Total	**$15.40**

Classmate 1	
Item	**Cost**
Chicken caesar salad	$8.95
Tomato and hot chilli pasta	$8.95
Brownie	$3.50
Total	**$21.40**

Classmate 2	
Item	**Cost**
Bruschetta	$4.50
Double beef burger with bacon	$6.45
Jumbo banana split	$4.50
Total	**$15.45**

Classmate 3	
Item	**Cost**
Garlic bread	$3.50
Rib eye steak	$18.95
Chocolate frozen yoghurt	$2.50
Total	**$24.95**

Item	Cost
Nachos with cheese, salsa and sour cream	$6.50
Chicken caesar salad	$8.95
Bruschetta	$4.50
Garlic bread	$3.50
Vegetable burger	$4.95
Tomato and hot chilli pasta	$8.95
Double beef burger with bacon	$6.45
Rib eye steak	$18.95
Cheesecake	$3.95
Brownie	$3.50
Jumbo banana split	$4.50
Chocolate frozen yoghurt	$2.50
Jug of orange juice	$4.60
Total	**$81.80**

6. Sample response:
You: $14.60
Classmate 1: $8.60
Classmate 2: $14.55
Classmate 3: $5.05

7. If the bill only showed the total amount and not the breakdown, then it would be very difficult to work out how much each person owes individually. To counter this, the restaurant could use a bill similar to that shown in the answer to question **5** above.

5.9 Review questions

1. a. 7 tenths   b. 7 hundredths   c. 7 thousandths

2. a. 7 ten thousandths
   b. 7 hundred thousandths
   c. 7 units

3. a. $(2 \times 1) + \left(6 \times \dfrac{1}{10}\right) + \left(4 \times \dfrac{1}{100}\right)$

   b. $(1 \times 10) + (8 \times 1) + \left(4 \times \dfrac{1}{10}\right) + \left(6 \times \dfrac{1}{1000}\right)$

   c. $(9 \times 10) + (6 \times 1) + \left(3 \times \dfrac{1}{10}\right) + \left(4 \times \dfrac{1}{100}\right) +$ $\left(2 \times \dfrac{1}{1000}\right) + \left(8 \times \dfrac{1}{10\,000}\right)$

4. a. 6.4   b. 0.943   c. 4.11

5. a. 0.459   b. 1.603   c. 2.793

6. a. >   b. <   c. <

7. a. <   b. <   c. >

8. a. 0.13, 0.22, 0.34, 0.71, 0.86
   b. 0.124, 0.247, 0.258, 0.274, 0.285

9. a. 0.826, 0.834, 0.859, 0.888, 0.891
   b. 0.217, 0.356, 0.358, 0.365, 0.385

10. a. $\dfrac{4}{5}$ b. $\dfrac{17}{100}$ c. $\dfrac{9}{25}$ d. $\dfrac{187}{1000}$

11. a. $\dfrac{1}{8}$ b. $\dfrac{71}{125}$ c. $\dfrac{41}{200}$ d. $\dfrac{19}{20}$

12. a. $1\dfrac{1}{2}$ b. $4\dfrac{3}{5}$ c. $3\dfrac{12}{25}$ d. $5\dfrac{1}{4}$

13. a. $2\dfrac{3}{4}$ b. $2\dfrac{5}{8}$ c. $1\dfrac{14}{25}$ d. $8\dfrac{8}{25}$

14. a. 2.05   b. 13.86   c. 17.90

15. a. 1.3   b. 0.048   c. 1.9260

16. a. 14   b. 29   c. 87   d. 100

17. a. 9.1   b. 10.09   c. 6.993

18. a. 15.647   b. 215.72   c. 41.737 45

19. a. 5.3   b. 9.07   c. 2.04

20. a. 90.659   b. 0.949   c. 1.575

21. a. 18.6   b. 42.03   c. 66.0180   d. 5.3494

22. a. 2.53   b. 667.48   c. 5122.32   d. 0.926 08

23. a. 2.6   b. 13.45   c. 6.45   d. 182.94

24. a. 14 662.81   b. 48.0643   c. 839.204   d. 368
25. a. 1.312   b. 0.516   c. 0.783   d. 0.000 18
26. a. 0.87   b. 2.7168   c. 14.9084   d. 1.023 36
27. a. 6.25   b. 0.0009   c. 0.4096   d. 0.000 006 25
28. a. 0.41   b. 2.08   c. 73.62   d. 0.16
29. a. 1.4623   b. 10.236   c. 9.612 347
30. a. 0.200 32   b. 264.983 002 6   c. 34.6294
31. a. 0.05   b. 0.625   c. 0.3125   d. 0.125
32. a. $4.\dot{5}$ b. $0.\overline{82}$ c. $19.\overline{2781}$ d. $83.01\dot{6}\dot{2}$
33. a. 8   b. 51   c. 11   d. 34.3
34. a. 894.73   b. 6200.762   c. 2.413   d. 21 426.78
35. a. 0.42   b. 0.05   c. 0.94   d. 1.39
36. a. 6%   b. 19.7%   c. 58.03%   d. 0.8%
37. a. 5   b. 3.72   c. 2.25   d. 6.9
38. a. 19.26   b. 38.76   c. 6.12   d. 40.6926
39. $53.30
40. 1.125 L
41. $3.70 (rounded to nearest 5c) or $3.68
42. a. $0.\overline{076 923}$ b. $0.\overline{384 615}$
43. $80
44. No. Overall, after two months your salary will decrease by 1% using either option. However, option 1 gives an extra income (10% of the original salary) in the first month.
45. 20
46. 7 min 12s

6 Ratios and rates

LESSON
6.1 Overview

6.1.1 Why learn this?

Ratios and rates are used to compare different quantities. Ratios compare quantities of the same kind. For example, in a class of Year 7 students, if 20% walk to school and 80% don't, this can be expressed as a ratio of 1 to 4. This means that for every 1 student who walks to school, 4 students don't. Ratios are used in design or architectural drawings, with a drawing being an exact 'miniature' or scale diagram of the item to be made. Plans for a new home would be drawn at a ratio of 1 : 1000, which means that every 1 mm on the plan represents 1000 mm in the new home.

In comparison, rates enable you to compare quantities expressed in different units. Speed is an example of a rate; it compares distance travelled with the time taken, and is measured in km/h or m/s. You will see speed signs whenever you go for a drive. Rates and ratios help you make choices, too, such as when you decide which box of chocolates to purchase. Civil engineers examine rates of water flow, which helps them to design pipes that ensure the water gets to your home. Electrical engineers examine rates for stepping up and stepping down voltage, which ensures electricity is able to get to your home. Business managers explore various types of costed rates to ensure they always get the 'best deal'. Carpenters evaluate different rates to ensure they order enough timber to complete a job, and doctors and pharmacists calculate different rates to ensure you take the right amount of medicine based upon your height and weight. By comparing price and quantity, you can choose the 'best buy', which is the item with the lowest price per unit. Architects, chefs, engineers, fashion designers and builders all use rates and ratios in their field of work.

Hey students! Bring these pages to life online

Watch videos

Engage with interactivities

Answer questions and check solutions

Find all this and MORE in jacPLUS

Reading content and rich media, including interactivities and videos for every concept

Extra learning resources

Differentiated question sets

Questions with immediate feedback, and fully worked solutions to help students get unstuck

1. A cake mixture contains 1 cup of sugar and 2 cups of flour. Express this statement as a ratio.

2. For the diagram shown, write the following ratios in simplest terms.
   a. shaded parts: unshaded parts
   b. shaded parts: total number of parts

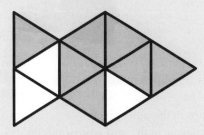

3. Express each of the following as a fraction.
   a. 5 pears to 7 oranges
   b. 500 m to 2 km
   c. 7 basketballs to 35 tennis balls
   d. 8 hours to 120 minutes

4. **MC** Select which of the following options is the simplest form of the ratio 8 : 32.
   A. 4 : 1   B. 1 : 4   C. 4 : 2   D. 4 : 16

5. A school has 2 students who wear glasses to every 5 students who don't wear glasses. A particular class where this ratio applies has 8 students who wear glasses. Calculate how many students don't wear glasses in this class.

6. **MC** Select which of the following are the same as the ratio 5 cm to 1 m. Select all answers that apply.
   A. 1 : 20   B. 5 : 100   C. 2 : 50   D. 25 : 500

7. **MC** Select which of the following is the best buy when buying a packet of cookies.

   A. 150 g for $1.75   B. 200 g for $2.30
   C. 250 g for $2.75   D. 350 g for $3.55

8. A bus is travelling at a speed of 60 km per hour. Calculate how long it takes to travel 145 km.

9. Hot chocolate can be bought in 250 g jars for $9.50 or in 100 g jars for $4.10. Calculate the cheaper way to buy the hot chocolate, and how much cheaper it is per 100 g.

10. Write 16 hours to 1 day as a simplified ratio.

11. Express the following as rates in the units given.
    a. 464 words written in 8 minutes, in the unit words/minute
    b. 550 litres of water flowing through a pipe in 12 minutes, in the unit kL/h

12. Yindi bought 3 dog treats for $21.60. Determine the price of 8 dog treats.

13. Out of 100 customers selected for a mobile internet usage survey, 63 customers were aged less than 25 years, 28 customers were aged 25–65 years and the rest were aged over 65 years.
    a. Write the ratio of customers aged 25–65 years to those aged less than 25 years.
    b. Write the ratio of customers aged over 65 years to those aged 25–65 years.
    c. Write the ratio of customers aged less than 25 years to the total number of customers surveyed.
    d. Write the number of customers aged less than 25 years as a fraction of the total number of customers surveyed.

14. If 12 oranges cost $6.00, calculate how much it will cost to buy 5 oranges.

15. If Eva takes 3 hours to clean a room and Will takes 5 hours to clean a room, determine how long it will take to clean a room if they work together.

LESSON
6.2 Ratios

LEARNING INTENTION

At the end of this lesson you should be able to:
- recognise and compare quantities using ratios
- simplify ratios.

▶ 6.2.1 Introduction to ratios

eles-3886

- **Ratios** are used to compare 2 or more quantities measured in the same units. That is, a ratio says how much of one value there is compared to another value. For example, in the diagram below, the ratio is 1 blue square to 4 pink squares.

1 : 4

Expressing ratios

- **A ratio of 1 to 4 is written as 1 : 4 (the quantities are separated by a colon) and is read as 'one to four'.**
- **Ratios can also be written in fraction form:**

$$1 : 4 \text{ is the same as } \frac{1}{4}$$

- **Ratios do not have names or units of measurement.**
- **A simplified ratio contains only whole numbers.**

WORKED EXAMPLE 1 Expressing quantities as ratios

Express each of the following statements as a ratio.
a. **The fertiliser contains 3 parts of phosphorus to 2 parts of potassium.**
b. **Mix 1 teaspoon of salt with 3 teaspoons of flour.**

THINK	WRITE
a. Write the 3 parts of phosphorus to 2 parts of potassium as a ratio (the first quantity should be written first).	a. $3 : 2$
b. Write the 1 teaspoon of salt with 3 teaspoons of flour as a ratio (first quantity should be written first).	b. $1 : 3$

DISCUSSION

Is the ratio 5:2 the same as 2:5? Does the order matter?

Express each of the following as a fraction.
a. **5 cm to 7 cm**
b. **3000 m to 2 km**

THINK	WRITE
a. 1. Write the 5 cm to 7 cm as a ratio by replacing 'to' with ':' (5 cm should be written first). As the two numbers have the same unit, units are not included.	a. $5:7$
2. Write the ratio as a fraction by placing the first number as a numerator and the second number as a denominator (no unit required).	$\dfrac{5}{7}$
b. 1. Express both quantities in the same units. Convert 3000 m to km or 2 km to m.	b. 3000 m to 2 km 3000 m to 2000 m
2. Write the 3000 m to 2000 m as a ratio by replacing 'to' with ':' (3000 m should be written first). Omit the units and simplify.	$3000:2000$ $3:2$
3. Write the ratio as a fraction by placing the first number as a numerator and the second number as a denominator.	$\dfrac{3}{2}$

▶ 6.2.2 Simplifying ratios

eles-3887

- Equivalent ratios are ratios that are the same when we compare them.
  For instance, the ratios $2:3$ and $4:6$ are equivalent, as the second ratio can be obtained by multiplying both parts of the first ratio by 2 and therefore they are the same ratios when compared.
  The ratios $10:5$ and $2:1$ are also equivalent, as the second ratio is obtained by dividing both parts of the first ratio by 5.
- Ratios are usually written in simplest form — that is, reduced to lowest terms. This is achieved by dividing each number in the ratio by the highest common factor (HCF).

Simplification of ratios

Equivalent ratios are always multiplied or divided by the same number. They work in the same way as equivalent fractions.

WORKED EXAMPLE 3 Simplifying ratios

Express the ratio 9 : 15 in simplest form.

THINK	WRITE
1. Determine the largest number by which both 9 and 15 can be divided (that is, the highest common factor, HCF). It is 3.	The HCF is 3.
2. To go from the ratio with the larger values to the ratio with the smaller values, divide both 9 and 15 by 3 to obtain an equivalent ratio in simplest form.	
3. Write the answer.	3 : 5

Digital technology

Since simplifying ratios follows a similar process to simplifying fractions, a calculator can be used to help check your answer. For example, in Worked example 3, the ratio 9 : 15 was simplified to 3 : 5.

Using a calculator, input 9 ÷ 15, then use the decimal to fraction button to S ↔ D display the answer as a fraction in simplest form.

WORKED EXAMPLE 4 Simplifying ratios of quantities expressed in different units

Write the ratio of 10 cm to 30 mm in simplest form.

THINK	WRITE
1. Write the question.	10 cm to 30 mm
2. Express both quantities in the same units by changing 10 cm into mm (1 cm = 10 mm).	100 mm to 30 mm
3. Omit the units and write the two numbers as a ratio.	100 : 30
4. Determine the HCF by which both 100 and 30 can be divided. It is 10.	The HCF is 10.
5. Simplify the ratio by dividing both sides by the HCF.	
6. Write the answer.	10 : 3

---

on Resources

🔹 **Interactivities** Introduction to ratios (int-3956)

Simplification of ratios (int-3949)

Exercise 6.2 Ratios

6.2 Quick quiz on	6.2 Exercise

Individual pathways

■ PRACTISE	■ CONSOLIDATE	■ MASTER
1, 4, 5, 10, 12, 15, 18	2, 6, 8, 11, 13, 16, 19	3, 7, 9, 14, 17, 20

Fluency

1. **WE1** Express each of the following statements as a ratio.
   a. The cake mixture contains 2 cups of flour and 5 cups of water.
   b. Mix 1 cup of sugar with 3 cups of milk.
   c. Sebastian Vettel's F1 car is twice as fast as Mick's delivery van.
   d. The Tigers team finished the season with a win to loss ratio of 5 to 2.

2. For the rectangle shown:
   a. write the ratio of unshaded parts to shaded parts
   b. write the ratio of unshaded parts to the total number of parts.

3. Explain why order is important when writing ratios.

4. **WE2** Express each of the following as a fraction.
   a. 5 wins to 7 losses
   b. 3500 m to 25 km
   c. 3 apples to 4 oranges
   d. 10 basketballs to 3 tennis balls

5. **WE3** Express the following ratios in simplest form.
   a. $75 : 60$ b. $154 : 24$ c. $36 : 6$ d. $48 : 42$

6. **WE4** Express the following ratios in simplest form.
   a. $7 \, \text{cm} : 9 \, \text{cm}$ b. $120 \, \text{minutes} : 2 \, \text{hours}$
   c. $240 \, \text{g} : 2 \, \text{kg}$ d. 5 weeks to 2 months

7. Express the following ratios in simplest form.
   a. $5 \, \text{cm} : 10 \, \text{mm}$ b. $600 \, \text{g} : 8 \, \text{kg}$
   c. $24 \, \text{hours} : 16 \, \text{minutes}$ d. 21 boys : 14 girls

8. Draw rectangles to represent the following ratios.
   a. Shaded area to unshaded area is $1 : 1$ b. Shaded area to unshaded area is $2 : 7$
   c. Unshaded area to shaded area is $2 : 3$ d. Unshaded area to shaded area is $3 : 1$

9. Complete the patterns of equivalent ratios.

a.	b.	c.	d.
$1 : 4$	$2 : 3$	$64 : 32$	$48 : 64$
$2 : 8$	$4 : 6$	$\_\_ : 16$	$\_\_ : 32$
$\_\_ : 12$	$8 : \_\_$	$\_\_ : 8$	$\_\_ : 16$
$\_\_ : 16$	$\_\_ : 15$	$8 : \_\_$	$6 : \_\_$
$5 : \_\_$	$\_\_ : 21$	$\_\_ : 1$	$\_\_ : 4$

Understanding

10. The length to width ratio of the Australian flag is $2:3$. If the length of the flag is $60\,\text{cm}$, determine the width of the flag.

11. Compare the following, using a mathematical ratio (in simplest form).

    a. The Hawks beat the Cats by 81 points to 60 points.

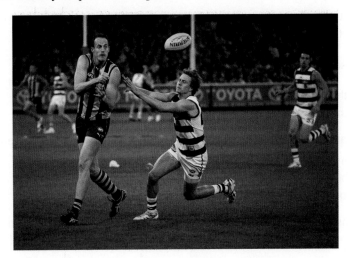

    b. The cost of a bag of Maltesers chocolate is $3, but Darrell Lea chocolate costs $5.
    c. In the movie audience, there were 160 children and 24 adults.
    d. During the race, Rebecca's average speed was $300\,\text{km/h}$ while Donna's average speed was $120\,\text{km/h}$.

12. Copy and complete each of the following equivalent ratios.

    a. $1:3 = \underline{}:12$ b. $2:3 = 4:\underline{}$ c. $66:48 = :\underline{}:8$ d. $36:72 = 12:\underline{}$

13. Copy and complete each of the following equivalent ratios.

    a. $20:12 = :\underline{}:3$ b. $\underline{}:6 = 8:24$ c. $8:6 = \underline{}:72$ d. $40:\underline{} = 5:6$

14. The table shows the nutritional information from a box of cereal.

Nutritional information	Average quantity per serving
Energy	660 kJ
Protein	4.5 g
Fat	0.1 g
Carbohydrates	22 g
Sugars	2 g
Fibre	3 g
Sodium	55 mg

Write the following ratios in simplest form.

a. Sugar to total carbohydrate
b. Fat to protein
c. Protein to fibre
d. Sodium to protein

Communicating, reasoning and problem solving

15. The cost of tickets to two different concerts is in the ratio 4 : 7. If the more expensive ticket costs $560, then determine the cost of the cheaper ticket.

16. A student received a score of $\dfrac{77}{100}$ for a Maths test.

    a. Write the ratio of the total marks received compared with the total marks lost.
    b. Write the ratio of the total marks lost compared with the total possible marks.

17. The total attendance at a concert was 62 400 people. Of this total, 24 800 people were male.

    a. Calculate how many females attended the concert.
    b. Compare the number of females to males by writing the ratio in simplest form.
    c. Write the number of females as a fraction of the number of males.

18. The ratio of gold coins to silver coins in a purse is 2 : 5. If there are 10 gold coins in the purse, determine the smallest number of coins that needs to be added so that the ratio of gold coins to silver coins changes to 3 : 4.

19. Out of 200 people selected for a school survey, 168 were students, 20 were teachers and the rest were support staff.

    a. Write the ratio of teachers to students.
    b. Write the ratio of support staff to teachers.
    c. Write the ratio of students to people surveyed.
    d. Write the number of teachers as a fraction of the number of people surveyed.

20. You drop a rubber ball from a height of 10 m. The ball rebounds to a height of 8 m and then to a height of 6.4 m. The ball continues to bounce in the same ratio.

    a. Calculate the height of the ball when the top of its bounce is less than 3 m.
    b. Determine on which number bounce this occurs.

LESSON
6.3 Rates

LEARNING INTENTION

At the end of this lesson you should be able to:
- recognise and express a rate from a written context
- simplify a rate and express the rate in its simplest form
- apply a rate to calculate the total value of an unknown quantity.

▶ 6.3.1 Identifying rates

eles-3888

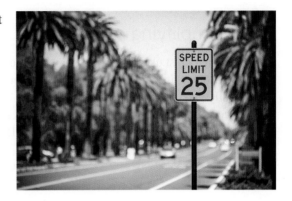

- A **rate** is used to compare two measurements of different kinds that are in different units. Unlike ratios, which are used to compare quanities measured in the same units.
- Rates have units. An example of a rate is speed (measured in km/h or m/s). A speed limit sign tells the driver that the maximum speed they are allowed to drive is 25 kilometres per hour.
- A forward slash (/), which is the mathematical symbol for 'per', is used to separate two different units.
- A rate is in its simplest form if it is per one unit.
- Rates are often written as one quantity **per** another quantity.

> **Expressing rates**
>
> **First quantity per second quantity**
>
> or
>
> $$\frac{\textbf{first quantity}}{\textbf{second quantity}}$$

WORKED EXAMPLE 5 Expressing a rate from a worded problem

Jennifer goes for a run every Monday morning and knows that she can run 3 kilometres in 15 minutes. Express, as a rate, the speed at which she can run.

THINK	WRITE
1. Speed is a rate the measures distance over time. From the question, determine the values for distance and time.	Distance = 3 km Time = 15 minutes
2. Express the rate as first quantity (distance) per second quantity (time) and write the answer.	3 km per 15 min This can also be written as 3 km/15 min.

WORKED EXAMPLE 6 Expressing a rate from a worded problem

Max notices a warning on his food packaging that tells him the sugar content in one serving is equal to 25 g. Rewrite this information in the form of a rate.

THINK	WRITE
1. Read the question carefully and determine two quantities. In this question, quantities are sugar and serving size.	Sugar = 25 g Serving size = 1 serve
2. Express rate as first quantity (sugar) per second quantity (serve) and write the answer.	25 g per 1 serve
3. As there is only 1 serving size, this can be written more simply as 25 g per serve.	25 g per serve

6.3.2 Simplifying rates

eles-3889

- It is often most useful to express rates in their simplest form.
  For instance, in Worked example 5, Jennifer knows she can run 3 kilometres in 15 minutes — speeds are usually given in kilometres per hour.
  To **convert** this rate of 3 km per 15 minutes into a rate of kilometres per hour (km/h), we multiply minutes by 4. That is because there are four lots of 15 minutes in every hour.

$$3 \text{ km} : 15 \text{ minutes}$$
$$\times 4$$
$$? \text{ km} : 60 \text{ minutes}$$

There are 60 minutes in an hour ($15 \times 4 = 60$).

To keep the rate the same, what we do to one side of the rate is what we also have to do to the other. This ensures that we do not change the rate's **value**.
This only works for multiplying and dividing, and not for adding or subtracting.

We also multiply the left side by 4 ($3 \times 4 = 12$).

$$3 \text{ km} : 15 \text{ minutes}$$
$$\times 4 \qquad\qquad \times 4$$
$$12 \text{ km} : 60 \text{ minutes}$$

Thus, the rate in kilometres per hour (km/h) is written as 12 km/h.

Converting rates

To convert rates, multiply or divide both quantities by the same number.

$$\div \left(\times \begin{array}{c} \text{First quantity : second quantity} \\ \\ \text{First quantity : second quantity} \end{array} \times \right) \div$$

WORKED EXAMPLE 7 Simplifying rates

Svetlana works for 4 hours at a local bakery. At the end of the week, she has earned $100 for her work. Calculate how much Svetlana earns in dollars per hour.

THINK

1. Read the question carefully and determine two quantities. In this question, the quantities are dollars and hours.

2. Express as a ratio of the first quantity (dollars) to the second quantity (hours). Remember to write out the units.

3. To convert the rate unit to dollars per hour we need to divide the hours by 4, to reduce 4 hours to 1 hour. Remember that what we do to one side of the rate, we must also do to the other side. Therefore, we also divide dollars by 4.

4. Write the answer as a rate in the correct units.

WRITE

Dollars = 100

Hours = 4

$\div 4$ ⌒ $100 : 4$ hours ⌒ $\div 4$ $25 : 1$ hour

Svetlana earns $25 per hour.

⊙ 6.3.3 Applications of rates

eles-3890

- Rates can be applied to find the total quantity of a measure being compared.
- Equivalent fractions can be used to calculate equivalent rates.

> ### Rates as equivalent fractions
>
> An equivalent fraction can be setup to calculate an equivalent rate.
>
> For example, to find how many beats in 3 minutes, the following equivalent fractions can be setup.
>
> $$\overset{\times 3}{\underset{\times 3}{\frac{60 \text{ beats}}{1 \text{ minute}} = \frac{? \text{ beats}}{3 \text{ minutes}}}}$$

WORKED EXAMPLE 8 Applying rate to calculate an unknown quantity

Rick has a job at the local cinema, where he is able to earn $22 per hour. If Rick works for 12 hours this week, calculate how much money he will earn in total.

THINK

1. Express the quantities given in the question as a rate in fraction form. The first quantity is dollars and the second quantity is hours.

2. Set up an equivalent fraction with the information given in the question.

3. To convert the amount of dollars in 12 hours, multiply both numerator and denominator by 12.

4. State your answer as a rate with correct units.

WRITE

$$\frac{22 \text{ dollars}}{1 \text{ hour}}$$

$$\overset{\times 12}{\underset{\times 12}{\frac{22 \text{ dollars}}{1 \text{ hour}} = \frac{?}{12 \text{ hours}}}}$$

$$\frac{22 \text{ dollars}}{1 \text{ hour}} = \frac{264 \text{ dollars}}{12 \text{ hours}}$$

Rick will earn $264 dollars in 12 hours.

Exercise 6.3 Rates

learn on

6.3 Quick quiz on	**6.3 Exercise**

Individual pathways

■ PRACTISE	■ CONSOLIDATE	■ MASTER
1, 4, 7, 10, 13	2, 5, 8, 11, 14	3, 6, 9, 12, 15

Fluency

1. **WE5 & 6** Identify the rates in the following circumstances and write them out in the form of one quantity per another quantity.

   a. Amir is watering the garden. He notices that he can fill his watering can up with water in one minute. The watering can is able to hold 7 litres of water. Express this as the rate at which the water is filling the can.

   b. James is training for his upcoming cricket game. If his current rate of runs per ball continues, he will make 33 runs from 7 balls. Write this as the rate of runs per ball.

   c. Tilly pays $6.00 for a bunch of bananas. State the cost of bananas per bunch.

   d. Cherese works at the local supermarket where, in a four-hour shift, she earns $80.00. Calculate the rate at which she is being paid.

   e. Dora swims every second morning. She knows that, in a 25-metre pool, she is able to swim 20 lengths in 15 minutes. Write her rate of swimming. Give your answer in metres per minute. You do not need to simplify your answer.

2. **WE7** Simplify the following rates, giving your answer as the unit rate where possible.

   a. 15 km per 30 minutes. Give your answer in the units km per hour.

   b. $150.00 for 6 hours of work. Give your answer in the units dollars per hour.

   c. A 14 m length of timber costs $21.00. Give your answer in the units dollars per metre.

3. Simplify the following rates, giving your answer as the unit rate where possible.

   a. 100 m was covered in 20 seconds. How many metres were covered per second?

   b. It cost $180.00 to wallpaper a room with a wall space of $50 \, m^2$. How much did the wallpaper cost per square metre?

   c. If a 5.5 kg bag of almonds cost $19.25, how much do almonds cost per kilogram?

   d. It costs $300.00 to carpet a room that is $8 \, m^2$. What is the cost of carpet per square metre?

4. **WE8** Patrick cycles for 2 hours, at a speed of 18 km/h. Calculate how far he has cycled.

5. Jackson works a 5-hour shift at the local bakery. If he is paid $85.00 for this shift, calculate his hourly pay rate.

6. A train is travelling at a speed of 50 km per hour. Calculate how long it takes to travel 150 km.

Understanding

7. Larah is working in the garden and buys 50 kg of premium potting mix for $75.00. Calculate how much the potting mix costs per kilogram.

8. Jacob was driving at a constant speed for 30 minutes. If he travelled for 35 km, calculate his speed.

9. Anthony was riding his bike at an average speed of 25 km/h. If he managed to cover a distance of 40 km, calculate how long his bike ride was (to the nearest minute).

Communicating, reasoning and problem solving

10. If 10 pears costs $15.00, explain how much it will cost to buy 5 pears.

11. If Michael can run 30 km in 3 hours, explain how long it will take him to run 20 km at a constant speed.

12. If Jaye can swim 400 metres in a time of 6 minutes and 45 seconds, calculate how long it will take her to swim 800 metres if she is swimming at a constant speed.

13. Water is poured into a canister at a rate of 6 L/ min. If the canister can hold 24 litres of water, calculate:
    a. how long will it take to fill the canister
    b. how long will it take to fill two canisters of the same size
    c. how long will it take until the canister is half full
    d. how long will it take to fill the canister with 18 litres of water.

14. Kelly is paid $14.25 per hour for her job working in retail. She is looking to save $240 to buy a new monitor for her computer and she knows she can only be rostered on for 4-hour shifts at work.
    a. Calculate how many hours she will need to work in order to buy the monitor. Round your answer to the nearest hour.
    b. Calculate how many 4-hour shifts she will need to complete in order to achieve this.
    c. After working enough 4-hour shifts, determine how much money is left over if she were to buy the monitor.

15. Charlee works two jobs, one at a supermarket and another as a basketball coach at weekends. She earns an hourly wage of $13.75 for her job at the supermarket and is paid a flat rate of $25.00 per week for coaching basketball. At the supermarket, the work roster is such that each shift goes for three hours and she cannot do more than one shift per day.
    a. If she is rostered on to work for 4 days this week, calculate how much money she will earn for the entire week, including her work as a basketball coach.
    b. At this rate, calculate how much money Charlee will earn for 3 weeks of work.
    c. Charlee worked for 24 hours at the supermarket over a 3-week period. Calculate how much money she earned in this time.

LESSON
6.4 The unitary method and best buys

LEARNING INTENTION

At the end of this lesson you should be able to:
- use the unitary method to determine the best value option
- apply the unitary method to solve problems.

▶ 6.4.1 Unit price

eles-3891

- The **unit price** is the price per unit. Examples include the price per 100 mL, the price per kg or the price of a single item.
- Supermarkets in Australia are now obliged to show the unit price on most products.
- To determine the 'best buy' when comparing similar items:
  - decide what unit you will use in your comparison
  - divide the price of each item by its number of units
  - identify the item with the smallest unit price. This is the best value item.

Price per unit

$$\text{Price per unit} = \frac{\text{price}}{\text{number of units}}$$

Remember to represent the same units when you are making comparisons.

WORKED EXAMPLE 9 Determining the best buy

Three shampoos are sold in the following quantities.

Brand A: 200 mL for $5.38
Brand B: 300 mL for $5.97
Brand C: 400 mL for $8.04

Determine which shampoo is the best buy.

	THINK		WRITE

THINK

WRITE

1. Determine a unit that all three brands can be reduced to. Here, 100-mL units can be found by dividing each quantity by 100.

$$\text{Brand A} = \frac{200}{100} = 2 \text{ units}$$

$$\text{Brand B} = \frac{300}{100} = 3 \text{ units}$$

$$\text{Brand C} = \frac{400}{100} = 4 \text{ units}$$

2. Determine the price per unit for each shampoo.

$$\text{Price per unit} = \frac{\text{price}}{100\text{-mL units}}$$

$$\text{Brand A} = \frac{5.38}{2} = \$2.69 \text{ per } 100 \text{ mL}$$

$$\text{Brand B} = \frac{5.97}{3} = \$1.99 \text{ per } 100 \text{ mL}$$

$$\text{Brand C} = \frac{8.04}{4} = \$2.01 \text{ per } 100 \text{ mL}$$

3. Write the answer.

Brand B: 300 mL of shampoo for $5.97 is the best buy.

6.4.2 The unitary method

eles-6120

- The **unit rate** is a rate per one of a quantity. Common unit rates are kilometres per 1 hour (km/h) or metres per 1 second (m/s).
- The **unitary method** is calculating the rate per one of a quantity and using that rate to calculate any other quantity.
- The unitary method is used because every value has a factor of one and can be easily multiplied or divided to determine an equivalent rate.

> ### The unitary method (unit rate)
>
> The unit rate is a special ratio that compares a quantity to one unit of another quantity. For example:
>
> $$\frac{60 \text{ kilometres}}{1 \text{ hour}} \text{ or } \$4 \text{ per kilogram}$$
>
> This is then used to calculate other equivalent rates.

WORKED EXAMPLE 10 Calculating the unit rate

A car travels 285 kilometres in 3 hours. Calculate how far the car travels per hour.

THINK

WRITE

1. Place the information given in a table.

Kilometres	285	?
Hours	3	1

2. To calculate kilometres per hour, 3 hours needs to be divided by 3. To keep the rate equivalent, divide the 285 kilometres by 3.

÷ 3

Kilometres	285	95
Hours	3	1

÷ 3

3. State the answer.

The car travels 95 kilometres per hour.

WORKED EXAMPLE 11 Solving problems using the unitary method

Three T-shirts cost \$84. Calculate the total cost of seven T-shirts.

THINK

1. Place the information given in a table as shown.

2. Because 3 is not a factor of 7, we need to first calculate the cost of 1 T-shirt by dividing by 3.

3. Using the cost of 1 T-shirt (unit price), multiply by 7 to calculate the cost of 7 T-shirts.

4. State the answer.

WRITE

Cost (\$)	84		
T-shirts	3		7

Cost (\$)	84	28	
T-shirts	3	1	7

Cost (\$)	84	28	196
T-shirts	3	1	7

Seven T-shirts cost \$196.

- Unit rate problems can be solved using a variety of methods. Worked example 11 could also have been solved using a number line or a bar model, shown below.

Resources

 Interactivities Unitary method (int-4154)

Exercise 6.4 The unitary method and best buys

learn on

6.4 Quick quiz	on	6.4 Exercise

Individual pathways

■ PRACTISE	■ CONSOLIDATE	■ MASTER
1, 4, 7, 9, 15, 19	2, 5, 10, 12, 13, 16, 20	3, 6, 8, 11, 14, 17, 18

Fluency

1. **WE9** Determine which of the following is the best buy.

Chocolate weight	Cost
150 g	$3.25
250 g	$4.75
325 g	$5.50

2. Determine which of the following is the best buy.

No. of pages	Cost
80	$1.98
160	$3.38
200	$3.98

3. Compare the cost of 400 g of biscuits for $2.48 and 500 g for $3.10 to identify which is the better buy.

4. **WE10** Amelia rides her bike 48 kilometres in 3 hours. Calculate how far Amelia travels per hour.

5. Ahmed earns $140 in a 5-hour shift at his part-time job. Calculate how much Ahmed earn per hour.

6. Nina and her brother Mauricio have made 196 cupcakes in 4 hours. Calculate how many cupcakes they made per hour.

7. Calculate the unit rate in each of the following tables.

a.
Onions	5	1
Cost ($)	$5.50	

b.
Gigabytes	8	1
Cost ($)	$44	

c.
Eggs	12	1
Cost ($)	$3.60	

8. Calculate how much you would pay for 4.5 kg of apples, given that 2 kg cost $3.80.

Understanding

9. **WE11** 10 litres of petrol costs $18. Determine the cost of 16 litres of petrol.

10. A 6-minute phone call costs $3.00. Determine the cost of an 8-minute phone call.

11. The winner of a 100-metre race finished in 12.5 seconds. Determine how far they had travelled after 8 seconds.

12. Compare the following to identify the best buy:
500 g of cheese for $12.80, 200 g of cheese for $5.62 or 80 g of cheese for $2.10.

13. Valeria can type 810 words in 15 minutes. Determine how many words she can type in 25 minutes.

14. Supermarkets are obliged by law to display the unit price when selling the same items in different quantities. Calculate the price per 100 g for the following amounts of chocolate. Give your answers correct to the nearest cent.
   a. 250 g for $5.95   b. 400 g for $8.50   c. 60 g for $2.00

Communicating, reasoning and problem solving

15. In some situations it is not always practical to go for the best buy. Suggest two reasons this might occur.

16. Explain some of the different situations in which you would use the unitary method.

17. While shopping in a sale, Carly spots a sign that says 'Buy 2 items, get the third half price (the least expensive item will be counted as the third item)'. If she finds a shirt for $25.99, a skirt for $87.99 and a belt for $15.97 and pays cash for the items, determine how much Carly will pay and justify your answer.

18. Penny loves to make bracelets. To make them, she needs to buy some beads. Brand A costs $7.50 for a box that contains 1000 beads; Brand B costs $17.50 for a box that contains 2500 beads. Determine which brand is the better buy.

19. Ice-cream sticks are sold in the following multi-packs. Compare the following to identify the best buy.

Cost ($)	Number of ice-cream sticks
$3.25	6
$4.99	10
$7.50	15

20. Some players from a soccer club went to a café to celebrate their win. Each had a burger and a drink. The bill came to $80. When it came time to pay the bill, 3 people had left without paying, so the remaining members had to pay an extra $2 each to cover the bill. If there were originally n people in the group, and the burger and drink deal cost b, write an equation using these pronumerals to show how the bill was settled. Do not attempt to solve your equation.

LESSON
6.5 Review

6.5.1 Topic summary

RATIOS AND RATES

Ratios

- Ratios compare quantities in the same unit.
- 1 : 3 is read as 'the ratio of 1 to 3'.
- Simplified ratios contain only integer values without units.
- They can be converted into fractions.

  e.g. 2 : 5 is the same as $\frac{2}{5}$

Equivalent ratios

- Ratios are equivalent if one can be converted into the other by multiplying or dividing by a factor.
  e.g.

  $\times 2 \left(\begin{array}{c} 4 : 6 \\ \\ 2 : 3 \end{array} \right) \times 2$

  4 : 6

  2 : 3

- Ratios are generally written in simplest form.

Rates

- A rate compares two different types of quantities in different units.
  e.g. 12 km/h or 12 cm/day
- A rate is in simplified form if it is per one unit.

  $\div 4 \left(\begin{array}{c} \$100 : 4 \text{ hours} \\ \\ \$25 : 1 \text{ hour} \end{array} \right) \div 4$

- A rate can be applied to find the total quantity of a measure being compared.
- An equivalent fraction can be set up to calculate an equivalent rate.

  $$\overset{\times 3}{\frac{60 \text{ beats}}{1 \text{ minute}} = \frac{? \text{ beats}}{3 \text{ minutes}}}_{\times 3}$$

Unitary method

- The unit rate is a rate per one of a quantity.
  eg. kilometres per hour or dollars per kilogram
- The unitary method involves calculating the rate per one and using that rate to calculate other quantities.

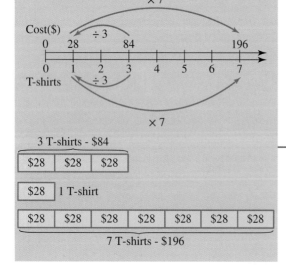

3 T-shirts - $84

$28	$28	$28

| $28 | 1 T-shirt |

$28	$28	$28	$28	$28	$28	$28

7 T-shirts - $196

Unit price

- The unit price is the price per single unit, such as kg, m or item.

  $$\text{Price per unit} = \frac{\text{price}}{\text{number of units}}$$

- The 'best buy' is the option with the lowest cost per unit.

6.5.2 Project

Successive discounts

Tony is a mechanic who has bought $250 worth of equipment at a tool store. Tony received 15% off the marked price of all items and received a further 5% trade discount. When two discounts are given on the same item or group of items, they are called successive discounts. In this case, Tony received successive discounts of 15% and 5%. Is this equal to a single discount of 20%?

Answer the following questions.
1. Calculate the amount that was due after Tony was given the first 15% discount.
2. From this amount, apply the trade discount of 5% to evaluate the amount that was due.
3. How much is the cash discount that Tony received?
4. Now calculate the amount that would have been due had Tony received a single discount of 20%. Is this the same answer?
5. Calculate the amount of cash discount that Tony received as a percentage of the original bill.
6. Explain whether the discount is the same if the 5% discount is applied before the 15% discount.

 Resources

Exercise 6.5 Review questions

 learnon

Understanding and fluency

1. For the diagram shown, write the following ratios in simplest terms.
   a. shaded parts : unshaded parts
   b. shaded parts : total number of parts

2. Express in simplest form.
   a. $8:10$ b. $14:21$ c. $16:64$ d. $24:18$

3. Express the following in simplest form.
   a. $\dfrac{1}{2}:4$

   b. $\dfrac{3}{5}:\dfrac{1}{5}$

   c. $1\dfrac{1}{2}:\dfrac{3}{4}$

   d. $0.8:10$

   e. $10^2:10^3$

   f. $\dfrac{3}{5}:\dfrac{3}{10}$

4. Rewrite each of the following ratios in simplest terms.
   a. $4\,\text{m}$ to $10\,\text{cm}$
   b. 3 oranges to 1 dozen (12) oranges
   c. 8 months to 1 year
   d. $3\,\text{cm}$ to $12\,\text{mm}$
   e. 70 cents to $2.00

5. Rewrite each of the following ratios in simplest terms.
   a. 3.5 kilograms to 500 grams
   b. 1.2 cm to 10 mm
   c. $\frac{1}{2}$ of a melon to 3 melons
   d. 2 hours and 40 minutes to 30 minutes
   e. $3\frac{1}{5}$ cm to $\frac{2}{5}$ m

6. Calculate the unit rate for each of the following.
   a. If 35 copies can be made in 7 minutes, how many copies can be made in one minute?
   b. If 9 crates can hold 216 bottles, how many bottles can one crate hold?
   c. If James can drive 180 km in 3 hours and 20 minutes, how many kilometres can he drive in one hour?
   d. If 6 movie tickets cost $50.40, how much does each movie ticket cost?

7. Calculate the given rates for each of the following.
   a. The total cost of 1.5 kg of potatoes at $2.99/kg
   b. The distance travelled at 65 km/h for 3.75 hours
   c. The time taken to travel 110 km at 80 km/h

8. When the AFL Grand Final was held in Queensland in 2020, 30 000 people attended. If there were 9872 women, 11 300 Richmond supporters, 8400 Geelong supporters and 2100 children in attendance, choose the best approximation of the ratio for each of the following.
   a. **MC** The number of Geelong supporters to the number of Richmond supporters
      A. 8400 : 1130
      B. 113 : 84
      C. 11 300 : 8400
      D. 84 : 113
   b. **MC** The number of children to the number of adults
      A. 2100 : 30 000
      B. 2100 : 9872
      C. 7 : 93
      D. 2100 : 19 700
   c. **MC** The number of women to the total number of people in attendance
      A. 1234 : 3750
      B. 9872 : 3000
      C. 9872 : 39 872
      D. 617 : 1870

9. Five flowers cost $45. Determine the cost of:
   a. 3 flowers
   b. 7 flowers
   c. 18 flowers.

10. Find the values of the pronumeral in the following ratios.
    a. $a : 7 = 3 : 14$
    b. $12 : 4 = 9 : b$

Communicating, reasoning and problem solving

11. Justin uses 375 grams of cream cheese to make a cheesecake that serves 12 people, using the recipe shown.
    Ingredients
    250 g sweet plain biscuits
    125 g unsalted butter
    375 g cream cheese, softened
    1 lemon, zested
    2 tsp vanilla essence
    a. Write the ratio of cream cheese to butter.
    b. Calculate how many grams of cream cheese Justin needs to make a cheesecake that serves 8 people.

12. If it takes 15 people 12 days to complete a job, explain how long it would take 20 people to complete the same job.

13. A video gamer has completed 8 levels in 1 hour 16 minutes. If there are 15 levels in the game, determine how long it will take them to complete all levels.

14. Identify which is the better value — 400 g of biscuits costing $2.98, or a pack of biscuits with 500 g costing $3.28.

15. The rate of ascent for the Discovery space shuttle was 114 000 m in 8.0 minutes.
a. Write this speed in km/min.
b. Write this speed in km/h.

16. Mitch is a very fast runner who can run 3 km in 9 minutes. If Mitch is able to maintain this rate, calculate how long it will take him to run 9 km.

17. The dimensions (length and width) of rectangle B are double those of rectangle A. Explain whether the area of rectangle B is double of the area of rectangle A.

| A |

| B |

18. A rectangle has dimensions (length and width) that are in the ratio 3 : 5. Complete the following table to find the dimensions of the rectangle if it has an area of 135 cm². The first row has been done for you.

Length (cm)	Width (cm)	Area (cm²)
3	5	15

19. Tammy walks to school at an average rate (speed) of 6 km/h.
a. It takes her 30 minutes to walk to school. Calculate the distance from her house to the school.
b. If the total time she spends walking to school and back home is 70 minutes, calculate the rate at which she is walking on her way home.

20. At a Year 7 social, the ratio of females to males was 3 : 2. Four additional females arrived and 4 males left, and the ratio was then 2 : 1. Determine the number of people at the social.

on To test your understanding and knowledge of this topic, go to your learnON title at www.jacplus.com.au and complete the **post-test**.

Answers

Topic 6 Ratios and rates

6.1 Pre-test

1. $1:2$

2. a. $7:3$ b. $7:10$

3. a. $\dfrac{5}{7}$ b. $\dfrac{1}{4}$ c. $\dfrac{1}{5}$ d. $\dfrac{4}{1}$

4. B

5. 20 students

6. A, B, D

7. D

8. 2 hours 25 minutes

9. The 250 g jar is cheaper by 30 cents per 100 g.

10. $2:3$

11. a. 58 words/minute   b. 2.75 kL/h

12. $57.60

13. a. $28:63$ b. $9:28$
    c. $63:100$ d. $\dfrac{63}{100}$

14. $2.50

15. $1\dfrac{7}{8}$ hours *or* 1 hour 52 minutes 30 seconds

6.2 Ratios

1. a. $2:5$ b. $1:3$ c. $2:1$ d. $5:2$

2. a. $7:5$ b. $7:12$

3. The order is very important in ratios. By changing the order of a ratio, the meaning changes. For example, if the ratio of losses to wins is $2:3$, the 2 corresponds to the losses and the 3 corresponds to the wins. If we change the order of the ratio to $3:2$ then the 3 corresponds to the losses and the 2 corresponds to the wins, which is not correct.

4. a. $\dfrac{5}{7}$ b. $\dfrac{7}{50}$ c. $\dfrac{3}{4}$ d. $\dfrac{10}{3}$

5. a. $5:4$ b. $77:12$ c. $6:1$ d. $8:7$

6. a. $7:9$ b. $1:1$ c. $3:25$ d. $5:8$

7. a. $5:1$ b. $3:40$ c. $90:1$ d. $3:2$

8. a.

   $1:1$

   b.

   $2:7$

   c.

   $2:3$

   d.

   $3:1$

9. a. $1:4$
   $2:8$
   $3:12$
   $4:16$
   $5:20$
   b. $2:3$
   $4:6$
   $8:12$
   $10:15$
   $14:21$
   c. $64:32$
   $32:16$
   $16:8$
   $8:4$
   $2:1$
   d. $48:64$
   $24:32$
   $12:16$
   $6:8$
   $3:4$

10. The width of the flag is 90 cm.

11. a. $27:20$ b. $3:5$ c. $20:3$ d. $5:2$

12. a. $1:3=4:12$ b. $2:3=4:6$
    c. $66:48=11:8$ d. $36:72=12:24$

13. a. $20:12=5:3$ b. $2:6=8:24$
    c. $8:6=96:72$ d. $40:48=5:6$

14. a. $1:11$ b. $1:45$ c. $3:2$ d. $11:900$

15. The cost of the cheaper ticket is $320.

16. a. $77:23$ b. $23:100$

17. a. 37 600 female   b. $47:31$ c. $\dfrac{47}{31}$

18. Fourteen coins need to be added to the purse. Three silver coins and 11 gold coins will change the ratio of gold to silver.

19. a. $5:42$ b. $3:5$ c. $21:25$ d. $\dfrac{1}{10}$

20. a. 2.6 m   b. Bounce 6

6.3 Rates

1. a. 7 litres per minute
   b. 4.71 runs per ball
   c. 6 dollars per bunch
   d. 20 dollars per hour
   e. $\dfrac{100}{3}$ metres per minute

2. a. 30 km/h
   b. $25.00 per hour
   c. $1.50 per metre

3. a. 5 m/s   b. $3.60 per m^2
   c. $3.50 per kg   d. $37.50 per m^2

4. 36 km

5. $17.00 per hour

6. 3 hours

7. $1.50 per kg

8. 70 km/h

9. 1 hour and 36 minutes

10. $7.50

11. 2 hours

12. 13.5 minutes or 13 minutes and 30 seconds

13. a. 4 minutes   b. 8 minutes
    c. 2 minutes   d. 3 minutes

14. a. 17 hours   b. 5 shifts   c. $45.00

15. a. $190.00   b. $570.00   c. $405.00

6.4 The unitary method and best buys

1. The 325-g option is the best buy.

2. The 200-page option is the best buy.

3. Neither is better (they are both 62 cents per 100 g).

4. 16 kilometres per hour

5. $28 per hour

6. 49 cupcakes per hour

7. a.

Onions	5	1
Cost ($)	$5.50	$1.10

b.

Gigabytes	8	1
Cost ($)	$44	$5.50

c.

Eggs	12	1
Cost ($)	$3.60	$0.30

8. $8.55

9. $28.80

10. $4.00

11. 64 metres

12. 500 g of cheese is the best buy.

13. 1350 words

14. a. $2.38   b. $2.13   c. $3.33

15. The amount required should always be considered, as the best buy will often be for the largest quantity. Waste due to spoilage and the best before date of the item should also be taken into account. The size of the item and storage space may also need to be considered. The total cost should also be considered, particularly if shopping on a budget.

16. There are various situations in which it would be used. Generally they involve comparing different types of a certain item.

17. Brand B. The cost for each bead for Brand B is cheaper (7c compared to 7.5c).
    a. $21.33   b. $18.13

18. $121.97

19. Options are 50 c per ice-cream stick, 49.9 c per ice-cream stick and 54.17 c per ice-cream stick. Therefore the $4.99 pack is the best buy.

20. $(n-3)(b+2)=80$

Project

1. $212.50

2. $201.88

3. $48.13

4. No

5. 19.25%

6. Yes

6.5 Review questions

1. a. $2:2=1:1$ b. $2:4=1:2$

2. a. $4:5$ b. $2:3$ c. $1:4$ d. $4:3$

3. a. $1:8$ b. $3:1$ c. $2:1$
   d. $2:25$ e. $1:10$ f. $2:1$

4. a. $40:1$ b. $1:4$ c. $2:3$
   d. $5:2$ e. $7:20$

5. a. $7:1$ b. $6:5$ c. $1:6$
   d. $16:3$ e. $2:25$

6. a. 5 copies   b. 24 bottles
   c. 54 km   d. $8.40

7. a. $\$4.485 \approx \$4.49$ b. 243.75 km
   c. 1.375 hours

8. a. D   b. C   c. A

9. a. $27   b. $63   c. $162

10. a. $\dfrac{3}{2}$ b. 3

11. a. $3:1$ b. 250 g

12. 9 days

13. 142.5 minutes or 2 hours 22.5 minutes

14. 500 g

15. a. 14.25 km/minutes   b. 855 km/h

16. It will take Mitch 27 minutes to run 9 km.

17. No, the area of rectangle B will not be double. It will be four times greater.

18.

Length (cm)	Width (cm)	Area (cm^2)
3	5	15
6	10	60
9	15	135

19. a. The school is 3 km away from her house.
    b. The rate is 4.5 km/h.

20. The total number of people at the social is 60.

7 Algebraic techniques

LESSON SEQUENCE

LESSON
7.1 Overview

Why learn this?

Algebra is a fundamental building block of mathematics. Algebra is, in fact, a mathematical language in which you are able to use the basic operations of addition, subtraction, multiplication and division. Using algebra, you are able to break down a problem or a puzzle into smaller parts and then find the solution or answer. Sometimes it is hard to imagine how we would apply algebra in our everyday lives, but we use algebra frequently without even realising it. The methods used to solve problems and the reasoning skills learned will be useful no matter what you do in life.

Every five years, on one particular night, every person in Australia completes a census form. The census collects information such as the number of people living at an address, their ages and their incomes. Statisticians use algebraic techniques to interpret the information, which can then be used in policymaking and to make predictions about future requirements for such things as schools, hospitals and aged care services. Scientists, architects, accountants and engineers all use algebra frequently in their work. Without algebra there would be no space travel, no electrical appliances, no television, no smartphones, no iPads, no computers and no computer games.

Hey students! Bring these pages to life online

▶ Watch videos

🧩 Engage with interactivities

🅰⁺ Answer questions and check solutions

Find all this and MORE in jacPLUS ▶

Reading content and rich media, including interactivities and videos for every concept

Extra learning resources

Differentiated question sets

Questions with immediate feedback, and fully worked solutions to help students get unstuck

1. **MC** Given the pattern 8, 13, 18, ..., the next three numbers are:

   A. 23, 28, 35   **B.** 23, 28, 33   **C.** 22, 26, 30   **D.** 22, 27, 32

2. Calculate the value of the expression $3p$ when $p = 8$.

3. In a tennis match, Charlie scored 12 more points than twice Lachlan's score. Lachlan scored 36 points. Calculate the number of points Charlie scored.

4. **MC** An expression that shows the difference between k and l, where k is bigger than l, is:

   A. $l - k$ **B.** $k \div l$ **D.** $k + l$ **E.** $k - l$

5. Simplify the expression $3x + 2y - x + 4y$ by collecting like terms.

6. Simplify the expression $4(2x + 3y) + 4(3x - 2y)$.

7. Identify two consecutive numbers with a product of 420.

8. Explain what happens when the result of $(5k + 3)$ is doubled.

9. Complete the following: $3x + 7 - \square = 2 + \square - x$.

10. Simplify the following expressions.

    a. $3a \times 5b \times 2c$ **b.** $4ab \times 13b^2$ **c.** $\dfrac{15pq}{3p}$ **d.** $\dfrac{36mn}{6n^2}$

11. If $a = 2$, $b = 5$ and $c = 3$, determine the value of the following expression: $c(4a - b) - 3a$.

12. Expand and simplify the following expression: $3(7p - 4q) - 2(2p - 5q)$.

13. A rectangle has a width of $3x$ and a length of $5y$. Determine the simplest expression for the perimeter of the rectangle.

14. Is the expression $a \times a = 2a$ True or False? Justify your answer.

15. Ashton has 2 pairs of basketball shoes and 5 pairs of runners, and Alyssa has 1 pair of basketball shoes and 3 pairs of runners. Given b is the number of basketball shoes and r is the number of runners, determine the simplified expression that shows the total number of basketball shoes and runners they have together.

LESSON
7.2 Introduction to algebra

LEARNING INTENTION

At the end of this lesson you should be able to:
- understand the use of pronumerals in algebra
- identify algebraic terminology
- construct expressions from given information.

▶ 7.2.1 Pronumerals

eles-3986

- A **pronumeral** is a letter or symbol that is used to represent one or more numerical values. For example, p could represent the number of pies in a shop.
- Any letter can be used as a pronumeral.
- A pronumeral that represents more than one numerical value can be called a **variable**.

> **Using pronumerals**
>
> - **Pronumerals are used to represent unknown quantities.**
> - **For example, without knowing the number of fish in a pond, this number could be represented by the pronumeral f.**

WORKED EXAMPLE 1 Use of pronumerals and constants

We don't know exactly how many lollies there are in a jar, but we know that each jar has the same number. We will refer to this unknown number as l, where l is the pronumeral.
a. **Draw a jar and label the number of lollies in the jar as l.**
b. **There is a jar of lollies plus one lolly. Represent this information using pictures and using symbols.**

THINK	WRITE
a. The number of lollies in the jar is unknown.	a. Let l be the number of lollies in the jar. l
b. Draw a jar of lollies and one lolly. Use 'l' in an algebraic expression to represent this information.	b. $l + 1$

▶ 7.2.2 Terms and expressions

eles-3987

- A **term** is a single mathematical expression.
  For example, $2, p, 3x$ and $6xy$ are all terms.

- In terms that contain numbers and pronumerals multiplied together, the number is placed at the front and the multiplication signs are removed.
  For example, $2 \times a \times c$ would be simplified to $2ac$.

- Pronumerals in a term are normally written in alphabetical order.
  For example, $5 \times y \times x = 5xy$. However, $5yx$ would still be correct, just as $2 \times 3 \times 5$ is the same as $2 \times 5 \times 3$.

- The number in front of the pronumeral(s) is known as the **coefficient**. For example, in the term $3t$, the coefficient of t is 3.

- Terms such as $m \times m$ are expressed as m^2. Similarly, $m \times m \times m$ is expressed as m^3, just as $2 \times 2 = 2^2$ and $2 \times 2 \times 2 = 2^3$.

- Numbers without pronumerals are called **constant** terms.

- **Expressions** are mathematical sentences made up of terms separated by $+$ or $-$ signs.
  For example, $3m + 5$, $9 - 2t$ and $a^2 + 2ab + b^2$ are all expressions. An expression can take on different values depending on the values of the pronumerals.

- The definitions are summarised in the following diagram.

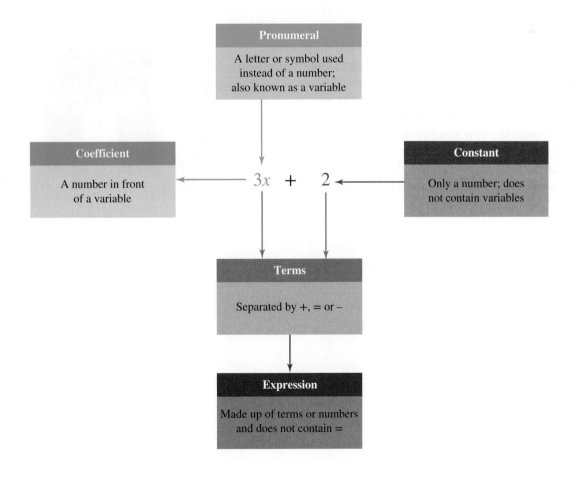

WORKED EXAMPLE 2 Writing terms and expressions

If y represents any number, write terms or expressions for:
a. 5 times that number
b. 2 less than that number
c. 8 more than that number
d. the number divided by 4 (or the quotient of y and 4)
e. the next consecutive number (that is, the counting number after y).

THINK

In each case, think about which operations are being used
and the order in which they occur.

a. An expression with '5 times' means 'multiply by 5'. When
multiplying, the multiplication sign is not shown. Remember
to put the number first.

b. An expression with '2 less' means 'subtract 2'.

c. An expression with '8 more' means 'add 8'.

d. An expression with 'divided by 4' means 'write as a quotient
(fraction)'.

e. An expression with 'the next consecutive number' means 'add 1
to the number'.

WRITE

a. $5 \times y = 5y$

b. $y - 2$

c. $y + 8$

d. $\dfrac{y}{4}$

e. $y + 1$

WORKED EXAMPLE 3 Writing simple algebraic expressions

Suppose there are d dogs currently at a lost dogs' home.
a. If three dogs find homes, write an expression to represent
the total number of dogs at the lost dogs' home.
b. Just after the three dogs find homes, the number of dogs
at the lost dogs' home is doubled. Write an expression to
represent the new number of dogs at the lost dogs' home.
c. The dogs are then sorted into 10 pens with the same
number of dogs in each pen. Write another expression to
represent the number of dogs in each pen.

THINK

a. The original number of dogs (d) is decreased by three.

b. The number of dogs left at the lost dogs' home is doubled.

c. The dogs are divided into 10 pens.

WRITE

a. $d - 3$

b. $2 \times (d - 3)$

c. $2 \times (d - 3) \div 10$ or $\dfrac{2d - 6}{10}$

WORKED EXAMPLE 4 Simplifying when terms or pronumerals are multiplied

If one jar of lollies has l lollies, then two jars of lollies will have double the number of lollies, that is $2 \times l = 2l$. In algebra, $2l$ means two lots of l, $3l$ means three lots of l, and so on.

a. There are two jars of lollies and three loose lollies. Each jar has the same number of lollies. Write an expression using symbols.

b. There are three jars of lollies, four packets of lollipops and four loose lollies. Each jar has the same number of lollies and each packet of lollipops has the same number of lollipops. Write an expression using symbols.

THINK

a. There are two jars of lollies (l) plus (+) the three that are loose.

b. There are three jars of lollies (l) and four packets of lollipops (p) plus (+) the four loose lollies.

WRITE

a. Let l be the number of lollies in the jar.

$$2 \times l + 3 = 2l + 3$$

b. Let l be the number of lollies in each jar and p be the number of lollipops in each packet.

$$3 \times l + 4 \times p + 4 = 3l + 4p + 4$$

COMMUNICATING — COLLABORATIVE TASK: How many shoes?

You own a certain number of pairs of shoes. Your friend owns five pairs more than you own, and your sister owns three pairs fewer than you own. Your brother owns four pairs fewer than your friend owns.

1. Write an expression to individually represent each of the following.
   a. The number of pairs of shoes you own
   b. The number of pairs of shoes your friend owns
   c. The number of pairs of shoes your sister owns
   d. The number of pairs of shoes your brother owns
2. Share your answers with a partner. Discuss any differences between your answers.
3. As a class, discuss any differences that you found.
4. Does it matter if each student in the class uses a different symbol? Why? Why not?

---

on Resources

Interactivities Terms and expressions (int-4004)

7.2 Quick quiz on	7.2 Exercise

Individual pathways

■ PRACTISE	■ CONSOLIDATE	■ MASTER
1, 3, 4, 7, 11, 12, 13, 18, 19, 24, 25, 26, 28, 31	2, 5, 6, 8, 14, 15, 20, 21, 27, 29, 32	9, 10, 16, 17, 22, 23, 30, 33, 34

Fluency

1. State what pronumerals are used for in algebra.

2. **WE1** If b represents the total number of bananas in a fruit bowl, write a short paragraph that could explain the following table.

Day	Number of bananas in the fruit bowl
Monday	b
Tuesday	$b-4$
Wednesday	$b-6$
Thursday	$b+8$
Friday	$b+5$

3. **WE2** Write an expression for the total number of coins in each of the following. The symbol represents a full piggy bank and ⬭ represents one coin. Use c to represent the number of coins in a full piggy bank.

a.

b.

c.

d.

4. Write an expression for the total number of coins in each of the following, using c to represent the number of coins in a piggy bank.

a.

b.

c.

d.

5. Christie and Jane each had two full piggy banks and seventeen coins. They combined their money and spent every cent on a day out in town.

$$2 \times \left(\text{🐷🐷} + \begin{array}{c} \text{⬭⬭⬭⬭⬭⬭⬭⬭⬭} \\ \text{⬭⬭⬭⬭⬭⬭⬭⬭} \end{array} \right)$$

Write an expression to represent the number of coins they had in total. Use c to represent the number of coins in a piggy bank.

6. Luke has three macadamia nut trees in his backyard. He saves takeaway containers to store the nuts in. He has two types of containers, rectangular and round .

Using *m* to represent the number of nuts in a rectangular container and *n* to represent the number of nuts in a round container, write expressions for the following.

a.

b.

c.

d. where ● represents one nut.

e.

7. Using *a* and *b* to represent numbers, write an expression for:
   a. the sum of *a* and *b*
   b. the difference between *a* and *b*
   c. the product of *a* and *b*
   d. *a* multiplied by 3 and the result divided by *b*.

8. **WE3** Using *a* and *b* to represent numbers, write an expression for:
   a. 15 divided by negative *a*
   b. five times *b* subtracted from three times *a*
   c. *b* divided by negative 5
   d. seven less than *a*.

9. Write the following expressions in ascending order.
   a. $m + 3, m - 2, m + 1, m - 8$
   b. $n - 15, n + 25, n - 10.3$

10. Write your own descriptions for what the following expressions could represent.
    a. $d + 2, d, d - 1$
    b. $t, 2t, t + 1$
    c. $y, 3y, 10y$

11. **WE4** Write an expression for each of the following.
    a. The sum of *b* and 2
    b. 3 less than *t*
    c. 6 added to *d*
    d. 5 taken away from *k*

12. Write an expression for each of the following.
    a. The sum of *g*, *n* and *w*
    b. *d* increased by *h*
    c. *n* increased by *n*
    d. *h* added to *c*

13. Write an expression for each of the following.
    a. *g* subtracted from 12
    b. The product of *d* and 4
    c. *b* multiplied by *f*
    d. *y* added to the product of 3 and *m*

14. If *a*, *b* and *c* represent any 3 numbers, write an expression for each of the following.
    a. The sum of all 3 numbers
    b. The difference between *a* and *c*
    c. The product of *a* and *c*
    d. The product of all 3 numbers

15. If *a*, *b* and *c* represent any 3 numbers, write an expression for each of the following.
    a. The quotient of *b* and *c* (that is, *b* divided by *c*)
    b. The sum of *a* and *c*, divided by *b*
    c. 3 more than *a*
    d. The difference between *c* and the product of *a* and *b*

16. Write expressions for the following rules.
    a. The number of students left in the class if *x* students leave for the canteen out of a total group of *t* students
    b. The amount of money earned by selling *b* cakes, where each cake is sold for $4.00
    c. The total number of sweets if there are *g* bags of sweets with 45 sweets in each bag
    d. The cost of one concert ticket if 5 tickets cost $*t*

Understanding

17. There are n mice in a cage. If the number of mice doubles each month, write an expression for the number of mice:

    a. 1 month from now
    b. 2 months from now
    c. 3 months from now.

18. **MC** m is used to represent an unknown whole number. Choose the answer that matches the description 'six more than the number'.

    A. $6 + m$ **B.** $6m$
    C. $m - 6$ **D.** $m \div 6$

19. **MC** m is used to represent an unknown whole number. Choose the answer that matches the description 'a fifth of the number'.

    A. $5m$ **B.** $m - 5$ **C.** $\dfrac{m}{5}$ **D.** $m + \dfrac{1}{5}$

20. **MC** m is used to represent an unknown whole number. Choose the answer that matches the description 'the number just before that number'.

    A. $m + 1$ **B.** $m - 1$ **C.** m **D.** $m + 2$

21. **MC** m is used to represent an unknown whole number. Choose the answer that matches the description 'eight more than the product of that number and 10'.

    A. $10(m + 8)$ **B.** $10m + 8$ **C.** $8m + 10$ **D.** $8(m + 10)$

22. **MC** m is used to represent an unknown whole number. Choose the answer that matches the description 'five more than three times the number'.

    A. $5m + 3$ **B.** $3m - 5$ **C.** $3m + 5$ **D.** $5m - 3$

23. Answer True or False for each of the statements below.

    a. $3x$ is a term.   **b.** $3mn$ is a term.
    c. $g = 23 - t$ is an expression.   **d.** $g = 5t - 6$ is an equation.

24. Answer True or False for each of the statements below.

    a. $rt = r \times t$ **b.** $5 + A = A + 5$
    c. $3d + 5$ is an expression.   **d.** $7 - B = B - 7$

25. Answer True or False for each of the statements below.

    a. $3x = 9$ is an expression.   **b.** The expression $g + 2t$ has two terms.
    c. $2f + 4 = 4 + 2f$ **d.** $5a - 7t + h$ is an equation.

26. There are a apples at your house. At your friend's house there are six more apples than there are at your house. Your friend uses eight apples to make an apple pie. Use this information to write an expression describing the number of apples at your friend's house:

    a. before making the apple pie
    b. after making the apple pie.

27. Your friend spends $\$d$ on a bus fare each day.

    a. If she uses the bus three days a week, state how much she spends on bus fares for the week.

    b. There are 40 school weeks in a year and she travels three times a week on the bus. Write an expression for how much it will cost her for the bus fare each year.

    c. It costs an adult an extra $\$1.50$ for the same bus trip. Write an expression for the cost of an adult catching the bus three days per week.

Communicating, reasoning and problem solving

28. Compare the expressions $\dfrac{x+1}{3}$ and $\dfrac{x}{3}+1$. Give possible examples for each expression.

29. Explain the difference between $y+y+y$ and $y \times y \times y$.

30. p represents a number between 1 and 2 as shown:

Show the location of the following on the number line:

    a. $p+1$ **b.** $p-2$ **c.** $3-p$

31. a. Ali earns $\$7$ more per week than Halit. If Halit earns $\$x$, write an expression for their total earnings per week.

    b. Sasha is twice as old as Kapila. If Kapila is x years old, write an expression for their total age.

    c. Frank has had 3 more birthdays than James. If James is $2x$ years old, write an expression for their total age.

32. a. Write a rule to describe how you would calculate the total cost of two hamburgers and one drink.

    b. Let h represent the cost of a hamburger. If the drink costs $\$3.75$ and the total costs $\$7.25$, write an equation that links all this information.

33. Two friends each had five $\$1$ coins and a piggy bank full of 20c coins. They combined their money and spent it all at a games arcade. Assume that both friends had the same amount of money in their piggy banks.

    a. Write an expression for the total amount of money that the two friends had.

    b. If there were 85 coins in each piggy bank, calculate the total amount of money spent.

34. Ten couples go out for dinner. If everyone shakes hands once with everyone else except their partner, determine how many handshakes there are.

LESSON
7.3 Substituting and evaluating

LEARNING INTENTION

At the end of this lesson you should be able to:
- substitute a number for a pronumeral to evaluate an algebraic expression or formula.

▶ 7.3.1 Substituting

eles-3988

- The value of an expression can be determined by substituting in values that are represented by the pronumeral.
- **Substitution** means replacing the pronumeral with the number it represents.

WORKED EXAMPLE 5 Substituting into an expression

If $n = 3$, evaluate the following expressions.

a. $n + 4$ **b.** $2n$ **c.** $4n - 2$

THINK	WRITE
a. 1. Substitute $n = 3$ into the expression.	**a.** $n + 4 = 3 + 4$
2. Perform the addition and write the answer.	$= 7$
b. 1. Substitute $n = 3$ into the expression.	**b.** $2n = 2 \times 3$
2. Perform the multiplication and write the answer.	$= 6$
c. 1. Substitute $n = 3$ into the expression.	**c.** $4n - 2 = 4 \times 3 - 2$
2. Perform the multiplication first.	$= 12 - 2$
3. Perform the subtraction and write the answer.	$= 10$

WORKED EXAMPLE 6 Substituting to complete a table of values

Generate a number pattern for the algebraic expression $3t - 6$ by substituting values of t:

t	1	2	5	12
$3t - 6$				

THINK	WRITE
1. Write the expression.	$3t - 6$
2. Substitute $t = 1$ into the expression.	When $t = 1$:
3. Evaluate.	$3(1) - 6$ $= 3 \times 1 - 6$ $= 3 - 6$ $= -3$
4. Substitute $t = 2$ into the expression.	When $t = 2$:
5. Evaluate.	$3(2) - 6$ $= 3 \times 2 - 6$ $= 6 - 6$ $= 0$
6. Substitute $t = 5$ into the expression.	When $t = 5$:
7. Evaluate.	$3(5) - 6$ $= 3 \times 5 - 6$ $= 15 - 6$ $= 9$
8. Substitute $t = 12$ into the expression.	When $t = 12$:
9. Evaluate.	$3(12) - 6$ $= 3 \times 12 - 6$ $= 36 - 6$ $= 30$
10. Complete the table.	

t	1	2	5	12
$3t - 6$	-3	6	9	31

WORKED EXAMPLE 7 Substituting into a formula

When researching holiday destinations, Jack saw that the temperature in Hawaii was 88 °F. The formula

$C = \dfrac{5}{9}(F - 32)$ is used to convert this temperature from Fahrenheit to Celsius.

Calculate the temperature in Hawaii in degrees Celsius, correct to 1 decimal place.

THINK	WRITE
1. Write the formula.	$C = \dfrac{5}{9}(F - 32)$
2. Substitute $F = 88$ into the equation.	$C = \dfrac{5}{9}(88 - 32)$
3. Evaluate the right-hand side.	$= 31.1111\ldots$
4. Round your answer to 1 decimal place.	$= 31.1\,°C$

Digital technology

Scientific calculators have a fraction button that can be used to compute calculations involving fractions. The S⇔D button can be used to convert your answer between fraction and decimal form.

$$\frac{5}{9}(88-32)$$

$$31.11111111$$

WORKED EXAMPLE 8 Substituting into a formula

Calculate the value of m by substituting $x = 4$ into the following formula.

$$m = 3(2x + 3)$$

THINK	WRITE
1. Write the formula.	$m = 3(2x + 3)$
2. Substitute $x = 4$ into the equation.	$m = 3(2 \times 4 + 3)$
3. Evaluate the right-hand side of the equation. Use BIDMAS to perform the operations in the correct order.	$m = 3(8 + 3)$ $m = 3(11)$ $m = 3 \times 11$ $m = 33$

---

on Resources

▶ **Video eLesson** Substitution (eles-2426)

✦ **Interactivities** Substitution (int-4003)

Exercise 7.3 Substituting and evaluating

learn on

7.3 Quick quiz on	7.3 Exercise

Individual pathways

■ PRACTISE	■ CONSOLIDATE	■ MASTER
1, 4, 7, 8, 12, 17, 20	2, 5, 9, 11, 13, 14, 18, 21	3, 6, 10, 15, 16, 19, 22

Fluency

1. **WE5** If $a = 4$, evaluate the following expressions.

   a. $a + 7$ b. $5a$ c. $2a - 10$ d. $3 + 3a$

2. If $b = 6$, evaluate the following expressions.

   a. $b - 7$ b. $11b$ c. $3b + 1$ d. $-8b + 7$

3. If $c = 3$, evaluate the following expressions.

   a. $c - 16$ b. $-4c$ c. $12 + 7c$ d. $11 - c$

4. **WE6** Generate a number pattern below for the algebraic expression $g-2$ by substituting values of g.

g	1	2	3	4
$g-2$				

5. Generate a number pattern below for the algebraic expression $2t-3$ by substituting values of t.

t	1	2	3	4
$2t-3$				

6. Generate a number pattern below for the algebraic expression $12f+7$ by substituting values of f.

f	2	4	6	8
$12f+7$				

7. Substitute the given values into each formula to determine the value of g if $g = 25 - 4w$.

   a. $w = 1$
   b. $w = 3$
   c. $w = 6$
   d. $w = 0$

8. **WE8** Calculate the value of h by substituting the given value of the pronumeral into the formula $h = 2(g + 1)$.

   a. $g = 1$
   b. $g = 0$
   c. $g = 12$
   d. $g = 75$

9. Calculate the value of i by substituting the given value of the pronumeral into the formula $i = 5(x - 2)$.

   a. $x = 6$
   b. $x = 10$
   c. $x = 11$
   d. $x = 2$

10. Calculate the value of j by substituting the given value of the pronumeral into the formula $j = 4(12 - p)$.

   a. $p = 2$
   b. $p = 3$
   c. $p = 12$
   d. $p = 11$

11. Calculate the value of k by substituting the given value of the pronumeral into the formula $k = 5(2g - 3)$.

   a. $g = 2$
   b. $g = 14$
   c. $g = 5$
   d. $g = 9$

Understanding

12. Calculate the value of l by substituting the given value of the pronumeral into the formula $l = 2(d + 2) - 3$.

   a. $d = 3$
   b. $d = 0$
   c. $d = 7$
   d. $d = 31$

13. Calculate the value of m by substituting the given value of the pronumeral into the formula $m = 3(f - 1) + 17$.

   a. $f = 1$
   b. $f = 3$
   c. $f = 6$
   d. $f = 21$

14. Calculate the value of n by substituting the given value of the pronumeral into the formula $n = 4s - s$.

   a. $s = 3$
   b. $s = 1$
   c. $s = 101$
   d. $s = 72$

15. Calculate the value of o by substituting the given value of the pronumeral into the formula $o = 3(y + 5) - 2$.

   a. $y = 1$
   b. $y = 5$
   c. $y = 0$
   d. $y = 12$

16. Calculate the value of p by substituting the given value of the pronumeral into the formula $p = 50 - 6v$.

   a. $v = 4$
   b. $v = 7$
   c. $v = 1$
   d. $v = 8$

17. The area of a triangle is given by $A = \dfrac{1}{2} \times w \times h$. Find the height h and width w that gives an area of 12 square units.

Communicating, reasoning and problem solving

18. **WE7** The formula $M = 0.62K$ can be used to convert distances in kilometres (K) into miles (M). Use the formula to convert the following distances into miles:

    a. 100 kilometres
    b. 250 kilometres
    c. 15 kilometres

19. The formula for finding the perimeter (P) of a rectangle of length l and width w is $P = 2l + 2w$. Use this formula to calculate the perimeter of the rectangular swimming pool shown.

20. Determine the values of x when the result of $5x - 2$ is a two-digit number.

21. In Australian rules football, a goal is worth 6 points and a behind is worth 1 point. The equation $S = 6g + b$ can be used to determine a team's total score (S) from the number of goals (g) and behinds (b) that it scores.

    a. The Giants kicked 16 goals and 13 behinds. Calculate their total score.
    b. The Swans kicked 12 goals and 21 behinds. Calculate their total score.
    c. The Swans had a total of 33 scoring shots. Had they been more accurate and kicked 15 goals instead of 12, would they have scored more than the Giants?

22. When he was a young man, the mathematician Carl Friedrich Gauss discovered that the sum (S) of the first n natural numbers is given by the formula $S = \dfrac{n(n + 1)}{2}$.

    Using this formula, find the answer to the sum of the following series of numbers:

    a. $1 + 2 + 3 + 4 + \ldots + 46 + 47 + 48 + 49$
    b. $54 + 55 + 56 + 57 + \ldots + 85 + 86 + 87 + 88$

LESSON
7.4 Simplifying expressions using like terms

LEARNING INTENTION

At the end of this lesson you should be able to:
- simplify expressions by adding or subtracting like terms.

▶ 7.4.1 Simplifying expressions containing like terms

eles-3989

- Terms that contain the exact same pronumerals are called **like terms**.
- Terms that do not contain the exact same pronumerals are called **unlike terms**. Examples of like and unlike terms are:

Like terms	Unlike terms
$2a$ and $5a$	$2a$ and $5b$
x and $13x$	$3x$ and $5xy$
$2bc$ and $4cb$	$4abc$ and $3cbd$
$3g^2$ and $45g^2$	$16g$ and $45g^2$

- Note that the order of pronumerals in a term does not matter. The terms *ab* and *ba* are like terms because they both contain the exact same pronumerals.
- Like terms can be combined by adding or subtracting coefficients. For example, $3a + 2a = 5a$.

- Unlike terms cannot be combined or simplified in any way. For example, $3a + 2b$ cannot be written in a simpler way.

WORKED EXAMPLE 9 Simplifying expressions

Where possible, simplify the following expressions by adding or subtracting like terms.
a. $4g + 6g$ **b.** $11ab - ab$ **c.** $6ad + 5da$ **d.** $4t + 7t - 5$ **e.** $8x + 3y$

THINK

a. The pronumerals are the same. The terms can be added.

b. The pronumerals are the same. The terms can be subtracted. Note: *ab* is the same as $1ab$ or $1ba$ or *ba*.

c. Although the order of the pronumerals is different, the terms are like terms and can be added.

d. The pronumerals are the same and the first two terms can be added. Do not subtract the constant term.

e. The pronumerals are **not** the same. These are not like terms and therefore cannot be simplified.

WRITE

a. $4g + 6g = 10g$

b. $11ab - ab = 10ab$

c. $6ad + 5da = 6ad + 5ad$
$\qquad\qquad\quad = 11ad$

d. $4t + 7t - 5 = 11t - 5$

e. $8x + 3y$ cannot be simplified.

WORKED EXAMPLE 10 Evaluating expressions containing like terms

First simplify the following expressions, then determine the value of the expression by substituting $a = 4$.
a. $5a + 2a$ **b.** $7a - a + 5$

THINK

a. 1. Add the like terms.

    2. Substitute $a = 4$ into the simplified expression.

b. 1. Subtract the like terms.

    2. Substitute $a = 4$ into the simplified expression.

WRITE

a. $5a + 2a = 7a$
$\qquad\qquad = 7 \times 4$
$\qquad\qquad = 28$

b. $7a - a + 5 = 6a + 5$
$\qquad\qquad\quad = 6 \times 4 + 5$
$\qquad\qquad\quad = 24 + 5$
$\qquad\qquad\quad = 29$

$3x^2 + 4x^2 = 7x^2$

The terms $3x^2$ and $4x^2$ are called *like terms*. Explain why.

$3x^2$ and $4x$ are called *unlike terms*. Explain why.

Equivalent algebraic expressions

- Terms that are identical but are displayed in different ways are called **equivalent expressions**. For example, you can simplify the expression $x + x + y + y - 2$ to $2x + 2y - 2$. This means that $x + x + y + y - 2$ and $2x + 2y - 2$ are equivalent expressions.
- The following are all pairs of equivalent expressions:

$$a + a + a + a = 4a \qquad b \times b = b^2 \qquad c \times d = cd \qquad e \div f = \frac{e}{f}$$

COMMUNICATING — COLLABORATIVE TASK: Silently collecting like terms

Equipment: $25 - 30$ cards, each containing an algebraic term. Across the set of cards there should be 5–6 different pronumerals, and each card should have different coefficients.

1. Each student chooses a card with a term written on it. They must not look at their cards or talk.
2. Each student displays their card facing outward so that the rest of the class can see their term, but the student holding the card cannot.
3. Without talking or looking at their own card, students then help each other to form groups around the classroom with other like terms.
4. Once the class is happy that they are all in like-term groups, students can look at their own card and talk.
5. Each group then simplifies by adding and subtracting like terms.
6. Each group then writes their part of the expression on the board to provide the fully simplified class expression.
7. Discuss any difficulties experienced when gathering and simplifying like terms.

on Resources

- **Video eLesson** Simplification of expressions (eles-1884)
- **Interactivities** Simplifying expressions containing like terms (int-4006)

Exercise 7.4 Simplifying expressions using like terms

7.4 Quick quiz on	7.4 Exercise

Individual pathways

■ PRACTISE	■ CONSOLIDATE	■ MASTER
1, 4, 7, 9, 12, 15, 18, 21, 22, 25	2, 5, 8, 10, 13, 16, 20, 23, 26	3, 6, 11, 14, 17, 19, 24, 27, 28

Fluency

1. **WE9** Where possible, simplify the following expressions by adding or subtracting like terms.
   a. $3a + 2a$
   b. $9y + 5y$
   c. $3c + 12c$
   d. $4u - 2u$
   e. $7e + 13e$

2. Where possible, simplify the following expressions by adding or subtracting like terms.
   a. $7t - 2t$
   b. $12ab + 2ab$
   c. $8fg - 2fg$
   d. $4e - e$
   e. $6t + t$

3. Where possible, simplify the following expressions by adding or subtracting like terms.
   a. $f + 4f$
   b. $6y - 6y$
   c. $4x + 14x$
   d. $3m + 16m - 7m$
   e. $6a + 4a - 2a$

4. Simplify the following expressions (if possible).
   a. $24ab + ab - 7$
   b. $5y + y - 3y$
   c. $5t + 5s$
   d. $4t + 8t - 3 + 2t$
   e. $7r + 2r + 5r - r$

5. Simplify the following expressions (if possible).
   a. $7y + 6$
   b. $4 + 3g - g$
   c. $6t - 5t$
   d. $18bg - 18bg$
   e. $11pq + 3qp$

6. Simplify the following expressions (if possible).
   a. $7t + 4t - 5$
   b. $32t - 31t$
   c. $7xy - 7x$
   d. $5t + 6t - 8$
   e. $5t - t + 3$

7. Simplify each of the following.
   a. $3t - 3t$
   b. $18r - 18r$
   c. $12ab - 12ab$
   d. $5x - 5x + 8$

8. Simplify each of the following.
   a. $6t + 7 - 6t$
   b. $13xyz - 13xyz$
   c. $5x + 7 - 5x$
   d. $5y + 2y - y$

9. **WE10** Simplify the following expressions, then determine the value of the expression by substituting $a = 7$.
   a. $3a + 2a$
   b. $7a + 2a$
   c. $6a - 2a$
   d. $9a + a$

10. First simplify the following expressions, then determine the value of the expression by substituting $a = 7$.
    a. $13a + 2a - 5a$
    b. $3a + 7a$
    c. $17 + 5a + 3a$
    d. $6a - a + 2$

11. First simplify the following expressions, then determine the value of the expression by substituting $a = 7$.
    a. $a + a$
    b. $7a - 6a$
    c. $7a - 7a$
    d. $12a + 5a - 16$

Understanding

12. Answer True or False to each of the following statements.
    a. $4t$ and $6t$ are like terms.
    b. $3x$ and x are like terms.
    c. $5g$ and $5t$ are like terms.
    d. $5a$ and $6a$ are like terms.
    e. $4a$ and $4ab$ are like terms.

13. Answer True or False to each of the following statements.
    a. $6gh$ and $7gk$ are like terms.
    b. yz and $45zy$ are like terms.
    c. $3acd$ and $6cda$ are like terms.
    d. $5g$ and $5fg$ are like terms.
    e. $8gefh$ and $3efgh$ are like terms.

14. Answer True or False to each of the following statements.

  a. $6ab$ and $3ba$ are like terms.   **b.** $8xy$ and $5xy$ are like terms.   **c.** $7eg$ and $7g$ are like terms.
  d. $7y$ and $18yz$ are like terms.   **e.** $12ep$ and $4pe$ are like terms.

15. **MC** Which one of the following is a like term for $7t$?

  A. $7g$ **B.** $5tu$ **C.** 2 **D.** $5t$

16. **MC** Which one of the following is a like term for $3a$?

  A. $5ab$ **B.** $6a$ **C.** $3w$ **D.** $3+a$

17. **MC** Which one of the following is a like term for $5ab$?

  A. $6abe$ **B.** $4b$ **C.** $4ba$ **D.** 5

18. **MC** Which one of the following is a like term for $7bgt$?

  A. $7t$ **B.** $5gt$ **C.** $2gb$ **D.** btg

19. **MC** Which one of the following is a like term for $2fgk$?

  A. $fg+k$ **B.** $5fg$ **C.** $2fgk$ **D.** $2f$

20. **MC** Which one of the following is a like term for $20m$?

  A. $11m$ **B.** $20mn$ **C.** $20f$ **D.** $m+2$

21. **MC** Which one of the following is a like term for $9xyz$?

  A. $9x$ **B.** $2yz$ **C.** $8xz$ **D.** xzy

Communicating, reasoning and problem solving

22. Answer True or False to each of the following statements and justify your answer.

  a. The expression $4x+3x$ can be simplified to $7x$.
  b. The expression $8y+4y$ can be simplified to $10y$.
  c. The expression $4x+3$ can be simplified to $7x$.

23. Answer True or False to each of the following statements and justify your answer.

  a. The expression $3a-a$ can be simplified to $2a$.
  b. The expression $7a+4d$ can be simplified to $11ad$.
  c. The expression $5x-3x$ can be simplified to $2x$.

24. Answer True or False to each of the following statements and justify your answer.

  a. The expression $7+2x$ can be simplified to $9x$.
  b. The expression $3h+12h+7$ can be simplified to $15h+7$.
  c. The expression $16g+6g-7g$ can be simplified to $15g$.

25. There are x chocolates in each box. Ahmed buys 7 boxes while Kevin buys 3 boxes. Write an expression for the total number of chocolates.

26. There are d mangoes in a kilogram. Nora buys 4 kg but they are so good she buys 2 kg more. Write an expression for the total number of mangoes Nora bought.

27. Find the simplest expression for the perimeter of the following shapes:

a.

b.

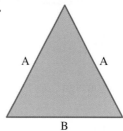

28. Given that OD bisects COE, that $\angle COD = (4x + 16)°$ and that $\angle DOE = (6x + 4)°$, explain if there is enough information for you to determine the angles for COD and DOE. If so, what are the measures for each?

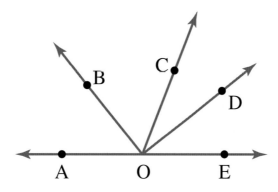

LESSON
7.5 Multiplying and dividing terms

LEARNING INTENTION

At the end of this lesson you should be able to:
- simplify expressions when multiplying, using indices where appropriate
- simplify expressions when diving by identifying common factors.

▶ ## 7.5.1 Multiplying terms

eles-3990

- We multiply algebraic terms using this process:
  - Split each term into its coefficient and its pronumerals.
  - Determine the product of the coefficient of the terms.
  - Combine like factors into a power. For example, $x \times x = x^2$.
  - Combine the coefficient and pronumerals into one term.

Unlike adding and subtracting, when we multiply algebraic terms, we can collect them into one term.

WORKED EXAMPLE 11 Simplify by using index notation where appropriate

Simplify:

a. $2 \times k$ b. $3 \times 2k$ c. $k \times 2kx$ d. $3k \times 2k$ e. $3k \times 2k^2 x$

THINK	WRITE
a. 1. Remove the multiplication sign.	a. $2 \times k$ $= 2k$
b. 1. Rewrite in expanded form. 2. Multiply the numbers (the coefficients). 3. Remove the multiplication sign.	b. $3 \times 2 \times k$ $= 6 \times k$ $= 6k$
c. 1. Rewrite in expanded form. 2. Multiply the like pronumerals. 3. Remove the multiplication signs.	c. $k \times 2 \times k \times x$ $= 2 \times k^2 \times x$ $= 2k^2 x$
d. 1. Rewrite in expanded form. 2. Multiply the numbers and write pronumerals with index. 3. Remove the multiplication sign.	d. $3 \times k \times 2 \times k$ $= 6 \times k^2$ $= 6k^2$
e. 1. Rewrite in expanded form. 2. Multiply the numbers and write pronumerals with index. 3. Remove the multiplication signs.	e. $3 \times k \times 2 \times k \times k \times x$ $= 6 \times k^3 \times x$ $= 6k^3 x$

Multiplying pronumerals

When we multiply pronumerals, order is not important. For example:

$$3 \times 6 = 6 \times 3 \qquad 6 \times w = w \times 6 \qquad a \times b = b \times a$$

- **The multiplication sign ($\times$) is usually removed to simplify the expression.**

$$3 \times g \times h = 3gh$$

- **Although order is not important, conventionally the pronumerals in each term are written in alphabetical order. For example:**

$$2 \times b \times a \times c = 2abc$$

▶ 7.5.2 Dividing terms

- We divide algebraic terms using this process:
  - Split each term into its coefficient and its pronumerals.
  - Determine the quotient of the coefficient of the terms.
  - Cancel any common factors. For example, $\dfrac{x}{x} = \dfrac{1\cancel{x}}{\cancel{x}} = 1$.
  - Combine the coefficient and pronumerals into one term.

Unlike adding and subtracting, when we divide algebraic terms, we can collect them into one term.

Simplify:

a. $\dfrac{16f}{4}$

b. $16f \div 2f$

c. $\dfrac{4f}{16f^2}$

d. $\dfrac{36fg}{40f^2g}$

THINK	WRITE
a. 1. Write in expanded form.	a. $\dfrac{16 \times f}{4}$
2. Divide numerator and denominator by the highest common factor (HCF), 4.	$= \dfrac{\overset{4}{\cancel{16}} \times f}{\underset{1}{\cancel{4}}}$
3. Write in simplest terms.	$= 4f$
b. 1. Write as a fraction in expanded form.	b. $\dfrac{16 \times f}{2 \times f}$
2. Divide numerator and denominator by HCF 2 and f.	$= \dfrac{\overset{8}{\cancel{16}} \times \cancel{f}}{\underset{1}{\cancel{2}} \times \cancel{f}}$
3. Write in simplest terms.	$= 8$
c. 1. Write in expanded form.	c. $\dfrac{4 \times f}{16 \times f \times f}$
2. Divide numerator and denominator by HCF, 4 and f.	$= \dfrac{\overset{1}{\cancel{4}} \times \cancel{f}}{\underset{4}{\cancel{16}} \times \cancel{f} \times f}$
3. Write in simplest terms. Omit the multiplication sign.	$= \dfrac{1}{4f}$
d. 1. Write in expanded form.	d. $\dfrac{36 \times f \times g}{40 \times f \times f \times g}$
2. Divide numerator and denominator by HCF 4, f and g.	$= \dfrac{\overset{9}{\cancel{36}} \times \cancel{f} \times \cancel{g}}{\underset{10}{\cancel{40}} \times \cancel{f} \times f \times \cancel{g}}$
3. Write in simplest terms. Omit the multiplication sign.	$= \dfrac{9}{10f}$

Exercise 7.5 Multiplying and dividing terms

7.5 Quick quiz	7.5 Exercise

Individual pathways

■ PRACTISE	■ CONSOLIDATE	■ MASTER
1, 2, 6, 9, 11, 14	3, 5, 7, 12, 15	4, 8, 10, 13, 16

Fluency

1. **WE11** Simplify the following.

   a. $2a \times a$
   b. $ab \times 7a$
   c. $7pq \times 3p$
   d. $3b^2 \times 2cd$
   e. $20m^{12} \times 2m^3$

2. Simplify the following.

   a. $4 \times 3g$
   b. $7 \times 3h$
   c. $4d \times 6$
   d. $3z \times 5$
   e. $6 \times 5r$

3. Simplify the following.

   a. $5t \times 7$
   b. $4 \times 3u$
   c. $7 \times 6p$
   d. $7gy \times 3$
   e. $2 \times 11ht$

4. Simplify the following.

   a. $4x \times 6g$
   b. $10a \times 7h$
   c. $9m \times 4d$
   d. $3c \times 5h$
   e. $9g \times 2x$

5. **WE12** Simplify the following.

   a. $\dfrac{8f}{2}$
   b. $\dfrac{6h}{3}$
   c. $\dfrac{15x}{3}$
   d. $9g \div 3$
   e. $10r \div 5$

6. Simplify the following.

   a. $4x \div 2x$
   b. $8r \div 4r$
   c. $\dfrac{16m}{8m}$
   d. $14q \div 21q$
   e. $\dfrac{3x}{6x}$

Understanding

7. Simplify the following.

   a. $\dfrac{12h}{14h}$
   b. $50g \div 75g$
   c. $\dfrac{8f}{24f}$
   d. $35x \div 70x$
   e. $24m \div 36m$

8. Simplify the following:

   a. $\dfrac{3 \times a}{a^2}$
   b. $\dfrac{25p^{12} \times 4q^7}{15p^2}$
   c. $\dfrac{8x^3 \times 7y^2}{6x \times 14y^3}$
   d. $w \times \dfrac{5}{w^2}$
   e. $\dfrac{12x^3 y^2}{6x^4 \times 14y^3}$

9. Simplify the following.

   a. $3a \times a$
   b. $-2p \times 2p$
   c. $5 \times 2x \times x$
   d. $ab \times 5a$

10. Simplify the following.

    a. $-3xy \times 2 \times 6x$
    b. $4pq \times 3p \times 2q$
    c. $5m \times n \times 2nt \times -t$
    d. $-7 \times xyz \times -3z \times -2y$

Communicating, reasoning and problem solving

11. A student shows the following working:

$$\frac{2 + 3x^1}{5x^1}$$

    Explain whether you agree with their working.

12. Consider the expression $\dfrac{24xy^2}{7ab} \times \dfrac{14a^2}{27y}$.

   a. Multiply the numerators and multiply the denominators, then cancel any common factors.
   b. Cancel the common factors in the numerators and denominators first, then multiply the remaining factors accordingly.
   c. Compare your answer to part **a** with your answer to part **b**. Which process do you feel is better to use?

13. Explain why $x \div (yz)$ is equivalent to $\dfrac{x}{yz}$ and $x \div (y \times z)$, but is not equivalent to $x \div y \times z$ or $\dfrac{x}{y} \times z$.

14. Calculate the areas of the following shapes.

   a.

   b.

   c.

15. Evaluate the expression $\dfrac{4ht}{3dk} \times \dfrac{-12hk}{9dt}$, by substituting the values of $h = 1$, $d = 2$, $k = 3$ and $t = 4$.

16. A rectangle has been cut out of a corner of a square.

   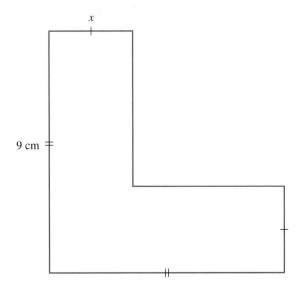

   a. Find the perimeter of the shape.
   b. Find the area of the shape if:
      i. $x = 2\,\text{cm}$ ii. $x = 3\,\text{cm}$.
   c. Find the area of the shape in terms of x.

LESSON
7.6 Number laws

LEARNING INTENTION

At the end of this lesson you should be able to:
- understand and apply the Commutative Law for addition and multiplication
- understand and apply the Associative Law for addition and multiplication
- understand that the Commutative and Associative laws do not apply to subtraction or division.

▶ 7.6.1 The Commutative Law

eles-3992

- The **Commutative Law** says we can swap numbers over and still get the same answer when adding and mutiplying.

The Commutative Law for addition

- $a + b = b + a$
  For example, $5 + 3 = 8$ and $3 + 5 = 8$.

- This law *does not hold for subtraction*.
  For example, $5 - 3 \neq 3 - 5$.

The Commutative Law for multiplication

- $a \times b = b \times a$
  For example, $2 \times 3 = 6$ and $3 \times 2 = 6$.

- This law *does not hold for division*.
  For example, $6 \div 2 \neq 2 \div 6$.

WORKED EXAMPLE 13 Testing the Commutative Law

Evaluate the following expressions if $a = 2$ and $b = 3$. Comment on the results obtained.

a. i. $a + b$ ii. $b + a$ b. i. $5a - 2b$ ii. $2b - 5a$

c. i. $a \times b$ ii. $b \times a$ d. i. $a \div b$ ii. $b \div a$

THINK	WRITE
a. i. 1. Substitute the correct value for each pronumeral.	a. i. $a + b = 2 + 3$
2. Evaluate and write the answer.	$= 5$
ii. 1. Substitute the correct value for each pronumeral.	ii. $b + a = 3 + 2$
2. Evaluate and write the answer.	$= 5$
3. Compare the result with the answer obtained in part **a i**.	The same result is obtained; therefore, order is not important when adding two terms.

b. i. 1. Substitute the correct value for each pronumeral.

  2. Evaluate and write the answer.

ii. 1. Substitute the correct value for each pronumeral.

  2. Evaluate and write the answer.

  3. Compare the result with the answer obtained in part **b i**.

c. i. 1. Substitute the correct value for each pronumeral.

  2. Evaluate and write the answer.

ii. 1. Substitute the correct value for each pronumeral.

  2. Evaluate and write the answer.

  3. Compare the result with the answer obtained in part **c i**.

d. i. 1. Substitute the correct value for each pronumeral.

  2. Evaluate and write the answer.

ii. 1. Substitute the correct value for each pronumeral.

  2. Evaluate and write the answer.

  3. Compare the result with the answer obtained in part **d i**.

b. i. $5a - 2b = 5 \times 2 - 2 \times 3$

  $= 4$

ii. $2b - 5a = 2 \times 3 - 5 \times 2$

  $= -4$

  Two different results are obtained; therefore, order is important when subtracting two terms.

c. i. $a \times b = 2 \times 3$

  $= 6$

ii. $b \times a = 3 \times 2$

  $= 6$

  The same result is obtained; therefore, order is not important when multiplying two terms.

d. i. $a \div b = 2 \div 3$

  $= \dfrac{2}{3} \ (\approx 0.67)$

ii. $b \div a = 3 \div 2$

  $= \dfrac{3}{2} \ (1.50)$

  Two different results are obtained; therefore, order is important when dividing two terms.

▶ 7.6.2 The Associative Law

eles-3993

- The **Associative Law** says that it does not matter how we group the numbers (i.e. which we calculate first) when adding or multiplying.

The Associative Law for addition

- $(a + b) + c = a + (b + c) = a + b + c$
  For example, $(9 + 3) + 7 = 12 + 7 = 19$ and $9 + (3 + 7) = 9 + 10 = 19$.

- This law *does not hold for subtraction*.
  For example, $(8 - 3) - 2 = 5 - 2 = 3$, but $8 - (3 - 2) = 8 - 1 = 7$.

The Associative Law for multiplication

- $(a \times b) \times c = a \times (b \times c) = a \times b \times c$
  For example, $(2 \times 3) \times 5 = 6 \times 5 = 30$ and $2 \times (3 \times 5) = 2 \times 15 = 30$.

- This law *does not hold for division*.
  For example, $(24 \div 6) \div 2 = 4 \div 2 = 2$, but $24 \div (6 \div 2) = 24 \div 3 = 8$.

WORKED EXAMPLE 14 Applying the Associative Law

Use the Associative Law to complete the following.

a. $2w + (8d + \_\_) = (\_\_ + 8d) + 5h$

b. $(x \times \_\_) \times 4g = \_\_ (d \times 4g)$

THINK

WRITE

a. The Associative Law states that the numbers can be regrouped. $2w$, $8d$ and $5h$ are numbers. $5h$ is missing from the left-hand side and $2w$ is missing from the right-hand side.

a. $2w + (8d + 5h) = (2w + 8d) + 5h$

b. The Associative Law states that the numbers can be regrouped. d is missing from the left-hand side and x is missing from the right-hand side.

b. $(x \times d) \times 4g = x \times (d \times 4g)$

WORKED EXAMPLE 15 Applying the associative and commutative laws

Use the associative and commutative laws to complete the following.

a. $4r + (6t + \_\_) = 5g + (\_\_ + \_\_)$

b. $a \times (3t \times 2w) = (2w \times \_\_) \_\_ a$

THINK

WRITE

a. The associative and commutative laws state that the numbers can be regrouped and rearranged. $5g$ is missing from the left-hand side. Both $4r$ and $6t$ are missing from the right-hand side.

a. $4r + (6t + 5g) = 5g + (4r + 6t)$

b. The associative and commutative laws state that the numbers can be regrouped and rearranged. $3t$ and a multiplication sign are missing from the right-hand side.

b. $a \times (3t \times 2w) = (2w \times 3t) \times a$

---

on Resources

 eWorkbook The associative law for multiplication (int-4008)

The commutative law for multiplication (int-4009)

Exercise 7.6 Number laws

learn on

7.6 Quick quiz on	7.6 Exercise

Individual pathways

■ PRACTISE	■ CONSOLIDATE	■ MASTER
1, 2, 5, 9, 12	3, 6, 7, 10, 13	4, 8, 11, 14

Fluency

1. State whether the following are True or False. *Hint:* If you are not sure, try substituting values for the pronumerals to help you decide.

   a. $2s + (3w + 5z) = (2s + 3w) + 5z$

   b. $x \times (d + y) = (x \times d) + y$

   c. $g(jk) = (gj)k$

   d. $4 \div (a \div c) = (4 \div a) \div c$

2. **WE14** Use the Associative Law to complete the following.

    a. $w + (r\_\_6y) = (\_\_ + \_\_) + 6y$ **b.** $6t \times (4r \times \_\_) = (6\_\_ \times 4r)\_\_3s$

    c. $(9y + \_\_) + 3w = \_\_ + (2r + 3w)$ **d.** $(z + 2p) + \_\_ = z\_\_(\_\_ + 6t)$

3. **WE15** Use the associative and commutative laws to complete the following.

    a. $6t + (3w + \_\_) = \_\_ + (6t + 7v)$ **b.** $s \times (9r \times \_\_) = 2c \times (\_\_ \times \_\_)$

    c. $(3c + w) + \_\_ = d + (\_\_ + \_\_)$ **d.** $(g \times \_\_) \times 2y = 2y \times (3b \times \_\_)$

4. State whether the following are True or False. *Hint:* If you are not sure, try substituting values for the pronumerals to help you decide.

    a. $3g + (k \div m) = (3g + k) \div m$ **b.** $4t - (p + 2b) = (4t - p) + 2b$

    c. $3r + (a + 4c) = a + (3r + 4c)$ **d.** $a + (b \times c) = (a \times c) + b$

Understanding

5. Calculate the values of the following expressions and comment on the results if $a = 3$, $b = 8$ and $c = 2$.

    a. $3a + (2b + 4c)$ **b.** $(3a + 2b) + 4c$

6. Calculate the values of the following expressions and comment on the results if $a = 3$, $b = 8$ and $c = 2$.

    a. $9a - (2b - 5c)$ **b.** $(9a - 2b) - 5c$

7. Determine the values of the following expressions and comment on the results if $a = 3$, $b = 8$ and $c = 2$.

    a. $a \times (b \times c)$ **b.** $(a \times b) \times c$

8. Calculate the values of the following expressions and comment on the results if $a = 3$, $b = 8$ and $c = 2$.

    a. $4a \div (3b \div 2c)$ **b.** $(4a \div 3b) \div 2c$

Communicating, reasoning and problem solving

9. The Commutative Law does not hold for subtraction. What can you say about the results of $x - a$ and $a - x$?

10. Explain whether the Associative Law holds when you are adding fractions.

11. Use the numbers $1, 5, 6$ and 7 with any arithmetic operations to give 21 as the result.

12. **a.** If $x = 1$, $y = 2$ and $z = 3$, find the value of:

      i. $(x - y) - z$ **ii.** $x - (y - z)$.

    b. Comment on the answers with special reference to the Associative Law.

13. Siblings Jaimi and Rohan want to buy a new video game together. Jaimi has $47 less than the cost of the video game and Rohan has $2 less than the cost. Both of them have at least $1 each. If they pool their money, they still do not have enough to buy the new video game. If the cost of the video game is a whole number, calculate how much it costs.

14. Andrew thinks of a number, adds 9 and multiplies the result by 3. Andrea thinks of a number, multiplies it by 3 and adds 9 to the result. Both Andrew and Andrea get an answer of 60. Explain whether Andrew and Andrea began by selecting the same number. Show your working to support your answer.

LESSON
7.7 Expanding brackets

LEARNING INTENTION

At the end of this lesson you should be able to:
• write expressions in an expanded form by using the Distributive Law.

▶ 7.7.1 The Distributive Law

eles-3994

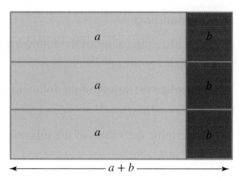

• The Distributive Law can be used to expand or remove brackets.
• Expanding brackets means writing an equivalent expression without brackets.
• In the diagram, the blue rectangles represent a, the pink squares represent b and the green rectangles represent $a + b$.
• As can be seen in the diagram, there are three groups of $a + b$, which can be written as $3(a + b)$. Alternatively, we can see that there are 3 as and 3 bs, so the expression could also be written as $3a + 3b$. Therefore $3(a + b) = 3a + 3b$.
• This can also be determined by collecting like terms:

$$3(a + b) = (a + b) + (a + b) + (a + b)$$
$$= a + a + a + b + b + b$$
$$= 3a + 3b$$

• Expressions such as this may be written in factorised form (with brackets) or expanded form (without brackets).

$$\overbrace{3(a + b)}^{\text{factorised form}} = \overbrace{3a + 3b}^{\text{expanded form}}$$

• Similarly, the Distributive Law can be determined using an area model. For example, $2(x + 3)$ can be modelled as:

• The area model shows that when multiplying the length and width of each rectangle, we would get the area of $2x + 6$. Therefore, $2(x + 3)$ is the same as $2x + 6$.

The Distributive Law

The Distributive Law provides a quick method of expanding expressions that are in factorised form.

$$a(b + c) = ab + ac$$

The term that is out at the front of the brackets gets distributed to (multiplied by) each of the terms inside the brackets.

The Distributive Law also applies when the terms inside the brackets are subtracted.

$$a(b - c) = ab - ac$$

- The Distributive Law holds true no matter how many terms there are inside the brackets. For example, $a(b + c - d) = ab + ac - ad$.
- The Distributive Law is not used when the terms inside the brackets are multiplied or divided. For example, $2(4 \times 5) = 2 \times 4 \times 5$ and not $(2 \times 4) \times (2 \times 5)$.

WORKED EXAMPLE 16 Applying the Distributive Law

Write the following expressions in expanded form, then simplify them.
a. $4(2a + d)$ b. $a(3x + 2y)$ c. $3(k + 4a - m)$

THINK

a. 1. According to the Distributive Law, the number 4 is distributed (or multiplied) to both $2a$ and d.

    2. Simplify the terms and write the answer.

b. 1. According to the Distributive Law, the number represented by the pronumeral a is distributed to both $3x$ and $2y$.

    2. Simplify the terms and write the answer.

c. 1. According to the Distributive Law, the number 3 is distributed to k, $4a$ and m.

    2. Simplify the terms and write the answer.

WRITE

a.
$$4(2a + d) = 4 \times 2a + 4 \times d$$
$$= 8a + 4d$$

b.
$$a(3x + 2y) = a \times 3x + a \times 2y$$
$$= 3ax + 2ay$$

c.
$$3(k + 4a - m) = 3 \times k + 3 \times 4a - 3 \times m$$
$$= 3k + 12a - 3m$$

WORKED EXAMPLE 17 Applying the Distributive Law

Use the Distributive Law to expand and simplify the following.
a. $4(2s + 3k) - 2k$ b. $a(3 - c) + a$

THINK

a. 1. Use the Distributive Law to expand $4(2s + 3k)$.

WRITE

a.
$$4(2s + 3k) - 2k = 4 \times 2s + 4 \times 3k - 2k$$

2. Simplify the terms. $= 8s + 12k - 2k$

3. Collect like terms and write the answer. $= 8s + 10k$

b. 1. Use the Distributive Law to expand $a(3-c)$.

b.
$$a(3-c) + a = a \times 3 - a \times c + a$$

2. Simplify the terms. $= 3a - ac + a$

3. Collect like terms and write the answer. $= 4a - ac$

 Resources

 Interactivity The distributive law (int-4007)

Exercise 7.7 Expanding brackets

learn

| 7.7 Quick quiz on | 7.7 Exercise |

Individual pathways

■ PRACTISE	■ CONSOLIDATE	■ MASTER
1, 3, 6, 8, 11, 14	4, 7, 9, 12, 15	2, 5, 10, 13, 16

Fluency

1. **WE16** Write the following expressions in expanded form, then simplify (if possible).

   a. $3(a + 2b)$ **b.** $5(x - 4z)$ **c.** $10(2g + 3h)$ **d.** $b(2c - 3a)$

2. Write the following expressions in expanded form, then simplify (if possible).

   a. $2c(8d - q)$ **b.** $3b(3b - 2g)$ **c.** $x(9a + 7b)$ **d.** $7k(t + 3m)$

3. **WE17** Use the Distributive Law to expand the following expressions.

   a. $5(a + 2b) - 3a$ **b.** $4(n - 2c) + n$ **c.** $9(b + 2c) - 9b$ **d.** $x(a - b) + x$

4. Use the Distributive Law to expand each of the following.

   a. $3(2d + 7e)$ **b.** $3h(6j - 5) + 4h$ **c.** $4(k + 5a - 3)$ **d.** $4x(x + 1) - 2(2x - 3)$

5. Use the Distributive Law to expand the following expressions.

   a. $y(3 + 2z) + 2y$ **b.** $3x(d - 3) + 4x$ **c.** $6u(2f - 1) + 20u$ **d.** $7n(10 + 3h) - 50n$

6. Expand each of the following expressions and then simplify by collecting like terms.

   a. $8(5r + 2) + 14r$ **b.** $3(t - 6) - 12$ **c.** $3y + 3(12 - 2y)$ **d.** $2x - (16 - 3x)$

7. Expand each of the following expressions and then simplify by collecting like terms.
   a. $5 - 3t + 6(2t - 8)$
   b. $14 + 3ry + 8(r + 2ry)$
   c. $12ty + 2t(7y - 8h) + 6h$
   d. $16r^2 + 12r(2 - r) - 6r + 3$

8. Expand each of the following expressions and then simplify by collecting like terms.
   a. $3(x + 2) + 2(3 - x)$
   b. $6(4h - 3) + 7h(4x - 8)$
   c. $5(4x + y) - 2(3y + 3)$
   d. $6(3 - t^2) - 8(t^2 + 3)$

Understanding

9. Expand each of the following and then simplify.
   a. $3(x - y) + 5(y + 2x)$
   b. $5(a - 2y) + a(7 + 6y)$
   c. $6c(g + 3d) + g(3c + d)$
   d. $y(6d - r) + 2y(3d + 2r)$

10. In expressions such as $a(b + c) = ab + ac$, the left-hand side is in *factor form* and the right-hand side is in *expanded form*. Write each of the following in factor form.
    a. $2x + 4$
    b. $6x - 6y$
    c. $4x + 12$
    d. $12fg - 16gh$

Communicating, reasoning and problem solving

11. The formula for the perimeter (P) of a rectangle of length l and width w is $P = 2(l + w)$. Use the rule to calculate the perimeter of a magazine cover with $l = 28.6$ cm and $w = 21$ cm.

12. For each of the following expressions, show three different ways to make the statements true by placing appropriate terms in the boxes.
    a. $\Box(\Box + 2) = 12ab + \Box$
    b. $\Box(\Box + 2) = 36kn + \Box$

13. Consider the expression $2(1 - 5x)$. If x is a negative integer, explain why the expression will be a positive integer.

14. Expanding and simplifying the expression $2(8c - 16a)$ gives $16c - 32a$. Determine three expressions that give the same result.

15. a. Expand and simplify $2(x + y) - 3x$.
    b. Verify that your simplified expression is correct by substituting $x = 2$ and $y = 3$ into both the original and simplified expressions.

16. The price of a pair of shoes is $50. During a sale, the price of the shoes is discounted by $$d$.

    a. Write an expression to represent the sale price of the shoes.
    b. If you buy three pairs of shoes during the sale, write an expression to represent the total purchase price:
       i. containing brackets
       ii. in expanded form (without brackets).
    c. Write an expression to represent the total change you would receive from $200 for the three pairs of shoes purchased during the sale.

LESSON
7.8 Review

7.8.1 Topic summary

The language of algebra

- **Pronumerals** are letters used to represent numbers.
  e.g. t: the number of tickets sold
- A pronumeral is also called a **variable**.
- The number in front of a pronumeral is called the **coefficient**.
  e.g. The coefficient of $7x$ is 7.
  The coefficient of p is 1.
- **Terms** can contain a combination of numbers and one or more pronumerals, or may consist of a number only.
  e.g. $3t$, $2y$, $5gh$, 7 and m are examples of terms.
- A **constant term** is a term containing only a number.
  e.g. 7, 10 and +4 are examples of constant terms.
- **Expressions** are made up of terms separated by + and − signs.
  e.g. $5x + 4 - 2y$ is an example of an expression.
  - This expression has 3 terms: $5x$, $+4$ and $-2y$ (+4 is called a constant term).
  - The coefficient of x is 5 and the coefficient of y is -2.

ALGEBRAIC TECHNIQUES

Like terms

- **Like terms** have identical pronumeral parts.
  e.g. $3x$ and x are like terms.
  $5ab$ and $-6ba$ are like terms.
  $-5xy$ and $11y$ are **not** like terms.
- Like terms can be added together or subtracted.

Substitution

- Substitution is the process of replacing a pronumeral with its known value.
- By using substitution, the value of an expression can be determined.
  e.g. $2x + 5$ when $x = 6$ becomes:
  $2(6) + 5 = 2 \times 6 + 5 = 12 + 5 = 17$

Multiplying and dividing expressions

- When multiplying expressions, multiply the numbers and write the pronumerals in alphabetical order. Remove the multiplication sign.
  e.g. $5b \times 2a = 10ab$
- When dividing, write the division as a fraction and simplify by cancelling.
  e.g. $\dfrac{16f}{4} = 4f$

Algebraic laws

- Only addition and multiplication are commutative. This means that the order in which we add or multiply does not matter. The **Commutative Law** for addition and multiplication states that:
  $$a + b = b + a \text{ and } a \times b = b \times a$$
- Only addition and multiplication are associative. This means that it does not matter how we group the numbers together when adding or multiplying. The **Associative Law** for addition and multiplication states that:
  $$(a + b) + c = a + (b + c) = a + b + c$$
  $$(a \times b) \times c = a \times (b \times c) = a \times b \times c$$
- The **Distributive Law** states that multiplying a number by a group of numbers added together is the same as doing each multiplication separately. The Distributive Law is useful for expanding brackets.
  The Distributive Law states that:
  $$a \times (b + c) = a \times b + a \times c$$
  $$= ab + ac$$

Equivalent expressions

- Equivalent expressions are always equal when pronumerals are substituted.
  e.g. $4 + 6x$ and $11 + 6x - 7$ are equivalent.

Landscaping

A town's local council wants to create more garden beds in the parklands around the town as part of its beautification program. All garden beds are to be square, but can vary in size. They will be surrounded by $1\,m^2$ pavers, as shown in the diagram.

Garden bed 1

Garden bed 2

Garden bed 3

- The blue areas represent the square garden beds.
- The white squares represent the $1\,m^2$ pavers.

1. Use these three diagrams of the garden beds to complete the following table.

Measurements	Garden bed 1	Garden bed 2	Garden bed 3
Side length of the garden bed (m)			
Area of the garden bed (m^2)			
Number of pavers used			
Total length around the outside of pavers (m)			

Let s represent the side length of the garden bed.
Let a represent the area of the garden bed.
Let p represent the number of pavers used.
Let l represent the total length around the outside of the pavers.

2. Examine the patterns formed in your table and write formulas linking:
   - a and s
   - l and s
   - p and s
   - p and l.

3. You just got a job as the council's new landscape gardener. Your first task is to design a $10\,m \times 10\,m$ square garden bed that is to be constructed in the same way as demonstrated in the previous diagram. How many pavers will be needed for this design?

4. The council has decided to put a large, square garden bed on the grounds near the entrance to its offices. The area of ground measures $18\,m \times 20\,m$. The council wants you to place the largest possible square garden bed here. You also have to put $1\,m^2$ pavers around it. Draw a design of the council's plan for this area of ground on a sheet of paper, with clear labels showing the measurements involved. Indicate the number of pavers that will be required for the project.

 Resources

Interactivities Crossword (int-2598)
Sudoku puzzle (int-3169)

Fluency

1. Simplify the following expressions by adding or subtracting like terms.

   a. $3g + 4g$ **b.** $8y - 2y$ **c.** $4h + 5h$ **d.** $7ag - 2ag$ **e.** $6gy - 3yg$

2. Simplify the following expressions by adding or subtracting like terms.

   a. $6y - 2y + y$ **b.** $8r - 8r$ **c.** $4t + 6 + 3t$ **d.** $12gh + 6hg$ **e.** $8t - 2m + 3t$

3. Simplify the following expressions by adding or subtracting like terms.

   a. $3m + m$ **b.** $7g + 8g + 8 + 4$ **c.** $7h + 4t - 3h$

   d. $2b + 7c + 8b$ **e.** $11axy - 3axy$

4. First simplify the following expressions, then find the value of the expression if $x = 5$.

   a. $7x + 3x$ **b.** $2x + 3x - 4$ **c.** $11x + 12x$ **d.** $x + 2x$

5. First simplify the following expressions, then find the value of the expression if $x = 5$.

   a. $3x - 2x + 16$ **b.** $21x - 13x + 7$ **c.** $11 + 2x + 5x$ **d.** $7x - 4x + 3x$

6. Use the Distributive Law to expand the following expressions.

   a. $7(m - 3k)$ **b.** $w(g + 9a)$ **c.** $2y(8h - 7)$

7. Use the Distributive Law to expand the following expressions.

   a. $10m(7r - 2p)$ **b.** $3g(2a + 2c)$ **c.** $4j(6x - 3y)$

8. Expand the following and simplify (if possible).

   a. $5(t - 2s) - 3t$ **b.** $4(a + 3g) - 2g$ **c.** $d(3f + 9) - 2d$

9. Expand the following and simplify (if possible).

   a. $4z(2 - a) + 7z$ **b.** $3p(f - 2q) + 7p$ **c.** $10b(a + 6) + 5b$

Understanding

10. Use the Associative Law to complete the following equations.

   a. $a + (3b + \underline{}) = (\underline{} + \underline{}) + 5c$ **b.** $\dfrac{1}{a} + (2b + \underline{}) = (\underline{} + 2b) \underline{} 7g$

11. Use the Associative Law to complete the following equations.

   a. $(d \times 3y) \times z = \underline{} (\underline{} \times \underline{})$ **b.** $\left(\dfrac{a}{b} \times c\right) \times \underline{} = \dfrac{a}{\underline{}} (\underline{} \times f)$

12. Use the associative and commutative laws to complete the following equations.

   a. $g + (3m + \underline{}) = (2k + \underline{}) + 3m$ **b.** $(w + \underline{}) + 4r = \underline{} + (5g + w)$

13. Use the associative and commutative laws to complete the following equations.

   a. $\dfrac{2}{3} \times (2d \times \underline{}) = (r \times 2d) \times \underline{}$ **b.** $(3z \times 5b) \times \underline{} = \underline{} \times (6m \times 5b)$

14. Write an expression for:
    a. the difference between m and c
    b. money earned by selling b cakes for \$3 each
    c. the product of x and y
    d. 15 more than g
    e. 1 more than d
    f. the cost of 12 bananas at h cents each
    g. t multiplied by 5.

15. Otto works in a warehouse, packing boxes of chocolates for distribution across the country. An order from a store in Wollongong fills 35 boxes, with 18 chocolates left over.

    a. Write an expression to represent the number of chocolates ordered, using the pronumeral x to represent a full box of chocolates.
    b. If each full box contains 30 chocolates, use your expression to calculate how many chocolates the store ordered.

16. Are the following equations True or False for all values of the pronumerals?
    a. $7x - 10y = 10y - 7x$
    b. $9x \times -y = -y \times 9x$
    c. $16 \times 2x \times x = 32x^2$

17. Simplify the following.
    a. $3g \times 7g$
    b. $6xy \times 3y$
    c. $7ad \times 6a^2$
    d. $-3z^2 \times 8zy$

18. Simplify each of the following.
    a. $\dfrac{16m}{8}$
    b. $\dfrac{13x}{13}$
    c. $\dfrac{20d}{60d}$
    d. $\dfrac{6th}{12h}$
    e. $\dfrac{63r}{27rt}$

19. Simplify each of the following.
    a. $\dfrac{15kt^3}{r^2}$
    b. $\dfrac{8m^2n}{18mn^2}$
    c. $\dfrac{72def^2}{-18d^2e}$
    d. $\dfrac{9a^2b}{12ab^2}$
    e. $\dfrac{33k^2m^2}{11m^2k}$

20. Place the appropriate terms in the boxes below to make the following statements true. Remember to simplify the equation by multiplying by the reciprocal.
    a. $\square \div 3pq = 4p^2$
    b. $21x^2 \div \square = \dfrac{3x}{y}$
    c. $\square \div \dfrac{2a}{5b} = \dfrac{35b^2}{2a^2}$
    d. $\dfrac{7x}{2y} \div \square = \dfrac{7y^2}{24}$

Communicating, reasoning and problem solving

21. Mary has y books. Tom has 4 more books than Mary. Cindy has 5 times as many books as Tom. Write down an expression, in terms of y, that can be used to work out the number of books that Cindy has. (Don't actually calculate the number of books.)

22. It's Bianca's birthday today. She is m years old. She asks the ages of people at her party. Write expressions to describe the ages of the following people.
    a. Bianca's cousin Paul is 8 years older than Bianca.
    b. Bianca's sister Kate is 5 years younger than Bianca.
    c. Bianca's aunt Delia is three times Bianca's age.
    d. Delia's friend Nic is 3 years older than Delia.

23. Use the pronumeral x to make a general statement in the form ___ × ___ = ___ from the following clues.

$$4 \times \frac{1}{4} = 1; \qquad 30 \times \frac{1}{30} = 1; \qquad 1.2 \times \frac{1}{1.2} = 1$$

24. A couple spends x on developing 80 small wedding photos and y for 60 large wedding photos. Determine how much more they need to spend if they want an extra dozen photos of each size.

25. It costs x for four adults' tickets at the cinema. A child's ticket costs $3 less than an adult's ticket. On family night, all the tickets go for half price. Develop a formula in terms of x to calculate how much it will cost for a family of 2 adults and 3 children on family night.

on To test your understanding and knowledge of this topic, go to your learnON title at www.jacplus.com.au and complete the **post-test**.

Answers

Topic 7 Algebraic techniques

7.1 Pre-test

1. B
2. 24
3. 84
4. D
5. $2x + 6y$
6. $20x + 4y$
7. 20, 21
8. $10k + 6$
9. 5, $4x$
10. a. $30abc$ b. $52\,ab^3$ c. $5q$ d. $\dfrac{13m}{3n}$
11. 3
12. $17p - 2q$
13. $6x + 10y$
14. False, $a \times a = a^2$
15. $3b + 8r$

7.2 Introduction to algebra

1. Pronumerals are used in algebra to represent unknown numbers.
2. As an example, on Monday there were b bananas in the bowl; on Tuesday 4 bananas were taken out; on Wednesday 2 bananas were taken out; on Thursday 14 bananas were added; on Friday 3 bananas were taken out.
3. a. $2 \times c + 4$ b. $3 \times c + 4$ c. $2 \times c + 2$
   d. $2 \times c + 3$
4. a. $c + 3$ b. $2c$ c. $2c + 5$ d. $4c + 3$
5. $4c + 34$
6. a. $2m$ b. $2m + n$ c. $4m + 3n$
   d. $m + 2n + 5$ e. $3m + 2n + 8$
7. a. $a + b$ b. $a - b$ c. ab d. $\dfrac{3a}{b}$
8. a. $\dfrac{-15}{a}$ b. $3a - 5b$ c. $\dfrac{-b}{5}$ d. $a - 7$
9. a. $m - 8, m - 2, m + 1, m + 3$
   b. $n - 15, n - 10.3, n + 25$
10. Examples are given here.
    a. Let d be the number of dogs at my house. My friend has 2 more dogs at her house than I do $(d + 2)$, while another friend has 1 dog less at his house than I do $(d - 1)$.
    b. Let t be the number of textbooks in my schoolbag. My friend has double that number in his schoolbag $(2t)$ while another friend has 1 more textbook than I do, in his schoolbag $(t + 1)$.
    c. Let y be the number of dollars in my wallet. My friend has 3 times the number of dollars as I have, in her wallet $(3y)$, while another friend has 10 times the number of dollars as I do, in her wallet $(10y)$.
11. a. $b + 2$ b. $t - 3$ c. $d + 6$ d. $k - 5$
12. a. $g + n + w$ b. $d + h$
    c. $2n$ d. $c + h$

13. a. $12 - g$ b. $4d$ c. bf d. $3m + y$
14. a. $a + b + c$ b. $a - c$
    c. ac d. abc
15. a. $\dfrac{b}{c}$ b. $\dfrac{a + c}{b}$ c. $a + 3$ d. $c - ab$
16. a. $t - x$ b. $\$4b$ c. $45g$ d. $\dfrac{\$t}{5}$
17. a. $2 \times n$ b. $4 \times n$ c. $8 \times n$
18. A
19. C
20. B
21. B
22. C
23. a. True b. True c. False d. True
24. a. True b. True c. True d. False
25. a. False b. True c. True d. False
26. a. $a + 6$ b. $a - 2$
27. a. $\$3d$ b. $\$120d$ c. $\$3(d + 1.5)$
28. In $\dfrac{(x + 1)}{3}$, 1 is added to x and then the result is divided by 3. In $\dfrac{x}{3} + 1$, x is divided by 3 first and then 1 is added.
29. $y + y + y = 3y$ and $y \times y \times y = y^3$
30.

31. a. $\$(2x + 7)$ b. $3x$ c. $4x + 3$
32. a. Cost of a hamburger + cost of a hamburger + cost of a drink = total cost (or cost of the drink + twice the cost of a hamburger = total cost)
    b. $\$3.75 + 2h = \$7.25$
33. a. $10 + \dfrac{2n}{5}$, where n is the number of coins in each piggy bank.
    b. $\$44$
34. 180

7.3 Substituting and evaluating

1. a. 11 b. 20 c. -2 d. 15
2. a. -1 b. 66 c. 19 d. -41
3. a. -13 b. -12 c. 33 d. 8
4.

g	1	2	3	4
$g - 2$	-1	0	1	2

5.

t	1	2	3	4
$2t - 3$	-1	1	3	5

6.

f	2	4	6	8
$12f + 7$	31	55	79	103

7. a. $g = 21$ b. $g = 13$ c. $g = 1$ d. $g = 25$

8. a. $h = 4$ b. $h = 2$ c. $h = 26$ d. $h = 152$

9. a. $i = 20$ b. $i = 40$ c. $i = 45$ d. $i = 0$

10. a. $j = 40$ b. $j = 36$ c. $j = 0$ d. $j = 4$

11. a. $k = 5$ b. $k = 125$ c. $k = 35$ d. $k = 75$

12. a. $l = 7$ b. $l = 1$ c. $l = 15$ d. $l = 63$

13. a. $m = 17$ b. $m = 23$ c. $m = 32$ d. $m = 77$

14. a. $n = 9$ b. $n = 3$ c. $n = 303$ d. $n = 216$

15. a. $o = 16$ b. $o = 28$ c. $o = 13$ d. $o = 49$

16. a. $p = 26$ b. $p = 8$ c. $p = 44$ d. $p = 2$

17. $w = 4$ and $h = 6$
    (There are many possible solutions for this question.)

18. a. 62 miles   b. 155 miles   c. 9.3 miles

19. 38 m

20. The values of x are between 3 and 20.

21. a. 109
    b. 93
    c. No, they would have scored 108.

22. a. 1255   b. 2485

7.4 Simplifying expressions using like terms

1. a. $5a$ b. $14y$ c. $15c$ d. $2u$ e. $20e$

2. a. $5t$ b. $14ab$ c. $6fg$ d. $3e$ e. $7t$

3. a. $5f$ b. 0   c. $18x$ d. $12m$ e. $8a$

4. a. $25ab - 7$
   b. $3y$
   c. Cannot be simplified
   d. $14t - 3$
   e. $13r$

5. a. Cannot be simplified
   b. $4 + 2g$
   c. t
   d. 0
   e. $14pq$

6. a. $11t - 5$
   b. t
   c. Cannot be simplified
   d. $11t - 8$
   e. $4t + 3$

7. a. 0   b. 0   c. 0   d. 8

8. a. 7   b. 0   c. 7   d. $6y$

9. a. $5a, 35$ b. $9a, 63$ c. $4a, 28$ d. $10a, 70$

10. a. $10a, 70$ b. $10a, 70$ c. $17 + 8a, 73$
    d. $5a + 2, 37$

11. a. $2a, 14$ b. $a, 7$ c. $0, 0$
    d. $17a - 16, 103$

12. a. True   b. True   c. False
    d. True   e. False

13. a. False   b. True   c. True
    d. False   e. True

14. a. True   b. True   c. False
    d. False   e. True

15. D

16. B

17. C

18. D

19. C

20. A

21. D

22. a. True   b. False   c. False

23. a. True   b. False   c. True

24. a. False   b. True   c. True

25. $10x$

26. $6d$

27. a. $2L + 2W$ b. $2A + B$

28. $\angle COD = \angle DOE = 40°$

7.5 Multiplying and dividing terms

1. a. $2a^2$ b. $7a^2b$ c. $21p^2q$
   d. $6b^2cd$ e. $40m^{15}$

2. a. $12g$ b. $21h$ c. $24d$ d. $15z$ e. $30r$

3. a. $35t$ b. $12u$ c. $42p$
   d. $21gy$ e. $22ht$

4. a. $24gx$ b. $70ah$ c. $36dm$
   d. $15ch$ e. $18gx$

5. a. $4f$ b. $2h$ c. $5x$ d. $3g$ e. $2r$

6. a. 2   b. 2   c. 2   d. $\dfrac{2}{3}$ e. $\dfrac{1}{2}$

7. a. $\dfrac{6}{7}$ b. $\dfrac{2}{3}$ c. $\dfrac{1}{3}$ d. $\dfrac{1}{2}$ e. $\dfrac{2}{3}$

8. a. $\dfrac{3}{a}$ b. $\dfrac{20p^{10}q^7}{3}$ c. $\dfrac{2x^2}{3y}$
   d. $\dfrac{5}{w}$ e. $\dfrac{1}{7xy}$

9. a. $3a^2$ b. $-4p^2$ c. $10x^2$ d. $5a^2b$

10. a. $-36x^2y$ b. $24p^2q^2$
    c. $-10mn^2t^2$ d. $-42xy^2z^2$

11. The student's working is not correct. Factors can only be simplified when in product form. This question has addition.

12. a. $\dfrac{16axy}{9b}$

    b. $\dfrac{16axy}{9b}$

    c. The answers to parts a and b are the same.

13. The order of operations needs to be applied when dealing with algebraic expressions. Also, terms that contain more than one pronumeral should be dealt with as a single term. The first expressions are all equivalent to $\dfrac{x}{yz}$ and the final two terms are equivalent to $\dfrac{xz}{y}$.

14. a. $3x^2$ **b.** $2bx$ **c.** $5(x+3)$

15. $-\dfrac{4}{9}$

16. a. $P = 4 \times 9 = 36$ cm

    b.   **i.** When $x = 2$, $A = 81 - (7 \times 7) = 32$ cm^2

        ii. When $x = 3$, $A = 81 - (6 \times 6) = 45$ cm^2

    c. $A = 81 - (9 - x)^2$ cm^2

7.6 Number laws

1. a. True   **b.** False   **c.** True   **d.** False

2. a. $w + \left(\underline{r + 6y} \right) = \left(\underline{w + r} \right) + 6y$

    b. $6t \times (4r) \times \underline{3s} = \left(\underline{6t \times 4r} \right) \times 3$

    c. $\left(9y + \underline{2r} \right) + 3w = \underline{9y} + (2r + 3w)$

    d. $(z + 2p) + \underline{6t} = z + \left(\underline{2p + 6t} \right)$

3. a. $6t + \left(3w + \underline{7v} \right) = \underline{3w} + (6t + 7v)$

    b. $s \times \left(9r \times \underline{2c} \right) = 2c \times \left(\underline{s \times 9r} \right)$

    c. $(3c + w) + \underline{d} = d + \left(\underline{3c + w} \right)$

    d. $\left(g + \underline{3b} \right) \times 2y = 2y \times \left(3b \times \underline{g} \right)$

4. a. False   **b.** False   **c.** True   **d.** False

5. a. 33   **b.** 33. They are the same.

6. a. 21   **b.** 1. They are different.

7. a. 48   **b.** 48. They are the same.

8. a. 2   **b.** $\dfrac{1}{8}$. They are different.

9. These expressions will have different results unless $x = a$.

10. The Associative Law does hold for fractions because, just like whole numbers, they are rational numbers. You can check this by adding a number of fractions in different orders.

11. $6 \div (1 - 5 \div 7) = 21$

12. a.   **i.** -4 **ii.** 2

    b. The answers are not equal. The Associative Law holds for addition but not for subtraction. The order of subtraction influences the outcome, whereas the order of addition has not affect on the outcome.

13. $48

14. They did not start with the same number. Andrew started with 11 and Andrea started with 17.

7.7 Expanding brackets

1. a. $3a + 6b$ **b.** $5x - 20z$ **c.** $20g + 30h$

    d. $2bc - 3ab$

2. a. $16cd - 2cq$ **b.** $9b^2 - 6bg$ **c.** $9ax + 7bx$

    d. $7kt + 21km$

3. a. $2a + 10b$ **b.** $5n - 8c$ **c.** $18c$

    d. $ax - bx + x$

4. a. $6d + 21e$ **b.** $18hj - 11h$ **c.** $4k + 20a - 12$

    d. $4x^2 + 6$

5. a. $5y + 2yz$ **b.** $3dx - 5x$ **c.** $12fu + 14u$

    d. $20n + 21hn$

6. a. $54r + 16$ **b.** $3t - 30$ **c.** $36 - 3y$ **d.** $5x - 16$

7. a. $9t - 43$ **b.** $19ry + 8r + 14$

    c. $26ty - 16ht + 6h$ **d.** $4r^2 + 18r + 3$

8. a. $x + 12$ **b.** $28hx - 32h - 18$

    c. $20x - y - 6$ **d.** $-14t^2 - 6$

9. a. $13x + 2y$ **b.** $12a - 10y + 6ay$

    c. $9cg + 18cd + dg$ **d.** $12dy + 3ry$

10. a. $2(x + 2)$ **b.** $6(x - y)$ **c.** $4(x + 3)$

    d. $4g(3f - 4h)$

11. Perimeter $= 99.2$ cm

12. Sample responses are given.

    a. $3a(4b + 2) = 12ab + 6a$

    b. $4n(9k + 2) = 36nk + 8n$

13. If x is a negative integer, then $-5x$ will be a positive integer. Adding this positive integer to 1 will result in another positive integer. The result will be positive.

14. $4(4c - 8a)$, $8(2c - 4a)$ and $16(c - 2a)$

15. a. $2y - x$ **b.** Both expressions equal 4.

16. a. $(50 - d)$

    b.   **i.** Sale price $(P) = 3(50 - d)$

        ii. $P = 150 - 3d$

    c. Amount of change $(C) = 200 - (150 - 3d) = 50 + 3d$

Project

1.

Measurements	Garden bed 1	Garden bed 2	Garden bed 3
Side length of the garden bed (m)	1	2	3
Area of the garden bed (m^2)	1	4	9
Number of pavers used	8	12	16
Total length around the outside of pavers (m)	12	16	20

2. $a = s \times s, l = 4s + 8, p = 4s + 4, p = l - 4$

3. 44

4. 68

7.8 Review questions

1. a. $7g$ **b.** $6y$ **c.** $9h$ **d.** $5ag$ **e.** $3gy$

2. a. $5y$ **b.** 0 **c.** $7t + 6$

    d. $18gh$ **e.** $11t - 2m$

3. a. $4m$ **b.** $15g + 12$ **c.** $4h + 4t$

    d. $10b + 7c$ **e.** $8axy$

4. a. $10x, 50$ **b.** $5x - 4, 21$
c. $23x, 115$ **d.** $3x, 15$

5. a. $x + 16, 21$ **b.** $8x + 7, 47$
c. $11 + 7x, 46$ **d.** $6x, 30$

6. a. $7m - 21k$ **b.** $gw + 9aw$ **c.** $16hy - 14y$

7. a. $70mr - 20mp$ **b.** $6ag + 6cg$ **c.** $24jx - 12jy$

8. a. $2t - 10s$ **b.** $4a + 10g$ **c.** $3df + 7d$

9. a. $15z - 4az$ **b.** $3fp - 6pq + 7p$
c. $10ab + 65b$

10. In this question, the order of the terms is important.
a. $a + (3b + 5c) = (a + 3b) + 5c$
b. t

11. In this question, the order of the terms is important.
a. $(d \times 3y) \times z = d \times (3y \times z)$

b. $\left(\dfrac{a}{b} \times c\right) \times f = \dfrac{a}{b}(c \times f)$

12. In this question, the order of the terms is important.
a. $g + (3m + 2k) = (2k + g) + 3m$

b. $(w + 5g) + 4r = 4r + (5g + w)$

13. In this question, the order of the terms is important.
a. $\dfrac{2}{3} \times (2d + r) = (r \times 2d) \times \dfrac{2}{3}$

b. $(3z + 5b) \times 6m = 3z \times (6m \times 5b)$

14. a. $m - c$ **b.** $\$3b$
c. xy **d.** $g + 15$
e. $d + 1$ **f.** $12h$ cents
g. $5t$

15. a. $35x + 18$ **b.** 1068

16. a. False **b.** True **c.** True

17. a. $21g^2$ **b.** $18xy^2$ **c.** $42a^3 d$ **d.** $-24yz^3$

18. a. $2m$ **b.** x **c.** $\dfrac{1}{3}$ **d.** $\dfrac{t}{2}$ **e.** $\dfrac{7}{3t}$

19. a. $15kt$ **b.** $\dfrac{4m}{9n}$ **c.** $-\dfrac{4f^2}{d}$ **d.** $\dfrac{3a}{4b}$ **e.** $3k$

20. a. $12p^3 q$ **b.** $7xy$ **c.** $\dfrac{7b}{a}$ **d.** $\dfrac{12x}{y^3}$

21. $5y + 20$

22. a. $m + 8$ **b.** $m - 5$ **c.** $3m$ **d.** $3m + 3$

23. $x \times \dfrac{1}{x} = 1$

24. $\$\dfrac{3x + 4y}{20} = \dfrac{3x}{20} + \dfrac{y}{5}$

25. $\$\dfrac{5x - 36}{8}$

Semester review 1

The learnON platform is a powerful tool that enables students to complete revision independently and allows teachers to set mixed and spaced practice with ease.

Student self-study

Review the **Course Content** to determine which topics and lessons you studied throughout the year. Notice the green bubbles showing which elements were covered.

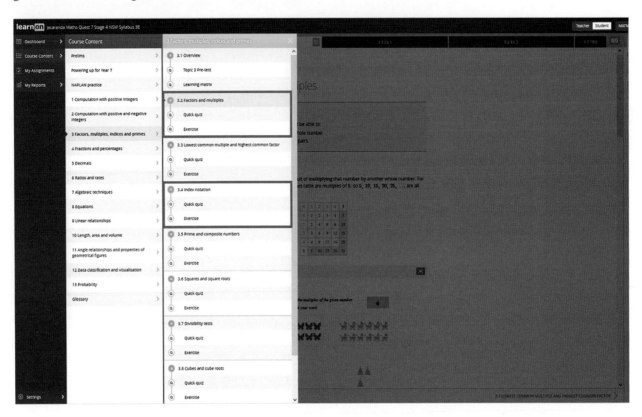

Review your results in **My Reports** and highlight the areas where you may need additional practice.

Use these and other tools to help identify areas of strengths and weakness and target those areas for improvement.

Teachers

It is possible to set questions that span multiple topics. These assignments can be given to individual students, to groups or to the whole class in a few easy steps.

Go to **Menu** and select **Assignments** and then **Create Assignment**. You can select questions from one or many topics simply by ticking the boxes as shown below.

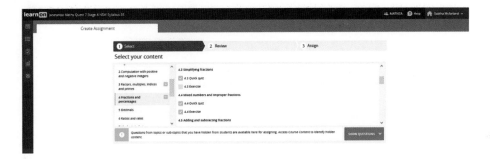

Once your selections are made, you can assign to your whole class or subsets of your class, with individualised start and finish times. You can also share with other teachers.

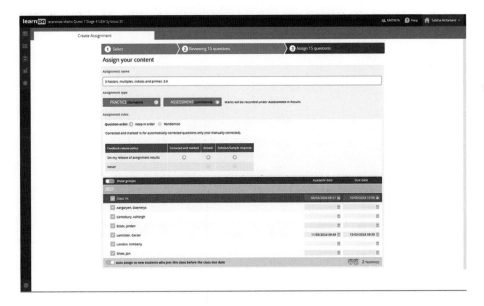

More instructions and helpful hints are available at www.jacplus.com.au.

8 Equations

LESSON
8.1 Overview

Why learn this?

We use equations to write sentences and information containing mathematical symbols and algebra. Many everyday situations can be represented by equations. Most likely you have been solving equations without even realising it. For example, if you pay for an item that costs $4.25 and you hand a $5.00 note to the shop assistant, you will expect to receive 75 cents in change. This answer is obtained using an equation. Solving equations means finding the numbers you didn't originally know.

Understanding how to solve equations is one of the most useful skills you will learn in algebra. To work out which mobile phone plan gives the best value requires you to solve equations. If you are travelling overseas to a country that uses a different currency, you can use the exchange rate and equations to work out the cost of an item in Australian dollars. The computer chips used in washing machines, dryers, laptops, cars and other items are based on mathematical equations and algorithms. Equations are also put to use in traffic control systems, space programs, aircraft, medicine and in many other areas. Scientists, architects, engineers and accountants are just some of the people who use algebra and equations frequently in their work.

Exercise 8.1 Pre-test

1. Write an equation that represents the following:
   A number, x, is multiplied by 6 and the answer is 72.

2. Solve, by inspection, the equation $\dfrac{t}{4} = 8$.

3. A tennis team won 4 more games than it lost. If the team played 18 games, how many did they win?

4. Calculate the output number for the following flowchart.

5. **MC** Select which of the following is a correct step to solve for m in the equation $\dfrac{m}{2} = -5$.

   A. $m = -5 - 2$ **B.** $m = -5 \times 2$ **C.** $m = -5 + 2$ **D.** $m = \dfrac{-5}{2}$

6. When 4 is added to a number and the result is doubled, it gives a value of 16. Determine the number.

7. If 5 less than a number divided by 4 gives a result of 3, determine the number.

8. **MC** In order, identify the steps required to backtrack the following to x.

   A. -5 then $\div\, 3$ **B.** -5 then $\times -3$ **C.** $+5$ then $\div\, 3$ **D.** $+5$ then $\times -3$

9. **MC** I think of a number, add 5 to it and then multiply the result by 3. I then divide this result by 2. The final value is 21. Select the equation that describes these steps.

   A. $\dfrac{5(x+3)}{2} = 21$ **B.** $\dfrac{3(x+5)}{2} = 21$ **C.** $\dfrac{3(x+5)}{3} = 21$ **D.** $\dfrac{3(x-5)}{2} = 21$

10. Kyle is 5 years older than Noah. If the sum of their ages is 29, calculate Noah's age.

11. Use backtracking to solve $4x - 5 = 7$.

12. Simplify the equation $8x - 15 - 3x - 7 = 28$ and then solve using backtracking.

13. The sum of three consecutive even numbers is 84. Determine the lowest number.

14. In a class of 27 students, there are 5 more boys than girls. Calculate the number of boys.

15. A triangle has side lengths of x, $4x + 3$ and $5x - 7$, and a perimeter of 51 cm. Determine the value of x.

LESSON
8.2 Introduction to equations

LEARNING INTENTION

At the end of this lesson you should be able to:
- write an equation to represent a worded problem
- solve equations by inspection
- solve equations using guess and check.

▶ 8.2.1 Equations

eles-4012

- **Equations** are mathematical statements that tell us that two expressions are equal. The left-hand side (LHS) is equal to the right-hand side (RHS).
- For a mathematical statement to be classified as an equation it must contain an equals sign ($=$).

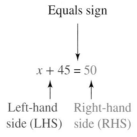

Equals sign

$$x + 45 = 50$$

Left-hand   Right-hand
side (LHS)   side (RHS)

Equations

- **An equation is a mathematical statement containing an equals sign ($=$).**
- **The LHS of an equation is always equal to (has the same value as) the RHS of the equation.**

- Remember from topic 7 that algebraic expressions, which are made up of terms and algebraic operations, do not contain an $=$ sign and do not have sides. For instance, $3t + 4$ is an algebraic expression, $3t + 4 = 6$ is an equation.
- In an algebraic expression, the pronumeral can represent any number. In an equation, we are able to determine the value of the pronumeral.

Writing an equation from words

- Mathematical equations can be created based on words or sentences.
- Some English words can be changed into mathematical operations.

Addition (+)	Subtraction (−)	Multiplication (×)	Division (÷)
sum	difference	product	quotient
more than	less than	multiply	divide
increased by	decreased by	times	shared equally
add *or* added to	minus	lots of	split
double *(multiplied by 2)*		squared *(multiplied by itself)*	

WORKED EXAMPLE 1 Writing an equation for a worded problem

Write an equation to represent each of these worded problems.
a. When I multiply a number by 8, the answer is 24.
b. When I divide a number by 5, the answer is 7.
c. When I divide 60 by a number, the answer is 10.
d. When I subtract 7 from a number, the answer is 25.
e. When a number is increased by 9, the answer is 35.
f. When I square a number, the result is 36.

THINK	WRITE
a. 1. Use a pronumeral to describe the number.	a. Let m be the number.
2. Multiply the number by 8 to form the LHS of the expression.	LHS: $m \times 8 = 8m$
3. Write the RHS of the expression.	RHS: 24
4. Write the equation by writing that the LHS is equal to the RHS.	$8m = 24$
b. 1. Use a pronumeral to describe the number.	b. Let t be the number.
2. Divide the number by 5 to form the LHS of the expression.	LHS: $t \div 5 = \dfrac{t}{5}$
3. Write the RHS of the expression.	RHS: 7
4. Write the equation by writing that the LHS is equal to the RHS.	$\dfrac{t}{5} = 7$
c. 1. Use a pronumeral to describe the number.	c. Let s be the number.
2. Divide 60 by the number to form the LHS of the expression.	LHS: $60 \div s = \dfrac{60}{s}$
3. Write the RHS of the expression.	RHS: 10
4. Write the equation by writing that the LHS is equal to the RHS.	$\dfrac{60}{s} = 10$
d. 1. Use a pronumeral to describe the number.	d. Let l be the number.
2. Subtract 7 from the number (that is, take 7 away from the number) to form the LHS of the expression.	LHS: $l - 7$
3. Write the RHS of the expression.	RHS: 25
4. Write the equation by writing that the LHS is equal to the RHS.	$l - 7 = 25$
e. 1. Use a pronumeral to describe the number.	e. Let a be the number.
2. The word *increased* refers to addition. To increase the number a by 9, add 9 to it.	LHS: $a + 9$
3. Write the RHS of the expression.	RHS: 35
4. Write the equation by writing that the LHS is equal to the RHS.	$a + 9 = 35$
f. 1. Use a pronumeral to describe the number.	f. Let z be the number.
2. Square the number (that is, multiply the number by itself) to form the LHS of the expression.	LHS: $z \times z = z^2$
3. Write the RHS of the expression.	RHS: 36
4. Write the equation by writing that the LHS is equal to the RHS.	$z^2 = 36$

▶ 8.2.2 Solving an equation by inspection

eles-4013

- To solve an equation you must determine the value of the pronumeral that makes the LHS of the equation equal to the RHS.
- Some basic equations can be solved by **inspection**. This process involves determining the answer by simply inspecting (or looking at) the equation.

WORKED EXAMPLE 2 Solving an equation by inspection

Solve the following equations by inspection.

a. $\dfrac{w}{3} = 4$

b. $h - 9 = 10$

THINK	WRITE
a. 1. Write the equation.	a. $\dfrac{w}{3} = 4$
2. Think of a number that, when divided by 3, gives 4. Try 12.	$12 \div 3 = 4$
3. Based on this, w must be 12.	$w = 12$
b. 1. Write the equation.	b. $h - 9 = 10$
2. Think of a number that equals 10 when 9 is subtracted from it. Try 19.	$19 - 9 = 10$
3. Based on this, h must be 19.	$h = 19$

DISCUSSION

With simple equations such as $\dfrac{w}{3} = 4$ could you use a visual representation, such as the one below, to help you solve the equation?

How would you use that diagram to help you solve $\dfrac{w}{4} = 3$?

▶ 8.2.3 Guess, check and improve

eles-4014

- Sometimes equations cannot be solved easily by inspection. In these cases, one possible method that could be used is called 'guess, check and improve'.
  For example, you can solve $4x + 6 = 46$ using the following method.

This logic puzzle uses 4 different equations. Can you solve the final equation?

$\square + \square + \square + \square = 28$

$\square + \square + \square = 17$

$\square + \square = 9$

$\square + \square + \square = ?$

Solve $4x + 6 = 46$			
Guess	**Check**		**Improve**
$x = 3$	LHS $= 4 \times 3 + 6$ $= 18$	RHS $= 46$	$x = 3$ makes the LHS too small. Since x appears on the LHS only, choose a larger value for x to make the LHS larger.
$x = 11$	LHS $= 4 \times 11 + 6$ $= 50$	RHS $= 46$	$x = 11$ makes the LHS too large, but closer than before. Choose a value for x that is less than 11 but greater than 3.
$x = 10$	LHS $= 4 \times 10 + 6$ $= 46$	RHS $= 46$	LHS $=$ RHS, so $x = 10$ is the solution.

WORKED EXAMPLE 3 Solving an equation by guess, check and improve

Use guess, check and improve to solve the equation $2x + 21 = 4x - 1$.

THINK

1. Set up a table with 4 columns displaying the value of x, the value of the LHS and RHS equations (after substitution) and a comment on how these two values compare to the equation.
2. Substitute the first guess, say $x = 1$, into the LHS and RHS equations and comment on the results.
3. Repeat step 2 for the other guesses until the correct answer is found.
4. Write the answer.

WRITE

	Check		
Guess x	**$2x + 21$**	**$4x - 1$**	**Comment**
1	23	3	$4x - 1$ is too small.
6	33	23	This is closer.
10	41	39	Very close.
11	43	43	That's it!

The answer is $x = 11$.

WORKED EXAMPLE 4 Using guess and check to determine unknown numbers

Use guess, check and improve to determine two numbers whose sum is 31 and whose product is 150.

THINK

1. The numbers add up to 31, so guess 2 numbers that do this. Then check by finding their product.
2. Guess 1 and 30, then calculate the product.
3. Guess 10 and 21, then calculate the product.
4. Try other numbers between 1 and 10 for the first number. Determine the product of the two numbers.
   Stop when the product of the two numbers is 150.
5. Write the answer.

WRITE

Guess the sum (small number first)	**Check the product (P)**	**Comment**
1 and 30	$1 \times 30 = 30$	P is too low.
10 and 21	$10 \times 21 = 210$	P is too high.
5 and 26	$5 \times 26 = 130$	P is too low.
8 and 23	$8 \times 23 = 184$	P is too high.
6 and 25	$6 \times 25 = 150$	That's it!

The two numbers that add up to 31 and have a product of 150 are 6 and 25.

DISCUSSION

Explore how Aboriginal and/or Torres Strait Islander peoples may use storytelling and/or dance to represent mathematical equations.

 Resources

Interactivities Writing equations (int-4041)
                    Guess, check and improve (int-4042)

Exercise 8.2 Introduction to equations   **learn** on

| 8.2 Quick quiz on | 8.2 Exercise |

Individual pathways

■ PRACTISE	■ CONSOLIDATE	■ MASTER
1, 4, 6, 8, 11, 14	2, 5, 9, 12, 15	3, 7, 10, 13, 16, 17

Fluency

1. **WE1** Write an equation to represent each of these worded problems.

   a. When I add 7 to a number, the answer is 11.
   b. When I add 12 to a number, the answer is 12.
   c. When I subtract 7 from a number, the answer is 1.
   d. When I subtract 4 from a number, the answer is 7.

2. Write an equation to represent each of these worded problems.

   a. When I multiply a number by 2, the answer is 12.
   b. When I multiply a number by 5, the answer is 30.
   c. When I divide a number by 7, the answer is 1.
   d. When I divide a number by 5, the answer is 2.

3. Write an equation to represent each of these worded problems.

   a. When I subtract a number from 15, the answer is 2.
   b. When I subtract a number from 52, the answer is 8.
   c. When I divide 21 by a number, the answer is 7.
   d. When I square a number, the answer is 100.

4. **WE2** Solve the following equations by inspection.

   a. $x + 7 = 18$ b. $y - 8 = 1$ c. $3m = 15$ d. $\dfrac{m}{10} = 3$

5. Solve the following equations by inspection.

   a. $\dfrac{k}{5} = 0$
   b. $b + 15 = 22$
   c. $b - 2.1 = 6.7$
   d. $\dfrac{c}{3} = 1.4$
   e. $5x = 14$

Understanding

6. **WE3** Use guess, check and improve to solve the following equations.

   a. $3x + 11 = 5x - 1$
   b. $5x + 15 = x + 27$
   c. $x + 20 = 3x$

7. Use guess, check and improve to solve the following equations.

   a. $12x - 18 = 10x$
   b. $6(x - 2) = 4x$
   c. $3(x + 4) = 5x + 4$

8. **WE4** Use guess, check and improve to find two numbers whose sum and product are given:

   a. sum $= 21$, product $= 98$
   b. sum $= 26$, product $= 165$
   c. sum $= 54$, product $= 329$
   d. sum $= 178$, product $= 5712$

9. Use guess, check and improve to find two numbers whose sum and product are given:

   a. sum $= 153$, product $= 4662$
   b. sum $= 242$, product $= 14\,065$
   c. sum $= 6.1$, product $= 8.58$
   d. sum $= 978$, product $= 218\,957$

10. Copy and complete this table by substituting each x-value into $x^2 + 4$ and $4x + 1$. The first row has been completed for you. Use the table to determine a solution to the equation $x^2 + 4 = 4x + 1$. (Remember, x^2 means $x \times x$.)

x	$x^2 + 4$	$4x + 1$
0	4	1
1		
2		9
3	13	
4		

Communicating, reasoning and problem solving

11. Explain the thought process you would use to determine the solution to $8x = 48$ by inspection.

12. Explain why the inspection method is not suitable to solve the equation $5x = 34$.

13. Annisa is trying to find the solution to the equation $3x - 7 = 17$ using the guess, check and improve method. Her first guess is $x = 5$, which is not the solution. Should her next guess be greater than or less than 5?

14. A football team won four more games than it lost. The team played 16 games. Determine how many games the team won.

15. A plumber cut a 20-metre pipe into two pieces. One of the pieces is three times as long as the other. Determine the lengths of the two pieces of pipe.

16. Lily is half the age of Pedro. Ross is 6 years older than Lily and 6 years younger than Pedro. Determine Pedro's age.

17. Angus and his grandfather share a birthday. Both their ages are prime numbers. Angus's age has the same two digits as Grandpa's but in reverse order. In 10 years' time, Grandpa will be three times as old as Angus. Determine how old both Grandpa and Angus will be in 10 years.

LESSON
8.3 Building up expressions and backtracking

LEARNING INTENTION

At the end of this lesson you should be able to:
- complete a flowchart to determine the output
- construct a flowchart from a given expression.

▶ 8.3.1 Flowcharts

eles-4016

- A flowchart is useful for keeping track of the steps in a sequence.

- In the preceding diagram, the instructions for each step are written above the arrows that join the boxes.
  - The first box shows what you start with.
  - The box in the middle shows what happens along the way as you complete the steps.
  - The last box shows what you finish with.
- Flowcharts can be used to keep track of expressions as operations are performed on them.
- In the following flowchart, the starting number is 8. This is called the **input number**. Performing the operations in the order displayed results in 1, the number in the last box. This is called the **output number**.

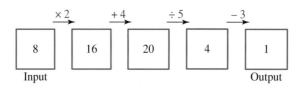

WORKED EXAMPLE 5 Completing a flowchart

Build up an expression by following the instructions on these flowcharts.

a.

b.

THINK

a. 1. The instruction above the first arrow says $\times 3$. The result is $m \times 3 = 3m$.

2. The instruction above the second arrow says $+5$. Adding 5 to $3m$ gives the result $3m + 5$.

3. State the expression (output).

WRITE

a.

The expression is $3m + 5$.

---

b.
  1. The instruction above the first arrow says $+5$.
     Adding 5 to m gives the result $m + 5$.
  2. The instruction above the second arrow says $\times 3$. That means multiply all of $m + 5$ by 3. The result is $3(m + 5)$.
  3. State the expression (output).

b.

The expression is $3(m + 5)$.

WORKED EXAMPLE 6 Drawing a flowchart

Draw a flowchart whose input number is m and whose output number is given by the following expressions.

a. $2m - 11$ **b.** $\dfrac{m + 9}{5}$ **c.** $4\left(\dfrac{m}{3} + 2\right)$

THINK

WRITE

a.
  1. The first step is to obtain $2m$. Do this by multiplying m by 2.

a.

  2. Next, subtract 11.

b.
  1. The expression $m + 9$ is grouped as though it is in a pair of brackets, so we must work out this part first. This means we add 9 to m first.

b.

  2. Next, the whole expression is divided by 5.

c.
  1. The brackets indicate we must first work from within the brackets, so:
     • divide m by 3
     • then add 2 to the result.

c.

  2. Multiply the result obtained in step 1 by 4.

▶ ## 8.3.2 Backtracking

eles-4017

• **Backtracking** is a method used to work backwards through a flowchart. It involves moving from the output towards the input.
• When working backwards through a flowchart, use **inverse** (or **opposite**) **operations**.

- A list of operations and their inverses (opposites) is highlighted in the following box.

Inverse operations

$-$ is the inverse operation of $+$

$+$ is the inverse operation of $-$

$\div$ is the inverse operation of $\times$

$\times$ is the inverse operation of $\div$

- Examine the following flowchart and the operations that were applied to the input to produce the output (working left to right). These are shown in purple on top.
  Using backtracking we can identify the opposite operations that need to be applied when working backwards through the flowchart. These are shown in pink on the bottom.

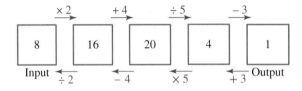

WORKED EXAMPLE 7 Backtracking a flowchart

Complete the following flowchart by writing in the operations that need to be carried out in order to backtrack to x.

THINK

1. Copy the flowchart and look at the operations that have been performed.
   Starting from the input, two operations have been performed: a multiplication by 5 and then the number 2 has been added.

2. Starting with the output, work backwards towards the input using inverse operations.
   The inverse operation of $+2$ is -2. Show this on the bottom of the flowchart by adding an arrow pointing towards the left (input).

3. The inverse operation of $\times 5$ is $\div 5$. Show this on the flowchart at the bottom, with an arrow pointing towards the left (input).

WRITE

COMMUNICATING — COLLABORATIVE TASK: Secret numbers

1. Think of a number by yourself and write it in your workbook. Don't tell anyone! Multiply your number by 4 and write down the answer. Next, add 20 and write down the answer. Then divide by 2 and write down the answer. Finally, subtract 5 and write down the answer.
2. As a class, write the flowchart for these instructions on the board.
3. Show how the flowchart works for different starting numbers.
4. Work out some of your classmates' secret numbers by working backwards through the flowchart.

Exercise 8.3 Building up expressions and backtracking learn on

8.3 Quick quiz on	**8.3 Exercise**

Individual pathways

■ PRACTISE	■ CONSOLIDATE	■ MASTER
1, 3, 7, 10, 13, 16	2, 5, 8, 11, 14, 17	4, 6, 9, 12, 15, 18, 19

Fluency

1. **WE5** Build up an expression by following the instructions on these flowcharts.

a.

b.

c.

d.

2. Build up an expression by following the instructions on these flowcharts.

a.

b.

c.

d.
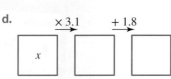

3. Build up an expression by following the instructions on the flowcharts. Use brackets or fractions as a grouping device, for example, $2(x+3)$ or $\dfrac{x-5}{4}$.

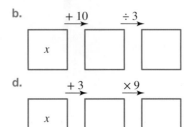

4. Build up an expression by following the instructions on the flowcharts. Use brackets or fractions as a grouping device, for example, $2(x+3)$ or $\dfrac{x-5}{4}$.

5. Copy and complete the following flowcharts by filling in the missing expressions.

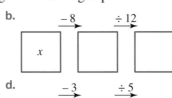

6. Copy and complete the following flowcharts by filling in the missing expressions.

Understanding

7. **WE6** Draw a flowchart whose input number is x and whose output is given by the following expressions.

   a. $5x + 9$

   b. $2(x + 1)$

   c. $\dfrac{x - 8}{7}$

8. Draw a flowchart whose input number is x and whose output is given by the following expressions.

   a. $12(x - 7)$

   b. $\dfrac{x}{5} - 2$

   c. $\dfrac{x + 6}{3}$

9. Draw a flowchart whose input number is x and whose output is given by the following expressions.

   a. $3(x + 7) - 5$

   b. $\dfrac{3x + 7}{2}$

   c. $4(3x + 1)$

10. **WE7** Complete the following flowcharts by writing in the operations that must be carried out in order to backtrack to x.

    a.

    b.

    c.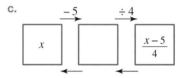

11. Add the operations to complete these flowcharts.

    a.

    b.

    c.

    d.

12. Add the operations to complete these flowcharts.

    a.

    b.

    c.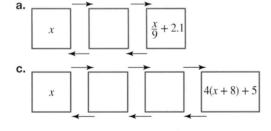

    d.

Communicating, reasoning and problem solving

13. Starting from p, the expression $2(p - 3)$ is built using the operations $\times 2$ and $- 3$. Identify which operation should be performed first.

14. I think of a number and add 8 to it. I multiply the result by 5 and then divide the result by 4. The answer is 30. Use a flowchart to build up an expression and write these steps in the form of an equation.

15. I think of a number, multiply it by 5 and add 15. The result is 3 less than 4 times the original number. If the original number is n, write an equation to show the relation. Show all your working.

16. Rachel and Jackson are 7 years apart in age. Jackson is older than Rachel. The sum of their ages is 51. Determine Rachel's age.

17. The sum of two numbers is 32 and their product is 247. Determine the two numbers.

18. Marcus and Melanie pooled their funds and purchased shares on the stock exchange. Melanie invested $350 more than Marcus. Together they invested $2780. Determine how much money Marcus invested.

19. The equation $1 + 2 + 3 + 4 + 5 + 6 + 7 + 8 + 9 = 100$ is clearly incorrect. By removing some + signs, including some − signs and combining some of the digits to make larger numbers, see if you can make the equation correct. Remember, only + and − signs are allowed and the equation must equal 100. One example is shown below. Determine as many other possible solutions as you can, showing full working.

$$123 + 45 - 67 + 8 - 9 = 100$$

LESSON
8.4 Solving linear equations using backtracking

LEARNING INTENTION

At the end of this lesson you should be able to:
- solve equations using backtracking.

▶ 8.4.1 Using backtracking to solve equations

eles-4018

- Backtracking is a process that can be used to solve equations.

Solving equations using backtracking

To solve an equation using backtracking, follow these steps.
1. **Use a flowchart to build up the equation.**
2. **Work backwards through the flowchart using inverse operations to determine the solution.**

For example, if we are asked to solve the equation $2x + 9 = 23$, we must first build up the equation and then use backtracking to determine the solution.

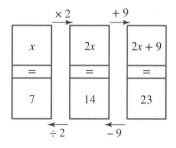

- The solution of the equation $2x + 9 = 23$ is $x = 7$.
- We can also check that we have found the correct answer by substitution.
  That is, LHS $= 2(7) + 9 = 2 \times 7 + 9 = 14 + 9 = 23 =$ RHS.

Draw a flowchart to represent the following worded problem and then solve it by backtracking.
I am thinking of a number. When I multiply it by 4 and then add 2, the answer is 14.

THINK	WRITE
1. Build an expression using x to represent the number. Start with x, multiply by 4 and add 2. The output number is 14.	
2. Backtrack to determine the value of x. • The inverse operation of $+2$ is -2 $(14-2=12)$. • The inverse operation of $\times 4$ is $\div 4$ $(12 \div 4 = 3)$.	
3. Write the solution.	The solution is $x = 3$. Therefore the number is 3.

Solve the following equations by backtracking.
a. $3(x + 7) = 24$

b. $\dfrac{x}{3} + 5 = 6$

THINK	WRITE
a. 1. Build the expression on the left-hand side of the equation. Start with x, then add 7 and then multiply by 3. The output number is 24.	a.
2. Backtrack to determine the value of x. • The inverse operation of $\times 3$ is $\div 3$ $(24 \div 3 = 8)$. • The inverse operation of $+7$ is -7 $(8 - 7 = 1)$.	
3. Write the solution.	The solution for the equation is $x = 1$.

b. **1.** Build the expression on the left-hand side of the equation. Start with x, then divide by 3 and then add 5. The output number is 6.

b.

2. Backtrack to determine the value of x.
- The inverse operation of $+5$ is -5, so subtract -5 from 6 ($6 - 5 = 1$).
- The inverse operation of $\div 3$ is $\times 3$, so multiply 1 by 3 ($1 \times 3 = 3$).

3. Write the solution.

The solution for the equation is $x = 3$.

WORKED EXAMPLE 10 Simplifying and solving equations by backtracking

Simplify and then solve the following equation by backtracking.

$5x + 13 + 2x - 4 = 23$

THINK

1. Simplify by adding the like terms together on the left-hand side of the equation. $5x + 2x = 7x$, $13 - 4 = 9$

2. Draw a flowchart and build the left-hand side of the equation $7x + 9$. Start with x, multiply by 7 and add 9. The output number is 23.

3. Backtrack to determine the value of x.
- The inverse operation of $+9$ is -9, so subtract 9 from 23 ($23 - 9 = 14$).
- The inverse operation of $\times 7$ is $\div 7$, so divide 14 by 7 ($14 \div 7 = 2$).

4. Write the solution.

WRITE

$5x + 13 + 2x - 4 = 23$
$7x + 9 = 23$

The solution for the equation is $x = 2$.

COMMUNICATING — COLLABORATIVE TASK: Think of a number

In pairs, discuss and investigate Mel's 'think of a number' puzzle, which is described below.

Mel loves playing 'Think of a number' with friends. Here's an example of one of her puzzles.

- Think of a number.
- Double it.
- Add 10.
- Divide by 2.
- Take away the number you first thought of.
- Your answer is . . . 5!

Let's investigate to see why the answer is always 5, whatever number you first think of. We can form expressions for each of the steps, using a variable as the starting value.

- Think of a number. $\qquad$ n
- Double it. $\qquad$ $n \times 2 = 2n$
- Add 10. $\qquad$ $2n + 10$
- Divide by 2. $\qquad$ $(2n + 10) \div 2 = n + 5$
- Take away the number you first thought of. $\qquad$ $n + 5 - n = 5$
- Your answer is . . . $\qquad$ 5

As you can see, the answer will always be 5 for any starting number.

Now you try

1. Write expressions for each step in the following, showing that you can determine the answer in each case.

a. Puzzle 1: Your birth year
- Take the year in which you were born.
- Subtract 500.
- Multiply by 2.
- Add 1000.
- Divide by 2.
- Your answer is . . . your birth year!

b. Puzzle 2: Your age
- Take your age (in years).
- Add 4.
- Multiply by 10.
- Subtract 10.
- Divide by 5.
- Subtract your age.
- Take away 6.
- Your answer will be your age!

c. Puzzle 3: Think of a number . . .
- Think of a number.
- Divide it by 2.
- Add 2.
- Multiply by 4.
- Take away your original number.
- Subtract your original number again.
- Your answer should be . . .

2. Write some 'Think of a number' puzzles yourself. Try them out on friends. They will marvel at your mystical powers!

Resources

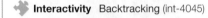 **Interactivity** Backtracking (int-4045)

Exercise 8.4 Solving linear equations using backtracking   learn

8.4 Quick quiz **on**   8.4 Exercise

Individual pathways

■ PRACTISE	■ CONSOLIDATE	■ MASTER
1, 4, 7, 11, 12, 15, 16	2, 5, 8, 10, 13, 17, 18, 19, 20	3, 6, 9, 14, 21, 22, 23, 24

Fluency

1. **WE8** Draw a flowchart to represent each of the following worded problems and then solve them by backtracking.

    a. When I multiply a number by 2 and then add 7 the answer is 11.
    b. When I add 3 to a number and then multiply by 5 the answer is 35.
    c. When I divide a number by 4 and then add 12 the answer is 14.
    d. When I add 5 to a number and then divide by 3 the answer is 6.
    e. When I subtract 7 from a number and then multiply by 6 the answer is 18.

2. Draw a flowchart to represent each of the following worded problems and then solve them by backtracking.

    a. When I subtract 4 from a number and then divide by 9 the answer is 7.
    b. When I divide a number by 11 and then subtract 8 the answer is 0.
    c. When I multiply a number by 6 and then add 4 the answer is 34.
    d. When I multiply a number by 5 and then subtract 10 the answer is 30.
    e. When I subtract 3.1 from a number and then multiply by 6 the answer is 13.2.

3. Draw a flowchart to represent each of the following worded problems and then solve them by backtracking.

    a. When I add $\dfrac{2}{5}$ to a number and then divide by 6 the answer is $\dfrac{4}{5}$.

    b. When I subtract $\dfrac{3}{4}$ from a number and then divide by $\dfrac{2}{3}$ the answer is $\dfrac{1}{6}$.

4. Draw a flowchart and use backtracking to determine the solution to the following equations.

    a. $5x + 7 = 22$ b. $9y - 8 = 1$ c. $4x + 12 = 32$ d. $8w + 2 = 26$ e. $4w + 5.2 = 28$

5. **WE9a** Solve the following equations by backtracking.

    a. $3(x + 7) = 24$ b. $2(x - 7) = 22$ c. $11(x + 5) = 99$ d. $6(x + 9) = 72$ e. $4(w + 5.2) = 26$

6. **WE9b** Solve the following equations by backtracking.

a. $\dfrac{x}{3} + 5 = 6$

b. $\dfrac{x}{9} - 2 = 3$

c. $\dfrac{x}{2} - 11 = 6$

d. $\dfrac{x}{7} - 5 = 6$

e. $\dfrac{x}{5} + 2.3 = 4.9$

Understanding

7. Solve the following equations by backtracking.

a. $\dfrac{x+4}{3} = 6$

b. $\dfrac{x-8}{7} = 10$

c. $\dfrac{x+11}{2} = 6$

d. $\dfrac{x-5}{7} = 0$

e. $\dfrac{x+2.21}{1.4} = 4.9$

8. Use backtracking to find the solution to the following equations.

a. $3x - 7 = 23$

b. $4(x+7) = 40$

c. $\dfrac{x}{5} - 2 = 8$

d. $5(x-3) = 15$

e. $6(x-4) = 18$

9. Use backtracking to find the solution to the following equations.

a. $\dfrac{x}{2.1} - 1.7 = 3.6$

b. $\dfrac{x+5}{3} - 3 = 7$

c. $4(x-2) + 5 = 21$

d. $3\left(\dfrac{x}{2} + 1\right) = 15$

e. $2(3x+4) - 5 = 15$

10. **WE10** Simplify and then solve the following equations by backtracking.

a. $2x + 7 + 3x + 5 = 27$

b. $3x + 9 + x - 4 = 17$

c. $3x + 5x + 2x = 40$

d. $6x + 6 - x - 4 = 37$

e. $7x - 4x + 8 - x = 10$

11. **MC** The correct flowchart required to solve the equation $6x + 5 = 8$ is:

A.

B.

C.

D.

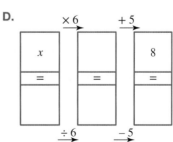

Communicating, reasoning and problem solving

12. Based on the scales in the diagram, identify the operations you need to apply to find the value of x. Calculate the value of x.

$$4x + 2 = 10$$

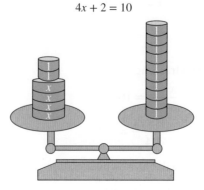

13. Chloe wrote the following explanation to solve the equation $3(x - 6) + 5 = 8$ by the backtracking method:
To solve this equation, work backwards and do things in the reverse order.
First add 5, then subtract 6, then finally multiply by 3.
Explain why her instructions are not correct.

14. The equation $4x - 20 = 8$ can be solved using backtracking by adding 20, then dividing by 4. Explain whether the equation could also be solved by dividing by 4 first.

15. Kevin is 5 years older than his brother Gareth. The sum of their ages is 31 years. Determine Gareth's age, letting x represent Gareth's age. Show your working.

16. The sum of three consecutive whole numbers is 51. Determine the numbers. (*Hint:* Let the smallest number equal x.)

17. Melanie and Callie went tenpin bowling together. Melanie scored 15 more pins than Callie. Their total score was 207. Determine what Callie scored.

18. The sum of three consecutive odd numbers is 27. Determine the 3 numbers.

19. The sum of 3 consecutive odd numbers is 39. Determine the 3 numbers.

20. In the high jump event Chris leapt 12 centimetres higher than Tim. Their two jumps made a total of 3 metres. Determine how high Chris jumped.

21. In three basketball games Karina has averaged 12 points each game. In the first game she scored 11 points. In the second she scored 17 points. In the third game she scored x points.

  a. From the given information, calculate the average of $11, 17$ and x.
  b. Write an equation using the answer to part **a**.
  c. Solve the equation.
  d. State how many points Karina scored in the third game.

22. Three consecutive multiples of 5 add up to 90. Determine the 3 numbers.

23. David is 5 years younger than his twin brothers. If the sum of their ages is 52, then how old is David?

24. Using the six consecutive numbers from 4 to 9, complete the magic square shown so that each row, column and diagonal totals 15.

	1	
3		
		2

LESSON
8.5 Solving linear equations using inverse operations

▶ 8.5.1 Keeping equations balanced
eles-4019

- A balance scale can be used to show whether things have equal mass.

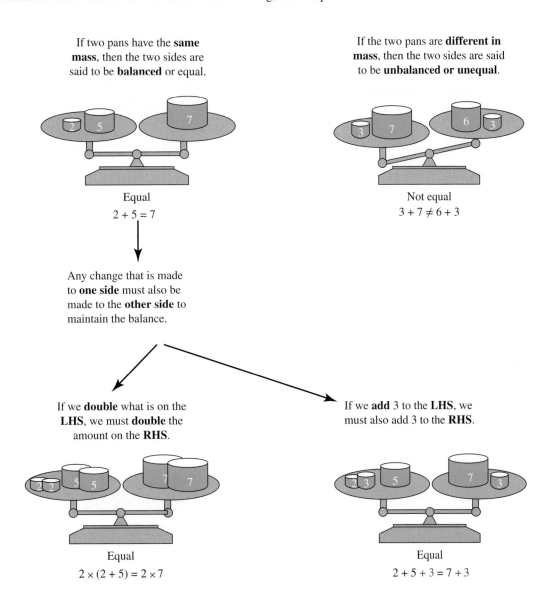

If two pans have the **same mass**, then the two sides are said to be **balanced** or equal.

Equal
$2 + 5 = 7$

If the two pans are **different in mass**, then the two sides are said to be **unbalanced or unequal**.

Not equal
$3 + 7 \neq 6 + 3$

Any change that is made to **one side** must also be made to the **other side** to maintain the balance.

If we **double** what is on the **LHS**, we must **double** the amount on the **RHS**.

Equal
$2 \times (2 + 5) = 2 \times 7$

If we **add** 3 to the **LHS**, we must also add 3 to the **RHS**.

Equal
$2 + 5 + 3 = 7 + 3$

- The scale in this diagram can be described by the equation $x + 3 = 7$.
- Making changes to both sides of the scale lets you work out how much the bag weighs (the value of the pronumeral).
- If we remove 3 weights from each side of the scale, we can see that the bag x weighs the same as 4 weights. Therefore, $x = 4$ in the equation $x + 3 = 7$.

Equal

▶ 8.5.2 Solving equations using balancing

eles-4020

- The balancing method can be used to solve equations.
- The solution to the equation can be found by changing the weights of both sides of the scales evenly until one side contains the pronumeral only.

WORKED EXAMPLE 11 Solving an equation by keeping equations balanced

For the following balance scale:
a. write the equation represented by the scale
b. calculate the value of the pronumeral.

Equal

THINK

a. 1. Examine the balance scale. On the LHS there are 2 bags, each weighing h, and 2 weights outside the bags. On the RHS there are 8 weights.

2. Write this as an equation.

b. 1. Remove 2 weights from both sides so that the balance will be maintained (-2). This leaves 2 bags on the LHS and 6 weights on the RHS.

2. To work out the value of h (the value of 1 bag), divide both sides by 2 and then simplify.
Note: Because both bags are labelled the same, they weigh the same amount.

WRITE

a.

Equal

$2h + 2 = 8$

b.

Equal

$2h + 2 - 2 = 8 - 2$
$2h = 6$

Equal

$\dfrac{(2h)}{2} = \dfrac{6}{2}$

$h = 3$

Each bag is equivalent to 3 weights.
That means $h = 3$.

Explain how you would use the balance scale method to solve $3x + 1 = 2x + 6$. Discuss a real-life scenario to represent this.

⊳ 8.5.3 Solving equations using inverse operations

eles-4021

- Both backtracking and the balance method of solving equations use inverse operations to calculate the value of an unknown.

  For example, the equation $2q + 3 = 11$ can be solved as follows.

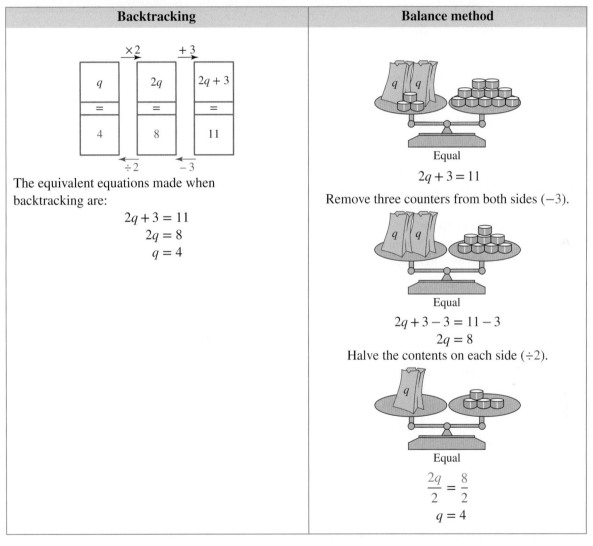

Backtracking	Balance method

Backtracking

The equivalent equations made when backtracking are:
$$2q + 3 = 11$$
$$2q = 8$$
$$q = 4$$

Balance method

Equal

$$2q + 3 = 11$$

Remove three counters from both sides (-3).

Equal

$$2q + 3 - 3 = 11 - 3$$
$$2q = 8$$

Halve the contents on each side $(\div 2)$.

Equal

$$\frac{2q}{2} = \frac{8}{2}$$
$$q = 4$$

- Both methods are equivalent, and use inverse operations to simplify the LHS of the equation until only the pronumeral remains.
- The operations that solve the equation and the order in which they are performed are the same.
- The equivalent equations created on the way to the solution are the same.
- The aim is to isolate the unknown value on one side of the equation.
- The last operation that was performed on the unknown value when building the equation is the first operation that is undone.

How to solve equations using inverse operations

To solve an equation using inverse operations:
1. determine the operations used to build the expression containing the pronumeral
2. apply the inverse operations in reverse order to both sides of the equation.

You can use a flowchart or a written list to determine the operations and their inverse.

Note: Building up expressions was covered in lesson 8.3.

WORKED EXAMPLE 12 Using inverse operations

Use inverse operations to solve the following equations.

a. $2y + 3 = 11$

b. $\dfrac{h}{5} + 1 = 3$

c. $2(k - 4) = 4$

THINK	WRITE
a. 1. Write the equation.	**a.** $2y + 3 = 11$
2. Note that on the LHS, y is multiplied by 2 and 3 is added to it. Apply inverse operations to: • remove 3 by subtracting 3 from both sides first • divide both sides by 2 to remove the 2 from y.	$2y + 3 - 3 = 11 - 3$ $2y = 8$ $\dfrac{2y}{2} = \dfrac{8}{2}$
3. Simplify.	$y = 4$
4. Write the solution.	The solution for the equation is $y = 4$.
b. 1. Write the equation.	**b.** $\dfrac{h}{5} + 1 = 3$
2. Note that on the LHS, h is divided by 5 and 1 is added to it. Apply inverse operations to: • remove 1 by subtracting 1 from both sides first • multiply both sides by 5 to remove the 5.	$\dfrac{h}{5} + 1 - 1 = 3 - 1$ $\dfrac{h}{5} = 2$ $\dfrac{h}{5} \times 5 = 2 \times 5$
3. Simplify.	$h = 10$
4. Write the solution.	The solution for the equation is $h = 10$.
c. 1. Write the equation.	**c.** $2(k - 4) = 4$
2. Note that on the LHS, 2 is multiplied by $k - 4$ and 4 is subtracted from k. Apply inverse operations to: • remove 2 by dividing both sides by 2 first • add 4 to both sides to remove the -4.	$\dfrac{2(k-4)}{2} = \dfrac{4}{2}$ $k - 4 = 2$ $k - 4 + 4 = 2 + 4$
3. Simplify.	$k = 6$
4. Write the solution.	The solution for the equation is $k = 6$.

• Not all solutions are positive integers. Equations can have decimal, fraction and even negative solutions too.

WORKED EXAMPLE 13 Solving equations with non-integer solutions

Solve the following equations.

a. $5d = 4$

b. $3c + 1 = 6$

THINK	WRITE
a. 1. Write the equation.	a. $5d = 4$
2. Note that on the LHS, d is multiplied by 5. Apply inverse operation to: • remove 5 by dividing both sides by 5.	$\dfrac{5d}{5} = \dfrac{4}{5}$
3. Simplify.	$d = \dfrac{4}{5}$
4. State the solution.	The solution for the equation is $d = \dfrac{4}{5}$.
b. 1. Write the equation.	b. $3c + 1 = 6$
2. Note that on the LHS, c is multiplied by 3 and 1 is added to it. Apply inverse operations to: • remove 1 by subtracting 1 from both sides first • divide both sides by 3 to remove 3.	$3c + 1 - 1 = 6 - 1$ $3c = 5$ $\dfrac{3c}{3} = \dfrac{5}{3}$
3. Simplify.	$c = \dfrac{5}{3}$
4. Write the solution. *Note:* We leave the solution in fraction form because it is a repeating decimal.	The solution for the equation is $c = \dfrac{5}{3}$ or $c = 1\dfrac{2}{3}$.

▶ 8.5.4 Solving equations using the cover-up method

eles-6121

- When an equation includes a pronumeral that is being subtracted (for example, $28 - 3x = 7$), an effective way to solve the equation is to use the cover-up method. This is a similar method to the guess, check and improve method used earlier in this topic.

WORKED EXAMPLE 14 Solving equations using the cover-up method

Solve the following equations.

a. $5 - v = 2$

b. $17 - 2a = 11$

THINK	WRITE
a. 1. Cover up the variable with your hand.	$5 - v = 2$
2. Think '5 minus what equals 2?'	$5 - \text{✋} = 2$
3. The missing value must be 3.	$\text{✋} = 3$
4. State your answer.	$v = 3$

b. 1. Cover up the variable with your hand.

$17 - 2a = 11$

2. Think '17 minus what equals 11?'

$17 - \text{✋} = 11$

3. The covered variable must have a value of 6.

$\text{✋} = 6$

4. Think '2 multiplied by what equals 6?'

$2a = 6$

5. State your answer.

$a = 3$

COMMUNICATING — COLLABORATIVE TASK: How many was that?

Work in pairs to answer the following questions and then share your ideas with others in your class.

1. A farmer told two friends that they could pick peaches from his tree but that they must not take more than 30 peaches each. They worked for a while, and then the taller one asked her friend, 'Have you picked your limit yet?'
He replied, 'Not yet, but if I had twice as many as I have now, plus half as many as I have now, I would have reached my limit.'
How many peaches did he have?

2. Your aunt from overseas sent an enormous box of chocolates for your mother's birthday. You were told not to eat any of the chocolates. Nevertheless, over the next five days you sneaked into the pantry when your mother was out and ate some of the chocolates. Each day you ate six more than the day before. After five days you were caught and your mother found out that you had eaten 100 chocolates. Your mother was extremely cross!
How many chocolates did you eat on *each* of the five days?

---

on Resources

Interactivities   Keeping equations balanced (int-4047)
Equations with rational number solutions (int-4048)
Negative integers (int-4049)

Exercise 8.5 Solving equations using inverse operations

learn on

Individual pathways

■ PRACTISE	■ CONSOLIDATE	■ MASTER
1, 2, 5, 8, 13, 16	3, 6, 9, 12, 14, 17	4, 7, 10, 11, 15, 18, 19, 20

Fluency

1. **WE11** For each of the following balance scales:

   i. write the equation represented by the scale

   ii. calculate the value of the pronumeral.

   a.

   Equal

   b.

   Equal

   c.

   Equal

2. **WE12** Use inverse operations to solve the following.

   a. $3g + 7 = 10$
   b. $4m - 6 = 6$
   c. $\dfrac{q}{3} + 8 = 11$

3. Use inverse operations to solve the following.

   a. $5b - 2 = 8$
   b. $2z + 13 = 37$
   c. $\dfrac{v}{4} + 31 = 59$

4. Use inverse operations to solve the following.

   a. $2(n - 5) = 8$
   b. $\dfrac{g - 2}{4} = 3$
   c. $4(y + 1) = 16$

5. **WE13** Solve the following equations.

   a. $3h = 7$
   b. $2k = 5$
   c. $2w + 1 = 8$

6. Solve the following equations.

   a. $3t - 4 = 9$
   b. $6h - 3 = 10$
   c. $3l + 4 = 8$

7. Solve the following equations.

   a. $5g + 3 = 11$
   b. $3h - 10 = 1$
   c. $8n - 4 = 0$

Understanding

8. **WE14** Solve the following equations.

   a. $6 - m = 2$
   b. $4 - d = 1$
   c. $12 - 3v = 6$

9. Solve the following equations.

   a. $13 - 2s = 7$
   b. $19 - 3g = 4$
   c. $30 - 5k = 20$

10. Solve the following equations.

   a. $2(x + 3) = 11$
   b. $p + 3 = 6\dfrac{1}{4}$
   c. $7 - x = 5$

11. Solve the following equations.

   a. $3(5 - y) = 12$
   b. $5(8 - 2h) = 15$
   c. $2(x + 3) = 7$

12. The formula $C = \dfrac{5(F-32)}{9}$ is used to convert degrees Fahrenheit to degrees Celsius. Use the formula to find:

   a. 45° Fahrenheit in degrees Celsius   **b.** 45° Celsius in degrees Fahrenheit.

Communicating, reasoning and problem solving

13. A taxi company charges a $3.60 flag fall (a fixed fee before the journey starts). An additional $2.19 per km is charged for the journey.

   a. Calculate the cost of a 4.6 km journey.
   b. If the journey cost $110.91, determine the distance travelled.

14. While shopping for music online, Olivia found an album she liked. She could buy the whole album for $17.95 or buy songs from the album for $1.69 each. Determine the number of individual songs Olivia could buy to make it cheaper to buy the whole album.

15. Explain why the order of inverse operations is important, using the equation $2x + 3 = 15$ as an example. In your explanation, use balanced scales to represent the equation.

16. A class of 25 students has 7 more boys than girls. Determine how many boys there are.

17. When 12 is subtracted from 7 times a number, the result is 9. Create an equation that represents this statement, then solve your equation to determine the number.

18. Given that the perimeter of the following triangle is 22 cm, calculate the value of x.

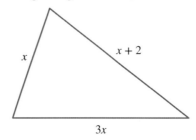

19. A number x becomes the fifth number in the set of numbers $\{3, 7, 9, 13\}$ to make the mean of the five numbers equal to their median. Determine all possible positive values for x.

20. These scales show that two bricks are equal to 2 kilograms plus half a brick. Determine the weight of each brick.

LESSON
8.6 Verifying solutions

▶ 8.6.1 Verifying solutions to equations by substitution

eles-4023
- Once the solution to an equation has been determined, the method of **substitution** can be used to verify whether that solution is correct.
- Verifying solutions to equations by substitution involves:
  - replacing the pronumeral with the numerical value of the solution
  - simplifying the equation to determine the value of the LHS and RHS of the equation.
- If the LHS = RHS after substitution, then the solution is correct.

WORKED EXAMPLE 15 Verifying solutions by substitution

For each of the following equations, use substitution to determine whether $x = 7$ is the solution to the equation.

a. $\dfrac{x+5}{3} = 4$

b. $2x - 8 = 10$

THINK	WRITE
a. 1. Write the equation.	a. $\dfrac{x+5}{3} = 4$
2. Write the LHS of the equation and substitute $x = 7$.	If $x = 7$, LHS $= \dfrac{x+5}{3}$ $= \dfrac{7+5}{3}$
3. Perform the calculation.	$= \dfrac{12}{3}$ $= 4$
4. Compare the LHS with the RHS of the equation.	LHS $= 4$, RHS $= 4$
5. Comment on the answer.	Since LHS $=$ RHS, $x = 7$ is the solution.
b. 1. Write the equation.	b. $2x - 8 = 10$
2. Write the LHS of the equation and substitute $x = 7$.	If $x = 7$, LHS $= 2x - 8$ $= 2(7) - 8$ $= 14 - 8$
3. Perform the calculation.	$= 6$
4. Compare the LHS with the RHS of the equation.	LHS $= 6$, RHS $= 10$
5. Comment on the answer.	$x = 7$ is not the solution, since LHS $\neq$ RHS.

Digital technology

Scientific calculators can be used to check solutions by substitution. Replace pronumerals with the desired value in brackets.

For example, the solution to the equation $3x - 15 = 12$ is $x = 9$.

$$3(9)-15 \qquad \boxed{D} \qquad \text{Math} \ \blacktriangle$$

$$12$$

WORKED EXAMPLE 16 Verifying solutions by substitution

A solution is given for each equation below. Verify whether the solution is correct.

a. $\dfrac{x+2}{3} = 2x - 12, \quad x = 10$

b. $3x - 7 = 2x + 3, \quad x = 10$

THINK	WRITE
a. 1. Write the equation.	a. $\dfrac{x+2}{3} = 2x - 12$
2. Write the LHS of the equation and substitute $x = 10$.	If $x = 10$, LHS $= \dfrac{x+2}{3}$
	$= \dfrac{10+2}{3}$
3. Perform the calculation.	$= \dfrac{12}{3}$
	$= 4$
4. Write the RHS of the equation and substitute $x = 10$.	If $x = 10$, RHS $= 2x - 12$
	$= 2(10) - 12$
5. Perform the calculation.	$= 20 - 12 = 8$
6. Comment on the answer.	$x = 10$ is not the solution, since LHS $\neq$ RHS.
b. 1. Write the equation.	b. $3x - 7 = 2x + 3$
2. Write the LHS of the equation and substitute $x = 10$.	If $x = 10$, LHS $= 3x - 7$
	$= 3(10) - 7$
3. Perform the calculation.	$= 30 - 7 = 23$
4. Write the RHS of the equation and substitute $x = 10$.	If $x = 10$, RHS $= 2x + 3$
	$= 2(10) + 3$
	$= 20 + 3$
5. Perform the calculation.	$= 23$
6. Comment on the answer.	Since LHS $=$ RHS, $x = 10$ is the correct solution.

 Resources

Interactivity Checking solutions (int-4046)

Individual pathways

■ PRACTISE	■ CONSOLIDATE	■ MASTER
1, 4, 6, 10, 13	2, 7, 8, 11, 14	3, 5, 9, 12, 15

Fluency

1. **WE15** Use substitution to determine whether each of the following is correct.
   a. $x = 3$ is the solution to the equation $x + 2 = 6$
   b. $x = 3$ is the solution to the equation $2x - 1 = 5$
   c. $x = 4$ is the solution to the equation $6x - 6 = 24$

2. Use substitution to determine whether each of the following is correct.
   a. $x = 5$ is the solution to the equation $4(x - 3) = 8$
   b. $x = 7$ is the solution to the equation $3(x - 2) = 25$
   c. $x = 8$ is the solution to the equation $5(x + 1) = 90$

3. Use substitution to determine whether each of the following is correct.
   a. $x = 81$ is the solution to the equation $3x - 53 = 80$
   b. $x = 2.36$ is the solution to the equation $5x - 7 = 4.8$
   c. $x = 4.4$ is the solution to the equation $7x - 2.15 = 18.64$.

4. **WE16** Verify whether the solution given for each of the following equations is correct.
   a. $5x + 1 = 2x - 7, x = 8$
   b. $5x = 2x + 12, x = 4$
   c. $4x = 3x + 8, x = 8$

5. Verify whether the solution given for each of the following equations is correct.
   a. $3x - 1.2 = x + 2.9, x = 1.9$
   b. $6x + 1.5 = 2x + 41.5, x = 10$
   c. $1.2(x + 1.65) = 0.2(x + 9.85), x = 0.45$

Understanding

6. Complete the following table to determine the value of $2x + 3$ when $x = 0, 1, 2, 3, 4$.

x	0	1	2	3	4
$2x + 3$					

   a. Identify the solution (that is, the value of x) for $2x + 3 = 11$.
   b. Identify the solution (that is, the value of x) for $2x + 3 = 5$.

7. Complete the following table to determine the value of $5(x - 2)$ when $x = 2, 3, 4, 5, 6$.

x	2	3	4	5	6
$5(x - 2)$					

   a. Identify the solution (that is, the value of x) for $5(x - 2) = 10$.
   b. Identify the solution (that is, the value of x) for $5(x - 2) = 20$.
   c. Guess the solution (that is, the value of x) to $5(x - 2) = 30$. Check your guess.

8. a. Copy and complete the following table.

x	$2x+1$	$3x-5$
3		
4		
5		10
6		
7		

b. Identify the solution to $2x+1=3x-5$.

9. a. Copy and complete the following table.

x	$\dfrac{x+3}{2}$	$2x-6$
3	3	
5		
7		8
9		
11		

b. Identify the solution to $\dfrac{x+3}{2}=2x-6$.

Communicating, reasoning and problem solving

10. Substitution is a method used to check that a solution to an equation is correct. Explain how you could use substitution to answer a multiple choice question that has potential solutions as the options.

11. Consider the diagram shown.

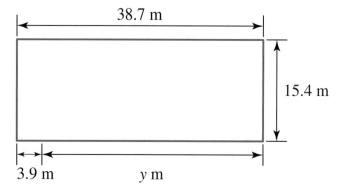

a. Explain whether you need to know both the length and width of this rectangle to be able to calculate the value of y.

b. Calculate the value of y.

c. Explain how you would check the solution.

12. Carol is making a quilt for her granddaughter. The quilt pattern requires that $\frac{1}{7}$ of the quilt is made of a pink fabric. Carol has $0.5\,\text{m}^2$ of pink fabric. She intends to use all the pink material in the quilt.
Calculate how much more fabric she needs for the entire quilt.

   a. Write an equation for this problem.
   b. Calculate the value of the unknown variable.
   c. Check your solution.

13. James and Alison share $125 between them, but Alison gets $19 more than James. Create an equation to describe this situation. Calculate how much money James and Alison each receive.

14. Aiko and Miyu are sisters who receive pocket money based on their age. Their parents pay them according to the equation $\$P = 3x - 25$, where $\$P$ represents their weekly pocket money and x is their age in years.

   a. Determine at what age Aiko and Miyu start receiving pocket money.
   b. Miyu is 2 years older than Aiko. Determine how much more pocket money Miyu receives each week compared to her sister. (Assume the sisters are both old enough to be receiving pocket money.)
   c. When Aiko is 11 years old, calculate how much pocket money each child receives.

15. Ivan is trying to find the solution to the equation $n + (n + 1) + (n + 2) = 393$. He believes the solution is $n = 133$.

   a. Show that Ivan is incorrect.
   b. Determine the correct solution, and prove that it is the solution using substitution.
   c. Write a worded problem that this equation represents.

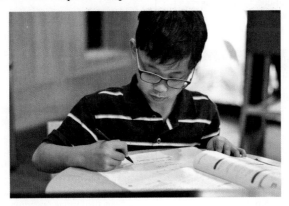

LESSON
8.7 Review

8.7.1 Topic summary

What is an equation?

- An equation is a mathematical statement that tells us that two expressions are equal.
  e.g. $7x + 9 = 23$ is an example of an equation.
- The left-hand side (LHS) is $7x + 9$.
- The right-hand side (RHS) is 23.
- This equation states that the LHS and the RHS are equal (have the same value).

EQUATIONS

Developing an equation from words

- When developing an equation from words, some English phrases or words can be turned into mathematical symbols.
  e.g.
  - Words such as *sum*, *more than*, *increased*, *add* or *added* imply addition and can be replaced with $+$.
  - Words such as *difference*, *less than*, *decreased* or *minus* imply subtraction and can be replaced with $-$.
  - Words such as *product* or *times* imply multiplication and can be replaced with $\times$.
  - Words such as *quotient* or *divide* imply division and can be replaced with $\div$.

Solving equations

- Solving an equation is the process of finding the value(s) of the pronumeral to make the equation true. e.g. The equation $7x + 9 = 23$ is true only when $x = 2$. Hence, the solution is $x = 2$.
- Equations can be solved using many different strategies such as the guess-and-check, balancing, cover-up and inverse methods.

Building expressions using flowcharts

- Flowcharts can be used to build an expression from words, steps, instructions or operations.
- In a flowchart, the starting number or position is the *input* and the final number or position is the *output*.
  e.g.

Input   Output

Solving equations using guess, check and improve

- Solving equations using the guess, check and improve method involves selecting a number for the pronumeral that you think might be a solution to the equation.
- Use substitution to check whether your guess is correct or not.
- After checking your guess, make improvements to your chosen number by either increasing or decreasing it until you find the solution.
  e.g. For $2x + 5 = 11$
  Guess: $x = 2$
  Check: LHS $= 2(2) + 5 = 4 + 5 = 9 \neq$ RHS
  Improve: Try $x = 3$
  LHS $2(3) + 5 = 6 + 5 = 11 =$ RHS
  Solution is $x = 3$.

Inverse operations

- Inverse (or opposite) operations are often used to help solve equations.
  $+$ is the inverse operation of $-$
  $-$ is the inverse operation of $+$
  $\times$ is the inverse operation of $\div$
  $\div$ is the inverse operation of $\times$

Solving equations using backtracking

- Backtracking uses inverse operations to determine the solution of an equation by moving backwards through a flowchart.
  e.g. Solving $7x + 9 = 23$ using backtracking gives:

Solution is $x = 2$.

Solving equations using balancing

- By performing the same operation on *both* sides of an equation, it remains balanced.
- Equations can be solved using inverse operations with balancing, that is, applying an inverse operation to both sides of the equation.
  e.g. $x - 20 = 12$
  $x - 20 + 20 = 12 + 20$
  $x = 32$

8.7.2 Project

Equations at the Olympic Games

The Olympic Games is held every four years between competing nations from all over the world. You might know that Sydney was the host city for the 2000 Olympic Games. A feature of this international sporting event is that records continue to be broken. Every Olympic Games sees competitors run faster, lift heavier weights, and so on. The desire to become better urges competitors to train harder in the hope that they will become record holders.

This project looks at running times for the men's and women's 100-metre running events. The running times for these events can be approximated using the following equations:

- Men's 100-metre event $\quad t = -0.0094y + 28.73$
- Women's 100-metre event $\quad t = -0.0173y + 45.31$

where t represents the running time (in seconds) and y represents the year.

For example, the men's time in 1996 can be approximated by substituting 1996 into the equation as shown.

$$t = -0.0094y + 28.73$$
$$= -0.0094 \times 1996 + 28.73$$
$$= 9.9676 \text{ seconds}$$

1. Use the equations provided to calculate the men's and women's running times for the 100-metre event in the following years.
   a. 1928　　　　b. 1968　　　　c. 1988　　　　d. 2016

2. The following table shows the winners and their running times for the 100-metre event final at four Olympic Games events. How well do these actual results compare with the times you calculated using the equations? What does this say about the equations?

Year	Men's winner	Time taken (seconds)	Women's winner	Time taken (seconds)
1928	Percy Williams	10.80	Elizabeth Robinson	12.20
1968	James Hines	9.95	Wyomia Tyus	11.08
1988	Carl Lewis	9.92	Florence Griffith Joyner	10.54
2016	Usain Bolt	9.81	Elaine Thompson	10.71

3. Predict the times for both men and women at the 2024 Olympic Games.

When making predictions about the future, we must remember thatthese predictions are based on the assumption that the patterns weobserve now will continue into the future.

Your answers to question 1 show you that men have historically runfaster times than women in the 100-metre event. However, closerinspection of the times shows that women are making greaterimprovements in their times over the years.

4. If running times continue to follow these patterns in future Olympic Games, decide whether it is possible that women's times will become equal to men's times. Discuss your answer.
5. Suggest reasons that running times are getting shorter. Do you think they will follow this pattern forever?

on Resources

Interactivities Crossword (int-2606)
Sudoku puzzle (int-3173)

Fluency

1. Solve these equations by inspection.

   a. $m + 7 = 12$
   b. $5h = 30$
   c. $s - 12 = 7$
   d. $\dfrac{d}{5} = 4$

2. Use guess, check and improve to find two numbers whose sum and product are given.

   a. Sum = 83, product = 1632
   b. Sum = 86, product = 1593

3. Calculate the output number for each of the following flowcharts.

   a.

   b.

   c.

   d.

4. Use backtracking and inverse operations to calculate the input number for each of the following flowcharts.

   a.

   b.

   c.

   d.

5. Build up an expression by following the instructions on the flowchart.

   a.

   b.

   c.

   d.

6. Build up an expression by following the instructions on the flowchart.

   a.

   b.

Understanding

7. Draw a flowchart that has an input of x and an output that is given by the expressions below.

   a. $5(x + 7)$
   b. $\dfrac{x}{4} - 3$
   c. $6x - 14$
   d. $\dfrac{x + 2}{5}$

8. Draw a flowchart and use backtracking to calculate the solution to the following equations.

   a. $7x + 6 = 20$

   b. $9(y - 8) = 18$

   c. $\dfrac{m}{5} - 3 = 9$

   d. $\dfrac{s + 7}{5} = 5$

9. Use backtracking to determine the solution to these equations.

   a. $3(d + 1) = 15$

   b. $\dfrac{t}{4} - 11 = 14$

   c. $6d - 3 = 15$

   d. $\dfrac{a + 6}{4} = 3$

10. Simplify the expression and then solve the equation for each of the following.

    a. $7v + 3 + 3v + 4 = 37$

    b. $6c + 15 - 5c - 8 = 19$

11. For each of the following equations there is a solution given. Is the solution correct?

    a. $5x - 7 = 2x + 2, x = 3$

    b. $\dfrac{x + 9}{2} = 2x - 7, x = 5$

12. For each of the following balance scales:
    i. write the equation represented by the scale
    ii. calculate the value of the pronumeral.

    a.

    Equal

    b.

    Equal

    c.

    Equal

13. Use inverse operations to solve the following equations.

    a. $2v + 1 = 7$

    b. $\dfrac{x}{2} + 3 = 4$

    c. $4w - 3 = 9$

    d. $2(g + 1) = 10$

14. Solve the following equations. Make sure you verify your solutions.

    a. $4k + 3 = 10$

    b. $2p - 3 = 8$

    c. $3q + 6 = 7$

    d. $5t + 1 = 3$

15. Solve the following equations. Make sure you verify your solutions.

    a. $3 - x = 1$

    b. $7 - 2p = 3$

    c. $10 - 4r = 2$

    d. $13 - 3z = 7$

Communicating, reasoning and problem solving

16. Sophie and Jackie each have a collection of football cards. Jackie has 5 more cards than Sophie. Together Jackie and Sophie have 67 cards. Create and solve an equation that shows how many cards Sophie owns.

17. Andreas has completed 2 more pieces of homework than Richard, who submitted x pieces of homework for the semester. If the total number of pieces of homework submitted by the two boys is 12, determine how many pieces of homework Andreas submitted.

18. Keith is 6 years younger than his twin brothers. If the sum of the brothers' ages is 48, determine Keith's age.

19. Margaret bought six 2-hour tickets and four daily tickets for $28. If the cost of one daily ticket is $4, write and solve an equation to calculate the cost of a 2-hour ticket.

20. Jetski hire is $10 per hour, plus a $30 deposit.
    a. Calculate the cost of hiring a jetski for:

       i. 1 hour   ii. 2 hours   iii. 3 hours.

    b. Write a rule that could be used to calculate the cost of hiring a jetski for h hours.
    c. Use your rule to calculate the cost of hiring a jetski for 8 hours.
    d. You have $85 to spend. Write an equation to help you work out how many hours of jetski hire you can afford. Solve this equation.
    e. Work out how much money (if any) you would have left over after you paid the hire charge and hire the jetski for 5 hours.
    f. After spending 5 hours on the water you return the jetski with no damage. Explain whether you would have enough money for a hamburger on the way home.

21. Suppose it costs $30 for an adult and $15 for a child to enter the Sydney Royal Easter Show.
    a. Calculate the entry cost for two adults and three children.
    b. Calculate the entry cost for one adult and five children.
    c. Determine the maximum number of children who could get into the Show for $100.
    d. Identify the possible combinations of adults and children that could get in for exactly $300.

22. Judy is watching her daughter play in the park. She notices that some children are riding tricycles and some are riding bicycles. Altogether, 19 children are on cycles. She counts that there are 43 wheels on the cycles. Determine how many children are on tricycles and how many are on bicycles.

23. Two angles in a triangle have the same magnitude. The sum of the magnitudes of these angles is the magnitude of the third angle. Determine the magnitudes of the angles and use this information to describe the triangle in this scenario.

24. A stage screen is 4 times as long as it is wide. If it were 5 metres wider and 4 metres shorter it would be a square. Determine the dimensions of the stage screen.

25. The largest angle in a triangle is 65 degrees more than the smallest angle. The third angle is 10 degrees more than the smallest angle. Calculate the size of the smallest angle.

on To test your understanding and knowledge of this topic, go to your learnON title at www.jacplus.com.au and complete the **post-test**.

Answers

Topic 8 Equations

8.1 Pre-test

1. $6x = 72$
2. 32
3. 11 games
4. 4
5. C
6. 4
7. 32
8. A
9. B
10. 12 years old
11. $x = 3$
12. $x = 10$
13. 26
14. 16 boys
15. 5.5 cm

8.2 Introduction to equations

1. a. $x + 7 = 11$ b. $x + 12 = 12$
   c. $x - 7 = 1$ d. $x - 4 = 7$
2. a. $2x = 12$ b. $5x = 30$
   c. $\dfrac{x}{7} = 1$ d. $\dfrac{x}{5} = 2$
3. a. $15 - x = 2$ b. $52 - x = 8$
   c. $\dfrac{21}{x} = 7$ d. $x^2 = 100$
4. a. $x = 11$ b. $y = 9$
   c. $m = 5$ d. $m = 30$
5. a. $k = 0$ b. $b = 7$
   c. $b = 8.8$ d. $c = 4.2$
   e. $x = 2.8$
6. a. $x = 6$ b. $x = 3$ c. $x = 10$
7. a. $x = 9$ b. $x = 6$ c. $x = 4$
8. a. 7 and 14   b. 11 and 15
   c. 7 and 47   d. 42 and 136
9. a. 42 and 111   b. 97 and 145
   c. 2.2 and 3.9   d. 347 and 631

10.

x	$x^2 + 4$	$4x + 1$
0	4	1
1	5	5
2	8	9
3	13	13
4	20	17

$x = 1$ and $x = 3$ are both solutions.

11. Think of the 8 times table. Determine which multiple of 8 is equal to 48. $6 \times 8 = 48$, so the answer is $x = 6$.

12. 34 does not appear in the 5 times table. Therefore the solution will not be a whole number.
13. Greater than 5
14. 10
15. 5 metres, 15 metres
16. 24 years old
17. Angus will be 27 and his grandfather will be 81.

8.3 Building up expressions and backtracking

1. a.

   b.

   c.

   d.

2. a.

   b.

   c.

   d.

3. a.

   b.

   c.

   d.

12. a.

b.

c.

d.

13. -3 should be performed first, then $\times 2$.

$$x \xrightarrow{+8} x+8 \xrightarrow{\times 5} 5(x+8) \xrightarrow{\div 4} \frac{5(x+8)}{4} = 30$$

14. $\dfrac{5(x+8)}{4} = 30$

15. $5n + 15 = 4n - 3$

16. 22 years old

17. The two numbers are 19 and 13.

18. $1215

19. Some examples are:
$12 + 3 - 4 + 5 + 67 + 8 + 9 = 100$
$123 + 4 - 5 + 67 - 89 = 100$
$123 - 45 - 67 + 89 = 100$

8.4 Solving linear equations using backtracking

1. a. 2   **b.** 4   **c.** 8   **d.** 13   **e.** 10

2. a. 67   **b.** 88   **c.** 5   **d.** 8   **e.** 5.3

3. a. $4\dfrac{2}{5}$ **b.** $\dfrac{31}{36}$

4. a. $x = 3$ **b.** $y = 1$ **c.** $x = 5$
   d. $w = 3$ **e.** $w = 5.7$

5. a. $x = 1$ **b.** $x = 18$ **c.** $x = 4$
   d. $x = 3$ **e.** $w = 1.3$

6. a. $x = 3$ **b.** $x = 45$ **c.** $x = 34$
   d. $x = 77$ **e.** $x = 13$

7. a. $x = 14$ **b.** $x = 78$ **c.** $x = 1$
   d. $x = 5$ **e.** $x = 4.65$

8. a. $x = 10$ **b.** $x = 3$ **c.** $x = 50$
   d. $x = 6$ **e.** $x = 7$

9. a. $x = 11.13$ **b.** $x = 25$ **c.** $x = 6$
   d. $x = 8$ **e.** $x = 2$

10. a. $x = 3$ **b.** $x = 3$ **c.** $x = 4$
   d. $x = 7$ **e.** $x = 1$

11. A

12. First subtract 2 from both sides, then divide both sides by 4; $x = 2$.

13. Inverse operations must be performed in the reverse order. The forward order is: First subtract 6, then multiply by 3 and finally add 5. The reverse order is: Subtract 5, divide by 3, then add 6.

14. Yes. Dividing both sides by 4 gives $x - 5 = 2$. By then adding 5 to both sides, the solution $x = 7$ is obtained.

15. Gareth is 13 years old.

16. 16, 17 and 18

17. Callie scored 96 pins.

18. 7, 9 and 11

19. 11, 13 and 15

20. Chris jumped 156 centimetres.

21. a. 12

   b. $\dfrac{x + 28}{3} = 12$

   c. $x = 8$

   d. Karina scored 8 points in the third game.

22. 25, 30 and 35

23. David is 14 years old.

24.

8	1	6
3	5	7
4	9	2

8.5 Solving linear equations using inverse operations

1. a. i. $4r = 12$ ii. $r = 3$
   b. i. $2n + 3 = 7$ ii. $n = 2$
   c. i. $3t + 5 = 8$ ii. $t = 1$

2. a. $g = 1$ **b.** $m = 3$ **c.** $q = 9$

3. a. $b = 2$ **b.** $z = 12$ **c.** $v = 112$

4. a. $n = 9$ **b.** $g = 14$ **c.** $y = 3$

5. a. $h = 2\dfrac{1}{3}$ **b.** $k = 2\dfrac{1}{2}$ **c.** $w = 3\dfrac{1}{2}$

6. a. $t = 4\dfrac{1}{3}$ **b.** $h = 2\dfrac{1}{6}$ **c.** $l = 1\dfrac{1}{3}$

7. a. $g = \dfrac{8}{5}$ **b.** $h = \dfrac{11}{3}$ **c.** $n = \dfrac{1}{2}$

8. a. $m = 4$ **b.** $d = 3$ **c.** $v = 2$

9. a. $s = 3$ **b.** $g = 5$ **c.** $k = 2$

10. a. $x = 2\dfrac{1}{2}$ **b.** $p = 3\dfrac{1}{4}$ **c.** $x = 2$

11. a. $y = 1$ **b.** $h = 2\dfrac{1}{2}$ **c.** $x = \dfrac{1}{2}$

12. a. $7\dfrac{2}{9}\,°\text{C}$ **b.** $113\,°\text{F}$

13. a. $13.67   b. 49 km

14. 11 songs

15. The removal of the three weights from the right-hand side needs to be done before the weights on each side are halved. If the division is performed first, the odd numbers of weights on each side of the scales would have to be halved, which will not result in whole numbers.

16. 16 boys

17. 3

18. 4 cm

19. 3, 8, 13

20. $1\frac{1}{3}$ kilograms

8.6 Verifying solutions

1. a. No   b. Yes   c. No

2. a. Yes   b. No   c. No

3. a. No   b. Yes   c. No

4. a. No   b. Yes   c. Yes

5. a. No   b. Yes   c. No

6.

x	0	1	2	3	4
$2x+3$	3	5	7	9	11

a. $x=4$ b. $x=1$

7.

x	2	3	4	5	6
$5(x-2)$	0	5	10	15	20

a. $x=4$ b. $x=6$ c. $x=8$

8. a.

x	$2x+1$	$3x-5$
3	7	4
4	9	7
5	11	10
6	13	13
7	15	16

b. $x=6$

9. a.

x	$\dfrac{x+3}{2}$	$2x-6$
3	3	0
5	4	4
7	5	8
9	6	12
11	7	16

b. $x=5$

10. You could substitute all of the options into the equation and see which one makes that equation true.

11. a. No. Length is enough.
    b. 34.8 m
    c. Adding 3.9 and the value of y. The answer should be 38.7 m.

12. a. $\frac{1}{7}x=0.5$ b. $x=3.5$

13. James $53 and Alison $72

14. a. 9
    b. $6
    c. Aiko $8, Miyu $14

15. a. $133+(133+1)+(133+2)=402$
    $\neq 393$
    b. $n=130$, $130+(130+1)+(130+2)=393$
    c. Sample response: The sum of three consecutive numbers is 393. What is the smallest of these numbers?

Project

1. Men's: 1928, 10.607 s; 1968, 10.231 s; 1988, 10.043 s; 2016, 9.780 s
   Women's: 1928, 11.956 s; 1968, 11.264 s; 1988, 10.918 s; 2016, 10.433 s

2. All calculated times are well within 0.4 seconds of the actual running times.

3. Using the formula: Men's time $=9.7044$ s; Women's time $=10.2948$ s.

4. Women's times are coming down faster than men's. It is possible only if these patterns continue.

5. Different training programs, fitter athletes, better shoes, etc. There must be some levelling out, as the times cannot keep coming down forever – it will be impossible to run 100 metres in 0 seconds.

8.7 Review questions

1. a. $m=5$
   b. $h=6$
   c. $s=19$
   d. $d=20$

2. a. 32, 51   b. 27, 59

3. a. 11
   b. 2
   c. 16
   d. 35

4. a. 16   b. 42   c. 11   d. 8

5. a. $7x+8$
   b. $\frac{x}{3}-5$
   c. $6(x+2)$
   d. $\frac{x-7}{5}$

6. a. $5(x+3)-9$ b. $\dfrac{4x+11}{8}$

7. a.

b.

c.

d.

$$x \xrightarrow{+2} x+2 \xrightarrow{\div 5} \dfrac{x+2}{5}$$

8. a. $x = 2$ **b.** $y = 10$ **c.** $m = 60$ **d.** $s = 18$

9. a. $d = 4$ **b.** $t = 100$ **c.** $d = 3$ **d.** $a = 6$

10. a. $v = 3$ **b.** $c = 12$

11. a. Yes   **b.** No

12. a.   i. $4x + 1 = 5$
     ii. $x = 1$
  b.   i. $3p + 5 = 11$
     ii. $p = 2$
  c.   i. $2q + 3 = 13$
     ii. $q = 5$

13. a. $v = 3$ **b.** $x = 2$ **c.** $w = 3$ **d.** $g = 4$

14. a. $k = 1\dfrac{3}{4}$ **b.** $p = 5\dfrac{1}{2}$ **c.** $q = \dfrac{1}{3}$ **d.** $t = \dfrac{2}{5}$

15. a. $x = 2$ **b.** $p = 2$ **c.** $r = 2$ **d.** $z = 2$

16. Sophie owns 31 cards.

17. 7 pieces of homework

18. 12 years old

19. $2

20. a.   i. $40   ii. $50   iii. $60
  b. Cost $= 30 + 10h$
  c. $110
  d. $85 = 30 + 10h; h = 5.5$ hours
  e. $5
  f. Your deposit is returned ($30), which means you could buy a hamburger on the way home.

21. a. $105
  b. $105
  c. 6 children
  d. 10 adults; 9 adults and 2 children; 8 adults and 4 children; 7 adults and 6 children; 6 adults and 8 children; 5 adults and 10 children; 4 adults and 12 children; 3 adults and 14 children; 2 adults and 16 children; 1 adult and 18 children; 20 children

22. 14 bicycles and 5 tricycles

23. 45°, 45°, 90°. This is a right-angled isosceles triangle.

24. Length 12 m, width 3 m

25. 35°

9 Linear relationships

LESSON SEQUENCE

LESSON
9.1 Overview

Why learn this?

Using a coordinate system enables us to pinpoint locations in different settings. When you go to the movies, a musical, a concert or a sporting grand final, your ticket shows where you will be seated. For example, a ticket to a concert may give the location of your seat as Section 47, Row G, Seat 5. A boarding pass on a flight always shows where you will be sitting. For example, a boarding pass that shows seat 30A means you will be seated in row 30, seat A, by the window.

Have you ever wondered where you are on a map, or globe, of the world? Sydney is 33.87°S, 151.21°E, which means it is in the southern hemisphere, 33.87° south of the equator and 151.21° east of the Royal Observatory in Greenwich, England. New York is 40.71°N, 74.01°W, so it is in the northern hemisphere and west of Greenwich. In coordinate geometry, we define a point by how far it is horizontally and vertically from the centre point. Scientists and geographers often use coordinates when they analyse data. Reading and interpreting maps involves knowing about coordinates. Computer programmers use coordinates when creating artwork. A computer screen is made up of tiny points of light called pixels. Programmers use coordinates to identify which pixels to light up when creating art. Understanding coordinates, and maps, assists you in many of your everyday activities.

Hey students! Bring these pages to life online

▶ **Watch videos**

🧩 **Engage with interactivities**

A+ **Answer questions and check solutions**

Find all this and MORE in jacPLUS

Reading content and rich media, including interactivities and videos for every concept

Extra learning resources

Differentiated question sets

Questions with immediate feedback, and fully worked solutions to help students get unstuck

Use the following map to answer questions **1** and **2**.

1. Wagga Wagga is located in D3. Use the grid map to give grid references for:

   a. Mt Kosciuszko   **b.** Port Campbell   **c.** The Grampians.

2. Using the grid map, name a town that is located within each of the following grid references.

   a. C1   **b.** B2   **c.** D3

3. Consider the chessboard shown.
   a. One of the black rooks (in chess, rook is another name for castle) is on a8. State the location of the white rooks.
   b. One of the white bishops is on c1. State the position of the second white bishop.
   c. A bishop can move diagonally any number of (unoccupied) squares at a time. With its next move, the bishop at b3 could capture the pawn at what position?

4. Identify the coordinates of point X on the Cartesian plane.

5. Calculate the area of the rectangle formed by connecting the points $(-2, 3), (3, 3), (3, -4)$ and $(-2, -4)$ on a Cartesian plane.

6. **MC** Identify which of the following points is located in the second quadrant.
   A. $(1, 5)$ **B.** $(1, -5)$ **C.** $(-1, -5)$ **D.** $(-5, 1)$

7. Determine the coordinates of the midpoint between $(-2, 6)$ and $(0, 2)$.

8. **MC** Identify the point that sits on the x-axis.
   A. $(1, 1)$ **B.** $(1, 0)$ **C.** $(3, 3)$ **D.** $(0, 1)$

9. Determine the value of point P if the points $(2, 5), (4, 11)$ and $(6, P)$ are to be colinear.

10. If the values in the following table are linear, determine the missing value.

x	2	3	5	7
y	5	9		25

11. **MC** Select which of the following points form a linear pattern when plotted.
   A. $(0, 2), (2, 4), (3, 6)$ **B.** $(-1, 0), (-3, 4), (-4, 6)$
   C. $(-1, 1), (1, 1), (2, 2)$ **D.** $(3, 3), (4, -4), (5, 5)$

Use the following information to answer questions **12** and **13**.

Teo plants a fast-growing hedge. The plant starts at a height of 45 cm, and for each of the next three weeks he measures the height to be 51 cm, 57 cm and 63 cm.

12. Explain whether the hedge is following a linear growth pattern.

13. If H represents the height of the hedge after x weeks, determine the equation that models the height of the plants.

14. **MC** The rule $2x - 5$ can be used to generate a number pattern, where x represents the position in the pattern. Using the rule, identify the 20th number in the pattern.
   A. 15 **B.** 27 **C.** 35 **D.** 45

15. Consider the following figures using matches.

   Determine the amount of matches in the 10th figure.

LESSON
9.2 The Cartesian plane

LEARNING INTENTION

At the end of this lesson you should be able to:
- plot points on a Cartesian plane
- identify the coordinates of points on a Cartesian plane.

▶ 9.2.1 The Cartesian plane

eles-4635

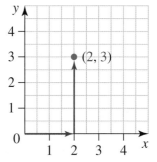

- The **Cartesian plane** is formed from two straight, perpendicular number lines called **axes**.
- The horizontal axis is referred to as the **x-axis**.
- The vertical axis is referred to as the **y-axis**.
- The point where the two axes intersect is called the **origin** and can be labelled with the letter O. The origin has the coordinate (0,0).
- Both axes must be marked with evenly spaced ticks and numbered with an appropriate scale (usually the distance between each tick is one unit).
- The x-axis must be labelled with an x at the far right of the horizontal axis and the y-axis must be labelled with a y at the top of the vertical axis.
- **Cartesian coordinates** are ordered pairs, written in brackets and separated by a comma, which give the position of a point on the plane. For example, (2,3) is a coordinate in the graph shown.
- In a coordinate, the first number is the x-coordinate and shows how far right (or left) from the origin a point is located. The second number is the y-coordinate and shows how far up (or down) from the origin a point is located.

Cartesian coordinates

A Cartesian coordinate is written in the form (x, y).

The Cartesian coordinate (2, 3) can be located by moving two units right from the origin and then three units up.

WORKED EXAMPLE 1 Creating a Cartesian plane and plotting points

Draw a Cartesian plane with axes extending from 0 to 5 units. Mark the following points with a dot, and label them.

a. $(2, 4)$ **b.** $(5, 0)$ **c.** $(0, 2)$ **d.** $\left(3\frac{1}{2}, 2\right)$

THINK

First rule and label the axes.

a. (2, 4) means starting at the origin, go across 2 units, and then up 4 units.

b. (5, 0) means go across 5 units and up 0 units. It lies on the x-axis.

c. (0, 2) means go across 0 units and up 2 units. It lies on the y-axis.

d. $\left(3\frac{1}{2}, 1\right)$ means go across $3\frac{1}{2}$ units and up 1 unit.

Label each point.

DRAW

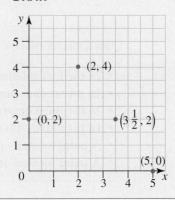

Identify the coordinates for each of the points A, B, C and D.

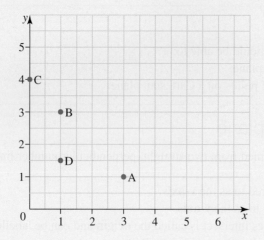

THINK	WRITE
Point A is 3 units across and 1 unit up.	A is at $(3, 1)$.
Point B is 1 unit across and 3 units up.	B is at $(1, 3)$.
Point C is 0 units across and 4 units up.	C is at $(0, 4)$.
Point D is 1 unit across and $1\frac{1}{2}$ units up.	D is at $\left(1, 1\frac{1}{2}\right)$.

9.2.2 The four quadrants of the Cartesian plane

eles-4636

- The axes on the Cartesian plane can extend infinitely in both directions.
- On the x-axis, the values to the left of the origin are negative.
- On the y-axis, the values below the origin are negative.
- By extending the axes in both directions, the Cartesian plane is divided into four sections called **quadrants**.
- The quadrants are numbered one to four, starting with the top right as quadrant 1, then numbering in an anti-clockwise direction.

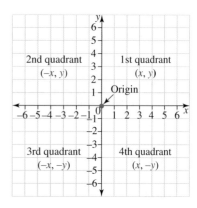

> ### Quadrants of the Cartesian plane
>
> - **In the first quadrant, the x and y values are both positive.**
> - **In the second quadrant, the x value is negative and the y value is positive.**
> - **In the third quadrant, both the x and y values are negative.**
> - **In the fourth quadrant, the x value is positive and the y value is negative.**
> - **A point located on the y-axis has an x-coordinate of 0.**
> - **A point located on the x-axis has a y-coordinate of 0.**

a. **Plot the following points on the Cartesian plane.**

$$A(-1, 2), B(2, -4), C(0, -3), D(4, 0), E(-5, -2)$$

b. **State the location of each point on the plane (that is, the quadrant, or the axis on which it sits).**

THINK

a. 1. Draw a set of axes, ensuring that they are long enough to fit all the values.

2. Plot the points. The first point is one unit to the left and two units up from the origin. The second point is two units to the right and four units down from the origin (and so on).

b. Look at the plane and state the location of each point. Remember that the quadrants are numbered in an anti-clockwise direction, starting at the top right. If the point is on the axis, specify which axis it is.

WRITE

a.

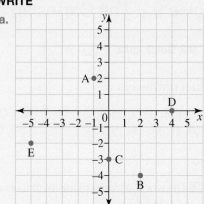

b. Point A is in the second quadrant.
Point B is in the fourth quadrant.
Point C is on the *y*-axis.
Point D is on the *x*-axis.
Point E is in the third quadrant.

COMMUNICATING — COLLABORATIVE TASK: Alphanumeric coordinates of Aboriginal languages in NSW and ACT

Similar to a Cartesian plane, grid references (often called area references) are used to locate places on a map. A grid reference map refers to a square region rather than a point on a map. The reference for each square is given by two coordinates — a letter and a number.

The map shown is of the different Aboriginal languages in NSW and ACT.

1. Using the map provided, give the grid reference(s) for each of the following locations.
   a. Baraba Baraba   b. Biripi
   c. Dharug   d. Karenggapa
2. Name the language(s) located in each of the following grid references.
   a. D3   b. F5
   c. A3   d. B4
3. Name the largest Aboriginal language area and write down all of its grid references.
4. Describe a disadvantage of using alphanumerical grid references.
5. Investigate other ways that locations are given on maps. How do pilots navigate to specific runways around the world?
6. With your classmates, discuss other uses for alphanumeric grid references.

Exercise 9.2 The Cartesian plane

learnon

9.2 Quick quiz on	9.2 Exercise

Individual pathways

■ PRACTISE	■ CONSOLIDATE	■ MASTER
1, 4, 5, 8, 11, 14, 16, 19, 23	2, 6, 9, 12, 13, 17, 20, 24, 25	3, 7, 10, 15, 18, 21, 22, 26, 27

Fluency

1. **WE1** Draw a Cartesian plane with axes extending from 0 to 5 units. Mark the following ordered pairs with a dot, and label them.

   a. $(4, 3)$ b. $(1, 4)$ c. $(3, 3)$ d. $(2, 0)$ e. $(0, 4)$ f. $(0, 0)$

2. Draw a Cartesian plane with axes extending from 0 to 10 units. Mark the following ordered pairs with a dot and label them.

   a. $(4, 2)$ b. $(3, 8)$ c. $(9, 9)$ d. $(5, 0)$ e. $(7, 2)$ f. $(0, 10)$

3. Draw a Cartesian plane with axes extending from 0 to 20 units using a scale of 2 units for both axes. Mark the following ordered pairs with a dot and label them.

   a. $(12, 16)$ b. $(2, 8)$ c. $(18, 4)$ d. $(10, 10)$ e. $(6, 0)$ f. $(14, 20)$

4. **WE2** Identify the coordinates for each of the points A to L.

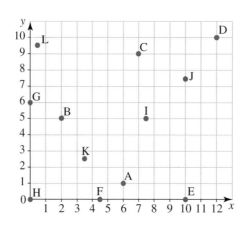

5. **WE3** a. Plot the following points on the Cartesian plane.

   $A(-1, -3)$, $B(-2, 5)$, $C(3, -3)$, $D(0, -4)$, $E(-2, -2)$, $F(-5, 0)$, $G(3, 1)$, $H(3, 0)$, $I(-4, -2)$, $J(4, -5)$

   b. State the location of each point on the plane (that is, the quadrant, or the axis it sits on).

6. **a.** Plot the following points on the Cartesian plane.

A $(-4, 3)$, B $(5, -7)$, C $(-5, -5)$, D $(0, -7)$, E $(-9, 0)$, F $(-2, 8)$, G $(-3, -6)$, H $(-1, 6)$, I $(2, 8)$, J $(-7, 7)$

   b. State the location of each point on the plane (that is, the quadrant or the axis it sits on).

7. **a.** Plot the following points on the Cartesian plane.

A $(-12, 7)$, B $(-7, 1)$, C $(-10, -10)$, D $(0, 7)$, E $(-3, 0)$, F $(-13, 5)$, G $(-9, -11)$, H $(1, -9)$, I $(2, 8)$, J $(3, -3)$

   b. State the location of each point on the plane (that is, the quadrant or the axis it sits on).

8. Identify the coordinates for each of the points A to F.

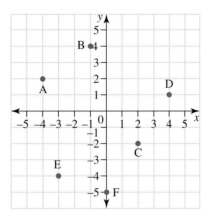

9. Identify the coordinates for each of the points A to F.

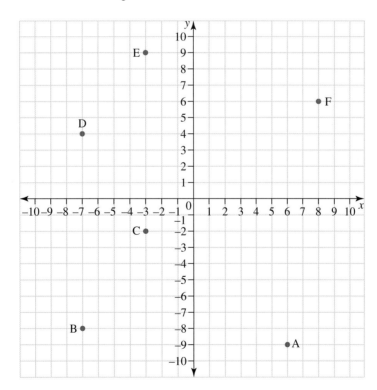

10. Identify the Cartesian coordinates for each of the points A to F.

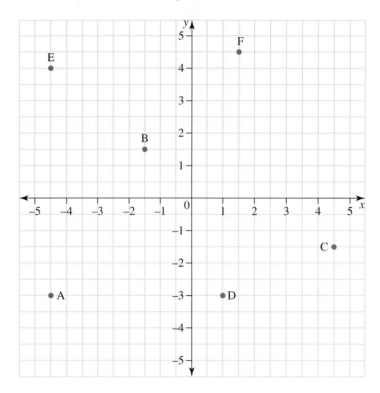

Understanding

11. Each of the following sets of Cartesian axes (except one) has something wrong with it. From the list below, match the mistake in each diagram with one of the sentences.

  i. The units are not marked evenly.
  ii. The y-axis is not vertical.
  iii. The axes are labelled incorrectly.
  iv. The units are not marked on the axes.
  v. There is nothing wrong.

a.

b.

c.

d.

e.

12. From the graph, write down the coordinates of 2 points that:

   a. have the same x-coordinate
   b. have the same y-coordinate.

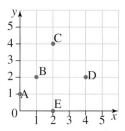

13. Messages can be sent in code using a grid like the one shown in the following graph, where the letter B is represented by the coordinates (2, 1).

   Use the graph to decode the answer to the following riddle:
   Q: Where did they put the man who was run over by a steamroller?
   A: (4, 2) (4, 3), (3, 2) (5, 3) (4, 4) (1, 4) (4, 2) (5, 4) (1, 1) (2, 3), (4, 2) (4, 3), (3, 5)
   (1, 1) (3, 4) (4, 1) (4, 4) (4, 4) (4, 2) (4, 5), (4, 4) (5, 1) (2, 5) (5, 1) (4, 3), (5, 1) (4, 2)
   (2, 2) (3, 2) (5, 4), (1, 1) (4, 3) (4, 1), (4, 3) (4, 2) (4, 3) (5, 1)
   The commas outside the brackets in the above indicate separate words. Do not use the commas in your answer; use spaces between your decoded words.

14. Rule a Cartesian plane with both axes extending from 0 to 10 units. Plot the following points and join them in the order given to make a geometric figure. Name each shape.

   a. (2, 2) − (5, 2) − (2, 6) − (2, 2)
   b. (4, 4) − (8, 4) − (6, 8) − (4, 4)
   c. (1, 1) − (10, 1) − (8, 9) − (2, 9) − (1, 1)
   d. (0, 0) − (8, 0) − (10, 10) − (2, 10) − (0, 0)

15. Here is an exercise that may require extra care and concentration. On graph paper or in your exercise book rule a pair of Cartesian axes. The x-axis must go from 0 to 26 and the y-axis from 0 to 24. Plot the following points and join them in the order given.

   $(0, 15) − (4, 17) − (9, 22) − (10, 21) − (12, 24) − (16, 22) − (15, 21) − (18, 19) − (20, 24) − (22, 18) − (26, 12)$

   $−(26, 10) − (23, 4) − (20, 3) − (18, 4) − (14, 7) − (11, 7) − (4, 6) − (2, 7) − \left(2\frac{1}{2}, 8\right) − (0, 15)$

   Complete the picture by joining $(19, 2) − (21, 2) − (20, 0) − (19, 2)$.

16. Consider a rectangle formed by connecting the points (3, 2), (9, 2), (9, 5) and (3, 5) on a Cartesian plane. Calculate:

   a. the area of the rectangle
   b. the perimeter of the rectangle.

17. Consider the following set of points:
   A(2, 5), B(−4, −12), C(3, −7), D(0, −2), E(−10, 0),
   F(0, 0), G(−8, 15), H(−9, −24), I(18, −18), J(24, 0).

   Identify which of the following statements are true.

   a. Points A and J are in the first quadrant.   b. Points B and H are in the third quadrant.
   c. Only point I is in the fourth quadrant.   d. Only one point is in the second quadrant.
   e. Point F is at the origin.   f. Point J is not on the same axis as point E.
   g. Point D is two units to the left of point F.

18. Consider the following triangle ABC.

   a. State the coordinates of the vertices of the triangle ABC.
   b. Calculate the area of the triangle.
   c. Reflect the triangle in the x-axis. (You need to copy it into your workbook first.) What are the new coordinates of the vertices?
   d. Now reflect the triangle you have obtained in part **c** in the y-axis, and state the new coordinates of the vertices.

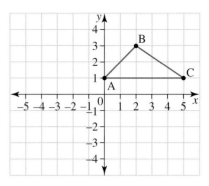

Communicating, reasoning and problem solving

19. Consider the square ABCD shown.

   a. State the coordinates of the vertices of the square ABCD.
   b. Calculate the area of the square ABCD.
   c. Extend sides AB and CD three squares to the right and three squares to the left.
   d. Calculate the area of the new shape.
   e. Extend sides AD and BC three squares up and three squares down.
   f. Calculate the area of the new shape.
   g. Compare the three areas calculated in parts **b**, **d** and **f**. Explain the changes of area in relation to the change in side length.

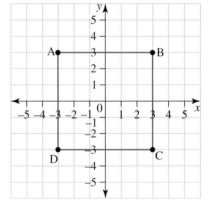

20. a. Explain what would happen to a shape in quadrant 3 if it was reflected in:

    i. the x-axis   **ii.** the y-axis.

   b. Explain what would happen to a shape in quadrant 4 if it was reflected in:

    i. the x-axis   **ii.** the y-axis.

   c. Write a rule about reflections:

    i. in the x-axis   **ii.** in the y-axis.

21. a. Determine the coordinates of the point that lies in the middle of the following pairs of points:

    i. $(2, 6)$ and $(4, 10)$ **ii.** $(9, 3)$ and $(3, 1)$
    iii. $(-7, 4)$ and $(-1, -6)$ **iv.** $(-2, -3)$ and $(8, -7)$

   b. Write a rule that could be used to find the middle of the points (a, b) and (m, n).

22. (x, y) is a point that lies somewhere on the Cartesian plane. State which quadrant/s the point could be located in, given the following conditions:

   a. $x < 0$ **b.** $y > 0$ **c.** $x < 0$ and $y < 0$ **d.** $x > 0$ and $y < 0$

23. A line AB passes through points $A(-5, 6)$ and $B(3, 6)$.

   a. Draw line AB on a Cartesian plane.
   b. Determine the length of AB.
   c. Determine the coordinates of the middle point of the line interval AB.
   d. Draw another horizontal line and determine the coordinates of the middle point.
   e. Is there a formula to calculate the x-coordinate of the middle point of a horizontal line?

24. A line CD passes through points C(1, 7) and D(1, −5).

    a. Draw line CD on a Cartesian plane.
    b. Determine the length of CD.
    c. Determine the coordinates of the middle point of the line interval CD.
    d. Draw another vertical line and determine the coordinates of the middle point.
    e. Explain whether there is a formula to calculate the y-coordinate of the middle point of a vertical line.

25. Consider the die shown. If we consider the centre dot to be the origin, give a possible set of coordinates for the other four dots.

26. Consider the following set of points:
(0, 3), (2, 7), (4, 11), (6, 15), (8, 19)

    a. Describe the pattern that connects these points to one another.
    b. Write the next three points in the sequence.
    c. Write a rule that connects the x-coordinate to the y-coordinate for each point in this sequence.

27. Consider a circle with a radius of 5 units that has its centre as the origin.

    a. Identify the coordinates of the four points where the circle cuts each axis.
    b. By drawing this circle on the Cartesian plane, explain whether you can identify any other points with integer coordinates that lie on this circle.

LESSON
9.3 Linear number patterns

LEARNING INTENTION

At the end of this lesson you should be able to:
- construct a number pattern from a rule
- identify a rule from a number pattern
- identify a number pattern that can be used to describe a geometric sequence
- express a number pattern algebraically.

 9.3.1 Describing relationships and patterns

eles-4637

- Mathematicians aim to describe relationships in the world around us. The relationships and rules they find can often be described using number patterns.
- Each number in a number pattern is called a **term**.
- Linear relationships are number patterns that increase or decrease by the same amount from one term to the next. For example, the sequence 1, 4, 7, 10, … increases by 3 from one term to the next, so it is a linear relationship.
- Number patterns can be described by rules. For example, the number pattern above can be described by the rule 'start with 1 and add 3 each time'.

WORKED EXAMPLE 4 Creating a sequence from a rule

Using the rule 'start with 32 and subtract 3 each time', write the first 5 terms of the number pattern.

THINK	WRITE
1. Start with 32 and subtract 3.	$32 - 3 = 29$
2. Keep subtracting 3 from the previous answer until five numbers have been calculated.	$29 - 3 = 26$ $26 - 3 = 23$ $23 - 3 = 20$
3. Write the answer.	The first five numbers are 32, 29, 26, 23 and 20.

WORKED EXAMPLE 5 Identifying a rule from a number pattern

Consider the number pattern: $4, 8, 12, \ldots$
a. Describe the pattern in words.
b. Write down the next 3 terms in the pattern.

THINK	WRITE
a. The next number is found by adding 4 to the previous number.	a. Next number = previous number $+ 4$
b. 1. Add 4 each time to get the next three numbers.	b. $12 + 4 = 16$ $16 + 4 = 20$ $20 + 4 = 24$
2. Write down the next three numbers.	The next three numbers are 16, 20 and 24.

- We can also describe a number pattern in words by looking at the value of any particular term and relating it to its position in the number pattern.
- For example, we could use a table of values to show the relationship from the previous worked example.

Position in pattern	1	2	3	4	5	6
Value of term	4	8	12	16	20	24

- This number pattern can be described as 'value of term = position in pattern $\times 4$'.
- After a rule in the format for a number pattern has been identified, you can use it to find any future value in the pattern.

WORKED EXAMPLE 6 Determining the rule of a number pattern

Consider the number pattern $6, 13, 20, \ldots$
a. Determine the rule for the number pattern by relating the value of a term to its position in the pattern.
b. Determine the 15th term in the pattern.

THINK	WRITE
a. 1. Draw a table relating the position in the pattern to the value of the term.	a. <table><tr><td>Position in pattern</td><td>1</td><td>2</td><td>3</td></tr><tr><td>Value of term</td><td>6</td><td>13</td><td>20</td></tr></table>
2. The value of the term is found by multiplying the position in the pattern by 7 and subtracting 1.	Value of term = position in pattern $\times 7 - 1$

b. 1. Substitute the position in the pattern into the formula and evaluate.

Value of term $= 15 \times 7 - 1$
$\qquad\qquad\quad = 104$

2. Write the answer.

The 15th term in the pattern is 104.

- When the terms in a number pattern are decreasing, the rule can be described by subtracting a multiple of the position in the pattern from a fixed number. For example, the number pattern 31, 25, 19, ... can be described by 'value of term $= 37 -$ position in pattern $\times 6$'.
- If you are given an algebraic expression, you can generate a number pattern by substituting values into the expression.

WORKED EXAMPLE 7 Calculating terms in a number pattern

The rule $x - 3$ can be used to generate a number pattern, where x represents the position in the pattern. Using the rule, calculate:
a. the first five terms of the number pattern
b. the 10th term of the number pattern
c. the 100th term of the number pattern.

THINK	WRITE
a. 1. Write the rule.	**a.** $x - 3$
2. Substitute $x = 1, 2, 3, 4$ and 5 into the rule for the first five tems.	When $x = 1$ $x - 3 = 1 - 3$ $\qquad = 2$ When $x = 2$ $x - 3 = 2 - 3$ $\qquad = -1$ When $x = 3$ $x - 3 = 3 - 3$ $\qquad = 0$ When $x = 4$ $x - 3 = 4 - 3$ $\qquad = 1$ When $x = 5$ $x - 3 = 5 - 3$ $\qquad = 2$
3. Write the answer.	The first five terms of the number pattern are $-2, -1, 0, 1$ and 2.
b. 1. Write the rule.	**b.** $x - 3$
2. Substitute $x = 10$ into the rule.	When $x = 10$ $x - 3 = 10 - 3$ $\qquad = 7$
3. Write the answer.	The 10th term of the number pattern is 7.
c. 1. Write the rule.	**c.** $x - 3$
2. Substitute $x = 100$ into the rule.	When $x = 100$ $x - 3 = 100 - 3$ $\qquad = 97$
3. Write the answer.	The 100th term of the number pattern is 97.

⏵ 9.3.2 Patterns using shapes

eles-6122

- We can often find number patterns when we look at shapes.
- For example, looking at the number of dots grown in the Goompi model, we can investigate the changes in each drawing.
- The pattern shows that it starts with 3 dots, and 4 new dots are added each time.
- By using a table of values, we can see a number pattern developing.

Diagram	0	1	2	3	4
Number of dots	3	7	11	15	19

- The pattern for the number of dots is 3, 7, 11… the rule here is:

$$\text{total number of dots} = 4 \times \text{diagram} + 3 \text{ dots}$$

WORKED EXAMPLE 8 Geometric number patterns

Consider a set of hexagons constructed according to the following pattern.

a. **Using matches, pencils or similar objects, construct these figures, then draw the next two figures in the series.**
b. **Draw a table showing the relationship between the number of hexagons in the figure and the number of matches used to construct the figure.**
c. **Devise a rule to describe the number of matches required for each figure in terms of the number of hexagons in the figure.**
d. **Use your rule to determine the number of matches required to make a figure consisting of 15 hexagons.**

THINK

a. Construct the given figures with matches, then draw the next two figures.

b. Draw a table showing the number of matches needed for each figure in terms of the number of hexagons. Fill it in by looking at the figures.

c. 1. Look at the pattern in the number of matches going from one figure to the next. It is increasing by 5 each time.

   2. If we take the number of hexagons and multiply this number by 5, it does not give us the number of matches. However, if we add 1 to this number, it does give us the number of matches in each shape.

WRITE

a. The next two figures are:

b.
Number of hexagons	1	2	3	4	5
Number of matches	6	11	16	21	26

c. The number of matches increased by 5 in going from one figure to the next.

Number of matches = number of hexagons × 5 + 1

d. 1. Use the rule to find the number of matches to make a figure with 15 hexagons.	**d.** Number of matches for 15 hexagons $= 15 \times 5 + 1$
2. Write the answer.	$= 76$
	76 matches would be required to construct a figure consisting of 15 hexagons.

9.3.3 Using algebra to describe patterns

eles-4639

- Once a pattern has been identified, we can use algebra to describe it.
- If we let $m =$ the number of matches and $h =$ the number of hexagons from the pattern in the previous worked example, the equation $m = 5h + 1$ would describe this pattern.

WORKED EXAMPLE 9 Using algebra to describe patterns

Consider the set of shapes constructed according to the following pattern.

a. **Draw a table of values showing the number of matches needed for each figure in the pattern.**
b. **Express the relationship between the number of matches (m) and the number of squares (s) in the pattern as an algebraic equation.**

THINK

a. Look at the figures in the pattern to complete a table of values.

b. 1. Look at the pattern in the number of matches going from one figure to the next. It is increasing by 3 each time.

2. If we take the number of squares and multiply this number by 3, it does not give us the number of matches. However, if we add 1 to this number, it does give us the number of matches in each shape.

3. Replace 'number of matches' with 'm' and 'number of squares' with 's' to create your algebraic equation, then simplify by removing the multiplication sign.

WRITE

a.
Number of squares	1	2	3
Number of matches	4	7	10

b. The number of matches increased by 3 in going from one figure to the next.

Number of matches = number of squares $\times 3 + 1$

$m = s \times 3 + 1$
$m = 3s + 1$

---

on Resources

Interactivities Identifying patterns (int-3801)
  Geometric patterns (int-3802)

Exercise 9.3 Linear number patterns

9.3 Quick quiz on	9.3 Exercise

Individual pathways

■ PRACTISE	■ CONSOLIDATE	■ MASTER
1, 3, 6, 8, 12, 14, 17	2, 4, 7, 9, 13, 15, 18, 19	5, 10, 11, 16, 20, 21, 22

Fluency

1. **WE4** Using the following rules, write down the first five terms of the number patterns.
   a. Start with 1 and add 4 each time.
   b. Start with 5 and add 9 each time.

2. Using the following rules, write down the first five terms of the number patterns.
   a. Start with 50 and subtract 8 each time.
   b. Start with 64 and subtract 11 each time.

3. Write down the next three terms for the following number patterns.
   a. 2, 4, 6, 8, ...
   b. 3, 8, 13, 18, ...

4. **WE5** Consider the number pattern 27, 24, 21, 18, ...
   a. Describe the pattern in words.
   b. Write down the next three terms in the pattern.

5. Consider the number pattern 48, 41, 34, 27, ...
   a. Describe the pattern in words.
   b. Write down the next three terms in the pattern.

6. Fill in the missing numbers in the following number patterns.
   a. 3, ___ , 9, 12, ___ , ___
   b. 8, ___ , ___ , 14, ___
   c. 4, 8, ___ , 32, ___

7. Fill in the missing numbers in the following number patterns.
   a. ___ , ___ , 13, 15, ___
   b. 66, 77, ___ , 99, ___ , 121
   c. 100, ___ , ___ , 85, 80, ___

8. **WE6** Consider the number pattern 5, 15, 25, ...
   a. Determine the rule for the number pattern by relating the value of a term to its position in the number pattern.
   b. Calculate the 15th term in the pattern.

9. Consider the number pattern 44, 41, 38, ...
   a. Determine the rule for the number pattern by relating the value of a term to its position in the number pattern.
   b. Calculate the 15th term in the pattern.

10. Consider the number pattern 102, 108, 114, ...
    a. Determine the rule for the number pattern by relating the value of a term to its position in the number pattern.
    b. Calculate the 35th term in the pattern.

11. Consider the number pattern 212, 199, 186, ...
    a. Determine the rule for the number pattern by relating the value of a term to its position in the number pattern.
    b. Calculate the 23rd term in the pattern.

Understanding

12. **WE7** The rule $x + 4$ is used to generate a number pattern, where x represents the position in the pattern.
    a. Generate the first five terms of the number pattern.
    b. Determine the 10th term of the number pattern.
    c. Determine the 100th term of the number pattern.

13. A number pattern is given by the rule $2x - 2$, where x represents the position in the pattern.
    a. Generate the first five terms of the number pattern.
    b. Generate the 20th term of the number pattern.
    c. Generate the 77th term of the number pattern.

Communicating, reasoning and problem solving

14. Explain why it is more useful to describe a number pattern in terms of relating the value of a term to its position in the pattern than to describe a number pattern by the rule between terms.

15. Are all patterns that you can describe by written descriptions also describable using algebraic symbols? Explain your answer.

16. Explain why it is useful to describe the rule for a pattern by drawing a table and looking for the connection between the top row and the bottom row of the table.

17. **WE8** a. Construct the following figures using matches, then draw the next two figures in the pattern.

    b. Draw a table showing the relationship between the number of shapes in each figure and the number of matchsticks used to construct it.
    c. Devise a rule that describes the pattern relating the number of shapes in each figure and the number of matchsticks required to construct it.
    d. Use your rule to calculate the number of matchsticks required to construct a figure made up of 7 such shapes.

18. a. Construct the following figures using matches, then draw the next two figures in the pattern.

    b. Draw a table showing the relationship between the number of shapes in each figure and the number of matchsticks used to construct it.
    c. Devise a rule that describes the pattern relating the number of shapes in each figure and the number of matchsticks required to construct it.
    d. Use your rule to calculate the number of matchsticks required to construct a figure made up of 7 such shapes.

19. **WE9** a. For the geometric pattern shown, draw a table of values showing the relationship between the number of shapes in each figure and the number of matchsticks used to construct it.

b. Express the relationship between the number of matches (m) and the number of shapes (n) in the pattern as an algebraic equation.

20. a. Construct the following figures using matches, then draw the next two figures in the pattern.

b. Draw a table showing the relationship between the number of shapes in each figure and the number of matchsticks used to construct it.

c. Devise a rule that describes the pattern relating the number of shapes in each figure and the number of matchsticks required to construct it.

d. Use your rule to calculate the number of matchsticks required to construct a figure made up of 7 such shapes.

21. a. For the geometric pattern shown, draw a table of values showing the relationship between the number of shapes in each figure and the number of matchsticks used to construct it.

b. Express the relationship between the number of matches (m) and the number of shapes (n) in the pattern as an algebraic equation.

22. Create a matchstick pattern that can be represented by the rule $m = 5n + 3$.

LESSON
9.4 Plotting simple linear relationships

LEARNING INTENTION

At the end of this lesson you should be able to:
- plot a simple linear relationship from a table of values
- identify whether a series of points forms a linear relationship.

▶ 9.4.1 Plotting simple linear relationships

eles-4640

- When a set of points is plotted on the Cartesian plane, a pattern may be formed. If a pattern forms a straight line, we call it a *linear pattern*. If the pattern forms a curve, we call it a *non-linear pattern*.
- The coordinates of the points that form a pattern can be presented in a table of values or as a set of points.

- If shown in a table, the coordinates of each point should be read 'in columns'; the top number in each column gives the *x*-coordinate and the bottom number gives the corresponding *y*-coordinate of the point.

x	0	1	2	3
y	8	6	4	2

$\quad\quad\quad\quad\quad$ (0, 8) $\quad\quad$ (1, 6) $\quad\quad$ (2, 4) $\quad\quad$ (3, 2)

- The Cartesian coordinates of the points are ordered pairs. That is, the first number always represents the *x*-coordinate of a point and the second always represents the *y*-coordinate. A set of ordered pairs forms a relationship between *x* and *y*.
- If the points form a linear pattern when plotted, we say that the relation between *x* and *y* is linear.
- a linear relationship can continue on and have an infinite number of ordered pairs.

WORKED EXAMPLE 10 Plotting a set of points to observe a pattern

Plot the following set of points on the Cartesian plane, and comment on any pattern formed.
$$(1, 3), (2, 4), (3, 5), (4, 6), (5, 7)$$

THINK

1. Look at the coordinates of the points in the set: the *x*-values range between 1 and 5, while the *y*-values range between 3 and 7. Draw a set of axes, ensuring they are long enough to fit all the values.

2. Plot the points.

WRITE

3. Comment on the pattern that the points form.

The points form a linear pattern.

WORKED EXAMPLE 11 Plotting points presented in a table of values

Plot the following points on a Cartesian plane and decide whether the relationship is linear.

x	0	1	2	3	4
y	8	7	6	5	4

THINK

1. The *x*-values range from 0 to 4; the *y*-values range from 4 to 8. Draw a set of axes that will include all these values. (The axes need not be the same length. In this case you can extend the y-axis a bit more than the *x*-axis, as the *y*-values are higher.)

2. The entries in the table can be rewritten as a set of points. The top number in each column gives the *x*-coordinate of the point and the bottom number gives the corresponding *y*-coordinate. Thus the first column gives the point (0, 8), the second (1, 7), and so on. Plot all points.

WRITE

3. Consider the pattern formed by the points. It is a straight line. Draw a suitable conclusion about the relationship between x and y.

The relationship between x and y is linear.

WORKED EXAMPLE 12 Constructing a table of values from a rule

Consider the relationship between x and y given by the rule $y = 3x - 4$.
a. Describe the set of points produced by this relationship between x and y.
b. Construct a table of values for the integers from -2 to 2.

THINK

a. When a rule is written in this form, it means that each y-coordinate must be found from a corresponding x-coordinate. In this case the y-value is four less than three times the x-value.

b. 1. The interval $-2 \le x \le 2$ represents the following x-coordinates:
$x = -2, -1, 0, 1, 2$
For each x-coordinate, calculate the corresponding y-coordinate.

2. Place this information into a table of values.

WRITE

a. The rule describes a set of points such that the y-coordinate of each point is four less than three times the value of the x-coordinate.

b. $y = 3 \times -2 - 4 = -10$
$y = 3 \times -1 - 4 = -7$
$y = 3 \times 0 - 4 = -4$
$y = 3 \times 1 - 4 = -1$
$y = 3 \times 2 - 4 = 2$

x	-2	-1	0	1	2
y	-10	-7	-4	-1	2

 Resources

Interactivities Plotting simple linear relationships (int-4138)

Exercise 9.4 Plotting simple linear relationships

learn on

9.4 Quick quiz on	9.4 Exercise

Individual pathways

■ PRACTISE	■ CONSOLIDATE	■ MASTER
1, 4, 7, 10, 12, 15	2, 5, 8, 9, 13, 16	3, 6, 11, 14, 17, 18

Fluency

1. WE10 Plot the following sets of points on the Cartesian plane, and comment on any pattern formed.
   a. $(0, 1), (1, 2), (2, 3), (3, 4), (4, 5)$
   b. $(0, 3), (1, 4), (2, 5), (3, 6), (4, 7)$
   c. $(2, 2), (3, 3), (4, 4), (5, 5), (6, 6)$
   d. $(0, 2), (1, 3), (2, 1), (3, 0), (4, 5)$
   e. $(1, 10), (2, 8), (3, 6), (4, 4), (5, 2)$
   f. $(2, 3), (3, 3), (4, 3), (5, 3), (6, 3)$

2. Plot the following sets of points on the Cartesian plane. (Note that some coordinates are negative, so you will need to use all four quadrants.) Comment on any pattern formed.

   a. $(-2, 0), (-1, 2), (0, 4), (1, 6), (2, 8)$
   c. $(-2, 7), (-1, 6), (0, 4), (1, 3), (2, 1)$

   b. $(-3, -3), (-2, 0), (-1, 3), (0, 6), (1, 9)$
   d. $(-6, 8), (-5, 6), (-4, 4), (-3, 2), (-2, 0)$

3. Plot the following sets of points on the Cartesian plane. Comment on any pattern formed.

   a. $(-3, 8), (-1, 5), (1, 2), (3, -1), (5, -4)$
   c. $(-4, 4), (-1, 3), (2, 1), (5, -3), (8, -11)$

   b. $(-5, 0), (-5, 5), (-5, 3), (-5, -2), (-5, -4)$
   d. $(7, 8), (1, -2), (-2, -7), (4, 3), (10, 13)$

4. **WE11** Plot the points in each table on a Cartesian plane, and decide whether or not the relationship is linear.

a.

x	0	1	2	3	4
y	3	4	5	6	7

b.

x	0	1	2	3	4
y	0	1	4	9	16

c.

x	1	2	3	4	5
y	7	6	5	4	3

d.

x	1	2	3	4	6
y	12	6	4	3	2

e.

x	0	1	2	3	4
y	0	3	6	9	12

f.

x	2	3	4	5	6
y	0	-2	-4	-6	-8

5. Plot the points in each table on a Cartesian plane and decide whether the relationship is linear or not.

a.

x	0	1	2	3	4
y	1	4	7	10	13

b.

x	-2	-1	0	1	2
y	10	5	0	-5	-10

c.

x	0	1	2	3	4
y	-1	0	2	6	14

d.

x	-3	-2	-1	0	1
y	10	-8	6	-4	2

6. Plot the points in each table on a Cartesian plane, and decide whether or not the relationship is linear.

a.

x	0	1	2	3	4
y	8	5	2	-1	-4

b.

x	-2	-1	0	1	2
y	5	3	1	-1	-3

c.

x	0	1	2	3	4
y	-2	4	-6	8	-10

d.

x	-2	-1	0	1	2
y	8	7	4	-1	-8

Understanding

7. **a.** Plot the following points on the set of axes: $(0, 10), (1, 8), (2, 6), (3, 4), (4, 2), (5, 0)$.
   b. Describe the pattern formed by the points.
   c. Extend the pattern by plotting the next two points.
   d. State the coordinates of the two points you have plotted.

8. The points in the following table form a linear pattern. The y-coordinate of the middle point is missing.

x	1	2	3	4	5
y	0	3		9	12

   a. Plot the four points (whose x- and y-coordinates are both known) on the Cartesian plane.
   b. Use your graph to predict the missing y-coordinate of the middle point.
   c. Add the middle point to your graph and check whether it fits the pattern.

9. Which of the following sets of points, when plotted, will form a linear pattern?

   a. $(0, 3), (1, 5), (2, 7), (3, 9), (4, 11)$
   c. $(1, 6), (2, 5), (3, 7), (4, 8), (5, 9)$
   e. $(0, 2), (1, 4), (2, 4), (3, 5), (4, 6)$

   b. $(-1, 8), (-2, 7), (-3, 6), (-4, 5), (-5, 4)$
   d. $(2, 3), (3, 5), (4, 7), (5, 9), (6, 11)$
   f. $(10, 7), (9, 4), (8, 3), (7, 1), (6, 0)$

10. Consider the linear pattern shown.

a. If the pattern is continued to the right, identify the coordinates of the next point.

b. If the pattern is extended to the left, identify the coordinates of the previous point.

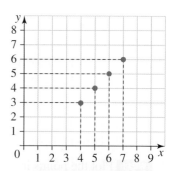

11. Rachel planted a cherry tomato plant in her vegetable patch. She measured the height of her plant every week for four consecutive weeks, obtaining the following data: 20 mm, 24 mm, 28 mm, 32 mm. Rachel forgot to measure her plant in week 5, but the following week she found the height of the cherry tomato to be 40 mm.

a. Fill in the table of values below.

Week						
Height (mm)						

b. Plot the heights of the cherry tomato plant on a set of axes.

c. Explain whether the relationship between the number of the week and the height of the plant is linear.

d. Assuming the pattern was the same for the first 6 weeks, determine the height of the tomato plant in week 5.

e. If the pattern continues, predict the height of the cherry tomato plant in week 8.

Communicating, reasoning and problem solving

12. **WE12** The relationship between the two variables x and y is described by the rule $y = x + 3$.

a. Complete the following statement: 'The rule describes a set of points such that the y-coordinate of each point is …'

b. Construct a table of values for whole numbers from -2 to 2, using the given rule.

c. Plot the points on a set of axes.

d. Is the relationship linear? Justify your answer.

13. The relationship between the two variables x and y is described by the rule $y = 2x - 3$.

a. Complete the following statement: 'The rule describes a set of points such that the y-coordinate of each point is three less than …'

b. Construct a table of values for whole numbers from -2 to 2, using the given rule.

c. Plot the points from the table on a set of axes.

d. Is the relationship linear? Justify your answer.

14. a. On the same set of axes, plot the following linear relationships: $y = x$, $y = 3x + 1$ and $y = 5x - 2$.

b. On the same set of axes, plot the following linear relationships: $y = -x$, $y = -3x + 1$ and $y = -5x - 2$.

c. Explain the differences and similarities between the two sets of graphs.

15. a. Use the following information to complete the table of coordinates.

Point A is two units to the left of the origin and five units below the origin.

Point B is one unit to the left of the origin and two units above point A.

Point C is one unit below the origin on the vertical axis.

Point D is one unit to the right of C and two units above point C.

Point E is two units to the right of the y-axis and three points above the x-axis.

	A	B	C	D	E
x					
y					

b. Plot these points on a Cartesian plane.

c. Determine the value of the horizontal step between two adjacent points.

d. Is the relationship linear? Explain your answer.

16. The table of values shown displays the relationship between time, in minutes, and parking fee, in dollars.

Time, minutes	30	60	90	120	
Parking fee, $	1.20	2.40		4.80	6.00

a. Write a rule for this relationship.

b. Use the rule found in part **a** to calculate the missing values.

17. The following table of values gives the relationship between the weight of coffee beans (in kg) and the cost per kg from a wholesale distributor.

Weight (kg)	1	5		20	25
Cost per kg ($)	49.00	46.00	42.25	34.75	

a. Write a rule for this relationship.

b. Use the rule found in part **a** to calculate the missing values.

c. Joel orders a 22 kg bag of beans each week for his coffee shop. Calculate how much he has to pay the wholesaler for his coffee.

d. The biggest bag that the wholesaler produces is a 30 kg bag of coffee beans. A chain coffee store orders a number of these bags each week. If the total cost of their order was $5722.50, determine the weight of coffee beans that was ordered.

18. Each row in my classroom has the same number of desks, and the rows are straight. My desk is fourth from the front and fourth from the back. It has one desk to its left and two desks to its right. Determine how many desks are in the classroom.

LESSON
9.5 Review

9.5.1 Topic summary

The Cartesian plane

- The Cartesian plane is made up of two perpendicular number lines put together. One line is horizontal and the other vertical. These lines are called axes.
- The horizontal axis is called the x-axis.
- The vertical axis is called the y-axis.

Extending the axis

- Each of the axes can continue infinitely in both directions.
- On the x-axis, values to the left of the origin are negative.
- On the y-axis, values below the origin are negative.
- The Cartesian plane can be split into four quadrants:

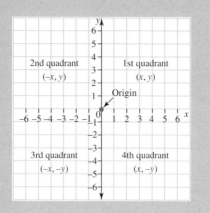

Coordinates of a point

- A coordinate is a pair of numbers, separated by a comma and in brackets, that specifies the location of a point: for example, (4, 2).
- The first number is the x-coordinate and the second is the y-coordinate.
- The coordinate (4, 2) represents the point found by moving 4 units to the right of the origin and then 2 units up.

LINEAR RELATIONSHIPS

Number patterns

- Geometric pattern:

- Number pattern: 6   11   16
- Table:

Number of shapes (s)	1	2	3
Number of matches (m)	6	11	16

- Rule: $m = 5 \times s + 1$

Linear relationships

- If a sequence of points forms a straight line when plotted, we call this a *linear relationship*.
- If the points form a line that is curved, we call this a *non-linear relationship*.

Linear relationship   Non-linear relationship

9.5.2 Project

Battleships game

The goal of this two-person game is to sink all of your opponent's ships before they do the same to you.

Equipment: Each player has two 10×10 grids. The grids are labelled with letters across the top (A to J) and numbers down the side (1 to 10). One of the grids is labelled 'My grid' and the other is labelled 'Opponent's grid'.

My grid									
A	**B**	**C**	**D**	**E**	**F**	**G**	**H**	**I**	**J**
1									
2									
3									
4									
5									
6									
7									
8									
9									
10									

Opponent's grid									
A	**B**	**C**	**D**	**E**	**F**	**G**	**H**	**I**	**J**
1									
2									
3									
4									
5									
6									
7									
8									
9									
10									

Set-up: Each player decides where to place four ships on their 'My grid': a five-square battleship, a four-square cruiser, a three-square submarine and a two-square destroyer. All ships must be placed in straight lines either horizontally or vertically, but not diagonally. It is fine (but not required) for two or more ships to be adjacent to each other.

How to play: Players take turns taking shots at each other. A shot is taken by calling out the coordinates of a square on the 10×10 grid, such as A1 or H4. Each player takes one shot at a time. If the player calls the coordinates of a square where one of the opponent's ships is located, the opponent calls out 'Hit' and says which ship was hit; if the shot misses, the opponent says 'Miss'.

Players should mark the shots they take on their 'Opponent's grid' and whether each shot was a hit or a miss, so that they don't call any square more than once. Players should also mark their 'My grid' with shots taken by their opponent.

A ship is sunk when all of its squares have been hit. When this happens, the player whose ship was sunk says, for example, 'You sank my battleship.'

The first player to sink all of their opponent's ships wins the game.

Exercise 9.5 Review questions

learnon

Fluency

1. The grid shown in part **b** shows a simplified map that uses alphanumeric grid references.
   a. Fill in the blank cells in the table.

Location	Grid reference
	B3
Alex's house	
	D1
Lena's house	

   b. A playground is to be built at A3. Mark its location on the grid with an X.

	1	2	3	4
D	Post office	Bus stop	Star cafe	Library
C		Apple street		
B	Alex's house	Maya's house	Joseph's house	Lena's house
A				

2. Study the graph to answer these questions.
   a. Identify the letter that represents the point $(-3, 4)$.
   b. Identify the points on the x–axis.
   c. Name all points whose x–coordinate is zero.
   d. Identify the points in the second quadrant.
   Give their coordinates.
   e. State the coordinates of C and F.

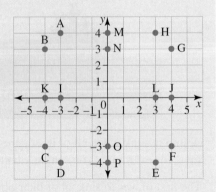

3. Use the graph to answer these questions.
   a. Fill in the blanks in the table.

Coordinates	Point
(0, 2)	
	C
	D

   b. Mark E (4, 6) on the graph. Join all the points with a straight line.

4. The following table gives the x and y coordinates of some points.

x	0	1	2	3
y	0	2	4	6

   a. Using the axes shown, plot the points given in the table and join them.
   b. Explain whether the relationship between x and y is linear.

5. Write down the next three terms for the following number patterns.
   a. 3, 6, 9, 12, …
   b. 6, 14, 22, 30, …

Understanding

6. Consider the number pattern 50, 65, 80, …
   a. Determine the rule for the number pattern by relating the value of a term to its position in the number pattern.
   b. Calculate the 15th term in the pattern.

7. The rule $3x + 4$ is used to generate a number pattern, where x represents the position in the pattern.
   a. Generate the first three terms of the number pattern.
   b. Determine the 10th term of the number pattern.
   c. Determine the 50th term of the number pattern.

8. Plot the points in each table on a Cartesian plane, and decide whether or not the relationship is linear.

   a.

x	0	1	2	3	4
y	2	4	6	8	10

   b.

x	2	4	6	8	10
y	3	5	9	15	23

Communicating, reasoning and problem solving

9. A game consists of a grid, and a dart that is thrown to land on a square of the grid.

A	B	C	D	E
F	G	H	I	J
K	L	M	N	O
P	Q	R	S	T
U	V	W	X	Y

The dart is thrown and lands on a letter that:
- has horizontal symmetry
- also has vertical symmetry
- and is not a vowel.

What are the possibilities for the landing position of the dart? Give instructions for locating any possibilities on the grid.

10. Rachel and Nathan are playing a battleship game. Rachel's grid (in progress, up to this moment) is shown below.
    a. For each of the following attempts, state whether Nathan will miss (M), sink (S) or hit (H) one of Rachel's ships.
       i. A7   ii. F9   iii. I10   iv. C3
    b. Specify the location of all of Rachel's smallest ships.
    c. State the set of grid references that Nathan needs if he is to sink Rachel's largest ship.

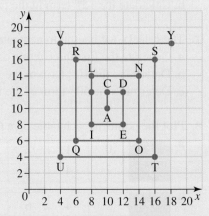

11. Answer the following questions, then translate each letter of your answer into an ordered pair by using the grid.
    a. In which Australian state is the Great Barrier Reef?
    b. Which Australian state is known as the Garden State?
    c. The Opera House is in which city?

12. The following diagram gives the key to a code by matching letters to coordinates. Use the key to write in code this famous statement by Descartes: 'I think, therefore I am.'

13. On a carefully ruled pair of Cartesian axes, join the following points in the order given. It reveals a symbol used by an ancient mathematical secret society.

$$(1, 0) - (6, 4) - (0, 4) - (5, 0) - (3, 6) - (1, 0)$$

14. ABCD is a square drawn in the first quadrant as shown.
    Given A(0, 3) and B(4, 0) and C(7, 4), find the coordinates of D.

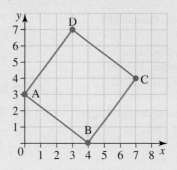

15. **a.** Plot the points A(3, 0), B(6, 7), C(10, 7), D(7, 0).
    i. What type of quadrilateral is ABCD?
    ii. Calculate its area.
    b. Draw a set of axes and on them plot points so that the encompassing triangle has an area of 20 square units. List the points and show why the area is 20 square units.

16. Four points — A, B, C and D — are drawn on a Cartesian plane and are joined in order to form a square. A is the point (−4, 0), B is (1, 5) and C is (6, 0). State the coordinates of point D.

17. Does this table represent a relationship between x and y that is linear?

x	1	2	3	4
y	1	4	9	16

Explain why or why not.

18. Consider the following statement: 'A linear relation, when plotted on a Cartesian plane, produces a straight line that cuts the x-axis at one point, and the y-axis at one point.'
    Discuss whether or not this statement is true.

19. The table of values shown displays the relationship between time, in minutes, and parking fee, in dollars.

Time, minutes	10	20	30		100
Parking fee, $	2.50	5.00	7.50	17.50	

    a. Write a rule for this relationship.
    b. Use the rule found in part a to calculate the missing values.

20. Create a matchstick pattern that can be represented by the rule $m = 3n + 1$

on   To test your understanding and knowledge of this topic, go to your learnON title at www.jacplus.com.au and complete the **post-test**.

Answers

Topic 9 Linear relationships

9.1 Pre-test

1. a. E2
   b. B1
   c. B2

2. a. Any of the following: Torquay, Geelong, Queenscliff, Sorrento, Wonthaggi, Berwick
   b. Any of the following: Warracknebeal, Dimboola, Wycheproof, Inglewood, Murtoa, Horsham, Navarre, Stawell, Avoca, Ararat, Clunes, Dunkeld, Penshurst, Derrinallum, Skipton, Clunes, Ballarat
   c. Any of the following: Griffith, Narrandera, Wagga Wagga

3. a. a1 and f1   b. b3   c. f7

4. (5, 3)

5. $35 \, \text{units}^2$

6. D

7. (−1, 4)

8. B

9. 17

10. 17

11. B

12. Yes, as each week the plant grows by 6 cm.

13. $H = 6x + 45$

14. C

15. 21 matches

9.2 The Cartesian plane

1.

2.

3.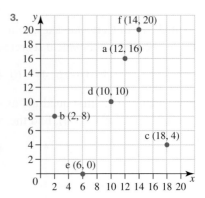

4. A(6, 1), B(2, 5), C(7, 9), D(12, 10), E(10, 0)
   $F\left(4\frac{1}{2}, 0\right)$, G(0, 6), H(0, 0), $I\left(7\frac{1}{2}, 5\right)$
   $J\left(10, 7\frac{1}{2}\right)$, $K\left(3\frac{1}{2}, 2\frac{1}{2}\right)$, $L\left(\frac{1}{2}, 9\frac{1}{2}\right)$

5. a.
   b. Quadrant I: point G; Quadrant II: point B; Quadrant III: points A, E, I; Quadrant IV: points C, J; x-axis: points F, H; y-axis: point D.

6. a.
   b. Quadrant I: point I; Quadrant II: points A, F, H, J; Quadrant III: point C, G; Quadrant IV: point B; x-axis: point E; y-axis: point D.

7. a.

b. Quadrant I: point I; Quadrant II: point A, B, F; Quadrant III: points C, G; Quadrant IV: points H, J; x-axis: points E; y-axis: point D.

8. A(−4, 2), B(−1, 4), C(2, −2), D(4, 1), E(−3, −4), F(0, −5)

9. A(6, −9), B(−7, −8), C(−3, −2), D(−7, 4), E(−3, 9), F(8, 6)

10. A(−4.5, −3), B(−1.5, 1.5), C(4.5, −1.5), D(1, −3), E(−4.5, 4)

11. a. B   **b.** C   **c.** D
   d. A   **e.** E   **f.** A

12. a. E, C   **b.** B, D

13. In hospital; in wards six, seven, eight and nine

14.

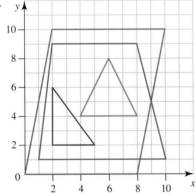

a. Right-angled triangle

b. Isosceles triangle

c. Trapezium

d. Parallelogram

15.

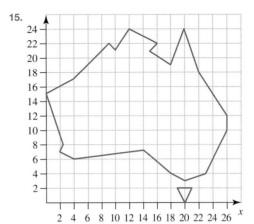

16. a. Area is 18 square units.
   b. Perimeter is 18 units.

17. b, d, e

18. a. A(0, 1) B(2, 3) C(5, 1)
   b. 5 square units
   c. A(0, −1) B(2, −3) C(5, −1)
   d. A(0, −1) B(−2, −3) C(−5, −1)

19. a. A(−3, 3), B(3, 3), C(3, −3), D(−3, −3)
   b. 36 square units
   c.

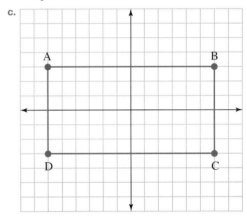

   d. 72 square units
   e.

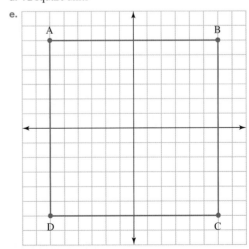

   f. 144 square units
   g. When the length is doubled, the area is doubled. When both the length and width are doubled, the area becomes four times larger.

20. a. i. It would move in quadrant 2.

ii. It would move in quadrant 4.

b. i. It would move in quadrant 1.

ii. It would move in quadrant 3.

c. i. Vertical reflection

ii. Horizontal reflection

21. a. i. $(3, 8)$

ii. $(6, 2)$

iii. $(-4, -1)$

iv. $(3, -5)$

b. $\left(\dfrac{a+m}{2}, \dfrac{b+n}{2}\right)$

22. a. Quadrant 2 or 3   **b.** Quadrant 1 or 2

c. Quadrant 3   **d.** Quadrant 4

23. a.

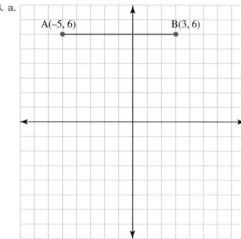

b. 8 units

c. $(-1, 6)$

d. Take care when determining the middle point of your second horizontal line.

e. $\dfrac{x_1 + x_2}{2}$

24. a.

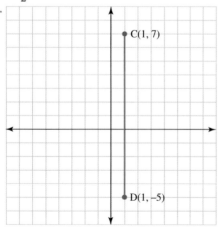

b. 12 units

c. $(1, 1)$

d. Take care when determining the middle point of your second vertical line.

e. $\dfrac{y_1 + y_2}{2}$

25. One possible answer is: $(2, 2), (-2, 2), (-2, -2), (2, -2)$.

26. a. The x-coordinate increases by 2 and the y-coordinate increases by 4. The y-coordinate is always 3 more than double the x-coordinate.

b. $(10, 23)$ $(12, 27)$ $(14, 31)$

c. $y = 2x + 3$

27. a. $(0, 5), (5, 0), (-5, 0), (0, -5)$

b. $(3, 4), (4, 3), (-3, 4), (-4, 3), (-3, -4), (-4, -3),$ $(3, -4), (4, -3)$ all lie on a circle of radius 5 units with centre at the origin.

9.3 Linear number patterns

1. a. $1, 5, 9, 13, 17$

b. $5, 14, 23, 32, 41$

2. a. $50, 42, 34, 26, 18$

b. $64, 53, 42, 31, 20$

3. a. $10, 12, 14$ **b.** $23, 28, 33$

4. a. Value of term $= 30 -$ position in pattern $\times 3$

b. $15, 12, 9$

5. a. Value of term $= 55 -$ position in pattern $\times 7$

b. $20, 13, 6$

6. a. $6, 15, 18$ **b.** $10, 12, 16$ **c.** $16, 64$

7. a. $9, 11, 17$ **b.** $88, 110$ **c.** $95, 90, 75$

8. a. Value of term $=$ position in pattern $\times 10 - 5$

b. 145

9. a. Value of term $= 47 -$ position in pattern $\times 3$

b. 2

10. a. Value of term $=$ position in pattern $\times 6 + 96$

b. 306

11. a. Value of term $= 225 -$ position in pattern $\times 13$

b. -74

12. a. $5, 6, 7, 8$ and 9

b. 14

c. 104

13. a. $0, 2, 4, 6$ and 8

b. 38

c. 152

14. When relating the value of a term to its position in the pattern, you can use that description to find any other future term in the number pattern.

15. Yes, you can substitute appropriate algebraic symbols for parts of the written descriptions to create an algebraic statement. For example, let the number of matches $= m$ and the number of shapes $= n$.

16. Once you have identified the connection between the top row and the bottom row in a table of values, you have identified the rule for the pattern.

17. a.

b.

Number of triangles	1	2	3	4	5
Number of matches	3	5	7	9	11

c. Number of matches = number of triangles × 2 + 1

d. 15

18. a.

b.

Number of houses	1	2	3	4	5
Number of matches	6	11	16	21	26

c. Number of matches = number of houses × 5 + 1

d. 36

19. a.

Number of shapes	1	2	3
Number of matches	5	10	15

b. $m = 5n$

20. a.

b.

Number of fence units	1	2	3	4	5
Number of matches	4	7	10	13	16

c. Number of matches = number of fence units × 3 + 1

d. 22

21. a.

Number of shapes	1	2	3
Number of matches	6	14	22

b. $m = 8n - 2$

22. See a sample pattern below. Other sample responses can be found in the worked solutions in the online resources.

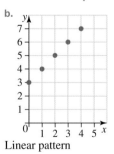

9.4 Plotting simple linear relationships

1. a.
   Linear pattern

b.
   Linear pattern

c.
   Linear pattern

d.
   Non-linear pattern

e.
   Linear pattern

f.
   Linear pattern

2. a.
   Linear pattern

b.
   Linear pattern

c.
   Non-linear pattern

d.
   Linear pattern

3. a.

Linear pattern

b.

Vertical line

c.

Non-linear pattern

d.

Linear pattern

4. a.

Linear relationship

b.

Non-linear relationship

c.

Linear relationship

d.

Non-linear relationship

e.

Linear relationship

f.

Linear relationship

c.

Non-linear pattern

5. a.

Linear pattern

d.

Non-linear pattern

b.

Linear pattern

6. a.

Linear pattern

b.

Linear pattern

c.

Non-linear pattern

d.

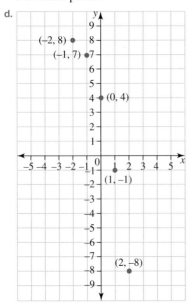

Non-linear pattern

7. a, c.

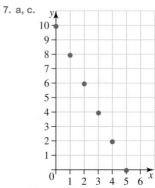

b. Linear pattern

d. (4, 2), (5, 0)

8. a, c.

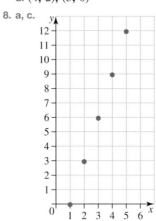

b. $y = 6$

9. a, b, d

10. a. (8, 7)   **b.** (3, 2)

11. a.

Week	1	2	3	4	5	6
Height (mm)	20	24	28	32		40

b.

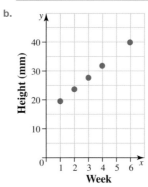

c. Yes, the points form a straight line.

d. 36 mm

e. 48 mm

12. a. The rule describes a set of points such that the y-coordinate of each point is 3 more than the x-coordinate.

b.

x	-2	-1	0	1	2
y	1	2	3	4	5

c.

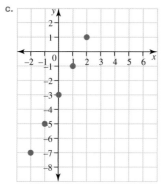

d. Yes, the relationship is linear as it produces a straight line.

13. a. The rule describes a set of points such that the y-coordinate of each point is three less than twice the x-coordinate.

b.

x	-2	-1	0	1	2
y	-7	-5	-3	-1	1

c.

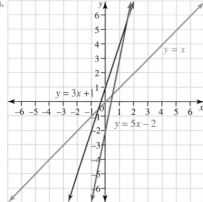

d. Yes; the points form a straight line.

e. $-7 \le y \le 1$

14. a.

b.

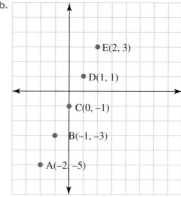

c. *Differences*: The coefficient of x is positive in the first set and negative in the second set. In the first set, the graphs are sloping up, while in the second set the graphs are sloping down. *Similarities*: They are all straight lines.

15. a.

x	-2	-1	0	1	2
y	-5	-3	-1	1	3

b.

E(2, 3)

D(1, 1)

C(0, −1)

B(−1, −3)

A(−2, −5)

c. 1 unit

d. Yes, the points form a straight line.

16. a. Parking fee $= 0.04 \times$ time

b.

Time, minutes	30	60	90	120	150
Parking fee, $	1.20	2.40	3.60	4.80	6.00

17. a. Cost per kg $= 49.75 - 0.75 \times$ weight in kg

b. 10 and 31

c. $731.50

d. Seven 30 kg bags, so the total weight ordered is 210 kg.

18. 28 desks

Project

Players should apply their knowledge of coordinates and the Cartesian plane and play the game to be the first to sink all their opponent's ships.

9.5 Review questions

1. a. Joseph's house: B1; Post office: B4

   b.

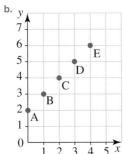

2. a. A

   b. K, I, L, J

   c. M, N, O, P

   d. A(−3, 4), B(−4, 3)

   e. C(−4, −3), F(4, −3)

3. a. A(2, 4), (3, 5)

   b.

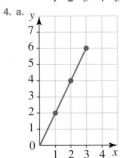

4. a.

   b. Yes, the points form a straight line.

5. a. 15, 18, 21   b. 38, 46, 54

6. a. Value of the term = position in the pattern × 15 + 35

   b. 260

7. a. 7, 10, 13   b. 34   c. 154

8. a.

   Linear

   b.

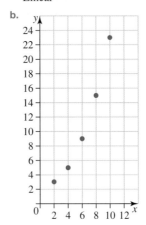

   Non linear

9. H or X

   H lies in column 3 and row 2.

   X lies in column 4 and row 5.

10. a. i. H   ii. H   iii. S   iv. M

    b. B9, C2, H6, I10

    c. A5 and A6 are already hit, so he needs A4 and A7 to sink it.

11. a. QUEENSLAND (6, 6), (4, 4), (12, 8), (12, 8), (14, 14), (16, 16), (8, 14), (10, 10), (14, 14), (12, 12)

    b. VICTORIA (4, 18), (8, 8), (10, 12), (16, 4), (14, 6), (6, 16), (8, 8), (10, 10)

    c. SYDNEY (16, 16), (18, 18), (12, 12), (14, 14), (12, 8), (18, 18)

12. (3, 1), (2, 3), (2, 1), (3, 1), (2, 2), (5, 1), (2, 3), (2, 1), (5, 0), (0, 3), (5, 0), (0, 1), (3, 2), (0, 3), (5, 0), (3, 1), (1, 0), (1, 2)

13.

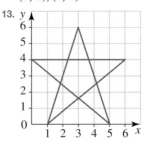

14. D is (3, 7).

15. a.

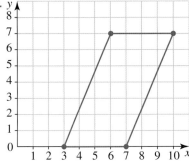

   i. Parallelogram

   ii. 28 square units

b.

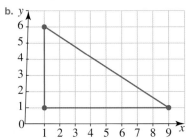

One example could be Triangle ABC where A is (1, 1), B is (1, 6) and C is (9, 1).

The area of this triangle is:

$\frac{1}{2} \times$ base $\times$ height

$= \frac{1}{2} \times 8 \times 5$

$= 20$ square units as required.

16. D$(1, -5)$

17. No; the relation is $y = x^2$, which is not in the form $y = mx + c$.

18. Not true, as a straight line could:
- be parallel to the y-axis and only cut the x-axis once
- be parallel to the x-axis and only cut the y-axis once
- pass through the origin, cutting the axes at only that point.

19. a. Parking fee $= 0.25 \times$ time

b.

Time, minutes	10	20	30	70	100
Parking fee, $	2.50	5.00	7.50	17.50	25.00

20.

10 Length, area and volume

LESSON
10.1 Overview

Why learn this?

Measurement, together with geometry, is important in our everyday lives. We are able to describe objects using numbers and units of measurement, such as millimetres, centimetres, grams and kilograms. How far is it from your home to the nearest beach? What is the size of a tennis court or a football ground? How long is a cricket pitch? How much sugar is in your favourite cookie? All of these questions and thousands more are answered using measurement.

When we measure objects we need to understand the units of measurement. Which is longer, a centimetre or a millimetre? Can you change centimetres to millimetres and millimetres to centimetres? From stump to stump, a cricket pitch is 20.12 metres or 2012 centimetres or 20120 millimetres!

Olympic swimming pools need to be 50 metres long and 25 metres wide. Builders must know these dimensions before constructing a pool. Before anything can be made, you need to decide how small or large it will be — that is, its dimensions or measurements. We use measurement to describe length, perimeter, area, volume and capacity every day. Can you imagine trying to build your home without first knowing what its size will be? Many professionals use measurement in their day-to-day work. Among these are property developers, builders, engineers, designers, dressmakers, chefs, architects and construction workers.

DISCUSSION

Why is it important to have an international standard metric system? Which countries don't follow this system?

Hey students! Bring these pages to life online

▶ **Watch videos**

🧩 **Engage with interactivities**

A+ **Answer questions and check solutions**

Find all this and MORE in jacPLUS ▶

Reading content and rich media, including interactivities and videos for every concept

Extra learning resources

Differentiated question sets

Questions with immediate feedback, and fully worked solutions to help students get unstuck

1. Convert 0.565 m into cm.

2. If the length of a cricket pitch is 20.12 m, calculate its length in km.

3. Luca went for a training run that was 1200 strides long. If each of his strides was 1.5 m, calculate the total distance he covered.

4. Determine the perimeter of this figure.

5. A rectangular pool has a perimeter of 34 m. If a concrete border of width 2 m was laid around the pool, calculate the perimeter around the outer section of the concrete border.

6. If the height of a city building is 36 m and it measures 10 cm in a photo, calculate how tall a building that measured 25 cm in a photo would be in real life.

7. Determine the area of the triangle shown.

8. A rectangular room with dimensions 3.5 m by 6 m is to be carpeted at a cost of $22.50 per m². Calculate the total cost to carpet the room.

9. Calculate the area of the figure at right.

10. A rectangular wall 6.2 m by 2.4 m is to be painted. A rectangular cabinet 1.8 m by 1.2 m rests against this wall. The area to be painted does not include the area of the cabinet. Calculate the area of the wall to be painted.

11. Calculate the volume of a rectangular prism with a length of 8 m, a width of 3 m and a height of 2 m.

12. A rectangular prism has a volume of 112 cm³. If it has a height of 2 cm and a width of 8 cm, determine its length.

13. A rectangular sink has a length of 45 cm, width of 40 cm and a depth of 30 cm. Calculate the sink's capacity, in litres.

14. Match the following three-dimensional shapes to their plan views.

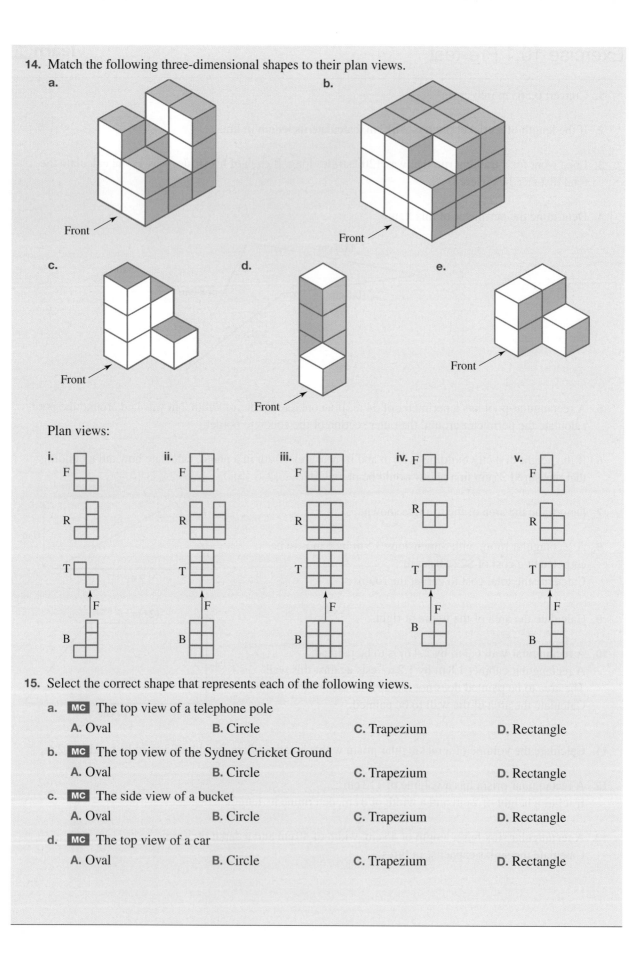

15. Select the correct shape that represents each of the following views.

a. **MC** The top view of a telephone pole

    A. Oval   **B.** Circle   **C.** Trapezium   **D.** Rectangle

b. **MC** The top view of the Sydney Cricket Ground

    A. Oval   **B.** Circle   **C.** Trapezium   **D.** Rectangle

c. **MC** The side view of a bucket

    A. Oval   **B.** Circle   **C.** Trapezium   **D.** Rectangle

d. **MC** The top view of a car

    A. Oval   **B.** Circle   **C.** Trapezium   **D.** Rectangle

LESSON
10.2 Units of length

▶ 10.2.1 Metric units of length

eles-4549

• The metric system is based on the number 10.
• The base unit of length of the metric system is the metre.
• The following figures show the most commonly used units of length along with their abbreviations and photos showing approximate examples.

1. Kilometre (km)

One kilometre is the distance travelled in one minute by a car travelling at a speed of 60 kilometres per hour.

2. Metre (m)

The length of an adult's stride

3. Centimetre (cm)

The width of each of your fingers

4. Millimetre (mm)

The width of a wire in this computer chip

DISCUSSION

Where do you use measurement in your everyday life?

You have been given the task of measuring the width of an A4 page. State which metric unit of length you would use.

THINK	WRITE
Estimate the length involved. In this case, it will be less than a metre and more than a centimetre. Consider which unit would be easiest to use. If you use metres, the width would be expressed as a decimal. If you use centimetres, the width would be expressed as a whole number. Therefore, centimetres is the best choice of unit.	Centimetres (or cm)

10.2.2 Converting units of length

eles-4550

- The main metric units of length are related as follows:

$$1\,km = 1000\,m \quad 1\,m = 100\,cm \quad 1\,cm = 10\,mm$$

Unit conversion

Units of length can be converted as shown in the following diagram. The numbers next to each arrow are called conversion factors.

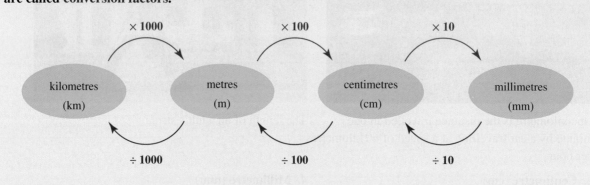

- When converting from a larger unit to a smaller unit, multiply by the conversion factor.
- When converting from a smaller unit to a larger unit, divide by the conversion factor.

Complete each of the following metric conversions.

a. **0.3285 km = _____ m**

b. **560 m = _____ mm**

c. **480 cm = _____ km**

d. **$2\frac{3}{5}\,m$ = _____ cm**

THINK	WRITE
a. To convert kilometres to metres, multiply by 1000. (Move the decimal point 3 places to the right.)	a. $0.3285\,km = 0.3285 \times 1000\,m$ $= 0.3285$ $= 328.5\,m$

b. To convert from metres to centimetres, multiply by 100 (since there is no decimal point, place two zeros after the final digit).

$\times 100$

b. $560\,\text{m} = 560 \times 100\,\text{cm}$
$= 56\,000\,\text{cm}$

To convert from centimetres to millimetres, multiply by 10 (place one zero after the final digit).

$\times 10$

$= 56\,000 \times 10\,\text{mm}$
$= 560\,000\,\text{mm}$

Note: Overall, we need to multiply by 100×10 or 1000.

c. To convert from centimetres to metres, divide by 100 (move the decimal point 2 places to the left).

$\div 100$

c. $480\,\text{cm} = 480 \div 100\,\text{m}$
$= 4.8\,\text{m}$

To convert from metres to kilometres, divide by 1000 (move the decimal point 3 places to the left).

$\div 1000$

$= 4.8 \div 1000\,\text{km}$
$= 0.0048\,\text{km}$

d. 1. Convert $2\dfrac{3}{5}\,\text{m}$ to an improper fraction.

d. $2\dfrac{3}{5}\,\text{m} = \dfrac{2 \times 5 + 3}{5}\,\text{m}$
$= \dfrac{13}{5}\,\text{m}$

2. Convert metres to centimetres by multiplying by 100.

$\times 100$

$= \dfrac{13}{5} \times 100\,\text{cm}$
$= \dfrac{1300}{5}\,\text{cm}$

Note: Remember when multiplying fractions by a whole number, the number multiplies the numerator.

3. Simplify the fraction.

$= 260\,\text{cm}$

---

on Resources

Interactivities Metric units of length (int-4010)

Converting units of length (int-4011)

10.2 Quick quiz on	10.2 Exercise

Individual pathways

■ PRACTISE	■ CONSOLIDATE	■ MASTER
1, 3, 4, 6, 9, 14, 16, 18, 20, 25, 28	2, 5, 7, 11, 13, 15, 17, 22, 23, 26, 29, 30	8, 10, 12, 19, 21, 24, 27, 31, 32, 33, 34

Fluency

1. **WE1** State which metric units (mm, cm, m or km) would be most suitable for measuring the real lengths marked in each photograph.

    a. The length of a large kangaroo

    b. Large distances between towns

    c. The diameter of a DVD

    d. The diameter of a planet

    e. The height of a tall building

2. State which metric units of length you would use for measuring the following.

    a. The length of a netball court
    b. The diameter of a netball
    c. The distance between Melbourne and Sydney
    d. The thickness of a magazine

3. **MC** Select which unit you would use to measure and compare the thicknesses of two different brands of chocolate biscuits.

    A. millimetres   **B.** kilometres   **C.** metres   **D.** centimetres

4. **WE2** Complete each of the following metric conversions.

    a. $2.0\,km = _____\,m$
    b. $7.0\,km = _____\,m$
    c. $5.3\,km = _____\,m$
    d. $0.66\,km = _____\,m$

5. Complete each of the following metric conversions.

   a. $0.25\,m = $ _____ cm
   c. $200.0\,cm = $ _____ mm

   b. $28.0\,cm = $ _____ mm
   d. $700.0\,m = $ _____ cm

6. Convert to the units indicated.

   a. $8000\,m = $ _____ km
   c. $700\,m = $ _____ km

   b. $6500\,m = $ _____ km
   d. $50\,m = $ _____ km

7. Convert to the units indicated.

   a. $6000\,cm = $ _____ m
   c. $45\,mm = $ _____ cm

   b. $57\,cm = $ _____ m
   d. $25\,600\,mm = $ _____ cm

8. Convert to the units indicated.

   a. $8\,km = $ _____ cm
   c. $72.33\,m = $ _____ mm

   b. $101\,m = $ _____ mm
   d. $30\frac{7}{20}\,mm = $ _____ m

9. **MC** Identify which of these distances is the same as $6.25\,km$.

   A. $625\,m$
   C. $625\,000\,mm$

   B. $0.006\,25\,m$
   D. $625\,000\,cm$

10. **MC** Identify which of these distances is the same as $7\,860\,000\,cm$.

   A. $78.6\,km$
   C. $786\,km$

   B. $786\,m$
   D. $786\,000\,cm$

11. Convert to the units indicated.

   a. $45\frac{1}{5}\,m$ to km
   b. $560\,mm$ to m
   c. $8\frac{3}{4}\,cm$ to mm

12. Convert to the units indicated.

   a. $0.0006\,km$ to cm
   b. $3\frac{9}{100}\,km$ to cm
   c. $48\,mm$ to cm

Understanding

13. Give an example of a length that each of the following people might measure in their jobs. (For example, a carpet layer would measure the length of a room.)

   A. Veterinary surgeon
   c. Carpenter
   e. Field athlete

   b. Costume designer
   d. Landscape gardener

14. The longest snake ever held in captivity was a female reticulated python named Colossus. She measured 8.68 m long. The shortest species of snake, the West Indian *Leptotyphlops bilineatus*, only grows to 108 mm. Write, in centimetres:

   a. the length of the longest snake
   b. the length of the shortest snake.

15. The world's highest mountains are Everest (8848 m) and K2 (8611 m). Convert these heights to kilometres.

16. Arrange the following in order from smallest to largest.

    a. 12.5 m, 150 cm, 0.02 km
    b. 350 cm, 0.445 m, 3000 mm
    c. 50 km, 500 m, 50 000 mm

17. Arrange the following in order from smallest to largest.

    a. 1700 cm, 1.7 m, 0.17 km
    b. 0.052 cm, 0.0052 mm, 0.000 052 m
    c. 990 cm, 0.909 m, 9000 mm

18. Add the following lengths, giving your answer in the specified unit.

    a. 75 cm and 3 m: cm
    b. 2700 m and 7.5 km: m
    c. 1.66 m and 58.2 cm: cm
    d. 0.000 675 km and 87.8 cm: cm

19. Calculate the difference between each of the following lengths, giving your answer in the specified unit.

    a. 72 km and 5600 m: m   b. 418 000 mm and 7.6 m: m

    c. $34\frac{3}{5}$ cm and $\frac{9}{20}$ m: cm   d. $2\frac{4}{5}$ km and 450 000 cm: km

20. The tallest building in Dubai is the Burj Khalifa, which stands 0.828 km high. Convert this height to metres.

21. A builder needs to build a wall 3.5 m high. If each layer of bricks adds 8 cm of height, calculate how many layers of bricks are in the wall.

22. Deanne buys a length of rope and cuts it into three smaller sections, each of length 7200 cm.
    Calculate how long the original piece of rope measured.

23. Norbert is 1.53 m tall in bare feet. If he wears shoes with soles that are 6.3 cm thick, calculate how tall he is when wearing these shoes, in metres.

24. Adrian is driving a truck with a rooftop 3.4 m above road level when he approaches an overpass bridge that has a clearance of 376 cm. Determine whether Adrian's truck will fit under the bridge. If so, state the room he has spare, in centimetres.

Communicating, reasoning and problem solving

25. Zvenglo is stacking identical boxes of height 330 mm. Calculate the height of a stack of six boxes, in centimetres.

26. Finita attaches a trellis (shown) that is 0.6 m high to the top of her 180 cm high fence. Calculate the height to the top of the new trellis from ground level. Give your answer in metres.

27. Astronomers use light-years as a measure of distance in the universe. A light-year is the distance travelled by light in one year. If light travels approximately 300 000 km in one second, calculate the distance travelled by light in:

   a. i. one minute
      iii. one day
   ii. one hour
   iv. one year (365 days).

   b. Explain why astronomers use this measurement for distance rather than kilometres.

28. A pin, 14 mm long, is made of wire. Determine how many whole pins could be made from 1 km of wire.

29. A licorice strap machine takes 3.80 m lengths of licorice and chops them into 10 cm long pieces. Determine how many pieces each 3.80 m length produces.

30. Waldo's noticeboard is 1.5 m long and 1.2 m wide. If he pins a calendar of length 70 cm and width 60 cm exactly in the middle of the board, determine the distance from the top of the calendar to the top of the noticeboard. (*Hint:* Draw a diagram of the situation.)

31. A childcare centre has three large cardboard boxes that the children stack up in various combinations. What stack heights are possible if the boxes' individual heights are 600 mm, 45 cm and 1.1 m? Give your answer in centimetres.

32. A mother and daughter are riding their bikes to the local market. The circumference of the mother's bike wheel is 2 m while the circumference of the daughter's bike wheel is 1 m.

   a. Calculate the number of rotations of the wheel there are in 100 m for the:
      i. mother
      ii. daughter

   b. Calculate the number of rotations of the wheel there are in 1 km for the:
      i. mother
      ii. daughter

33. In a 'log cabin' quilt, a pattern is created from a series of squares and rectangles. From a centre square, 4 congruent rectangles are placed to build a larger square. This process is repeated a number of times to build larger and larger squares.

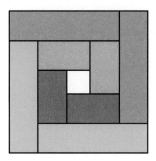

   Laurel starts with a centre square of side length 2 cm. If each rectangle has a width of 2 cm and a length that depends on its position in the block, give the dimensions of the rectangles that will need to be cut if the final block is to be 18 cm square. (Don't worry about seam allowance.)

34. It is said that the average person walks the equivalent of four times around the Earth in a lifetime. The circumference of the Earth is about 40 000 km. If you lived to age 80, approximate the number of kilometres per week you would walk to achieve this distance.

LESSON
10.3 Perimeter

LEARNING INTENTION

At the end of this lesson you should be able to:
• calculate the perimeter of a given shape
• calculate the perimeter of rectangles and squares.

▶ 10.3.1 Calculating the perimeter

eles-4552

• A **perimeter** is the distance around the outside (border) of a shape.
• To calculate the perimeter of the shape, convert all lengths to the same unit and then add all the lengths.

Perimeter = 4 + 5 + 3 + 2 + 10 + 10
= 34 cm

WORKED EXAMPLE 3 Calculating the perimeter of shapes

Calculate the perimeter of each shape.

a.

b. 13 cm
12 cm   10 cm

c. 30 mm
6 cm   45 mm

Note: **The dots are 1 unit apart.**

THINK		WRITE
a. 1. Count the number of unit intervals around the outside of the shape (16).	a.	
2. Write the answer.		The perimeter is 16 units.
b. 1. Check that the measurements are in the same unit. All are in centimetres.	b.	
2. Add the measurements to determine the perimeter.		12 13 + 10 ⎯⎯ 35
3. Write the answer with the correct unit.		The perimeter is 35 cm.

---

c. 1. Notice the measurements are not all the same. Convert to the smaller unit (6 cm = 60 mm).

c. 6 cm = 60 mm

2. Add the measurements that now have the same unit (mm) to determine the perimeter.

$$\begin{array}{r} 60 \\ 30 \\ +\,45 \\ \hline 135 \end{array}$$

3. Write the answer with the correct unit.

The perimeter is 135 mm.

WORKED EXAMPLE 4 Calculating perimeters of squares and rectangles

Calculate the perimeter of:
a. a rectangle that is 49.3 cm long and 22.0 cm wide
b. a square whose side length is 28 cm.

THINK

WRITE

a. 1. Draw a diagram and write its measurements.

a.

2. Check that the measurements are in the same unit. All are in centimetres. The perimeter is the distance around the rectangle, so add all the distances together.

$P = 49.3 + 22.0 + 49.3 + 22.0$
$\quad = 142.6$

3. Write the answer with the correct unit.

The perimeter is 142.6 cm.

b. 1. Draw a diagram and write its measurements.

b.

```
        28 cm
    ┌──────────┐
28 cm│          │28 cm
    └──────────┘
        28 cm
```

2. Check that the measurements are in the same unit. The perimeter is the distance around the square, so add all the distances together.

$P = 28 + 28 + 28 + 28$
$\quad = 112$

3. Write the answer with the correct unit.

The perimeter is 112 cm.

DISCUSSION

What information does the perimeter give you about a shape? Is it enough to draw the shape?

▶ 10.3.2 Calculating the perimeter of a rectangle and a square

• The perimeters of rectangles and squares can be found using the following formulas.

Perimeter of a rectangle

For a rectangle, the perimeter P is:

$$P = 2(l + w)$$

where l is its length and w is its width.

Perimeter of a square

For a square, the perimeter P is:

$$P = 4l$$

where l is its side length.

DISCUSSION

Consider the shape below.

How many ways of calculating its perimeter can you think of? Is there a method you think is more efficient?

COMMUNICATING — COLLABORATIVE TASK: Same perimeter, different shape

Equipment: a ball of string, A3 paper, marker pens, glue or sticky tape, ruler

1. Cut a piece of string into four equal lengths.
2. Measure and record the lengths of your pieces of string.
3. Make an enclosed shape with one of your pieces of string and stick it onto an A3 sheet of paper.
4. Write a caption about the perimeter of the shape. For example, you might write, 'It is 25 cm around this shape'.
5. Repeat steps **3** and **4** with your other pieces of string. Each time, make a different shape and write a different caption.
6. As a class, discuss the answer to the following question: What information does the perimeter give you about a shape?

on Resources

▶ **Video eLesson** Perimeter (eles-1874)

✦ **Interactivities** Perimeter (int-4013)
Perimeter of composite shapes (int-4014)

Exercise 10.3 Perimeter

learn on

10.3 Quick quiz **on**	10.3 Exercise

Individual pathways

■ PRACTISE	■ CONSOLIDATE	■ MASTER
1, 2, 4, 7, 10, 13, 16, 19, 22	3, 5, 8, 11, 14, 17, 20, 23	6, 9, 12, 15, 18, 21, 24

Fluency

1. **WE3a** Calculate the perimeter of each shape. The dots are 1 unit apart.

a.   b.   c.

    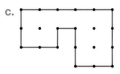

d.   e.

2. **WE3b** Calculate the perimeter of each of the following.

a.

12 cm

8 cm

b. 11 mm

35 mm

3. Calculate the perimeter of each of the following.

a.

b.

4. Calculate the perimeter of the shape shown.

5. Calculate the perimeter of the shape shown.

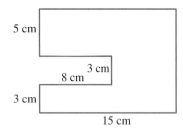

6. Calculate the perimeter of the shape shown.

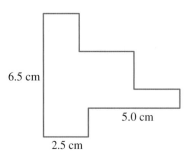

7. **WE3c** Calculate the perimeter of each shape, giving answers in the smaller unit in each case.

a.

b.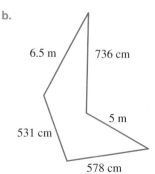

8. Calculate the perimeter of each shape, giving answers in the smaller unit in each case.

a.

6.8 cm

136 mm

b.
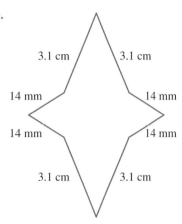
3.1 cm 3.1 cm

14 mm 14 mm

14 mm 14 mm

3.1 cm 3.1 cm

9. Calculate the perimeter of each shape, giving answers in the smaller unit in each case.

a.

3 km

4863 m

2.7 km

3.87 km

b.

2.88 km

2.16 km 4016 m

1982 m

6541 m

10. **MC** Identify the perimeter of the shape shown.

A. 9 units
B. 18 units
C. 16 units
D. 14 units

Note: The dots are 1 unit apart.

11. **WE4** Calculate the perimeter of:

a. a rectangle of length 45 cm and width 33 cm
b. a rectangle of length 2.8 m and width 52.1 cm
c. a square of side length 3.7 cm.

12. a. Calculate the perimeter of a rectangle of length $4\frac{1}{4}$ m and width $2\frac{1}{5}$ m.

b. Calculate the perimeter of a square of side length $8\frac{1}{5}$ mm.

Understanding

13. Calculate the length of party lights needed to decorate the perimeter of a rectangular tent with dimensions 15.5 m by 8.75 m.

14. Allowing an extra 30 cm for waste, calculate the length of picture frame required to frame the artwork shown in the photo.

63 cm

48 cm

15. Zedken wishes to install three strands of barbed wire at the top of the fences around a rectangular work site. The length of the site is 34.5 m, and its width is 19.8 m. Calculate the length of wire needed.

16. A new game, Bop-ball, is played on a triangular field, where each side of the triangle measures 46.6 m. A greenkeeper is marking the field's perimeter using a chalk-dispensing trundle wheel. Determine how far the greenkeeper will walk to mark the field.

17. Phang's property boundary dimensions are shown in the diagram.

    a. Calculate how many metres of fencing are needed to fence all but the front boundary.
    b. If the fencing costs $19 per metre, calculate the total cost.

18. Lucille has received a quote of $37 per metre for new fencing for a tennis court. The tennis court is 23.77 m long and 10.97 m wide. There should be a 3.6 m space between each side of the court and the fence, and a 6.4 m gap between each end of the court and the fence.

    a. Draw a diagram showing all given measurements.
    b. Determine how many metres of fencing are needed.
    c. Based on the quote, calculate the total cost of the fencing.

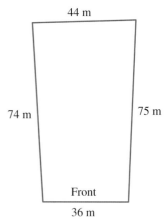

Communicating, reasoning and problem solving

19. Suggest a type of work where people need to calculate perimeters on a daily basis.

20. A piece of paper is 22 cm long, including a 2.2 cm sticky strip at each end. Five strips of paper are stuck together so that the sticky parts overlap exactly to make a loop of paper. Determine the circumference of (distance around) the loop.

21. Marc and Cathy are seeking quotes for the cost of building a fence on three sides of their property. They want to calculate approximate costs for each item to decide whether the quotes supplied sound reasonable.

The new fence requires three rails, a kickboard, posts and palings.

    a. Calculate the length of timber needed for the kickboard.
    b. Calculate the cost of the kickboard if the timber required for this costs $1.90 per metre.
    c. If the timber for the rails costs $2.25 per metre, determine the total cost of the timber for the railings.
    d. Determine how many posts will be needed for the new fence if each post is to be 5 metres apart and there needs to be a post at the end of each straight section of fence.
    e. Calculate the cost of the posts if the price of each post is $13.65.

Palings are 9 cm wide and are nailed so they overlap each other by 15 mm on each side.

    f. i. Calculate the approximate number of palings needed for the fence.
       ii. Palings cost $1.05 each. Determine how much money should be allowed for the total cost of the palings.

    g. Write an itemised list of all the costs involved. Include an amount to cover labour charges and miscellaneous items like the cost of nails. This amount is around $1000 (two people for two days at approximately $30.00 per hour for an eight-hour day). Estimate the cost of the new fence. This will provide Marc and Cathy with information to use when comparing builders' quotes.

22. Kelly has a piece of string 32 cm long. He uses it to form rectangular shapes that each have a length and a width that are a whole number of centimetres.

   a. Identify how many different rectangular shapes he can form.
   b. Predict what the sides of the rectangle will be to give the largest possible area. Check by drawing different rectangles to see if your prediction is correct.

23. Write an equation for the perimeter of the diagram shown in terms of a and b.

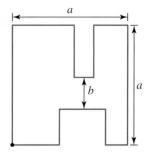

24. The rectangle shown is made up of three identical, smaller rectangles that fit inside (as shown). If its length measures 9 cm, work out its width without using a ruler.

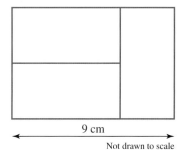

9 cm

Not drawn to scale

LESSON
10.4 Circles and circumference

▶ 10.4.1 Circumference

eles-4044

• The distance around a circle is called the **circumference** (C).
• The **diameter** (d) of a circle is the straight-line distance across a circle though its centre.
• The **radius** (r) of a circle is the distance from the centre to the circumference.
• The diameter and radius of a circle are related by the formula $d = 2r$. That is, the diameter is twice as long as the radius.

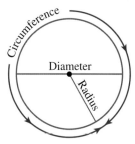

Measure the diameters and circumferences of a variety of different circles and cylinders (for example, coins and drink bottles), and record your results in a table. What do you notice about the value of $\frac{C}{d}$ for each circle you measured?

Pi (π)

- The ratio of acircle's circumference to its diameter, $\frac{C}{d}$ gives the same value for any circle no matter how large or small the circle is.
- This special numberis known as pi (π). That is, $\pi = \frac{C}{d}$.
- Pi (π) is an irrational number and cannot be written as a fraction.
- When pi is written as a decimal number, the decimal places continue forever with no repeated pattern.
- Pi written to eight decimal places is 3.141 592 65.
- All scientific calculators have a π button and this feature can be used when completing calculations involving pi.

Digital technology

Scientific calculators can be used to assist with calculations involving pi. The π button can be found on most calculators by pressing SHIFT + $\times 10^x$.

STAT   D

ㅠ

3.141592654

Calculating circumference

The circumference, C of a circle can be determined using one of the following formulas.

$$C = 2\pi r, \text{ where } r \text{ is the radius of the circle}$$

or

$$C = \pi d, \text{ where } d \text{ is the diameter of the circle}$$

Note: $2r = d$, where the diameter is twice the length of the radius

Calculate the circumference of each of the following circles, giving answers:

i. **in terms of π**

ii. **correct to 2 decimal places.**

a.

b.

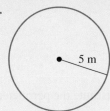

THINK

a. i. 1. Write the formula for the circumference of a circle.
Note: Since the diameter of the circle is given, use the formula that relates the circumference to the diameter.

2. Substitute the value $d = 24$ cm into the formula.

3. Write the answer and include the units.

ii. 1. Write the formula for the circumference of a circle.

2. Substitute the value $d = 24$ cm into the formula.

3. Evaluate the multiplication using a calculator and the π button.

4. Round correct to 2 decimal places and include the units.

b. i. 1. Write the formula for the circumference of a circle.
Note: Since the radius of the circle is given, use the formula that relates the circumference to the radius.

2. Substitute the value $r = 5$ m into the formula.

3. Write the answer and include the units.

ii. 1. Write the formula for the circumference of a circle.

2. Substitute the value $r = 5$ m into the formula.

3. Evaluate the multiplication using a calculator and the π button.

4. Round the answer correct to 2 decimal places and include the units.

WRITE

a. i. $C = \pi d$

$= \pi \times 24$

$= 24\pi$ cm

ii. $C = \pi d$

$= \pi \times 24$

$= 75.398\ldots$

$= 75.40$ cm

b. i. $C = 2\pi r$

$= 2 \times \pi \times 5$

$= 10\pi$ m

ii. $C = 2\pi r$

$= 2 \times \pi \times 5$

$= 31.415\ldots$

$= 31.42$ m

▶ 10.4.2 Perimeter of a semicircle

eles-6095

- We can use the formula for circumference to calculate the length of the curved section of a semicircle.
- Firstly calculate the full circumference and then divide by 2.
- You must remember to add the length of the diameter for the total circumference of the semicircle.

WORKED EXAMPLE 6 Calculating the perimeter of a semicircle

Calculate the perimeter of the following shape, correct to 2 decimal places.

THINK	WRITE
1. Identify the parts that constitute the perimeter of the given shape.	$P = \dfrac{1}{2}$ circumference $+$ straight-line section
2. Write the formula for the circumference of a circle. *Note:* If the circle were complete, the straight-line segment shown would be its diameter. So the formula that relates the circumference to the diameter is used.	$P = \dfrac{1}{2}\pi d +$ straight-line section
3. Substitute the value $d = 12$ cm into the formula.	$= \dfrac{1}{2} \times \pi \times 12 + 12$
4. Evaluate the multiplication using a calculator and the π button.	$= 18.849\ldots + 12$ $= 30.849\ldots$
5. Round the answer correct to 2 decimal places and include the units.	$= 30.85$ cm

DISCUSSION

Consider a cross-section of a yidaki (didgeridoo), as illustrated below.

Outer circumference of 24 cm

Thickness of 7 mm

Do you have enough information to determine its inner circumference?

- If the circumference of a circle is known, it is possible to determine its radius or diameter using one of the formulas below.
  If $C = 2\pi r$, then $r = \dfrac{C}{2\pi}$.
  If $C = \pi d$, then $d = \dfrac{C}{\pi}$.

WORKED EXAMPLE 7 Determining the radius of a circle from its circumference

Calculate the radius of a cylindrical water tank with a circumference of 807 cm, correct to 2 decimal places.

THINK	WRITE
1. Write the rule relating circumference and radius.	If $C = 2\pi r$, then $r = \dfrac{C}{2\pi}$.
2. Substitute the value of the circumference into the formula.	$r = \dfrac{807}{2\pi}\,\text{cm}$
3. Evaluate the expression using a calculator.	$r = 128.438\ldots\,\text{cm}$
4. Round the answer correct to 2 decimal places.	$r = 128.44\,\text{cm}$

 Resources

Interactivity   Circumference (int-3792)

Exercise 10.4 Circumference

learn

10.4 Quick quiz on	10.4 Exercise

Individual pathways

■ PRACTISE	■ CONSOLIDATE	■ MASTER
1, 4, 7, 9, 10, 13, 16, 19	2, 5, 8, 11, 12, 14, 17, 20,	3, 6, 15, 18, 21

Fluency

1. **WE5a** Calculate the circumference of each of these circles, giving answers in terms of π.

   a.

   2 cm

   b.

   10 cm

   c.

   7 mm

2. Calculate the circumference of each of these circles, giving answers correct to 2 decimal places.

   a.

   0.82 m

   b.

   7.4 km

   c.

   34 m

3. **WE5b** Determine the circumference of each of the following circles, giving answers in terms of π.

a.

4 m

b.

17 mm

c.

8 cm

4. Calculate the circumference of each of the following circles, giving answers correct to 2 decimal places.

a.

1.43 km

b.

0.4 m

c.

10.6 m

5. **WE6** Determine the perimeter of each of the shapes below. (Remember to add the lengths of the straight sections.) Give your answers to 2 decimal places.

a.

10 cm

b.

16 mm

c.
24 m

d.

11 mm

e.

20 cm

6. Calculate the perimeter of each of the shapes below. (Remember to add the lengths of the straight sections.) Give your answers to 2 decimal places.

a.

18 cm

b.

1.4 m
1.2 m

c.

48 m
50 m

d.
75 cm

30 cm

7. **MC** The circumference of a circle with a radius of 12 cm is:

A. $\pi \times 12$ cm
B. $2 \times \pi \times 12$ cm
C. $2 \times \pi \times 24$ cm
D. $\pi \times 6$ cm

8. **MC** The circumference of a circle with a diameter of 55 m is:

A. $2 \times \pi \times 55$ m
B. $\pi \times \dfrac{55}{2}$ m
C. $\pi \times 55$ m
D. $\pi \times 110 \times 2$ m

Understanding

9. Calculate the circumference of the seaweed around the outside of this sushi roll, correct to 2 decimal places.

19 mm

10. A scooter tyre has a diameter of 32 cm. Determine the circumference of the tyre. Give your answer to 2 decimal places.

11. Calculate the circumference of the Ferris wheel shown at right, correct to 2 decimal places.

12. In a Physics experiment, students spin a metal weight around on the end of a nylon thread. How far does the metal weight travel if it completes 10 revolutions on the end of a 0.88 m thread? Give your answer to 2 decimal places.

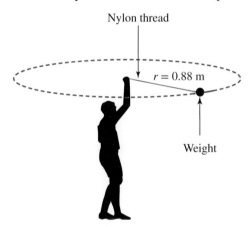

Nylon thread

$r = 0.88$ m

Weight

13. **WE7** Calculate the radius of a tyre with a circumference of 135.56 cm. Give your answer to 2 decimal places.

14. Calculate the diameter of a circle (correct to 2 decimal places where appropriate) with a circumference of:
    a. 18.84 m
    b. 64.81 cm
    c. 74.62 mm.

15. Calculate the radius of a circle (correct to 2 decimal places where appropriate) with a circumference of:
    a. 12.62 cm
    b. 47.35 m
    c. 157 mm.

Communicating, reasoning and problem solving

16. To cover a total distance of 1.5 km, a student needs to run around a circular track three times. Calculate the radius of the track correct to the nearest metre.

17. Determine the total length of metal pipe needed to assemble the wading pool frame shown. Give your answer in metres to 2 decimal places.

18. Nathan runs around the inside lane of a circular track that has a radius of 29 m. Rachel runs in the outer lane, which is 2.5 m further from the centre of the track. How much longer is the distance Rachel runs each lap. Give your answer to 2 decimal places.

19. A shop sells circular trampolines of four different sizes. Safety nets that go around the trampoline are optional and can be purchased separately. The entrance to the trampoline is via a zip in the net.

The following table shows the diameters of all available trampolines and their net lengths. Which safety net matches which trampoline?

	Diameter of trampoline		Length of safety net
a	1.75 m	i	6.03 m
b	1.92 m	ii	9.86 m
c	2.46 m	iii	5.50 m
d	3.14 m	iv	7.73 m

20. In *Around the world in eighty days* by Jules Verne, Phileas Fogg boasts that he can travel around the world in 80 days or fewer. This was in the 1800s, so he couldn't take a plane. What average speed is needed to go around the Earth at the equator in 80 days? Assume you travel for 12 hours each day and that the radius of the Earth is approximately 6390 km. Give your answer in km/h to 2 decimal places.

21. Liesel's bicycle covers 19 m in 10 revolutions of her bicycle wheel while Jared's bicycle covers 20 m in 8 revolutions of his bicycle wheel. Determine the difference between the radii of the two bicycle wheels. Give your answer in cm to 2 decimal places.

LESSON
10.5 Area

▶ 10.5.1 Metric units of area

eles-4554

- The **area** of a shape is the amount of flat surface enclosed by the shape.
- Area is measured in square units such as square centimetres, square metres and square kilometres.
- Commonly used metric units of area, with their abbreviations and examples, are shown in the following figures.

1. Square kilometres (km^2)

The area of a country or a large city like Sydney is given in square kilometres.

2. Square metres (m^2)

Square metres are used to measure the area of an object such as a classroom floor, whiteboard or a window.

3. Square centimetres (cm^2)

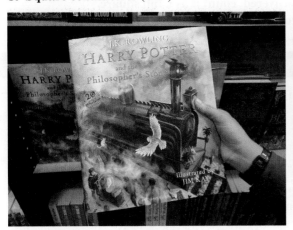

Small areas, such as the area of a sheet of A4 paper or a book cover, are measured in square centimetres.

4. Square millimetres (mm^2)

Very small areas, such as the area of a button or a postage stamp, are measured in square millimetres.

- The area of a farm or a large city park is measured in hectares (ha). $1\,\text{ha} = 10\,000\,\text{m}^2$.
- 1 square metre is the area enclosed by a square with sides that are 1 m long. The same is true for any other unit.
- If a shape is drawn on 1 cm grid paper, its area can be found by counting the number of squares that the shape covers.

WORKED EXAMPLE 8 Determining area by counting squares

The following figures are drawn on centimetre grid paper. Determine the area of each one.

a.

b.

THINK

a. Count the squares. Remember to include the correct unit (cm^2) in the answer.

b. Some of the squares are cut in half by the diagonal line. It takes two of these to make up one square centimetre. Count the squares. Remember to include the correct unit (cm^2) in the answer.

WRITE

a. $8\,\text{cm}^2$

1	4	
2	5	7
3	6	8

b. $8\,\text{cm}^2$

1/2	4	1/2
2	5	7
3	6	8

Estimating shaded areas

If a square of the grid paper is not completely covered by the shape, use the following as a guide to obtain an estimate of the area.
- **If more than half the square is covered, count it as a full square.**
- **If less than half the square is covered, do not count it at all.**

WORKED EXAMPLE 9 Estimating the shaded area

Estimate the shaded area of the following diagram drawn on centimetre grid paper.

THINK

1. Tick the squares that are more than half covered and count them.

2. State the answer, with the correct unit.

WRITE

$6\,\text{cm}^2$

⏵ 10.5.2 Calculating the area of a rectangle

eles-4555

• The areas of rectangles and squares can also be calculated by using the following formulas.

Area of a rectangle

For a rectangle, the area A is:

$$A = l \times w$$
$$= lw$$

where l is the length and w is the width of the rectangle.

Area of a square

For a square, the area A is:

$$A = l \times l$$
$$= l^2$$

where l is the length of the square's side.

Digital technology

A quick way to calculate the area of a square on a calculator is to use the 'square' button. The button has the symbol x^2 and will square the number that you type in.

WORKED EXAMPLE 10 Calculating areas of rectangles and squares

Calculate the areas of the following shapes.

a.

4 cm

3 cm

b.

4.5 m

THINK

a. 1. Write the formula for the area of a rectangle.

2. Substitute the value 4 for l and 3 for w and calculate the area.

WRITE

a. $A = lw$

$l = 4$ cm

$w = 3$ cm

$A = 4 \times 3$

$= 12 \text{ cm}^2$

3. State the answer in the correct unit (cm²).   The area of the rectangle is 12 cm².

b. 1. Write the formula for the area of a square.   b. $A = l^2$
      Note: All side lengths of a square are equal.

4.5 m

4.5 m

2. Substitute the value 4.5 for l and calculate the $A = 4.5^2$
   area.
                                                         $= 20.25 \text{ m}^2$

3. State the answer in the correct unit (m^2).   The area of the square is 20.25 m².

COMMUNICATING — COLLABORATIVE TASK: Rectangles with the same areas and perimeters

Work in pairs to answer the following questions.

Compare the perimeters of two different rectangles, each with an area of 60 cm². Which rectangles have the smallest perimeters and which have the largest perimeters?

Compare the areas of different rectangles, each with a perimeter of 24 cm. Which rectangles have the smallest areas and which have the largest areas?

▶ 10.5.3 Calculating the area of a triangle

eles-4556

- Triangles can be formed by cutting rectangles in half.

3 cm

4 cm

2.5 cm

5 cm

2 cm

3 cm

- The area of a triangle is therefore equal to half the area of a rectangle of the same base length and height.

Area of a triangle

To calculate the area A of a triangle:

$$A = \frac{1}{2}b \times h$$

$$= \frac{1}{2}bh$$

h

b

where b is the base and h the perpendicular height.

- To measure the perpendicular height of a triangle, the line must form a right angle with the base of the triangle. Sometimes this line will have to be drawn outside the triangle, as shown in the following diagram.

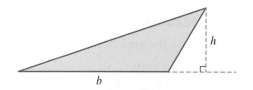

WORKED EXAMPLE 11 Calculating the area of a right-angled triangle

Calculate the area of the following shape.

THINK	WRITE

Method 1: Using the formula for the area of a triangle

1. Write the formula for the area of a triangle.

$$A = \frac{1}{2}bh$$

2. Identify the values of b and h.

$b = 8\,\text{cm}, h = 6\,\text{cm}$

3. Substitute the values of b and h into the formula.

$$A = \frac{1}{2} \times 8 \times 6$$

4. Evaluate. (Since one of the values is even, halve it first to make calculations easier.) Remember to include the correct unit (cm^2).

$$= 4 \times 6$$
$$= 24\,\text{cm}^2$$

Method 2: By cutting the rectangle in half

1. Draw an imaginary rectangle that contains the triangle. The rectangle should have the same base length and height as the triangle. Notice that the triangle forms half of the rectangle.

2. Use the formula $A = lw$ to calculate the area of this rectangle.

$$A = lw$$
$$= 8\,\text{cm} \times 6\,\text{cm}$$
$$= 48\,\text{cm}^2$$

3. Halve the area of the rectangle to get the area of the triangle. Remember to answer in the correct unit (cm^2).

$\text{Area of triangle} = \frac{1}{2} \text{ of area of rectangle}$

$$\text{Area of triangle} = \frac{1}{2} \times 48$$

$$= 24\,\text{cm}^2$$

Calculate the area of each of these triangles.

a.

b.

THINK	**WRITE**
a. 1. Write the formula for the area of a triangle.	a. $A = \frac{1}{2}bh$
2. Identify the values of b and h.	$b = 10, h = 4.5$
3. Substitute the values of b and h into the formula.	$A = \frac{1}{2} \times 10 \times 4.5$
4. Evaluate. (Since one of the values is even, halve it first to make calculations easier.) Remember to include the correct unit (cm²).	$= 5 \times 4.5$ $= 22.5 \text{ cm}^2$
b. 1. Write the formula for the area of a triangle.	b. $A = \frac{1}{2}bh$
2. Identify the values of b and h.	$b = 15, h = 7$
3. Substitute the values of b and h into the formula.	$A = \frac{1}{2} \times 15 \times 7$
4. Evaluate. (Since neither value is even, multiply 15 and 7 first, and then divide by 2.) Remember to include the correct unit (cm²).	$= \frac{1}{2} \times 105$ $= 52.5 \text{ cm}^2$

▶ 10.5.4 Area of a parallelogram

eles-4557

- A **parallelogram** is a quadrilateral having each pair of opposite sides parallel.

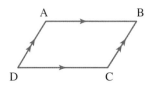

- To calculate the area of the parallelogram, first draw a diagonal from B to D to form two triangles, ABD and BDC. Label the base b and the perpendicular height h.

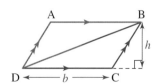

- The area of triangle BCD $= \dfrac{1}{2}b \times h$ and the area of triangle ABD $= \dfrac{1}{2}b \times h$.
  So, the area of the parallelogram:

$$\text{ABCD} = \dfrac{1}{2}b \times h + \dfrac{1}{2}b \times h$$
$$= b \times h$$

  where b is the base and h is the perpendicular height.
- The formula for the area of a parallelogram is similar to the formula for the area of a rectangle.
- If we dissect any parallelogram and rearrange its pieces, we can see it can be made into a rectangle.

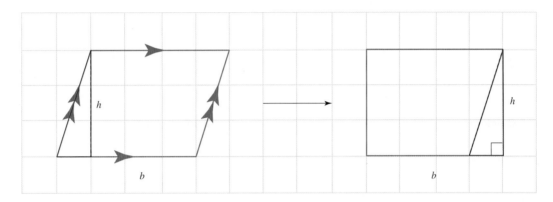

- Similarly, we can also bisect a pair of opposite sides of a parallelogram and rotate the triangles to also form a rectangle.

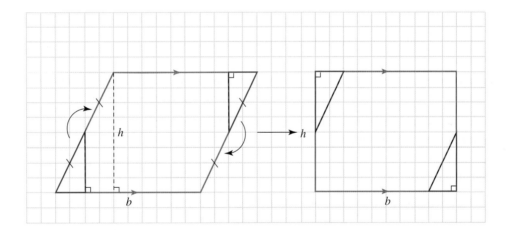

Area of a parallelogram

The area of a parallelogram is:

$$A = bh$$

where b is the base and h is the perpendicular height.

- For example, the area of the parallelogram shown is:

$$ABCD = b \times h$$
$$= 6 \times 4$$
$$= 24 \, cm^2$$

Exercise 10.5 Area

10.5 Quick quiz **on**	10.5 Exercise

Individual pathways

■ PRACTISE	■ CONSOLIDATE	■ MASTER
1, 4, 5, 8, 9, 10, 14, 17, 20, 22, 25	2, 6, 11, 13, 15, 18, 21, 23, 26, 27, 28, 29	3, 7, 12, 16, 19, 24, 30

Fluency

1. Identify which unit would be most suitable to measure the following areas.
   Choose from mm^2, cm^2, m^2, ha or km^2.

   a. A computer screen
   b. The Melbourne Cricket Ground
   c. A shirt button
   d. The Brisbane metropolitan area

2. Identify which unit would be most suitable to measure the following areas.
   Choose from mm^2, cm^2, m^2, ha or km^2.

   a. A house block
   b. The state of Queensland
   c. A basketball court
   d. A dairy farm

3. **WE8a** The following figures are drawn on centimetre grid paper. Determine the area of each one.

   a.
   b.
   c.

   d.
   e.

4. Estimate the areas of the following figures, which are drawn on centimetre grid paper.

a.

b.

c.

d.

e.

5. **MC** Select from the following an estimate for the area of the shape drawn on centimetre grid paper.

A. $18\,\text{cm}^2$
B. $21\,\text{cm}^2$
C. $16\,\text{cm}^2$
D. $23\,\text{cm}^2$

6. **MC** Select from the following an estimate for the area of the shape drawn on centimetre grid paper.

A. $16\,\text{cm}^2$
B. $20\,\text{cm}^2$
C. $25\,\text{cm}^2$
D. $28\,\text{cm}^2$

7. **MC** Select from the following an estimate for the area of the shape drawn on centimetre grid paper.

A. $15\,\text{cm}^2$
B. $17\,\text{cm}^2$
C. $19\,\text{cm}^2$
D. $21\,\text{cm}^2$

8. **WE9** Determine the areas of the following rectangles. (*Hint:* Use the formula $A = lw$.)

a.
5 cm
3 cm

b.
7 km
2 km

c.
2 m
5 m

d.
5 mm

9. **WE10** Calculate the areas of the following triangles.

a.
2 cm
4 cm

b.
3 cm
2 cm

c.
9 cm
9 cm

d.
5 cm
20 cm

10. **MC** Identify the height of the triangle shown.

A. 60 mm
B. 30 mm
C. 27 mm
D. 53 mm

30 mm
53 mm
27 mm
60 mm

11. **MC** Identify the height of the triangle shown.

A. 2.6 cm
B. 4.0 cm
C. 3.0 cm
D. 4.2 cm

4.0 cm
2.6 cm
3.0 cm

12. **MC** Identify the base length of the triangle shown.

A. 39 cm
B. 58 cm
C. 29 cm
D. 25 cm

39 cm
58 cm
29 cm
25 cm

13. **WE11** Calculate the area of each of these triangles.

a.

b.

c.

d.

14. Calculate the area of each of these triangles.

a.

b.

c.

d.

15. Calculate the area of the following parallelogram.

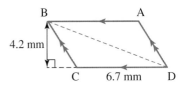

Understanding

16. **MC** Select the closest estimate of the area of this butterfly if a centimetre grid is placed over it as shown.

   A. $10 \, cm^2$
   B. $18 \, cm^2$
   C. $27 \, cm^2$
   D. $38 \, cm^2$

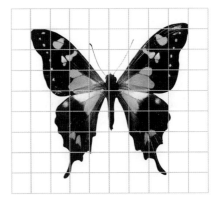

17. One of the largest open-air shopping centres is the Ala Moana centre in Honolulu, Hawaii, USA with more than 200 shops covering an area of 20 hectares. Convert this area to square metres.

18. The Taupo volcanic eruption that occurred in New Zealand nearly 2000 years ago is estimated to have flattened an area of $16\,000 \, km^2$.

   a. Convert this area to square metres.
   b. Convert this area to hectares.

19. Calculate how many square metres of carpet are needed to cover a rectangular room of length 5 m and width 3.5 m.

20. Determine the area of material needed to make a rectangular rug that is 4.2 m long and 230 cm wide. (Give your answer in square metres.) (*Hint:* Convert the width measurement into metres first.)

21. Toby is tiling his bathroom with ceramic tiles that cost $35.20 per box.

   a. Calculate how many square metres of tiles he will need if the rectangular room has a width of 2.5 m and a length of 3 m.
   b. Calculate how many boxes of tiles he should order if each box contains enough tiles to cover $0.5 \, m^2$.
   c. Determine the cost of the tiles.

Communicating, reasoning and problem solving

22. Alana, who works for Fast Glass Replacements, has been asked for a quote to replace three windows. Each window is 1.8 m long and 0.8 m wide. Determine the price Alana should quote if the glass costs $27 per square metre. (Include a delivery cost of $25 in the quote.)

23. A floor tiler charged $640.00 to tile a rectangular room. Her next job is to tile the floor of a rectangular room twice as long and twice as wide. Determine how much she should charge for the larger room. (The answer is not $1280.00.) Justify your answer.

24. Glen wants to pave his back courtyard. His courtyard is 12 m × 12 m. He has a choice of 10 cm × 25 cm clay pavers that cost $2.30 each, or concrete pavers that are 70 cm × 70 cm and cost $42 each. Compare the costs of each to determine the cheaper option.

25. For the sailboat shown, calculate the approximate area of the triangular mainsail when the sail is flat. Use the dimensions shown.

26. Jane is a landscape gardener who is laying a new lawn. The rectangular lawn is 13 m long and 8 m wide. Calculate how many square metres of turf Jane should order.
Calculate the total cost of the turf if it costs $12.50 per square metre.

27. Determine the area of cloth required to make the kite shown, given that the length BD is 80 cm and the length AC is 1.5 m.

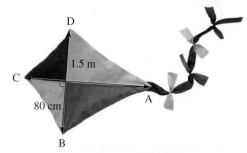

28. A church spire has six identical triangular faces that have the dimensions shown. Determine the area of copper roofing required to cover all six faces of the spire.

29. Calculate the total wing area of the following delta-winged jet aircraft.

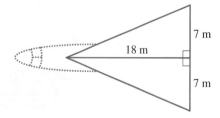

30. A rectangle has an area of 36 cm² and a perimeter of 26 cm. Each side of the rectangle is a whole number of centimetres. Determine the length and width of this rectangle. Show your working.

LESSON
10.6 Area of composite shapes

▶ 10.6.1 Calculating the area of composite shapes

eles-4558
- **Composite shapes** can be separated into parts, each of which is a simple shape (such as, for example, a rectangle, square or a triangle).
- The area of a composite shape can be found by adding the areas of each of the separate parts.

WORKED EXAMPLE 13 Calculating the area of rectangles in composite shapes

Calculate the areas of the following shapes.

a.

b.

THINK

WRITE

a. 1. Divide the shape into two rectangles.

a. $l = 1$ cm

2. Calculate the area of each rectangle separately by substituting the correct values of l and w into the formula $A = lw$.

Area of rectangle $1 = l \times w$
$= 4 \times 2$
$= 8 \, \text{cm}^2$

Area of rectangle $2 = l \times w$
$= 1 \times 2$
$= 2 \, \text{cm}^2$

3. Add the two areas. Remember to answer in the correct unit (cm²).

Area of shape $=$ Area of rectangle 1 $+$ Area of rectangle 2
$= 8 \, \text{cm}^2 + 2 \, \text{cm}^2$
$= 10 \, \text{cm}^2$

b. 1. Divide the shape into three rectangles.

b.

l = 10.2 mm

① *w* = 9.7 mm

21.7 mm ②

← 15.9 mm →

③ *w* = 7.3 mm

l = 18.5 mm

2. Calculate the area of each rectangle separately by substituting the correct values of *l* and *w* into the formula $A = lw$.

Area of rectangle $1 = l \times w$

$= 10.2 \times 9.7$

$= 98.94 \text{ mm}^2$

Area of rectangle $2 = l \times w$

$= (21.7 - 9.7 - 7.3) \times (18.5 - 15.9)$

$= 4.7 \times 2.6$

$= 12.22 \text{ mm}^2$

Area of rectangle $3 = l \times w$

$= 18.5 \times 7.3$

$= 135.05 \text{ mm}^2$

3. Add the three areas. Remember to answer in the correct unit (mm²).

Area of shape $=$ Area $1 +$ Area $2 +$ Area 3

$= 98.94 \text{ mm}^2 + 12.22 \text{ mm}^2 + 135.05 \text{ mm}^2$

$= 246.21 \text{ mm}^2$

- The area of the following pentagon can be found by adding the areas of the triangle and rectangle.

WORKED EXAMPLE 14 Calculating the area of a composite shape using addition

Calculate the area of the following shape.

7 cm

4 cm

12 cm

THINK	WRITE
1. Divide the shape into a rectangle and a triangle.	
2. Calculate the area of shape 1 using the formula $A = lw$. The value of $l = 7$ cm and $w = 4$ cm.	Area $1 = lw$ $l = 7$ cm, $w = 4$ cm
3. Substitute the values of l and w in the formula to calculate the area of the rectangle.	$= 7 \times 4$ $= 28$ cm^2
4. Calculate the area of shape 2 using the formula $A = \dfrac{1}{2}bh$. The value of $b = 5$ cm and $h = 4$ cm.	Area $2 = \dfrac{1}{2}bh$ $b = 5$ cm, $h = 4$ cm
5. Substitute the values of b and h in the formula to calculate the area of the triangle.	$= \dfrac{1}{2} \times 5 \times 4$ $= 10$ cm^2
6. Add the two areas to get the total area.	Total area $= A_1 + A_2$ $= 28 + 10$ $= 38$ cm^2
7. Write the answer with the correct units.	The area of the composite shape is 38 cm^2.

- In many situations the area of a given shape can be found by subtracting individual areas from each other.

WORKED EXAMPLE 15 Calculating the area of a composite shape using subtraction

Calculate the shaded area in the following figure.

THINK	WRITE
1. The figure shows a large square with the blue shaded area and a small square.	Area of a square $= l^2$
2. Calculate the area of the large square.	Area of the large square: $A = 5^2$ $= 5 \times 5$ $= 25$ m^2

3. Calculate the area of the small square.

Area of the small square: $A = 2^2$
$= 2 \times 2$
$= 4\,m^2$

4. To calculate the area of the shaded region, subtract the area of the small square from that of the large one.

Shaded area $= 25 - 4$
$= 21\,m^2$

5. Write the answer with the correct unit.

The shaded area is $21\,m^2$.

 Resources

▶ **Video eLesson** Composite area (eles-1886)

✦ **Interactivity** Area of composite shapes (int-4020)

Exercise 10.6 Area of composite shapes

learnon

10.6 Quick quiz on

10.6 Exercise

Individual pathways

■ PRACTISE	■ CONSOLIDATE	■ MASTER
1, 5, 9, 11, 13, 15, 17, 20	2, 3, 6, 10, 14, 18, 21	4, 7, 8, 12, 16, 19, 22

Fluency

For questions **1** to **4**, calculate the areas of the following shapes. (*Hint:* Divide the shapes into rectangles and squares before using the formula $A = lw$)

1. **WE13** a.

b.

2. a.

b.

3. a.

b.

4. a.

b.

5. **WE14** Calculate the area of the following composite shapes.

a.

b.

c.

6. Calculate the area of the following composite shapes.

a.

b.

c.

7. **WE15** Calculate the shaded area of the following shapes.

a.

b.

8. Calculate the area of the following garage wall.

9. Determine the area of bricks needed to cover the following courtyard.

Understanding

10. Determine the area of the hotel lobby in the following diagram.

11. a. Calculate the area of carpet needed for the floor plan shown.
    b. Determine the cost of carpeting the room if the carpet costs $25 per square metre.

12. Michael is paving a rectangular yard that is 15.5 m long and 8.7 m wide. A square fishpond with side lengths of 3.4 m is in the centre of the yard. Calculate the cost of paving Michael's yard if the paving material costs $17.50 per square metre.

13. Members of the Lee family want to pave the area around their new swimming pool. The pool is set into the corner of the yard as shown in the diagram. Calculate the area of paving (in square metres) required to cover the yard around the pool (shaded in the diagram).

14. In order to construct an advertising sign, the following letters are cut from plastic sheets. Calculate the total area of plastic required to make all 3 letters.

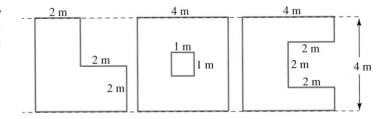

Communicating, reasoning and problem solving

15. Calculate the total area of the cattle station shown in the diagram, which has 3 large paddocks. (Give your answer in square kilometres.)

16. Calculate the total floor area of a concrete slab for a house as shown in the following diagram.

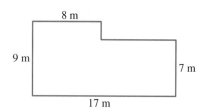

17. a. Calculate the area of the diagram. Show all your working.
    b. Calculate the area of the figure using an alternative method. Show all your working.
    c. Explain the differences between the two methods.

18. When determining the area of a composite shape, explain how you would decide whether to use addition or subtraction.

19. The stylised diagram of a set of traffic lights has been constructed using basic shapes, as shown.

    a. State the shapes used to draw the diagram and their dimensions in cm.
    b. Explain the steps required in calculating the black area surrounding the traffic lights shown.
    c. If the area of one of the large semicircles is closest to 353 cm², and the area of one of the small coloured circles is closest to 177 cm², calculate the area of the black surface. State the answer in cm².
    d. Explain why it is more appropriate to express the area in cm² rather than mm².

20. Ellen wants to frame her cross-stitch work. Her cross-stitch work is 90 cm long and 75 cm wide. The frame's border, shown in the diagram, measures 2.5 cm wide. One company charges $120 for the frame. Determine the cost of the frame per square centimetre. Round your answer to 2 decimal places.

21. A square of side length 1 metre is made up of smaller internal squares that are formed by joining the midpoints of the outer square as shown in the diagram. Determine how much of the square the shaded area represents.

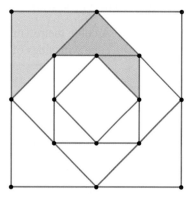

22. The diagram of the letter M shown has been constructed using basic shapes.

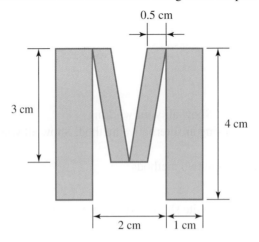

0.5 cm

3 cm

4 cm

2 cm 1 cm

a. Explain two methods of calculating the area of the letter using:
  i. addition
  ii. subtraction.

b. Using your preferred method, calculate the area of the letter.

LESSON
10.7 Volume of rectangular prisms

LEARNING INTENTION

At the end of this lesson you should be able to:
  • calculate the volume of a rectangular prism.

eles-4559

⊙ 10.7.1 Volume

  • The **volume** of a **3-dimensional object** is the amount of space it occupies.
  • Volume is measured in cubic units such as mm^3, cm^3 and m^3.
  • A cubic centimetre (cm^3) is the space occupied by a cube with sides of 1 cm.

A sugar cube has a volume of about $1\,cm^3$.

A cubic metre (m^3) is the space occupied by a cube with sides of 1 m.

1 m

1 m

1 m

Calculate how many cubic centimetres are in each solid shape. (Each small cube represents $1\,cm^3$.)

a.

b.

THINK	WRITE
a. 1. Count the cubes. There are 10 cubes in each layer. There are three layers altogether.	a. Volume $= 10\,cm^3 \times 3$
2. Give the answer in cubic centimetres.	$= 30\,cm^3$
b. 1. Count the cubes. There are 12 cubes in the first layer. There are 6 cubes in the second layer. There are 3 cubes in the third layer.	b. Volume $= 12\,cm^3 + 6\,cm^3 + 3\,cm^3$
2. Give the answer in cubic centimetres.	$= 21\,cm^3$

COMMUNICATING — COLLABORATIVE TASK: The blue cube

Work in pairs to answer the following questions.

1. The outside of a cube with 3 cm sides is painted blue. The cube is then cut into smaller cubes, each with sides 1 cm long. How many of the 1 cm cubes will have paint on:
   a. 1 side only
   b. 2 sides only
   c. 3 sides only
   d. no sides?

2. For a cube with 4 cm sides, made from 1 cm cube blocks and painted blue on the outside, determine the number of cubes painted on:
   a. 1 side only
   b. 2 sides only
   c. 3 sides only
   d. no sides.

3. Repeat question 2 for a cube with sides of 5 cm.

4. Is there a pattern that you can use to predict the numbers of cubes with paint on one, two, three or no sides if you know the number of cubes along the side of the large cube?

10.7.2 Calculating the volume of a rectangular prism

eles-4560

- A cube is a 6-sided 3-dimensional shape in which all sides are squares.
- A cuboid is a 6-sided 3-dimensional shape in which all sides are rectangles or squares.
- Cuboids are also known as rectangular **prisms**.

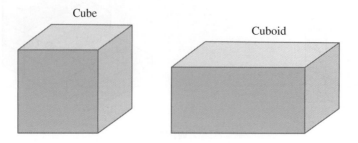

Cube

Cuboid

Volume of a rectangular prism

For a rectangular prism, the volume V is:

$$V = l \times w \times h$$
$$= lwh$$

where l is its length, w its width and h its height.

- We know that the area of a two-dimensional rectangle is length × width. If we imagine the two-dimensional rectangle being stretched to make a three-dimensional rectangle, the volume would then be its area (length × width) × height.
- Therefore, the volume of a rectangular prism can also be written as $V = Ah$.

WORKED EXAMPLE 17 Calculating the volume of a rectangular prism

Use the formula $V = lwh$ to calculate the volume of the following rectangular prism.

THINK	WRITE
1. State the formula for the volume of a rectangular prism.	$V = lwh$
2. Identify the length ($l = 7$ cm), width ($w = 4$ cm), and height ($h = 2$ cm) of the prism and substitute the values into the formula.	$= 7 \, \text{cm} \times 4 \, \text{cm} \times 2 \, \text{cm}$
3. Calculate the volume, stating the answer in the appropriate units (cm^3).	$= 56 \, \text{cm}^3$

Use the information given in the diagram to calculate the volume of the composite rectangular prism.

THINK	WRITE
1. To calculate the volume, first write the formula for the volume of a rectangular prism.	$V = lwh$

2. Split the shape into 2 rectangular prisms and calculate the volume of each rectangular prism.

Volume of rectangular prism $1 = 3 \times 2 \times 3$

$$= 18\,\text{cm}^3$$

Volume of rectangular prism $2 = 2 \times 1 \times 3$

$$= 6\,\text{cm}^3$$

3. Add the volumes to determine the volume of the composite rectangular prism.

$V = 18\,\text{cm}^3 + 6\,\text{cm}^3$
$ = 24\,\text{cm}^3$

on Resources

Interactivities Volume (int-4021)

Volume of a rectangular prism (int-4022)

Exercise 10.7 Volume of rectangular prisms

10.7 Quick quiz on	10.7 Exercise

Individual pathways

■ PRACTISE	■ CONSOLIDATE	■ MASTER
1, 2, 6, 9, 13, 16, 21, 24	3, 5, 7, 10, 11, 14, 17, 19, 22, 25	4, 8, 12, 15, 18, 20, 23, 26

Fluency

1. **WE16a** Calculate how many cubic centimetres are in each solid shape. (Each small cube represents $1 \, cm^3$.)

   a.
   b.
   c.
   d.

2. **MC** Select the volume of the prism shown.

   A. $4 \, cm^3$
   B. $12 \, cm^3$
   C. $1 \, cm^3$
   D. $8 \, cm^3$

   Note: Each cube has a volume of $1 \, cm^3$.

3. **WE16b** Calculate the volumes of the following solids. (Each small cube represents $1 \, cm^3$.)

   a.
   b.

   c.
   d.

4. Calculate the volumes of the following solids. (Each small cube represents $1\,\text{cm}^3$.)

a.

b.

c.

d.

5. **MC** Select the volume of the prism shown.

   A. $20\,\text{cm}^3$
   B. $16\,\text{cm}^3$
   C. $10\,\text{cm}^3$
   D. $22\,\text{cm}^3$

   Note: Each cube has a volume of $1\,\text{cm}^3$.

6. **WE17** Use the formula $V = lwh$ to calculate the volumes of the following rectangular prisms.

a.

2 cm
3 cm
2 cm

b.

4 m
1 m
3 m

c.

24 cm
16 cm
16 cm

7. Determine the volume of each of the following. (Each cube has a volume of $1\,\text{cm}^3$.)

a.

b.

c.

8. Use the formula $V = lwh$ to calculate the volumes of the following rectangular prisms.

a.

4 m

6 m   8 m

b.

15 cm

3 cm   6 cm

c.

3 cm

7.5 cm   4.5 cm

9. **MC** Select the volume of the prism shown.

A. $10 \, \text{cm}^3$
B. $20 \, \text{cm}^3$
C. $30 \, \text{cm}^3$
D. $40 \, \text{cm}^3$

Note: Each cube has a volume of $1 \, \text{cm}^3$.

10. **MC** Select the volume of the prism shown.

A. $16 \, \text{cm}^3$
B. $48 \, \text{cm}^3$
C. $32 \, \text{cm}^3$
D. $40 \, \text{cm}^3$

Note: Each cube has a volume of $1 \, \text{cm}^3$.

11. **WE18** Calculate the volumes of the following composite prisms.

a.

5 cm   6 cm

6 cm

1 cm   1 cm

b.

10 cm

4 cm

5 cm   2 cm

2 cm

12. Calculate the volumes of the following composite prisms.

a.

b.

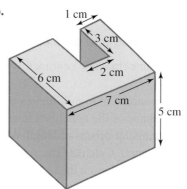

Understanding

13. Calculate the volume of the shoe box shown. (Give your answer in cm^3.)

14. The inside dimensions of a refrigerator are shown. Calculate the volume available for food storage inside the refrigerator.

15. a. A rectangular prism has a length of 30 cm and a width of 15 cm. If its volume is $9000\,cm^3$, determine its height.

b. A rectangular prism has a length of 13 cm and a width of 17 cm. If its volume is $1989\,cm^3$, determine its height.

c. A rectangular prism has a length of 10 cm and a width of 15 cm. If its volume is $1800\,cm^3$, determine its height.

16. The lunch box shown is a rectangular prism.

a. Calculate the volume of the lunch box in cubic centimetres.

b. Change each measurement to millimetres and hence calculate the volume of the lunch box in cubic millimetres.

17. Determine the volume of concrete (in cubic metres) that would be needed to make the base for a garage that is 6.5 m wide and 3 m long. The concrete base is 0.25 m deep.

Communicating, reasoning and problem solving

18. **a.** Calculate the volume (in cubic centimetres) of the matchbox shown.
    b. Change each measurement to millimetres and hence determine the volume of the matchbox in cubic millimetres.
    c. Matches are rectangular prisms of length 44 mm, width 2 mm and height 2 mm. Calculate the volume of a match (in cubic millimetres). (Ignore the red substance on the end of each match.)

   d. If a matchbox contains 50 matches, determine how much space is left in a full box.

19. Calculate how many cubic metres of water would be needed to fill a diving pool that has a length of 16 m, a width of 12 m and a depth of 4 m.

20. Heather wishes to cover a rectangular lawn with topsoil to a depth of 0.1 m. If the lawn is 24 m long and 17 m wide, calculate the volume of soil (in cubic metres) she needs to order.

21. The fruitcake shown in the photo is to be divided equally among 100 people at a wedding reception. Determine the volume of cake (in cubic centimetres) each guest will receive.

22. A swimming pool is rectangular and its width is exactly half its length. Determine the volume of water needed to fill it if the swimming pool is 50 metres long and has a constant depth of 2 metres.

23. A new rectangular patio has been built on the end of your house. It measures 3.8 m by 1.9 m. You want to plant a garden that is 1.5 m wide around the patio on three sides. The garden beds are to be 0.5 metres deep. If you order 7.3 cubic metres, explain with a diagram whether enough soil has been ordered.

24. Calculate the volume of clay in a house brick that has a length of 23 cm, a width of 11 cm and a height of 9 cm.

25. The areas of the three sides of a rectangular box are shown in the diagram. Determine the volume of the box, showing all your working.

26. A big cargo box measures 12 m by 6 m by 5 m. Determine how many small boxes of 40 cm by 25 cm by 10 cm can fit in the cargo box.

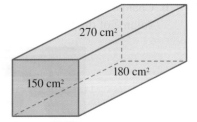

LESSON
10.8 Capacity

LEARNING INTENTION

At the end of this lesson you should be able to:
- choose an appropriate unit to measure capacity
- convert between units of volume and capacity.

▶ 10.8.1 Capacity

eles-4561

- The **capacity** of a container is the volume of liquid that it can hold.
- Capacity is commonly measured in millilitres (mL), litres (L), kilolitres (kL) and megalitres (ML).

WORKED EXAMPLE 19 Appropriate unit to measure capacity

Identify the most appropriate unit to measure the capacity of:
a. a drinking glass
b. a backyard swimming pool
c. a bathtub.

THINK	WRITE
a. A drinking glass is relatively small, so it would hold less than a litre.	a. millilitres
b. A backyard swimming pool is very large, so the most appropriate unit of measure would be kilolitres or megalitres (depending on the size).	b. kilolitres or megalitres
c. A bathtub would hold several litres of water but less than 1000 L or 1 kilolitre. The most appropriate unit of measure would be litres.	c. litres

---

- The following chart can be used to convert between megalitres, kilolitres, litres and millilitres.

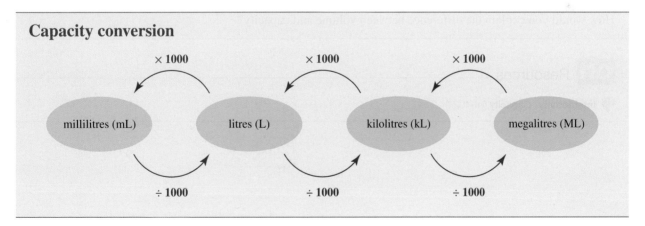

Capacity conversion

- Capacity can be measured in the same units as volume (e.g. cm^3 or m^3).
- The metric volume unit $1\ cm^3$ has a capacity of $1\ mL$. The metric capacity unit $1\ L$ is equivalent to $1000\ cm^3$.

WORKED EXAMPLE 20 Converting between different units of capacity

Complete the following unit conversions.

a. $6\ L =$ _____ mL   b. $700\ mL =$ _____ L   c. $0.45\ L =$ _____ cm^3

THINK

a. Check the conversion chart. To convert litres to millilitres, multiply by 1000. Since 6 is a whole number, do this by adding three zeros to the end of the number.

b. Check the conversion chart. To convert millilitres to litres, divide by 1000. To do this, move the decimal point 3 places to the left.

c. Check the conversion chart. To convert litres to millilitres, multiply by 1000. To do this, move the decimal point 3 places to the right. Note that $1\ mL = 1\ cm^3$.

WRITE

a. $6\ L = 6 \times 1000\ mL$
    $= 6000\ mL$

b. $700\ mL = 700 \div 1000\ L$
    $= 0.7\ L$

c. $0.45\ L = 0.45 \times 1000\ mL$
    $= 450\ mL$
    $= 450\ cm^3$

DISCUSSION

How would you explain the difference between volume and capacity?

 Resources

Interactivity Capacity (int-4024)

Exercise 10.8 Capacity

10.8 Quick quiz on	10.8 Exercise

Individual pathways

■ PRACTISE	■ CONSOLIDATE	■ MASTER
1, 2, 5, 8, 11, 14, 16, 18	3, 6, 9, 12, 15, 19, 21, 22	4, 7, 10, 13, 17, 20, 23, 24, 25, 26

Fluency

1. **WE19** Identify the most appropriate unit to measure the capacity of:

   a. Sydney Harbour
   b. a pond
   c. a small drink bottle
   d. a car fuel tank.

2. **WE20a,b** Complete the following unit conversions.

   a. $2\,L =$ _____ mL
   c. $7000\,mL =$ _____ L

   b. $3000\,mL =$ _____ L
   d. $5500\,mL =$ _____ L

3. Complete the following unit conversions.

   a. $2\dfrac{1}{2}\,L =$ _____ mL
   c. $0.035\,L =$ _____ mL

   b. $32\,000\,mL =$ _____ L
   d. $420\,L =$ _____ mL

4. Complete the following unit conversions.

   a. $1\dfrac{87}{100}\,L =$ _____ mL
   c. $\dfrac{1}{10}\,L =$ _____ mL

   b. $22\,500\,mL =$ _____ L
   d. $25\,L =$ _____ kL

5. **WE20c** Complete the following unit conversions.

   a. $750\,cm^3 =$ _____ mL
   c. $2.45\,L =$ _____ m^3

   b. $2500\,m^3 =$ _____ kL
   d. $78\,000\,cm^3 =$ _____ L

6. Complete the following unit conversions.

   a. $40\,000\,cm^3 =$ _____ mL = _____ L
   c. $5200\,L =$ _____ kL = _____ m^3

   b. $6\,L =$ _____ mL = _____ cm^3

7. **MC** A capacity of $25\,L$ is equal to:

   A. $0.025\,mL$ B. $250\,mL$ C. $0.25\,mL$ D. $25\,000\,mL$

8. **MC** A capacity of $35\,400\,mL$ is equal to:

   A. $35\,400\,000\,L$ B. $0.354\,L$ C. $3.5400\,L$ D. $35.4\,L$

9. Arrange in order from smallest to largest:

   a. $2\dfrac{1}{2}\,L, 25\,000\,mL, \dfrac{1}{4}\,L, 2.45\,L$
   b. $760\,mL, 0.765\,L, 7.65\,mL, 7.60\,L$

   c. $110\,mL, 0.1\,L, 0.011\,L, 1.1\,L$

Understanding

10. A water bottle has a capacity of 2 litres. Calculate how many 125 mL bottlefuls are required to fill it.

11. A bottle contains 250 mL of orange juice concentrate. Calculate how much water should be added to make up 2 L of juice from the concentrate.

12. A scientist dilutes (waters down) an acid solution by adding 120 mL of acid to 1.5 L of water. Calculate how much of the diluted solution this will make.

13. Most sports drinks are sold in 500 mL bottles. Calculate how many litres of sports drink are in one dozen (12) bottles.

14. A medicine bottle contains 125 mL of cough syrup. Calculate how many 2.5 mL doses could be administered from this bottle, assuming that none is spilt.

15. Anthea runs a market stall selling detergent. Calculate how many 200 mL bottles of detergent she could fill from a 45 L bulk container.

16. A milk bar sells 55 small bottles of lemon drinks in one week. If each bottle contains 180 mL, calculate how many litres are sold each week.

17. Petrov is working as a school laboratory technician. If there are 12 groups of students and each group requires 400 mL of salt solution for an experiment, calculate how many litres of solution should be prepared.

Communicating, reasoning and problem solving

18. a. Claire has made her favourite green cordial with 0.25 L of water and 30 mL of cordial. Determine whether a glass with a capacity of 250 mL will hold this volume.
b. Sharmila has made fresh juice with 0.3 L of apple juice and 40 mL of carrot juice. Determine whether a glass with a capacity of 350 mL will hold this volume.

19. A 185 mL container of hair conditioner is sold at the special price of $3.70. A 0.5 L container of the same conditioner costs $11.00. Compare them to identify the better buy. (*Hint:* Calculate the cost of 1 mL of hair conditioner in each case.)

20. Laurie connected a water tank to his 5 m × 3 m roof in order to collect rain water. In four hours, 2 cm of rain fell. The water tank holds 280 litres. Determine whether the tank overflowed. *Hint:* 1000 L = 1 m³.

21. Determine the number of millilitres of milk the container in the picture holds.

22. Water is to be poured into this fish tank. How many litres of water are needed to fill the tank to a depth of 28 cm?

30 cm

75 cm

24 cm

23. Using the measurements shown in the photo of the sinks, estimate how many litres of water each sink will hold if filled to the top. Both sinks have the same dimensions. (*Hint:* First convert the measurements to centimetres.)

180 mm

430 mm

350 mm

24. One litre of paint covers an area of 20 square metres. Calculate how many millilitres of paint are required to cover an area of 6 square metres. Show your working.

25. A tap drips water at a rate of one drop every three seconds. If it takes 750 drops of water to fill a 100 mL container, determine how many litres of water are lost in one year (365 days).

26. To achieve a world record, a 16.4 kL strawberry milkshake was made in the UK in 1996. To understand how large this is, consider the size of a rectangular prism that would have this capacity. Suggest three possible sets of measurements for this container.

LESSON
10.9 Drawing solids

LEARNING INTENTION

At the end of this lesson you should be able to:
- draw the cross-section of a given prism
- draw the views, or elevations, of a given solid
- construct a solid given its views or elevations.

▶ 10.9.1 Prisms

eles-4562

- A prism is a 3-dimensional figure that can be cut into parallel slices or cross-sections that have the same size and shape.
- All prisms have a uniform **cross-section** that is equal to the base area.

Draw the cross-section of the following prisms.

a.

b.

THINK

a. 1. Identify the cross-section of the prism that is uniform along the length.

DRAW

a.

2. Draw the cross-section as a 2-dimensional shape.

b. 1. Identify the cross-section of the prism that is uniform along the length.

b.

2. Draw the cross-section as a 2-dimensional shape.

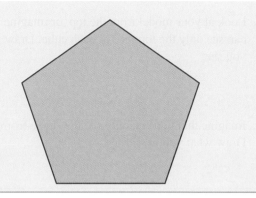

▶ 10.9.2 Plans and views

eles-4563

- An object can be viewed from different angles.
- Architects and draftspersons draw plans of buildings viewed from the front, the side or the top.
- The **front view**, or **front elevation**, is what you see if you are standing directly in front of an object.
- The **side view**, or **side elevation**, is what you see if you are standing directly to one side of the object. You can draw the left view or the right view of an object.
- The **top view**, or **bird's eye view**, is what you see if you are hovering directly over the top of the object looking down on it.
- The **back view**, or **back elevation**, is what you see if you are standing directly behind an object.

WORKED EXAMPLE 22 Drawing elevations or views

The following object is made from 4 cubes.
Draw plans of it showing:
a. the front view   b. the right view
c. the top view   d. the back view.

THINK

a. Make this shape using cubes. Place the shape at a considerable distance and look at it from the front (this way you can see only the front face of each cube). Draw what you see. (Or simply imagine looking at the shape from the front and draw what you see.)

b. Look at your model from the right, or imagine that you can see only the right face of each cube and draw what you see.

DRAW Front

a. Front view

b. Right view

c. Look at your model from the top, or imagine that you can see only the top face of each cube. Draw what you see.

d. Imagine that you can only see the shape from behind. Draw what you would see.

c. Top view

Front

d. Back view

- As you can see from Worked example 22, the back view is a mirror image of the front view. In fact, for any solid, the back view will always be the mirror image of the front view.
- Figures such as the one in Worked example 22 can be drawn using isometric dot paper. This will help to give the 3-dimensional perspective of the object.

WORKED EXAMPLE 23 Drawing a view of a solid

Draw plans of this solid showing:
a. **the front view**
b. **the right view**
c. **the top view.**

THINK

1. Determine an object of similar shape, or visualise the object in your head.

2. Whether viewed from the front, or from the right of the object, the rectangular shaft will appear as a long thin rectangle. The square prisms will also be seen as a pair of identical rectangles. So the front view and the right view are the same.

3. When the object is viewed from above, all we can see is the flat surface of the top square prism: that is, a large and a small square with the same centre.

DRAW

Front view   Right view

Top view

▶ 10.9.3 Isometric drawing

eles-4564

- When working with 3-dimensional models and designs, it is often useful to have the design or model drawn on paper (that is, in 2 dimensions).
- An **isometric drawing** is a 2-dimensional drawing of a 3-dimensional object.
- Isometric dot paper can be used to help with these drawings.

WORKED EXAMPLE 24 Completing an isometric drawing of an object

First copy the incomplete figure (Figure 1) onto isometric dot paper. Complete the isometric drawing of the object (Figure 2).

Figure 1

Figure 2

THINK

Study the object and identify the lines that have already been drawn. Fill in the missing lines on your isometric drawing to match the object.

DRAW

WORKED EXAMPLE 25 Drawing an object on isometric dot paper

Draw the following object on isometric dot paper. (You could construct it from a set of cubes.)

THINK

1. Use cubes to make the object shown (optional). Draw the front face of the object. The vertical edges of the 3-dimensional object are shown with vertical lines on the isometric drawing; the horizontal edges are shown with the lines at an angle (by following the dots on the grid paper).

2. Draw the left face of the object.

3. Add the top face to complete the isometric drawing of the object.

DRAW

WORKED EXAMPLE 26 Constructing a solid from given information

The front, right and top views of a solid are shown. Use cubes to construct the solid, then sketch the solid.

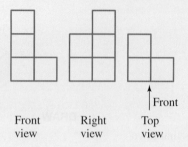

Front view   Right view   Top view   Front

THINK	CONSTRUCT
1. Use cubes to construct the solid.	
2. Check carefully that your solid matches each of the 3 views you are given. Make adjustments if necessary.	Front

---

COMMUNICATING — COLLABORATIVE TASK: Build it!

Equipment: building blocks

For this activity, you will work in groups of three or four as directed by your teacher.

1. Stack ten building blocks in any formation. An example is shown.
2. Draw diagrams of the formations you have assembled to show:
   a. the top view
   b. the front view
   c. the side view.
3. Dismantle your formation. Swap your drawings with another group and ask them to build your formation by using your diagrams.

Note: This could also be done using digital tools such as Desmos at desmos.com/geometry.

DISCUSSION

In your small groups from the previous activity, discuss some of the uses of these types of drawings. In which professions might these types of drawings be used?

---

on Resources

Interactivities Prisms (int-4148)
Plans and views (int-4149)

Exercise 10.9 Drawing solids

10.9 Quick quiz on	10.9 Exercise

Individual pathways

■ PRACTISE	■ CONSOLIDATE	■ MASTER
1, 3, 7, 8, 10, 12, 16, 19, 20	2, 4, 5, 9, 11, 14, 17, 21	6, 13, 15, 18, 22, 23

Fluency

1. **WE21** Draw the cross-sections of the following prisms.

   a.   b.

2. Draw the cross-sections of the following prisms.

   a.   b.

3. **WE22** The following objects are made from cubes. For each object draw the plans, showing the front view, the right view, the top view and the back view. (You may wish to use a set of cubes or building blocks to help you.)

   a.   b.   c.

4. The following objects are made from cubes. For each object draw the plans, showing the front view, the right view, the top view and the back view. (You may wish to use a set of cubes or building blocks to help you.)

   a.   b.   c.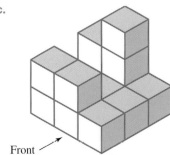

5. Draw the front, right and top views of these objects.

a.

Front

b.

Front

c.

Front

6. Draw the top, front and side views of the object shown in the following isometric drawing.

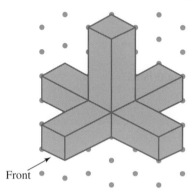

Front

7. **WE23** Draw the front, right and top views of each solid shown.

a.

Front

b.

Front

8. **WE24** Copy the following figures onto isometric dot paper and complete the isometric drawing of the objects shown.

a.

b.

c.

d.

9. **WE25** Draw each of the following objects on isometric dot paper. (You might wish to make them first from a set of cubes.)

a.   b.   c.   d.

Understanding

10. **WE26** The front, right and top views of a solid are shown. In each case, use cubes to construct the solid, then sketch the solid.

a.
Front view

Right view

Top view

Front

b.
Front view

Right view

Top view

Front

11. The front, right and top views of a solid are shown. In each case, use cubes to construct the solid, then sketch the solid.

a.
Front view

Right view

Top view
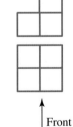

Front

b.
Front view

Right view

Top view
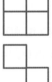

Front

12. **MC** The front, right and top views of a solid are shown. Select the correct solid represented by these views.

Front view Right view Top view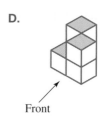

Front

A. B. C. D.

Front Front Front Front

13. Construct the following letters using cubes, and then draw the solids on isometric dot paper:
    a. the letter T with 5 cubes
    b. the letter L with 7 cubes
    c. the letter E with 10 cubes
    d. the letter H with 7 cubes.

14. Draw the following figure on isometric dot paper.

15. Draw a selection of buildings from this photograph of a city skyline on isometric dot paper.

Communicating, reasoning and problem solving

16. Consider the shapes with the following front and right views.
    Determine the minimum number of cubes that could be used to construct the shape with the following views. Determine the maximum number of cubes that could be used. Explain your reasoning.

Front view Right view

17. A shape is made using only 4 cubes. Its front view, right view and top view are shown.

Front view 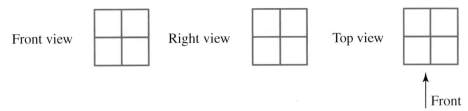 Right view Top view

Front

a. Explain whether it is possible to construct this solid.
b. Describe or draw what this solid would look like.

18. How many objects can you construct using the top view shown? Explain your answer with the use of diagrams.

19. On the same isometric dot paper, draw the two wooden boxes shown.

20. Draw the plans of the following diagram, showing the front view, right view and top view.

Front

21. Jonah is building a brick wall as shown in the photograph. Draw this part of the wall on isometric dot paper.

22. Draw top, front and side views for the 3D objects shown.

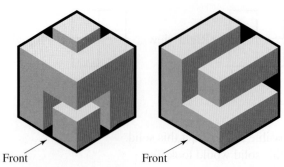

Front　　　Front

23. Sam has a set of wooden cube blocks he uses to build different shapes. His sister Chris challenged him to see how many different ways he could stack 4 cubes. He can only stack them on top of each other or side by side — not one behind the other. The arrangements must be different and not simply be a mirror image or rotation of another shape. Draw the different arrangements possible.

LESSON
10.10 Review

10.10.1 Topic summary

Area

- The area of a shape is the amount of 2-dimensional space within it.
- Area is measured in square units: mm^2, cm^2 m^2 and km^2.
- The area formulas for some common shapes are listed below.

Shape	Area formula
Square	$A = l \times l$ $= l^2$
Rectangle	$A = l \times w$
Parallelogram	$A = b \times h$
Triangle	$A = \frac{1}{2} \times b \times h$

- The hectare, ha, is a non-SI metric unit of area equal to a square with sides of 100 m.
$1 \text{ ha} = 10\,000 \text{ m}^2$

Capacity

- The amount of liquid an object can hold is known as *capacity*.
- Capacity can be measured in cubic units (cm^3 or m^3); however, it is more commonly measured in millilitres (mL), litres (L), kilolitres (kL) or megalitres (ML).
- You can convert between these units by using the diagram below.

$$1000 \text{ mL} = 1 \text{ L}$$
$$1000 \text{ L} = 1 \text{ kL}$$
$$1 \text{ cm}^3 = 1 \text{ mL}$$
$$1000 \text{ cm}^3 = 1 \text{ L}$$

Units of length

- The modern metric system in Australia is defined by the International System of Units (SI).
- The most common units of length are millimetres (mm), centimetres (cm), metres (m) and kilometres (km).
- You can convert between these units by using the diagram below.

e.g.
$3.4 \text{ m} = 3.4 \times 100 \times 10 \text{ mm}$
$= 3400 \text{ mm}$

Perimeter

- The perimeter, P, is the distance around a closed shape.
- When calculating the perimeter of a shape, make sure all lengths are in the same unit before adding them together.

4 cm   3.2 cm
4.3 cm   3 cm

$P = 4 + 3.2 + 3 + 4.3$
$= 14.5 \text{ cm}$

LENGTH, AREA AND VOLUME

Volume

- Volume is the amount of 3-dimensional space an object takes up.
- Volume is measured in cubic units: mm^3, cm^3, m^3 and km^3.
- A prism is a solid object with identical ends, flat faces and a uniform cross-section along its length.
- The volume of a rectangular prism is given by:

$V = l \times w \times h$

10.10.2 Project

Old units of length

When studying length and perimeter, you learned about the common metric units of length: millimetre, centimetre, metre and kilometre. It's only in the last few decades that these units have been commonly used in Australia. Prior to this, the units of length were based on the imperial system. Most countries in the world use units based on the metric system, but a few continue to use imperial units.

The origins of the imperial units are far and wide. Many have been adapted from ancient times or translated into English from other languages. Other units, as strange as it may seem, are based on parts of the body.

The ancient Egyptians built the Great Pyramid of Giza over four and a half thousand years ago. It was constructed with side lengths of 440 cubits and was 280 cubits high.

The *cubit* was a unit equal to the 'length of the forearm of the bent elbow to the tip of the extended middle finger'.

1. The metric height of the Great Pyramid of Giza is 146.5 metres; its base length is 230 metres. Use these to determine the value of the cubit to 3 decimal places.

2. Below are some units of length from the imperial system. Select five units, and calculate their equivalent in the metric system. Find out where and how they were used (what sorts of objects they measured, in what situations they were used — for example, for small or large distances, for racetracks, for the depth of water or in land surveying).

league	hand	line	mile
furlong	chain	inch	link
rod	foot	yard	fathom

Although Australians have used units of length from the metric system for over 3 decades, many institutions have stuck with units common to the imperial system. For example, a cricket pitch is still 22 yards long.

3. Find another sport that still uses lengths associated with the imperial system and state the unit(s) involved.

on **Resources**

 Interactivities Crossword (int-2600)
                                     Sudoku puzzle (int-3170)

Exercise 10.10 Review questions

learn on

Fluency

1. Identify which metric unit would be most suitable for measuring:
   a. the distance walked by a teacher at school during a week
   b. the length of a piece of spaghetti
   c. the width of a pencil.

2. Complete the following conversions.
   a. $560\,mm = $ _____ m
   b. $2300\,cm = $ _____ km
   c. $17\,m = $ _____ cm
   d. $\frac{3}{4}\,km = $ _____ mm
   e. $2.09\,m = $ _____ mm

3. Complete the following conversions.
   a. $6\frac{4}{5}\,cm = $ _____ m
   b. $22.5\,mm = $ _____ cm
   c. $\frac{63}{10\,000}\,km = $ _____ m
   d. $82\,000\,000\,m = $ _____ km
   e. $5\frac{9}{10}\,mm = $ _____ cm

4. Arrange from smallest to largest: $44.5\,m$, $455\,cm$, $455\,000\,mm$, $0.004\,45\,km$.

5. a. Add $45.6\,km$ to $5600\,m$.
   b. Calculate the difference between $80\,m$ and $4300\,cm$.

6. Identify the reading in kilograms on this bathroom scale. *Note:* The scale for kilograms is the inner circle.

7. Estimate the height of the front tree in the photograph, given that the person standing is 1.7 m tall.

8. Identify which unit of area would be the most appropriate for measuring the following. (Choose from mm^2, cm^2, m^2, ha and km^2.)
   a. The floor area of your classroom
   b. The area of the city of Brisbane
   c. A pin head
   d. A market garden
   e. The area of the continent of Antarctica

9. Calculate the circumference of each of these circles correct to 2 decimal places.
   a.

   11 cm

   b.

   44 mm

   c.
   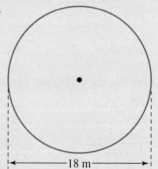
   18 m

10. Estimate the shaded area in the figure.
   Note: Each square on the grid has an area of $1 cm^2$.

Understanding

11. Use the formula $A = lw$ to calculate the areas of the following rectangles.

a.

15 m

10 m

b.

6 cm

1 cm

c.

47 mm

62 mm

12. Calculate the areas of the following shapes by first dividing them into rectangles.

a.

3 cm

1 cm

1 cm

4 cm

b.

5 cm

4 cm

2 cm   1 cm   2 cm

c.

3 cm

8 cm

4 cm

10 cm

13. Calculate the areas of the following shapes by first dividing them into rectangles.

a.

2 cm

4 cm

1 cm

3 cm

b.

20 cm

10 cm   8 cm

24 cm

10 cm

16 cm

c.

4 cm

2 cm

4 cm

1.5 cm

14. Calculate the areas of the following triangles.

a.

22 cm

57 cm

b.

16 m

12 m

c.

42 cm

64 cm

15. Calculate the areas of the following shapes.

a.

5 m

3 m

4 m

b.

16 mm

12 mm

20 mm

12 mm

16. Calculate the areas of the following shapes.

a.

5 cm

10 cm

5 cm

10 cm

b.

3 cm

4 cm

8 cm

17. Calculate the volume of each shape in cubic centimetres (cm^3). Each cube has a volume of 1 cm^3.

a.

b.

c.

d.

18. Draw the front, right and top views of these objects.

a.

b.

19. Draw isometric views of the objects, whose front (F), right (R) and top (T) views are given here.

a.

b.

20. Use the information given in the diagram to calculate the volume of the shape.

21. Use the information given in the diagram to determine the volume of the shape.

22. Complete the following conversions.

  a. $8\,L =$ _____mL

  b. $\dfrac{21}{50}L =$ _____mL

  c. $3300\,mL =$ _____L

  d. $1\dfrac{3}{250}L =$ _____mL

23. Complete the following conversions.

  a. $4\dfrac{3}{10}\,kL =$ _____L

  b. $0.0034\,kL =$ _____L

  c. $4755\,L =$ _____kL

  d. $432\,mL =$ _____L

Communicating, reasoning and problem solving

24. During a rescue operation in calm seas, a 16.5 m rope is dangled from a helicopter hovering 20 m above sea level. A man who is 175 cm tall is standing on the deck of a boat when he reaches 50 cm above his head for the rope. By how much does he fail to reach the rope if the deck is 1 m above sea level?

25. Calculate the perimeter of each shape.

  a.

  b.

Note: The dots are 1 cm apart.

  c.

  d.

26. Michelle rides three laps of the following dirt-bike track. Calculate how far she rides altogether.

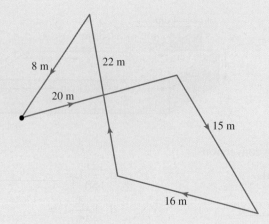

27. Determine the length of satin ribbon required to edge a rectangular blanket on all four sides if the blanket is 240 cm long and 195 cm wide. (Assume there is no overlap.)

28. Determine the cost of paving a rectangular courtyard that is 6.5 m long and 3.2 m wide. The courtyard is to be paved with concrete paving blocks, which cost $28 per square metre.

29. Calculate the area of card needed to make the triangular display sign shown.

30. When making a box kite, Katie cuts a square piece of material diagonally, as shown by the dotted line in the diagram. What is the area of each triangular piece of material?

31. a. A rectangular prism has a length of 40 cm and a width of 26 cm. If its volume is 29 120 cm³, calculate the height of the rectangular prism.
    b. A rectangular prism has a length of 30 cm and a width of 15 cm. If its volume is 3825 cm³, calculate its height.

32. In a food technology class, each of the 14 groups of students uses 350 mL of fresh milk to make pancakes. Calculate how many litres of milk should be ordered for the whole class.

33. Mario makes up raspberry cordial by pouring 275 mL of concentrate into a 2-litre container and filling the container with cold water.
    a. Calculate how much cold water Mario needs to add.
    b. Calculate how many 250 mL glasses of cordial Mario can pour from the large container.

34. A 6×16 rectangle is cut into two pieces as indicated. The two pieces are rearranged to form a right triangle. Calculate the perimeter of the resulting triangle.

35. Each small square in the following diagram measures $1\,\text{cm} \times 1\,\text{cm}$. Calculate the area of triangle ABC (in square cm).

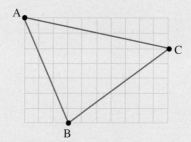

36. Stewart is having a swimming pool built in his back yard. The hole for the pool is 6 metres long, 4 metres wide and 1.5 metres deep. The excavated soil becomes aerated as it is dug out, and its volume is increased by $\dfrac{1}{10}$. Determine the volume of the excavated soil.

37. A rectangular cake is three times as long as it is wide. A layer of icing that is 1 cm thick is spread on top of the cake. If the total volume of icing used is $300\,\text{cm}^3$, determine the dimensions of the top of the cake.

38. A gardener is employed to pave a path that is 2 metre wide around a $10\,\text{m} \times 5.5\,\text{m}$ rectangular swimming pool.
    a. Determine the area of the paved path.
    b. Determine the minimum number of bricks that should be ordered if there are 60 bricks per square metre.

39. The perimeter of a rectangle is 20 cm. Investigate the shape of all such rectangles if the sides are whole numbers. Identify the dimensions of the rectangle with the largest area.

40. The following rectangle is divided into three identical smaller rectangles. The length of the large rectangle is 12 cm.
Determine the width of the large rectangle.

41. The front, side and top views of a solid are shown. Construct this solid using blocks, then sketch the solid.

a.
Front view

Right view

Top view

↑ Front

b.
Front view

Right view

Top view

↑ Front

on To test your understanding and knowledge of this topic, go to your learnON title at www.jacplus.com.au and complete the **post-test**.

Answers

Topic 10 Measurement

10.1 Pre-test

1. 56.5 cm
2. 0.020, 12 km
3. 90 m
4. 1.8 km
5. 46.8 cm
6. 50 m
7. 27 m^2
8. $472.50
9. 112 cm^2
10. 12.72 m^2
11. 48 m^3
12. 7 cm
13. 54 L
14. a. ii   b. iii   c. v   d. i   e. iv
15. a. B   b. A   c. C   d. D

10.2 Units of length

1. a. m   b. km   c. cm   d. km   e. m
2. a. m   b. cm   c. km   d. mm
3. A
4. a. 2000 m   b. 7000 m
   c. 5300 m   d. 660 m
5. a. 25 cm   b. 280 mm
   c. 2000 mm   d. 70 000 cm
6. a. 8 km   b. 6.5 km
   c. 0.7 km   d. 0.05 km
7. a. 60 m   b. 0.57 m
   c. 4.5 cm   d. 2560 cm
8. a. 800 000 cm   b. 101 000 mm
   c. 72 330 mm   d. 0.030 35 mm
9. D
10. A
11. a. 0.0452 km   b. 0.56 m   c. 87.5 mm
12. a. 60 cm   b. 309 000 cm   c. 4.8 cm
13. Sample responses are given here:
    a. the length of tube required to allow a cat to breathe through while in surgery
    b. the circumference of an actor's head so that an old-fashioned hat of the correct size can be made
    c. the length of a piece of milled hardwood required for the tabletop of a new table for a large family
    d. the height of a hill that requires terracing
    e. the length, in paces, of the runway leading to a long jump pit
14. a. 868 cm   b. 10.8 cm
15. Everest: 8.848 km, K2: 8.611 km

16. a. 150 cm, 12.5 m, 0.02 km
    b. 0.445 m, 3000 mm, 350 cm
    c. 50 000 mm, 500 m, 50 km
17. a. 1.7 m, 1700 cm, 0.17 km
    b. 0.0052 mm, 0.000 052 m, 0.052 cm
    c. 0.909 m, 9000 mm, 990 cm
18. a. 375 cm   b. 10 200 m
    c. 224.2 cm   d. 155.3 cm
19. a. 66 400 m   b. 410.4 m
    c. 10.4 cm   d. 1.7 km
20. 828 m
21. 44
22. 216 m
23. 1.593 m
24. Yes, 36 cm
25. 198 cm
26. 2.4 m
27. a. i. 18 000 000 km
       ii. 1 080 000 000 km
       iii. 25 920 000 000 km
       iv. 9 460 800 000 000 km
    b. Distances in the universe are very big and it would be inconvenient to use kilometres to measure them. The numbers in kilometres would be too large and would make interpreting and even reading distances difficult.
28. 71 428 pins
29. 38
30. 40 cm
31. 215 cm, 105 cm, 155 cm, 170 cm
32. a. i. 50 rotations
       ii. 100 rotations
    b. i. 500 rotations
       ii. 1000 rotations
33. 2 cm × 4 cm, 2 cm × 8 cm, 2 cm × 12 cm, 2 cm × 16 cm
34. 38.5 km

10.3 Perimeter

1. a. 8 units   b. 18 units   c. 18 units
   d. 22 units   e. 12 units
2. a. 40 cm   b. 92 mm
3. a. 30 m   b. 44 cm
4. 74 cm
5. 68 cm
6. 28 cm
7. a. 173 mm   b. 2995 cm
8. a. 408 mm   b. 180 mm
9. a. 14 433 m   b. 17 579 m
10. B
11. a. 156 cm   b. 664.2 cm   c. 14.8 cm
12. a. $12\frac{9}{10}$ m or 12.9 m   b. $32\frac{4}{5}$ mm or 32.8 mm

13. 48.5 m

14. 252 cm

15. 325.8 m

16. 139.8 m

17. a. 193 m   b. $3667

18. a.

b. 109.48 m

c. $4050.76

19. For example, fence construction companies would need to calculate perimeters on a daily basis.

20. 99 cm

21. a. 120 m

b. $228

c. $810

d. 25 posts

e. $341.25

f. i. 1600   ii. $1680

g. $4059.25

22. a. 8 different rectangles

b. Square sides 8 cm

23. $P = 6a - 2b$

24. 6 cm

10.4 Circles and circumference

1. a. 2π cm   b. 10π cm   c. 7π mm

2. a. 2.58 m   b. 23.25 km   c. 106.81 m

3. a. 8π m   b. 34π mm   c. 16π cm

4. a. 8.98 km   b. 2.51 m   c. 66.60 m

5. a. 25.71 cm   b. 82.27 mm   c. 61.70 m
   d. 39.28 mm   e. 71.42 cm

6. a. 120.82 cm   b. 5.88 m
   c. 250.80 m   d. 252.81 cm

7. B

8. C

9. 119.38 mm

10. 100.53 m

11. 100.53 cm

12. 25.13 m

13. 21.58 m

14. a. 6.00 m   b. 20.63 cm   c. 23.75 mm

15. a. 2.01 cm   b. 7.54 m   c. 24.99 mm

16. 80 m

17. 19.19 m

18. 15.71 m

19. a. iii   b. i   c. iv   d. ii

20. 41.82 km/h

21. 9.55 cm

10.5 Area

1. a. cm^2 b. m^2 c. mm^2 d. km^2

2. a. m^2 b. km^2 c. m^2 d. ha

3. a. 14 cm^2 b. 10 cm^2 c. 10 cm^2
   d. 16 cm^2 e. 8 cm^2

4. a. 9 cm^2 b. 6 cm^2 c. $9\frac{1}{2}$ cm^2
   d. 12 cm^2 e. 10 cm^2

5. A

6. C

7. C

8. a. 15 cm^2 b. 14 km^2 c. 10 m^2
   d. 25 mm^2

9. a. 4 cm^2 b. 3 cm^2 c. 40.5 cm^2 d. 50 cm^2

10. C

11. A

12. A

13. a. 416 cm^2 b. 3.125 m^2 c. 2.4 m^2 d. 493 mm^2

14. a. 583 cm^2 b. 600 km^2
    c. 337.5 cm^2 d. 152 cm^2

15. 28.14 mm^2

16. C

17. 200 000 m^2

18. a. 16 000 000 000 m^2
    b. 1 600 000 hectares

19. 17.5 m^2

20. 9.66 m^2

21. a. 7.5 m^2 b. 15 boxes   c. $528

22. $141.64

23. $2560

24. It would be cheaper to use the 70 cm × 70 cm concrete pavers.

25. 88.8 m^2

26. 104 m^2, $1300

27. 6000 cm^2 (0.6 m^2)

28. 12.6 m^2

29. 126 m^2

30. Length 9 cm, width 4 cm

10.6 Area of composite shapes

1. a. 20 cm^2 b. 48 m^2

2. a. 56 m^2 b. 400 mm^2

3. a. 72 cm^2 b. 76 m^2

4. a. 135 mm^2 b. 686 mm^2

5. a. 81 cm^2 b. 5 m^2 c. 16 cm^2

6. a. $56\,\text{cm}^2$ b. $588\,\text{cm}^2$ c. $13.5\,\text{m}^2$

7. a. $48\,\text{m}^2$ b. $325\,\text{m}^2$

8. $13.05\,\text{m}^2$

9. $39.28\,\text{m}^2$

10. $14.25\,\text{m}^2$

11. a. $27\,\text{m}^2$ b. \$675

12. \$2157.60 (rounded to the nearest 5 cents)

13. $75\,\text{m}^2$

14. $39\,\text{m}^2$

15. $4920\,\text{km}^2$

16. $135\,\text{m}^2$

17. a. $22\,\text{cm}^2$

    b. $22\,\text{cm}^2$

    c. Sample responses can be found in the worked solutions in the online resources.

18. If your composite shape is made up of two or more simple shapes, it is best to determine the area by addition. If the area of a shape has a section not required (a shaded area), it is best to calculate the total area and then subtract the area not required.

19. a. Square of side length $30\,\text{cm}$
    Three circles of diameter $15\,\text{cm}$
    Two semicircles of diameter $30\,\text{cm}$

    b. Using addition of areas: square + circle of diameter $30\,\text{cm}$ (the two equal semicircles form one circle). Using subtraction of areas: area calculated in previous step $-3 \times$ area of the smaller (coloured) circle.

    c. $1075\,\text{cm}^2$

    d. The area in mm^2 would be $107\,500\,\text{mm}^2$. This is more inconvenient to use because it is a larger number, which is harder to interpret and use in further calculations if needed.

20. $\$0.15/\text{cm}^2$

21. The shaded area is $\dfrac{7}{32}$ of the square.

22. a. i. Two rectangles with length $4\,\text{cm}$ and width $1\,\text{cm}$ and two parallelograms with length $3\,\text{cm}$ and width $0.5\,\text{cm}$

    ii. One square of side length $4\,\text{cm}$, one rectangle with length $2\,\text{cm}$ and width $1\,\text{cm}$, one triangle with height $3\,\text{cm}$ and base $1\,\text{cm}$ and two right-angled triangles with height $3\,\text{cm}$ and base $0.5\,\text{cm}$. The last two triangles form a triangle with base $1\,\text{cm}$ and height $3\,\text{cm}$.

    b. $11\,\text{cm}^2$

10.7 Volume of rectangular prisms

1. a. $8\,\text{cm}^3$ b. $6\,\text{cm}^3$ c. $12\,\text{cm}^3$ d. $48\,\text{cm}^3$

2. B

3. a. $4\,\text{cm}^3$ b. $24\,\text{cm}^3$ c. $15\,\text{cm}^3$ d. $13\,\text{cm}^3$

4. a. $7\,\text{cm}^3$ b. $35\,\text{cm}^3$ c. $14\,\text{cm}^3$ d. $17\,\text{cm}^3$

5. A

6. a. $12\,\text{cm}^3$ b. $12\,\text{m}^3$ c. $6144\,\text{cm}^3$

7. a. $12\,\text{cm}^3$ b. $10\,\text{cm}^3$ c. $10\,\text{cm}^3$

8. a. $192\,\text{m}^3$ b. $270\,\text{cm}^3$ c. $101.25\,\text{cm}^3$

9. D

10. B

11. a. $186\,\text{cm}^3$ b. $136\,\text{cm}^3$

12. a. $189\,\text{cm}^3$ b. $180\,\text{cm}^3$

13. $9690\,\text{cm}^3$

14. $477\,750\,\text{cm}^3$

15. a. $20\,\text{cm}$ b. $9\,\text{cm}$ c. $12\,\text{cm}$

16. a. $2700\,\text{cm}^3$ b. $2\,700\,000\,\text{mm}^3$

17. $4.875\,\text{m}^3$

18. a. $28.875\,\text{cm}^3$ b. $28\,875\,\text{mm}^3$
    c. $176\,\text{mm}^3$ d. $20\,075\,\text{mm}^3$

19. $768\,\text{m}^3$

20. $40.8\,\text{m}^3$

21. Each guest receives $267\,\text{cm}^3$ of cake.

22. $2500\,\text{m}^3$

23. $0.65\,\text{m}^3$ more soil is needed (or $2.075\,\text{m}^3$).

24. $2277\,\text{cm}^3$

25. Length $18\,\text{cm}$, width $15\,\text{cm}$ and height $10\,\text{cm}$. Volume is $2700\,\text{cm}^3$.

26. $36\,000$

10.8 Capacity

1. a. Megalitres   b. Kilolitres
    c. Millilitres   d. Litres

2. a. $2000\,\text{mL}$ b. $3\,\text{L}$ c. $7\,\text{L}$ d. $5.5\,\text{L}$

3. a. $2500\,\text{mL}$ b. $32\,\text{L}$ c. $35\,\text{mL}$ d. $420\,000\,\text{mL}$

4. a. $1870\,\text{mL}$ b. $22.5\,\text{L}$ c. $100\,\text{mL}$ d. $0.025\,\text{kL}$

5. a. $750\,\text{mL}$ b. $0.0025\,\text{kL}$
    c. $2450\,\text{cm}^3$ d. $78\,\text{L}$

6. a. $40\,000\,\text{mL} = 40\,\text{L}$ b. $6000\,\text{mL} = 6000\,\text{cm}^3$
    c. $5.2\,\text{kL} = 5.2\,\text{m}^3$

7. D

8. D

9. a. $0.25\,\text{L}$, $2.45\,\text{L}$, $2.5\,\text{L}$, $25\,000\,\text{mL}$
    b. $7.65\,\text{mL}$, $760\,\text{mL}$, $0.765\,\text{L}$, $7.60\,\text{L}$
    c. $0.011\,\text{L}$, $0.1\,\text{L}$, $110\,\text{mL}$, $1.1\,\text{L}$

10. 16 bottles

11. $1750\,\text{mL}\,(1.75\,\text{L})$

12. $1620\,\text{mL}\,(1.62\,\text{L})$

13. $6\,\text{L}$

14. 50 doses

15. 225 bottles

16. $9.9\,\text{L}$

17. $4.8\,\text{L}$

18. a. No   b. Yes

19. The $185\,\text{mL}$ container is the better buy.

20. The tank overflowed.

21. $616\,\text{mL}$

22. $50.4\,\text{L}$

23. Approximately $27\,\text{L}\,(27.09\,\text{L})$

24. $300\,\text{mL}$

25. $1401.6\,\text{L}$

26. $16.4\,\text{kL} = 16.4\,\text{m}^3$. An example could be length $2\,\text{m}$, width $2\,\text{m}$, height $4.1\,\text{m}$.

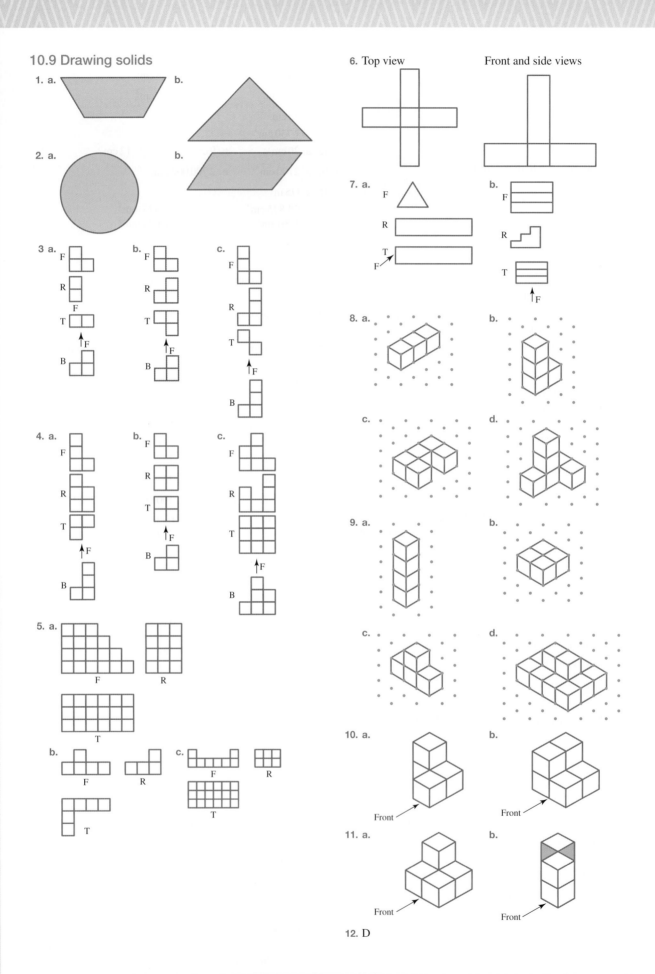

10.9 Drawing solids

1. a. [trapezium shape] b. [triangle shape]

2. a. [circle] b. [parallelogram]

3. a.
F, R, T, B views (F arrow)

b. F, R, T, B views (F arrow)

c. F, R, R, T, B views (F arrow)

4. a. F, R, T, B views (F arrow)

b. F, R, T, B views (F arrow)

c. F, R, T, B views (F arrow)

5. a. F, R, T views

b. F, R, T views

c. F, R, T views

6. Top view   Front and side views

7. a. F, R, T, F (arrow) b. F, R, T, F (arrow)

8. a. b. c. d.

9. a. b. c. d.

10. a. Front b. Front

11. a. Front b. Front

12. D

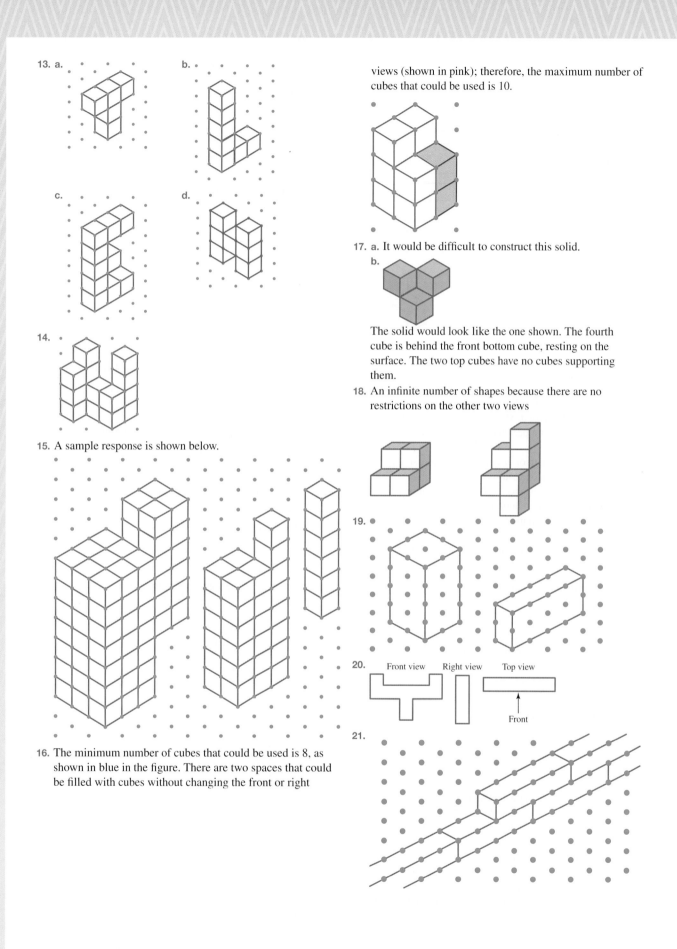

13. a.

b.

c.

d.

14.

15. A sample response is shown below.

16. The minimum number of cubes that could be used is 8, as shown in blue in the figure. There are two spaces that could be filled with cubes without changing the front or right

views (shown in pink); therefore, the maximum number of cubes that could be used is 10.

17. a. It would be difficult to construct this solid.

b.

The solid would look like the one shown. The fourth cube is behind the front bottom cube, resting on the surface. The two top cubes have no cubes supporting them.

18. An infinite number of shapes because there are no restrictions on the other two views

19.

20. Front view   Right view   Top view

Front

21.

22.

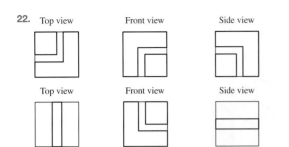

Top view   Front view   Side view

Top view   Front view   Side view

23. The arrangements are shown below.

Project

1. 1 cubit $= 0.523$ metres

2. A sample response is given here.

Inch — a unit of length used as a measure for display screens, the height of a person or sometimes the thickness of an object

Mile — a unit of length or distance used to measure the distance usually seen on road signs in the US

Furlong — a unit of distance used to measure dimensions of farmland and distance in horse racing

Foot — a unit of length used to measure the height of a person or an object

Fathom — a unit of length used to measure the depth of the water, or the length of a fisher's setline

3. A sample response is given here.

Basketball:

Baskets:	Inches
	Rings are 18 inches in inside diameter, with white cord 12-mesh nets, 15 to 18 inches in length.
Height of basket:	Feet
	10 feet (upper edge)
Circumference of ball:	Inches
	Not greater than 30 inches and not less than $29\frac{1}{2}$ inches.

Soccer:

Goals:	Yards/feet/inches
	Distance between posts is 8 yards. Distance from crossbar to the ground is 8 feet.
Penalty area:	Yards
	Two lines drawn at right angles to the goal line, 18 yards from the inside of each goalpost. Lines extend into playing field for 18 yards and are joined by a line drawn parallel with the goal line.

10.10 Review questions

1. a. Kilometres or metres

  b. Centimetres

  c. Millimetres

2. a. 0.56 m   **b.** 0.023 km   **c.** 1700 cm

  d. 750 000 mm   **e.** 2090 mm

3. a. 0.068 m   **b.** 2.25 cm   **c.** 6.3 m

  d. 82 000 km   **e.** 0.59 cm

4. 0.004 45 km, 455 cm, 44.5 m, 455 000 mm

5. a. 51 200 m (51.2 km)

  b. 3700 cm (37 m)

6. 80 kg

7. 7 m

8. a. m^2 **b.** km^2 **c.** mm^2

  d. ha   **e.** km^2

9. a. 69.12 cm   **b.** 138.23 mm   **c.** 56.55 m

10. $15 \, cm^2$

11. a. $150 \, m^2$ **b.** $6 \, cm^2$ **c.** $2914 \, mm^2$

12. a. $9 \, cm^2$ **b.** $19 \, cm^2$ **c.** $52 \, cm^2$

13. a. $9 \, cm^2$ **b.** $380 \, cm^2$ **c.** $11 \, cm^2$

14. a. $627 \, cm^2$ **b.** $96 \, m^2$ **c.** $1344 \, cm^2$

15. a. $16 \, m^2$ **b.** $432 \, mm^2$

16. a. $87.5 \, cm^2$ **b.** $41 \, cm^2$

17. a. $18 \, cm^3$ **b.** $30 \, cm^3$ **c.** $20 \, cm^3$ **d.** $28 \, cm^3$

18. a.

    F   R   T

  b.

    F   R   T

19. a.

  b.

20. $72 \, cm^3$

21. $112 \, cm^3$

22. a. 8000 mL   **b.** 420 mL

  c. 3.3 L   **d.** 1012 mL

23. a. 4300 L   **b.** 3.4 L

  c. 4.755 kL   **d.** 0.432 L

24. 25 cm

25. a. 46 cm   **b.** 20 cm

  c. 191 mm (19.1 cm)   **d.** 21 cm (210 mm)

26. 243 m

27. 870 cm

28. $582.40

29. 3.6 m^2

30. 0.72 m^2

31. a. 28 cm   b. 8.5 cm

32. 4.9 L

33. a. 1725 mL (1.725 L)
    b. 8 glasses

34. The perimeter is 46.

35. 32 cm^2

36. 39.6 cubic metres

37. 10 cm by 30 cm

38. a. Area of path is 78 m^2.
    b. 4680 bricks are needed.

39. Possible side dimensions: $9 \times 1; 8 \times 2; 7 \times 3; 6 \times 4; 5 \times 5$.
    Largest area is 25 square units.

40. 8 cm

41. a.   b.

11 Angle relationships and properties of geometrical figures

LESSON SEQUENCE

LESSON
11.1 Overview

Why learn this?

The word geometry comes from the Greek terms *geo-* (meaning *earth*) and *metron* (meaning *measurement*). Geometry is one of the oldest areas of mathematics, and it allows us to explore our world in a very precise way. Have you ever wondered how early sailors such as Captain Cook navigated around the world? They didn't have computers to guide them. They navigated using the stars and their knowledge of geometry. The famous Greek philosopher Thales of Miletus, who lived around 600 BC, used geometry to calculate the height of the pyramids in Egypt.

The geometry we study today can be traced back to around the year 300 BC and the work of another Greek mathematician, Euclid. Builders, architects, surveyors and engineers all use their knowledge of geometry to ensure buildings are stable and visually pleasing. Navigation, both in the air and at sea, is based on geometric principles. Travel in outer space involves the use of complex geometry. Geometry is also involved in sport. You need to carefully estimate the angle required when you are shooting for a goal in hockey or football. If you have played or watched someone playing pool, you will know that understanding angles is vital to play well. Many professions, including landscape gardening and interior design, use geometry every day.

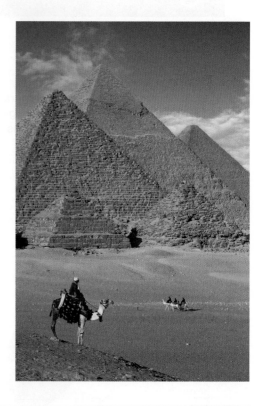

Hey students! Bring these pages to life online

Watch videos

Engage with interactivities

Answer questions and check solutions

Find all this and MORE in jacPLUS

Reading content and rich media, including interactivities and videos for every concept

Extra learning resources

Differentiated question sets

Questions with immediate feedback, and fully worked solutions to help students get unstuck

1. **MC** An angle of 125° is:

   A. an acute angle.
   B. a right angle.
   C. an obtuse angle.
   D. a reflex angle.

2. If an analogue clock is showing 1:30 pm, determine the smaller angle between the big and small hands.

3. State how many degrees are in a right angle.

4. **MC** A reflex angle is between:

   A. 45° and 180°
   B. 0° and 90°
   C. 90° and 120°
   D. 180° and 360°

5. **MC** Two acute angles that sum to be an acute angle are:

   A. 180° and 360°
   B. 90° and 180°
   C. 30° and 70°
   D. 50° and 30°

6. **MC** An analogue clock is showing 4 am. The obtuse and reflex angles that the big and small hands make between them respectively are:

   A. 90° and 180°
   B. 240° and 120°
   C. 120° and 240°
   D. 20° and 40°

7. Name the acute angle in the following diagram.

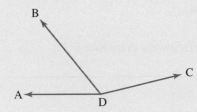

8. **MC** Identify the values of p and q in the following diagram.
   A. $p = 110°$, $q = 70°$
   B. $p = 10°$, $q = 170°$
   C. $p = 100°$, $q = 80°$
   D. $q = 110°$, $p = 70°$

9. **MC** If you measure the three angles of a triangle and then add the three angles together, what is the result?

   A. 90°
   B. 120°
   C. Between 90° and 180°
   D. 180°

10. Classify each of the following triangles, based on side length and angle type:

   i. Isosceles, right-angled triangle   **ii.** Isosceles, acute-angled triangle
   iii. Scalene, right-angled triangle   **iv.** Scalene, obtuse-angled triangle

a.

b.

c.

d.
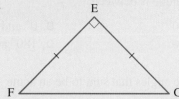

11. Calculate the value of *x* in the following diagram.

12. ΔABC is an isosceles triangle with AB = AC and ∠ABC = 64°.
Calculate the magnitude of ∠BAC.

13. Calculate the value of *x* in the following diagram.

14. **MC** Three angles in a quadrilateral are 135°, 72° and 38°. Identify the fourth angle.

  A. 135°   **B.** 38°   **C.** 72°   **D.** 115°

15. A driveway inclined at an angle of 18° from the horizontal has vertical posts along the driveway.
Determine the acute angle between the posts and the driveway.

LESSON
11.2 Measuring angles

LEARNING INTENTION

At the end of this lesson you should be able to:
- understand what an angle is
- estimate angles
- measure angles using a protractor.

11.2.1 Understanding angles

eles-4089

- An angle is a measurement of space between two lines that meet at a point.
- The point where the two lines meet is called a **vertex** and the lines are the arms of the angle.
- For *angle*, we use the symbol ∠.
- Angles are measured in degrees. The symbol for degrees is °.

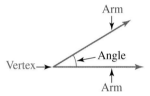

Estimating angles

- There are four important angles: 90°, 180°, 270° and 360°.

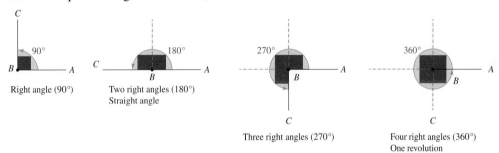

Right angle (90°) Two right angles (180°) Three right angles (270°) Four right angles (360°)
Straight angle One revolution

Estimating the size of an angle

- **First: determine which important angle it would measure closest to**
- **Second: decide whether it is greater or less than that angle**
- **Third: make an estimate.**

WORKED EXAMPLE 1 Estimating the angle

Estimate the size of the following angles.

a.   b.   c.

THINK

a. 1. The angle is closest to 90°.

  2. The angle is less than 90°.

  3. Make an estimate.

WRITE

a. The estimate is 60°.

---

b. 1. The angle is closest to 90°.

   2. The angle is more than 90°.

   3. Make an estimate.

c. 1. The angle is roughly halfway between 270° and 360°.

   2. Halfway between 270° and 360° is 270° + 45° = 315°
     Note: You add 45° because 45° is half of 90°.

   3. Write your estimate.

b. The estimate is 120°.

c. The estimate is 315°.

▶ 11.2.2 Measuring angles

eles-4090

- The tool for measuring angles is called a **protractor**.
- There are two types of protractors: circular and semicircular.

Circular protractor

Semicircular protractor

Using a protractor

To use a protractor, follow these steps.

1. Place the protractor on the angle so that:
- the zero line is along one line of the angle
- the centre of the protractor lines up with where the two lines meet (the vertex).

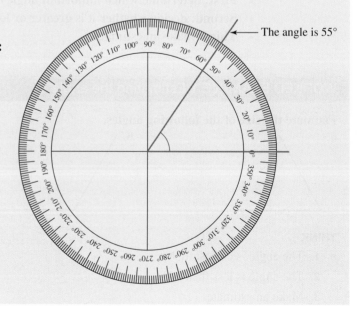

The angle is 55°

2. Read the angle by looking at where the other line crosses the protractor numbers.

The angle is 65°
(the other arm is at 0° on the inside scale, so we use the inside scale)

WORKED EXAMPLE 2 Measuring angles

Measure the size of each of the following angles.

a.

b.

THINK

a. **1.** Place the centre of the protractor on the vertex of the angle.

    2. Match the line that passes through the centre of the protractor and points to 0° with one of the arms of the angle.

    3. Read the size of the angle indicated by the other arm. (You may need to extend the line so that you can read it on the protractor.)

WRITE

a.

The size of this angle is 60°.

b. **1.** Place the centre of the protractor on the vertex of the angle.

    2. Match the line that passes through the centre of the protractor and points to 0° with one of the arms of the angle.

    3. Read the size of the angle indicated by the other arm. (You may need to extend the line so that you can read it on the protractor.)

b.

The size of this angle is 120°.

- A semicircular protractor has angles only up to 180°. To measure angles larger than 180°:
  1. measure the part of the angle that is bigger than 180°
  2. determine the total angle by adding the measured angle to 180°.

WORKED EXAMPLE 3 Measuring angles using a semicircular protractor

Measure the size of this angle using a semicircular protractor.

THINK	WRITE
1. Measure the angle after 180°.	 The angle after 180° is 130°.
2. Add the measured angle to 180°.	130° + 180° = 310°
3. Write the answer.	The size of the angle is 310°.

 Resources

 Interactivities The circular protractor (int-3952)
The semicircular protractor (int-3953)

Exercise 11.2 Measuring angles

learn on

11.2 Quick quiz on	11.2 Exercise

Individual pathways

■ PRACTISE	■ CONSOLIDATE	■ MASTER
1, 2, 7, 10	3, 5, 8, 11	4, 6, 9, 12

Fluency

1. **WE1** Answer the following.

   a. **MC** The size of the angle shown is closest to:

   A. 20°   B. 30°   C. 40°   D. 150°   E. 140°

b. The size of the angle shown is closest to:

    A. 42°   **B.** 138°   **C.** 142°   **D.** 145°

2. **WE2,3** Measure the size of each of the following angles.

  a.   **b.**   **c.**   **d.**

3. Measure the size of each of the following angles.

  a.   **b.**

  c.   **d.**

4. Measure the size of each of the following angles.

  a.   **b.**

  c.   **d.**

  e. Comment on the accuracy of your measurement compared with the answer in the text.

Understanding

5. In each of the photographs, measure the indicated angle.

  a.   **b.**

c.

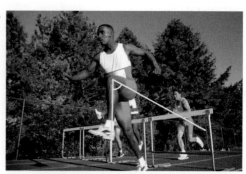

d.

6. The recommended slope for wheelchair access is 4° from the horizontal. An angle greater than this makes it more difficult for the person in the wheelchair to travel up the ramp.

Second section

Bottom section

a. Would you rate the slope of the bottom section of this ramp as 'difficult', 'as recommended' or 'easy'? Justify your rating.

b. Describe how you would rate the second section of the ramp.

Communicating, reasoning and problem solving

7. Consider the four different angles shown.

a. Carefully measure each angle and write the measurements.
b. Identify which two of these angles combined add to 90°.
c. Identify which two of these angles combined add to 180°.
d. Suggest one way of combining these angles to add to 360°. You can repeat angles.

i.

ii.

iii.

iv.

8. Sal is designing a new slide for her water park. She wants the slide to be both fun and safe. Below are side views of four possible designs:

Design 1   Design 2   Design 3   Design 4

a. State which design will give riders the fastest ride. What is the angle of that slide?
b. State which design will give riders the slowest ride. What is the angle of that slide?
c. The safety requirements are that slides for people aged 12 years and older must have an angle of no more than 35°. Identify which slides fit the safety requirements for Sal's park.
d. The safety requirements are that slides for people aged less than 12 years must have an angle of no more than 20°. Identify which slides fit the safety requirements for Sal's park.

Sal needs to choose the designs that give the fastest ride but also fit the safety requirements.

e. Identify which design Sal should use for the slide for people aged 12 and older.
f. Identify which design Sal should use for the slide for people aged less than 12 years.

9. Study the cartoon of the golfer to help you answer the following questions.

Copy the following table to record your answers.

Diagram	Estimate of angle	Measured angle	Difference
1			
2			
3			
4			

a. Without using a protractor, estimate the size of each angle in the first four diagrams of the golf sequence above and complete the second column of the table.
b. Explain how you obtained an estimate. Discuss the steps you followed.
c. Complete the third column by measuring each angle with a protractor.
d. Calculate the difference between your estimate and the actual value of each angle.
e. Describe how you could improve your skills in estimating the size of an angle.
f. Try these improvements for estimating the size of the angles in the remaining four diagrams. Copy and complete the following table.

Diagram	Estimate of angle	Measured angle	Difference
5			
6			
7			
8			

g. Has the difference between the estimated and actual value for each angle become smaller? Comment on whether your estimating skills have improved.

10. Sean is flying a kite. Depending on the position of the kite, the angle between the string and the horizontal line changes.

Complete the following statements:

a. As the kite moves from point A to point B, the angle increases from _____ to _____.

b. As the kite moves from point B to point C, the angle decreases from _____ to _____.

11. **a.** Using a protractor, measure the angles between the two clock hands shown in the diagrams.

i.

ii.

iii.

iv.

v.

vi.

b. Calculate the angle between the two arms without using a protractor.

12. Abby and Brinda are having an argument. Abby says there isn't a pattern when you measure the angles inside triangles and add them together. Brinda says there is a pattern.

Three triangles are shown.

Triangle 1

Triangle 2

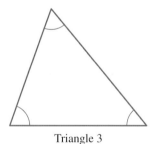

Triangle 3

a. Measure the angles in the corners of each triangle.

b. Add the angles of each triangle.

c. Explain who is right, Abby or Brinda.

LESSON
11.3 Constructing angles with a protractor

LEARNING INTENTION

At the end of this lesson you should be able to:
- draw an angle using a protractor.

▶ 11.3.1 Constructing angles with a protractor

eles-4092

- Protractors are used to draw angles.
- To draw an angle, start by drawing a straight line (arm) and putting a dot (vertex) at one end.

Draw a dot   Draw a straight line

- Next, line up the protractor so that the dot (vertex) is in the centre and the line points to 0°.
- Place a dot for the angle you want to measure (for example 60°).

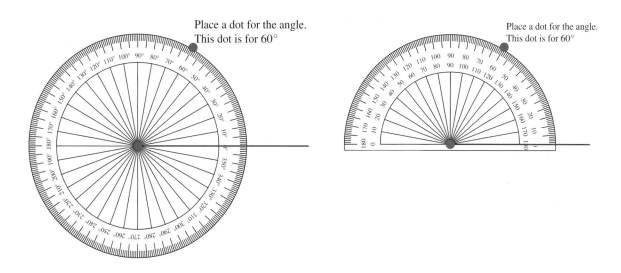

Place a dot for the angle. This dot is for 60°

Place a dot for the angle. This dot is for 60°

- Lastly, join the dots, and label the angle by drawing an arc between the two lines.

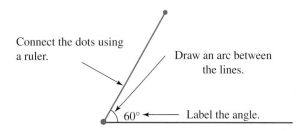

Connect the dots using a ruler. Draw an arc between the lines.

60° Label the angle.

Construct each of the following angles.

a. **50°**

b. **152.5°**

THINK	WRITE

a. 1. Draw a horizontal line. Put a dot at one end. This line is the arm and the dot is the vertex.

2. Place the protractor so that its centre is on the vertex and 0° is on the line.

3. Find 50° on the scale and mark a dot at the edge of the protractor.

a.

4. Join the dot and the vertex with a straight line using a ruler.

5. Draw the arc and label the angle.

50°

b. 1. Repeat steps 1 and 2 from part a.

2. Find 152.5° on the scale and mark a dot at the edge of the protractor. Remember to start at 0° and use the inside scale.

b.

3. Join the dots with a straight line using a ruler.

4. Draw the arc and label the angle.

152.5°

▶ 11.3.2 Constructing angles greater than 180° using a semicircular protractor

eles-4093

- To draw an angle greater than 180°, recall that 180° is a straight line.
- Subtract 180° from the given angle to determine how much larger than 180° the angle is.
  For example, for an angle of 220°: $220° - 180° = 40°$
  220° is 40° more than 180°.

- To draw the angle, draw the arm (line) and the vertex (dot) as described in section 11.3.1.
- Use the semicircular protractor upside down and start counting at 0° (inner scale) or 180° (outer scale) to measure the given angle (40° in the above example).

Start counting from 0° here.

Place a dot for the angle at 40°.

- Lastly, join the dots and label the outer angle by drawing the arc.

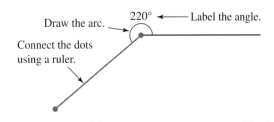

Draw the arc.

220° ◄——— Label the angle.

Connect the dots using a ruler.

WORKED EXAMPLE 5 Constructing angles using a semicircular protractor

Construct an angle of 210° using a semicircular protractor.

THINK	WRITE
1. Subtract 180° from 210°.	$210° - 180° = 30°$
2. Draw a horizontal line. Put a dot at one end. The line is the arm and the dot is the vertex.	Draw a dot Draw a straight line
3. Place the protractor upside down on the line so that its centre is on the dot.	
4. Start at 0° on the left-hand side. Mark the position of 30° with a dot.	
5. Remove the protractor and join the dots using a ruler.	
6. Draw the arc and label the outer angle as 210°.	210°

- *Note:* It is possible to draw an angle greater than 180° with the protractor upright. This diagram shows the angle of 210° (the same as in Worked example 5) constructed by placing the protractor in the upright position.

210°

 Resources

 Interactivities Construction of angles using a circular protractor (int-3954)
Construction of angles using a semicircular protractor (int-3955)

Exercise 11.3 Constructing angles with a protractor

learn on

11.3 Quick quiz **on**	11.3 Exercise

Individual pathways

■ PRACTISE	■ CONSOLIDATE	■ MASTER
1, 4, 9, 11, 14	2, 5, 7, 10, 12, 15	3, 6, 8, 13, 16

Fluency

1. **WE4a** Construct each of the following angles.
   a. 15° b. 9° c. 53° d. 75°

2. Construct each of the following angles.
   a. 45° b. 40° c. 88° d. 76°

3. **WE4b** Construct each of the following angles.
   a. 96° b. 92° c. 165° d. 143°

4. Construct each of the following angles.
   a. 140° b. 156° c. 127° d. 149°

5. **WE5** Construct each of the following angles using a semicircular protractor.
   a. 185° b. 210.5° c. 235° d. 260°

6. Construct each of the following angles using a semicircular protractor.
   a. 243° b. 192° c. 249° d. 214°

7. Construct each of the following angles.
   a. 295° b. 269° c. 307° d. 349°

8. Construct each of the following angles.
   a. 328° b. 300° c. 345° d. 358°

Understanding

9. a. Draw the following angles.

     i. 45°
     ii. 90°
     iii. 180°
     iv. 270°
     v. 360°

  b. Use the angles you have drawn to help fill in the gaps.
     i. The angle _____ is half of the angle 90°.
     ii. The angle _____ is half of the angle 180°.
     iii. The angle _____ is half of the angle 360°.
     iv. The angle _____ is a third of the angle 270°.
     v. The angle _____ is an eighth of the angle 360°.

10. a. Construct a 20° angle.
  b. Draw a 45° angle on one side of this angle.
  c. On the second arm of the angle in part b draw a 108° angle.
  d. What is the remaining reflex angle? *Hint:* A reflex angle is more than 180° but less than 360°.
  e. What is the total sum of the angles?

Communicating, reasoning and problem solving

11. Explain the method you would use to construct a reflex angle if you only had a 180° protractor.

12. It is possible to draw angles larger than 360°.
  a. Draw an angle of 400°.
  b. What smaller (0° to 90°) angle is the same as 400°?

13. We see the objects surrounding us because light reflects off the surface of the objects, as shown in the first diagram. The two angles shown are always equal.
Imagine you are in a room lit by a light globe as shown in the second diagram.

  a. Using a protractor and a ruler, draw the next four paths followed by the ray of light shown in the diagram.
  b. Draw another ray of light in a different direction.
  c. Use the same procedure as in part a to draw the first four paths of the line drawn in part b.

14. a. Draw a horizontal line that is 3 cm long. Use that line as the base to draw a triangle with angles of 60° in each corner.
  b. Draw a horizontal line that is 6 cm long. Use that line as the base to draw a triangle with angles of 60° in each corner.
  c. Compare the two triangles and state the similarities and differences.
  d. If you drew a triangle with base length 10 m and two corner angles of 60° each, determine the angle of the third corner.

15. On an A4 sheet of paper, draw a 20 cm long straight horizontal line close to the bottom of the page. Label this line AB.

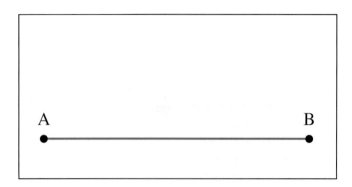

a. Using a protractor construct an angle of 30° at point A and an angle of 30° at point B. Extend the two lines until they intersect.
b. Cut out the triangle and compare it with the triangles constructed in this way by other students.
c. Write a statement about your findings.

16. Draw any triangle on an A4 sheet of paper.
a. Use a protractor to measure the three angles.
b. Cut out the triangle and then cut the angles as shown in the diagram.

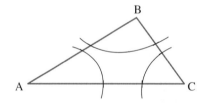

c. Place the angles next to each other so that their arms touch (see diagram).

d. Measure the new angle formed.
e. Compare your results with those of other students and write a statement to summarise your findings.

LESSON
11.4 Types of angles and naming conventions

LEARNING INTENTION

At the end of this lesson you should be able to:
- classify different types of angles according to size
- define, label and name points, rays, lines and intervals using capital letters
- name angles using three points.

▶ 11.4.1 Types of angles

eles-4094

- Angles can be classified according to their size.

An **acute angle** is greater than 0°, but less than 90°.

A **right angle** is an angle that equals exactly 90°.

An **obtuse angle** is greater than 90° but less than 180°.

A **straight angle** equals exactly 180°.

A **reflex angle** is greater than 180° but less than 360°.

An angle of **complete revolution** or a **perigon** is an angle of a full 360° circle.

WORKED EXAMPLE 6 Classifying angles

Classify each of the following angles according to its size.
a. **115°**　　　　　　　　　b. **27°**　　　　　　　　　c. **300°**

THINK	WRITE
a. The given angle is larger than 90°, but smaller than 180°, so classify it accordingly.	a. 115° is an obtuse angle.
b. The given angle is between 0° and 90°, so classify it accordingly.	b. 27° is an acute angle.
c. The given angle is larger than 180°, but less 360°, so classify it accordingly.	c. 300° is a reflex angle.

▶ 11.4.2 Naming points, rays, lines and intervals

eles-6096

- Points, rays, lines and intervals can be named using capital letters.
- A **point** is a specific location and is labelled with a single capital letter.

Point

- An **interval** can be labelled with two capital letters, one at each end of the interval.

Interval

- A **ray** has a fixed starting point but no end point with an arrow on one end.

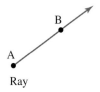

Ray

- A **line** is labelled by two points that sit on the line. The arrows at each end of the line indicate that the line continues without ending.

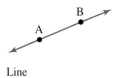

Line

WORKED EXAMPLE 7 Points, rays, lines and intervals

Draw:

a. **A ray from point D passing through point E**
b. **An interval from point F to G**
c. **A line through points H and I.**

THINK	WRITE
a. A ray with a fixed starting point at D passing through E with no end point and an arrow on one end.	a. Ray
b. An interval can be labelled with two capital letters, one at each end of the interval. Label one end F and the other G.	b.
c. A line is labelled by two points that sit on the line. The arrows at each end of the line indicate that the line continues without ending.	c.

⊙ 11.4.3 Naming angles

eles-4095

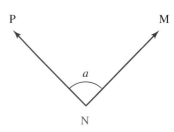

- Angles are named by:
  - using three letters: two letters to represent the arms of the angle, and a third letter to represent its vertex. The letter representing the vertex is always placed in the middle. For example, the angle shown can be named either ∠PNM or ∠MNP.
  - writing the letter that represents the vertex only. For example, the angle shown could be named ∠N.
  - writing the pronumeral that represents the angle inside the arms. For example, the angle shown could be named *a*.

WORKED EXAMPLE 8 Naming angles

Name each of the following angles.

a.

b.

THINK

a. Name the angle by starting either with X or with Y, then placing O (the letter representing the vertex) in the middle. You could also simply use the letter at the vertex. Remember to place the angle symbol (∠) before the letters.

b. Name the angle by starting either with A or with C, then placing B (the letter representing the vertex) in the middle. You could also simply use the letter at the vertex. Remember to place the angle symbol (∠) before the letters.

WRITE

a. ∠XOY or ∠YOX or ∠O

b. ∠ABC or ∠CBA or ∠B

WORKLED EXAMPLE 9 Drawing angles

Draw an acute angle ∠ADG.

THINK

1. Construct any acute angle (an angle greater than 0°, but less than 90°).

2. Letter D is in the middle, so place it at the vertex.

3. Place letter A at the end of one arm of the angle and letter G at the end of the other arm.
   (*Note:* It does not matter which arm is represented by AD or GD.)

WRITE

---

on Resources

Interactivities   Types of angles (int-3961)
Names of angles (int-3962)

11.4 Quick quiz on	**11.4 Exercise**

Individual pathways

■ PRACTISE	■ CONSOLIDATE	■ MASTER
1, 4, 7, 10, 12, 15, 20, 23	2, 5, 8, 11, 14, 16, 18, 21, 24	3, 6, 9, 13, 17, 19, 22, 25

Fluency

1. **WE6** Classify each of the following angles according to its size.

   a. 12°   b. 215.3°   c. 156°   d. 180°   e. 355.2°

2. Classify each of the following angles according to its size.

   a. 90°   b. 4.8°   c. 360°   d. 45°   e. 270°

3. The following list gives values of particular angles.

   3°, 45°, 65°, 123°, 69°, 234°, 90°, 360°, 300°, 270°, 165°, 210°, 180°

   a. List the acute angles.
   b. List the obtuse angles.
   c. List the reflex angles.

4. **WE7** Draw:
   a. A ray from point A passing through point B.
   b. An interval from point C to D.
   c. A line through points X and Y.

5. Consider the following diagrams and write down the type of angle shown in each case.

   a.

   b.

   c.

   d.

   e.

   f.

6. **WE8** Name each of the following angles.

a.

b. P
Q
R

c. S
D
X

d. P
G
Z

e. P ← F → H

Understanding

7. **WE9** Draw each of the following angles as specified in the brackets.

a. ∠CDE (acute)

b. ∠TRE (obtuse)

c. ∠QAS (straight)

8. Draw each of the following angles as specified in the brackets.

a. ∠FGH (reflex)

b. ∠KJF (right)

c. ∠VBN (acute)

9. Name the type of angle shown in this photograph.

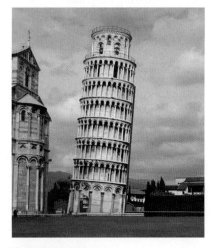

10. Name the type of angle shown in this photograph.

11. Name the type of angle shown in this photograph.

12. Name each acute and obtuse angle in the following diagram.

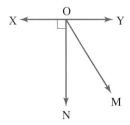

13. Name each acute and obtuse angle in the following diagram.

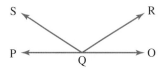

14. Name each acute and obtuse angle in the following diagrams.

a.

b.

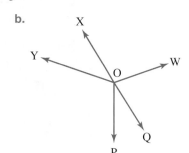

15. State the type of angle that the minute hand of a clock makes when it moves for the following lengths of time.

    a. 10 minutes   b. 15 minutes   c. 20 minutes

16. State the type of angle that the minute hand of a clock makes when moving for the following lengths of time.

    a. 30 minutes   b. 45 minutes   c. 1 hour

17. At various times of the day the hands of an analogue clock form two angles: the inside angle (the smaller one) and the outside angle (the larger one).
    State the type of the inside angle formed by the hands of a clock at the following times.

    a. 1 o'clock   b. 20 minutes to 12   c. 6 o'clock
    d. 9 o'clock   e. quarter to 4

18. At various times of the day the hands of an analogue clock form two angles: the inside angle (the smaller one) and the outside angle (the larger one).
    Write two different times when the inside angle of the clock is:

    a. acute　　　　　　　　　　b. obtuse　　　　　　　　　　c. right.

19. In this photograph, acute, right-angled and obtuse angles have been used.

    a. Name an acute angle.
    b. Name two right angles.
    c. Name an obtuse angle.
    d. State the name of a reflex angle.

Communicating, reasoning and problem solving

20. Give reasons for the following statements.

    a. The sum of two acute angles can be an acute angle.
    b. The sum of two acute angles can be a right angle.
    c. The sum of two acute angles can be an obtuse angle.
    d. The sum of two acute angles can be a reflex angle.

21. a. State whether it is possible to construct a complete revolution using one 90° angle, one straight angle and two acute angles.
    b. Construct a diagram to represent the complete revolution in part a.

22. Watchmakers take particular care that the angles between the hands of an analogue clock represent the time accurately. Determine the size of the smaller angle between the hour and minute hands when an analogue clock displays the time as 10 past 9.

23. Explain why it is important to always put the letter representing the vertex in the middle of an angle's name.

24. Label each angle marked in this photograph with the word describing the type of angle it is.

25. **a.** Consider the figure and identify how many angles are shown.

   b. At midday, the hour hand, the minute hand and the second hand on an analogue clock are all pointing to the 12.
   Determine the size of the smaller angle between the hour hand and the minute hand after the second hand has completed 30 revolutions.
   Hint: It is not 180°— draw a separate diagram to help.

LESSON
11.5 Geometrical properties of angles

LEARNING INTENTION

At the end of this lesson you should be able to:
- identify and determine the size of vertically opposite angles
- identify and determine complementary and supplementary angles
- identify and determine the size of angles formed when a transversal intersects parallel lines.

▶ 11.5.1 Vertically opposite, adjacent, complimentary and
eles-4102 supplementary angles

- When two straight lines intersect they form four angles with a common vertex.
- The angles that are opposite each other at the vertex are called **vertically opposite angles**.
- Vertically opposite angles are equal in size.
- In this diagram there are two pairs of vertically opposite angles:
  ∠AOB = ∠COD and ∠BOC = ∠AOD.
- **Adjacent angles** share a common arm and a common vertex. In this diagram ∠AOB is adjacent to ∠BOC.

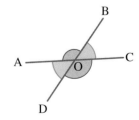

- **Complementary angles** are adjacent angles that add up to 90°.
  In this diagram, $x + y = 90°$.

- **Supplementary angles** are adjacent angles that add up to 180°.
  In this diagram, $x + y = 180°$.

WORKED EXAMPLE 10 Calculating a complementary angle

Calculate the value of x in the diagram.

THINK	WRITE
1. These angles are supplementary because they add up to 90°.	$x + 30° = 90°$
2. Subtract 30° from 90° to calculate the value of x.	$x = 90° - 30°$ $x = 60°$
3. Write the answer.	The value of x is 60°.

WORKED EXAMPLE 11 Calculating a supplementary angle

Calculate the value of x in the diagram.

THINK	WRITE
1. These angles are supplementary because they add up to 180°.	$x + 85° = 180°$
2. Subtract the known angle from 180° to calculate the value of x.	$x = 180° - 85°$ $x = 95°$
3. Write the answer.	The value of x is 95°.

Determine the value of the pronumerals in the diagram.

THINK	WRITE
1. The angle y and $120°$ are vertically opposite, so they are equal.	$y = 120°$
2. The angle z and the angle marked $120°$ form a straight line, so they are supplementary angles (that is, they add to $180°$). Subtract $120°$ from $180°$ to find the value of z.	$z = 180° - 120°$ $= 60°$
3. The angles x and z are vertically opposite, so they are equal.	$x = 60°$

⏵ 11.5.2 Parallel and perpendicular lines

eles-4103

- **Parallel lines** are two or more lines that are simple translations of each other — that is, moved up or down without being rotated or flipped.
- The distance between parallel lines is the same across their entire lengths. They never intersect each other. The two rails of the train tracks in this picture are parallel to each other.
- To write 'the line segments AB and CD are parallel' using mathematical symbols, simply write AB ∥ CD.
- The symbol ∥ means 'is parallel to'.
- The term 'transverse' means 'crossways'. A line that intersects with a pair of parallel lines is called a **transversal**.

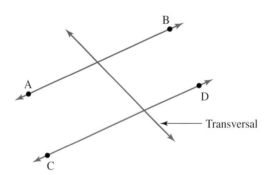

- When a transversal cuts a set of parallel lines, a number of angles are created. The following table shows how pairs of angles are related.

Angle identification	Relationship	Example
Corresponding angles are positioned on the same side of the transversal and are either both above or both below the parallel lines. Think of them as F-shaped.	$\angle a = \angle b$	
Co-interior angles are positioned 'inside' the parallel lines, on the same side of the transversal. Think of them as C-shaped. Co-interior angles are also called **allied angles**.	Angles are supplementary: $\angle a + \angle b = 180°$	
Alternate angles are positioned 'inside' the parallel lines on alternate sides of the transversal. think of them as Z-shaped.	$\angle a = \angle b$	
Vertically opposite angles are created when two lines intersect. The angles opposite each other are equal in size. Think of them as X-shaped.	$\angle a = \angle a$ $\angle b = \angle b$	

DISCUSSION

What shape does each of the angles above make? As a class, discuss ways to remember each angle type and the name given.

Perpendicular lines

- **Perpendicular** means 'at right angles'.
- In this diagram, the angles $\angle AOX$ and $\angle XOB$ are both right angles, which means the lines AB and OX are perpendicular.
- To write 'the line segments AB and OX are perpendicular' using mathematical symbols, simply write $AB \perp OX$.
- The symbol $\perp$ means 'is perpendicular to'.

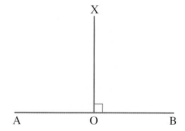

⊳ 11.5.3 Calculating angles associated with parallel lines

eles-4104

- Angle relationships can be used to calculate the size of missing angles.
- Remember that the Z, F or C shapes may be upside-down or backwards in some diagrams.

WORKED EXAMPLE 13 Calculating angles

For the diagram shown:
a. state the type of angle relationship
b. calculate the value of the pronumeral.

THINK

a. 1. Study the diagram: which shape, Z, F or C, would include both angles that are shown? Copy the diagram into your workbook and highlight the appropriate shape.

2. Write the answer by stating the name of the angles suggested by a C shape.

b. 1. Co-interior angles add up to 180°. Write this as an equation.

2. Solve for m by subtracting 45° from 180°.

3. Write the answer.

WRITE

a.

The highlighted angles are co-interior.

b. $m + 45° = 180°$

$m = 180° - 45°$
$m = 135°$

The value of m is 135°

WORKED EXAMPLE 14 Calculating angles

Calculate the value of the pronumeral in the diagram shown, giving reasons.

THINK

1. The two angles shown are neither C, Z nor F angles. So we must find some other angle first that will enable us to determine the size of angle x.

2. This other angle must be in a certain relationship with *both* given angles. Draw an F shape that includes a 120° angle.

3. The other angle in this F shape is related to both given angles: it corresponds to the 120° angle and is supplementary to angle x. Call this angle y.

4. State the size of angle y, specifying the reason (the angle relation).

5. The angles x and y are supplementary (add to 180°) as they lie on a straight line. State this as an equation.

WRITE

$y = 120°$ (because corresponding angles are equal)

$x + y = 180°$ (because the angle sum of a straight line is 180°)

6. Substitute the value of y into the equation.	$x + 120° = 180°$
7. Solve for x by subtracting 120° from 180°	$x = 180° - 120°$ $x = 60°$
8. Write the answer.	The value of x is 60°.

▶ 11.5.4 Testing for parallel lines

eles-4105

- Two lines are parallel if:
  - corresponding angles are equal
  - alternate angles are equal
  - co-interior angles are supplementary.

WORKED EXAMPLE 15 Determining parallel lines

Are the following lines A and B parallel? Explain your answer.

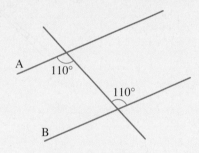

THINK

1. The given angles are equal to each other. They are both 110°.
2. Identify the type of angles provided by considering the Z, F or C shapes.
   The Z shape is most appropriate. Therefore the angles are alternate.

3. Since the two angles are alternate and equal, then the lines A and B must be parallel.

WRITE

$A \parallel B$

 Resources

▶ **Video eLesson** Parallel lines (eles-2309)

Interactivities Vertically opposite and adjacent angles (int-3968)
Corresponding angles (int-3969)
Co-interior angles (int-3970)
Alternate angles (int-3971)
Calculation of angles associated with parallel lines (int-3972)
Testing for parallel lines (int-3973)

11.5 Quick quiz on	**11.5 Exercise**

Individual pathways

■ PRACTISE	■ CONSOLIDATE	■ MASTER
1, 4, 7, 8, 11, 15, 18, 20, 23	2, 5, 9, 12, 14, 17, 21, 24	3, 6, 10, 13, 16, 19, 22, 25

Fluency

1. **WE10** Calculate the value of the pronumeral in the following diagram.

2. Calculate the value of the pronumeral in each of the following diagrams.

   a.

   b.

3. **WE11** Calculate the size of ∠ BOC.

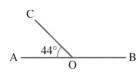

4. Calculate the value of the pronumeral in each of the following diagrams.

   a.
   b.
   c.

   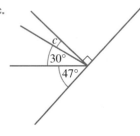

5. **WE12** Calculate the value of the pronumerals in each of the following diagrams.

   a.
   b.
   c.

6. Determine the value of the pronumerals in each of the following diagrams.

a.

b.

c.

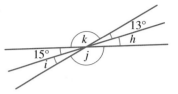

7. **MC** In the diagram:

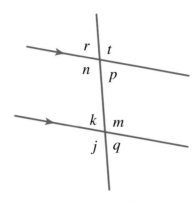

a. which angle is vertically opposite to angle p?

   A. k **B.** m **C.** r **D.** q

b. which angle is corresponding to angle p?

   A. k **B.** m **C.** r **D.** q

c. which angle is co-interior to angle p?

   A. k **B.** m **C.** r **D.** q

d. which angle is alternate to angle p?

   A. k **B.** m **C.** r **D.** q

8. **WE13** For each of the following diagrams:

  a. state the type of angle relationship

  b. determine the value of the pronumeral.

  i.

  ii.

  iii.

  iv.

  v.

  vi.

9. **WE14** Calculate the value of the pronumeral in each of the following diagrams. Give reasons for your answers.

a.

b.

c.

10. Calculate the value of the pronumerals in each of the following diagrams. Give reasons for your answers.

a.

b.

Understanding

11. Identify which of the following pairs of angles are complementary.
    a. 32° and 60°
    b. 70° and 30°
    c. 45° and 45°
    d. 28° and 72°
    e. 30° and 150°
    f. 42° and 48°

12. Identify which of the following pairs of angles are supplementary.
    a. 30° and 60°
    b. 70° and 110°
    c. 145° and 45°
    d. 118° and 72°
    e. 35° and 145°
    f. 42° and 138°

13. a. Determine the complement of 65°.
    b. Determine the supplement of 123°.

14. a. If the angle co-interior to x is 135°, calculate the size of angle x.
    b. If the angle corresponding to y is 55°, calculate the size of angle y.

15. Copy and complete this table.

Diagram	Type of angle relation	Associated shape	Rule
	Corresponding		Are of equal size
		Z	Are of equal size

16. Match each diagram with the appropriate angle name from the four options listed.

a.

b.

c.

d.

A. Co-interior angles (C)
C. Alternate angles (Z)
B. Corresponding angles (F)
D. None of the above

17. Look at the diagram. List all pairs of:

    a. vertically opposite angles
    b. corresponding angles
    c. co-interior angles
    d. alternate angles.

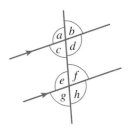

18. Calculate the value of the pronumerals in the following diagrams.

    a.

    b.

    c.

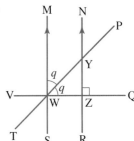

19. Calculate the value of the pronumerals in the following diagrams.

    a.

    b.

    c.

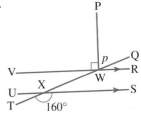

Communicating, reasoning and problem solving

20. **WE15** Is the line AB parallel to the line CD? Explain your answer.

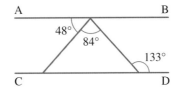

21. a. Calculate the size of angle a. Explain how you found it.
    b. Is the line KL parallel to MN? Explain your answer.
    c. Calculate all of the remaining internal and external angles of the triangle in the centre of the diagram.

22. Look at the following optical illusions. Are any of the lines parallel? Explain your answer.

    a.

    b.

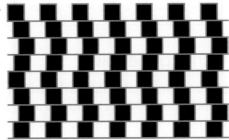

23. A hill is at an angle of 30° to the horizontal. A fence is put in, consisting of a railing parallel to the ground and a series of vertical fence posts. Calculate the angle, *p*, between the top of the fence post and the rail.

24. Two gates consist of vertical posts, horizontal struts and diagonal beams. Calculate the angle, *a*, as shown in the following gates.

a.

b.

25. In the diagram shown, calculate the magnitude of the acute angle ∠PQR and the reflex angle ∠PQR.

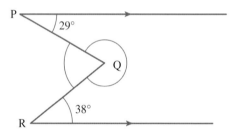

LESSON
11.6 Triangles

LEARNING INTENTION

At the end of this lesson you should be able to:
- classify triangles according to the lengths of their sides or the sizes of their angles
- calculate both interior and exterior angles of a triangle.

▶ 11.6.1 Naming triangles

eles-4096

- **Triangles** are closed shapes that have three edges (or sides) and three vertices (or angles).
- Capital letters are used at each vertex to name a triangle.

- For example, the triangle in the following bicycle frame can be named △PQR. (It can also be named △RPQ or △QRP, and so on.)

Classifying triangles

- Triangles can be classified according to:
  - the lengths of their sides
  - the sizes of their angles.

Lengths of sides		Sizes of angles	
Triangle name and distinguishing features	**Shape**	**Triangle name and distinguishing features**	**Shape**
Equilateral triangle All edges are the same length and all angles are the same size. This is shown by the identical marks on each edge and each angle.		**Acute-angled triangle** All interior angles are less than 90°.	
Isosceles triangle Two edges are the same length. The third edge is a different length. It is often called the base of the triangle. The two base angles are the same size.		**Right-angled triangle** One angle is exactly 90°. This angle is marked by a small square. The edge opposite the right angle is called the hypotenuse.	Hypotenuse
Scalene triangle All edges are different lengths and all angles are different sizes. This is shown by the unique markings on each edge and each angle.		**Obtuse-angled triangle** One angle is greater than 90°.	

- A triangle may belong to more than one class. For example, a right-angled triangle can also be a scalene triangle and an obtuse angle triangle can also be an isosceles triangle.

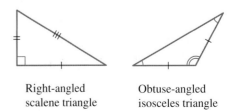

Right-angled scalene triangle   Obtuse-angled isosceles triangle

DISCUSSION

1. In the triangle shown, count the number of:
   a. equilateral triangles
   b. right-angled triangles
   c. isosceles triangles.

2. Count the number of triangles in this shape.

WORKED EXAMPLE 16 Classifying triangles by sides

Classify each of these triangles according to the lengths of its sides.

a.

b.

c.

THINK	WRITE
a. Sides AB and AC have identical markings on them, which means that they are equal in length. So △ABC has 2 equal sides. Classify this triangle based on that infomation.	a. △ABC is an isosceles triangle.
b. The 3 sides of △MNP have identical markings on them, which means that all 3 sides are equal in length. Classify this triangle based on that information.	b. △MNP is an equilateral triangle.
c. All 3 sides of △PRS are marked differently, which means no sides in this triangle are equal in length. Classify this triangle based on that information.	c. △PRS is a scalene triangle.

WORKED EXAMPLE 17 Classifying triangles by angles

Classify each of the triangles in Worked example 9 according to the size of its angles.

THINK	WRITE
a. In △ABC, ∠CAB is marked as the right angle. Classify this triangle based on that information.	a. △ABC is a right-angled triangle.
b. In △MNP all angles are less than 90°. Classify this triangle based on that information.	b. △MNP is an acute-angled triangle.
c. In △PRS, ∠PRS is greater than 90°; that is, it is obtuse. Classify this triangle based on that information.	c. △PRS is an obtuse-angled triangle.

ⓡ 11.6.2 Angles in a triangle

eles-4097

- The sum of the three angles in any triangle is always 180°.
- This can be proved by looking at the diagram shown.
- We know from the previous section that alternate angles are equal, so α, β and γ (the angles in the triangle) would always create a straight line or straight angle.
- A straight angle always equals 180°.
- We can see that as the angles α and β change to different-sized angles, it would change the size of γ at the top of the triangle but would still create a straight line.

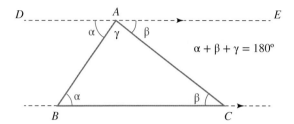

$$\alpha + \beta + \gamma = 180°$$

Sum of angles in a triangle

$$a° + b° + c° = 180°$$

The three angles of a triangle add up to 180°. That is, $a° + b° + c° = 180°$.

WORKED EXAMPLE 18 Calculating angles in triangles

Calculate the value of the pronumeral in each of the following triangles.

a.

b.

THINK	WRITE
a. 1. The sum of the 3 angles (b, 35° and 58°) must be 180°. Write this as an equation.	**a.** $b + 35° + 58° = 180°$
2. Simplify by adding 35° and 58° together.	$b + 93° = 180°$
3. Use inspection or backtracking to solve for b.	$b = 180° - 93°$
	$b = 87°$

| **b. 1.** The sum of the three angles (d, 37° and 42°) must be 180°. Write this as an equation. | **b.** $d + 37° + 42° = 180°$ |

2. Simplify by adding 37° and 42° together.

$d + 79° = 180°$

3. Use inspection or backtracking to calculate the value of d.

$d = 180° − 79°$
$d = 101°$

WORKED EXAMPLE 19 Calculating angles in triangles

Calculate the value of the pronumeral in the following triangle.

THINK

1. The markings on the diagram indicate that $\triangle ABC$ is isosceles because $AB = BC$. Therefore, the angles at the base are equal in size; that is, $\angle BCA = \angle BAC = 74°$. Write this as an equation.

2. All 3 angles in a triangle must add up to 180°. Write this as an equation.

3. Simplify by adding 74° and 74° together.

4. Subtract 148° from 180° to calculate the value of h.

WRITE

$\angle BCA = \angle BAC = 74°$

$\angle ABC + \angle BAC + \angle BCA = 180°$
$h + 74° + 74° = 180°$

$h + 148° = 180°$

$h = 180° − 148°$
$h = 32°$

WORKED EXAMPLE 20 Calculating angles in triangles

Calculate the value of the pronumeral in the following triangle.

THINK	WRITE

THINK

1. The markings on the diagram indicate that $\triangle MNP$ is isosceles because MN = NP.
   Therefore the angles at the base are equal in size.
   Write this as an equation.

2. All 3 angles in a triangle must add up to 180°.
   Write this as an equation.

3. Simplify by collecting like terms. In this example $a + a = 2a$.

4. Use inspection or backtracking to calculate the value of a.

WRITE

$\angle NPM = \angle NMP = a$

$\angle NMP + \angle NPM + \angle MNP = 180°$
$a + a + 40° = 180°$
$2a + 40° = 180°$

$2a = 180° - 40°$
$2a = 140°$
$a = \dfrac{140°}{2}$
$a = 70°$

COMMUNICATING — COLLABORATIVE TASK: Sketching triangles from a verbal description

Pair up with a classmate and each of you draw a triangle without the other person seeing. Take turns describing your triangles using only verbal descriptions and asking your partner to draw the triangle you have described. How accurate is their drawing compared to the triangle you described?

Investigating the angles

Pair up with a classmate and cut out a triangle. Tear off the three angles (the points of your triangle) and pass them to your partner. Can you make a straight line with the three angles your partner has given you?

▶ 11.6.3 Interior and exterior angles of a triangle

eles-4098

- The angles inside a triangle are called **interior angles**.
- When a side of a triangle is extended outwards, the angle formed is called an **exterior angle**.
- The exterior angle and the interior angle adjacent (next) to it add up to 180°.
- The sum of the two interior opposite angles is equal to the exterior angle.

Exterior angle of a triangle

The sum of the two interior opposite angles is equal to the size of the exterior angle.

$\angle BAC + \angle ABC = \angle BCD$

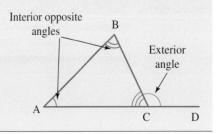

Determine the value of the pronumerals in the following diagram.

THINK

1. $\angle DAB = 125°$. $\angle BAC$ (angle p) and its adjacent exterior angle ($\angle DAB$) add up to $180°$. Write this as an equation.

2. Subtract $125°$ from $180°$ to determine the value of p.

3. The interior angles of $\triangle ABC$ add up to $180°$. Identify the values of the angles and write this as an equation.

4. Simplify by adding $83°$ and $55°$.

5. Subtract $138°$ from $180°$ to determine the value of n.

WRITE

$\angle BAC = p;\ \angle DAB = 125°$
$\angle BAC + \angle DAB = 180°$
$p + 125° = 180°$

$p = 180° - 125°$
$p = 55°$

$\angle BCA + \angle BAC + \angle ABC = 180°$
$\angle BCA = 83°;\ \angle BAC = p = 55°;\ \angle ABC = n$
$83° + 55° + n = 180°$

$n + 138° = 180°$

$n = 180° - 138°$
$n = 42°$

COMMUNICATING — COLLABORATIVE TASK: External angle of a triangle

Pair up with a classmate and draw a triangle with an external angle. Cut out the two angles opposite the external angle and place them on top of the external angle. What do you notice about the two opposite angles and the external angle?

Compare with the rest of the class. Does this work in every case?

 Resources

Interactivities Classification of triangles — sides (int-3963)
Classification of triangles — angles (int-3964)
Angles in a triangle (int-3965)
Interior and exterior angles of a triangle (int-3966)

11.6 Quick quiz on	11.6 Exercise

Individual pathways

■ PRACTISE	■ CONSOLIDATE	■ MASTER
1, 2, 5, 10, 13, 15, 17, 20	3, 6, 7, 9, 11, 14, 18, 21	4, 8, 12, 16, 19, 22

Fluency

1. **WE16** Name each of the following triangles using the capital letters, then classify each triangle according to the lengths of its sides.

a.

b.

c.

d.

e.
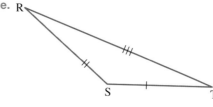

2. **WE17** Classify each of the triangles in question **1** according to the size of its angles.

3. **MC** Select which of these triangles is an equilateral triangle.

A.

B.

C.

D.

4. **MC** Select which of these triangles is not a scalene triangle.

A.

B.

C.

D.

5. **MC** Select the triangle that is both right-angled and scalene.

A.
B.
C.
D.

6. **MC** Select which of these triangles is both acute-angled and isosceles.

A.
B.
C.
D.

7. **WE18** Calculate the value of the pronumeral in each of the following triangles.

a.
b.
c.

d.
e.
f.

8. Calculate the value of the pronumeral in each of the following right-angled triangles.

a.
b.
c.

9. **WE19** Calculate the value of the pronumeral in each of the following triangles, giving reasons for your answer.

a.
b.
c.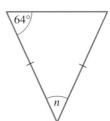

10. **WE20** Calculate the value of the pronumeral in each of the following triangles, giving reasons for your answer.

a.

b.

c.

d.

e.

f.

11. **WE21** Calculate the value of the pronumerals in each of the following diagrams.

a.

b.

c.

d.

12. Calculate the value of the pronumerals in each of the following diagrams.

a.

b.

c.

d.

Understanding

13. Add side and angle markings to these diagrams to show that:

a. ΔRST is an equilateral triangle

b. ΔUVW is an isosceles triangle

c. ΔPQR is a scalene triangle.

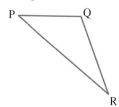

14. Add side and angle markings to these diagrams to show that:

a. ΔMNP is a right-angled triangle

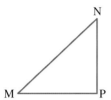

b. ΔABC is a right-angled and isosceles triangle

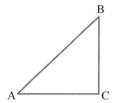

c. ΔMNO is a right-angled and scalene triangle.

15. Use your ruler, pencil and protractor to accurately draw:

a. an equilateral triangle with side lengths 6 cm and all angles 60°
b. an isosceles triangle with two sides that are 6 cm each with a 40° angle between them
c. a right-angled triangle whose two short sides are 6 cm and 8 cm (How long is the longest side?)
d. a scalene triangle with two of the sides measuring 4 cm and 5 cm and an angle of 70° between the two sides.

16. Calculate the missing angle in each of the triangles marked on the following photographs.

a.

b.

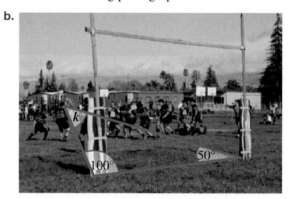

Communicating, reasoning and problem solving

17. Calculate the magnitudes of the unknown angles in the following triangles.

a. ΔMNP is an isosceles triangle with MN = MP and ∠MNP = 72°. Calculate the sizes of ∠MPN and ∠NMP.
b. ΔMNP is an isosceles triangle with MN = MP and ∠NMP = 72°. Calculate the sizes of ∠MPN and ∠MNP

18. a. Explain in your own words why it is impossible to construct a triangle with two obtuse angles.
b. Explain why the longest side of a triangle is always opposite the largest angle of the triangle.
c. Explain why the sum of the lengths of two sides of a triangle must be greater than the length of the third side.

19. a. Construct the isosceles triangle ABC with ∠B = ∠C = 80°.
b. Extend the line BC to the right and label the other end D where BD = AB and form the ΔABD.
c. Calculate the size of ∠D.
d. Extend the line CD to the right and label the other end E where DE = AD and form the ΔADE.
e. Calculate the size of ∠E.
f. Repeat this process again and discuss any pattern you observe.

20. **a.** An isosceles triangle has two angles of 55° each. Calculate the third angle.
    b. An isosceles triangle has two angles of 12° each. Calculate the third angle.
    c. Two angles of a triangle are 55° and 75°. Calculate the third angle.
    d. Two angles of a triangle are 48° and 68°. Calculate the third angle.

21. For each of the following sets of three angles, state whether it is possible to construct a triangle with these angles. Give a reason for your answer.
    a. 40°, 40°, 100°
    b. 45°, 60°, 70°
    c. 45°, 55°, 85°
    d. 111°, 34.5°, 34.5°

22. Determine the value of the pronumerals in the following shapes.

a.

b.

LESSON
11.7 Quadrilaterals and their properties

LEARNING INTENTION

At the end of this lesson you should be able to:
- classify quadrilaterals from a given diagram
- calculate the angles in a quadrilateral.

▶ 11.7.1 Naming quadrilaterals

eles-6097

- A **quadrilateral** is a 2-dimensional closed shape with four straight sides.
- Similar to triangles, quadrilaterals can be named by their vertices. For example, the quadrilateral shown can be named ABCD, BCDA or DCBA and so on.
- It cannot be named ACBD because those vertices are not connected by edges.

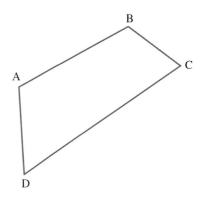

▶ 11.7.2 Classifying quadrilaterals

eles-4100

- Quadrilaterals can be either **convex** or non-convex (also called concave).
- In a convex quadrilateral, each of the interior angles is less than 180° and the two diagonals lie inside the quadrilateral.
- In a non-convex quadrilateral, one interior angle is more than 180° and one of the two diagonals lies outside of the quadrilateral.
- Quadrilaterals can be divided into two major groups: **parallelograms** and other quadrilaterals.

Convex quadrilaterals

Non-convex quadrilaterals

- Parallelograms are quadrilaterals with:
    1. both pairs of opposite sides equal in length and parallel
    2. opposite angles equal.

A *rectangle*, a *square* and a *rhombus* (diamond) are all examples of parallelograms. Therefore, some quadrilaterals can be classified as more than one type of quadrilateral.

Parallelograms	Shape	Properties
Parallelogram		Opposite sides are equal in length. Opposite angles are equal in size.
Rectangle		Opposite sides are equal in length. Adjacent sides are perpendicular. All angles are the same and equal 90°.
Rhombus		All sides are equal in length. Opposite angles are equal in size.
Square		All sides are equal in length. Adjacent sides are perpendicular. All angles are the same and equal 90°.

- Examples of other quadrilaterals (non-parallelograms) include *trapeziums*, *kites* and *irregular quadrilaterals*.

Other quadrilaterals	Shape	Properties
Trapezium		One pair of opposite sides is parallel.
Kite		Two pairs of adjacent (next to each other) sides are equal in length. One pair of opposite angles (the ones that are between the sides of unequal length) are equal.
Irregular quadrilateral		This shape does not have any special properties.

WORKED EXAMPLE 22 Naming quadrilaterals

Name the following quadrilaterals, giving reasons for your answers.

a.

b.

THINK

a. The markings on this quadrilateral indicate that all sides are equal in length and all angles equal 90°. Name the quadrilateral by finding the matching description in the table.

b. The arrows on the sides of this quadrilateral indicate that there are two pairs of parallel sides. Therefore it is a parallelogram. Check the descriptions in the table to see if it is a particular type of parallelogram.

WRITE

a. This quadrilateral is a square, since all sides are equal and all angles are 90°.

b. This quadrilateral is a parallelogram, since it has two pairs of parallel sides.

11.7.3 Angles in a quadrilateral

eles-4101

To investigate the sum of the angles in a quadrilateral, consider the following:

- All quadrilaterals have four internal angles.

- All quadrilaterals are made up of two triangles.

- From earlier work, we know that the sum of angles in a triangle = 180°.

Therefore:

$$2 \text{ triangles} = 2 \times 180°$$
$$= 360°$$

Sum of angles in a quadrilateral

The four angles of a quadrilateral add up to 360°.

$$a° + b° + c° + d° = 360°$$

WORKED EXAMPLE 23 Determining angles in a quadrilateral

Calculate the value of the pronumeral in the diagram.

THINK	WRITE
1. The sum of the angles in a quadrilateral is 360°. Write this as an equation.	$d + 100° + 136° + 65° = 360°$
2. Simplify by adding 100°, 136° and 65°.	$d + 301° = 360°$
3. Subtract 301° from 360° to calculate the value of d.	$d = 360° - 301°$ $d = 59°$

Determine the value of the pronomerals in the photograph, giving a reason for your answer.

THINK	WRITE
1. According to the markings, the opposite sides of the quadrilateral in the photograph are parallel and equal in length. Therefore, this quadrilateral is a parallelogram. In a parallelogram, opposite angles are equal. Write down the value of the pronumerals.	Opposite angles in a parallelogram are equal in size. Therefore, $j = 65°$ and $i = k$.
2. The interior angles of a quadrilateral add up to 360°. Form an equation by writing the sum of the angles on one side and 360° on the other side of an equals sign.	$65 + j + i + k = 360°$
3. Replace j in the equation with 65°.	$65 + 65 + i + k = 360°$
4. Simplify, by adding 65° and 65° together.	$130 + i + k = 360°$
5. Subtract 130° from 360° to determine the value of $i + k$.	$i + k = 360 - 130°$ $i + k = 230°$
6. We know that $i = k$, so each angle must be half of the remaining 230°. Divide 230° by 2.	$\dfrac{230°}{2} = 115°$
7. Angles i and k are both equal to 115°.	$i = 115°$ and $k = 115°$
8. Write the answer.	$i = 115°, j = 65°, k = 115°$

Calculate the value of the pronumerals in the diagram.

THINK

1. Form an equation by writing the sum of the angles on one side, and 360° on the other side of an equals sign.

2. The quadrilateral shown in the diagram is a kite. Referring to the description of a kite in the table, angle t and angle 136° are the angles between unequal sides and therefore must be equal in size.

3. Replace t in the equation with 136°.

4. Simplify by adding 136°, 50° and 136° together.

5. Subtract 322° from 360° to calculate the value of k.

WRITE

$k + t + 50° + 136° = 360°$

The figure is a kite.
$t = 136°$

$k + 136° + 50° + 136° = 360°$

$k + 322° = 360°$

$k = 360° - 322°$
$k = 38°$

COMMUNICATING — COLLABORATIVE TASK: Sketching quadrilaterals from a verbal description

Pair up with a classmate and each of you draw a quadrilateral without the other person seeing. Take turns describing your quadrilaterals using only verbal descriptions. Your partner must try to draw the quadrilateral you describe. How accurate is the drawing compared to the quadrilateral you described?

 Resources

Interactivities Types of quadrilaterals (int-4025)
Angles in a quadrilateral (int-3967)

11.7 Quick quiz on	11.7 Exercise

Individual pathways

■ PRACTISE	■ CONSOLIDATE	■ MASTER
1, 3, 6, 11, 12, 15, 18, 19, 22	2, 4, 7, 10, 13, 16, 20, 23	5, 8, 9, 14, 17, 21, 24

Fluency

1. State whether the quadrilaterals shown are convex or non-convex.

   a.

   b.

   c.

   d.

2. **WE22** Name the following quadrilaterals, giving reasons for your answers.

   a. i.

   ii.

   iii.

   iv.

   v.

   vi.

b. Name each of the quadrilaterals in the photographs, giving reasons for your answers.

i.

ii.

iii.

iv.

3. **MC** Identify the quadrilateral shown.
   A. Square
   B. Kite
   C. Parallelogram
   D. Rhombus

4. **MC** Identify the quadrilateral shown.
   A. Trapezium
   B. Kite
   C. Parallelogram
   D. Rhombus

5. **MC** Identify the quadrilateral shown.
   A. Trapezium
   B. Parallelogram
   C. Rhombus
   D. Kite

6. **WE23** Calculate the value of the pronumeral in each of the following diagrams.

a.

b.

c.

d.

e.

f.

7. **WE24** Calculate the value of the pronumeral in each of the following diagrams, giving reasons for your answers.

a.

b.

c.

d.

e.

f.

8. Use the following photographs to determine the values of the unknown angles.

a.

b.

9. **WE25** Calculate the value of the pronumerals in each of the following diagrams.

a.

b.

c.

d.

e.

f.

10. **MC** Identify the value of *t* in the diagram.
A. 360°
B. 112°
C. 222°
D. 138°

11. **MC** Identify the value of *r* in the diagram.
A. 117°
B. 180°
C. 63°
D. 126°

Understanding

12. State whether each of the following statements is true or false.

a. All squares are rectangles.
b. All squares are rhombuses.
c. All rectangles are squares.
d. Any rhombus with at least one right angle is a square.
e. A rectangle is a parallelogram with at least one angle equal to 90°.

13. **MC** A rectangle is a quadrilateral because:

A. it has 4 right angles
B. it has 2 pairs of parallel sides
C. its opposite sides are equal in length
D. it has 4 straight sides

14. The photograph shows the roof of a gazebo. Calculate the value of p.

15. Determine the size of the obtuse angle in the following photograph of a kite.

16. Two angles in a parallelogram are 45° and 135°. Calculate the values of the other 2 angles.

17. Tom measures 2 angles of a kite at 60° and 110°, but forgets which angle is which. Draw three different kites that Tom may have measured, showing the size of all angles in each diagram.

18. For each of the following sets of four angles, decide whether it is possible to construct a quadrilateral. Explain your answer.

 a. 25°, 95°, 140°, 100°

 b. 40°, 80°, 99°, 51°

Communicating, reasoning and problem solving

19. Three angles of a quadrilateral are 60°, 70° and 100°.

 a. Calculate the size of the fourth angle of this quadrilateral.
 b. Determine how many quadrilaterals with this set of angles are possible.
 c. Construct a quadrilateral with the given angle sizes. The choice of the length of the sides is up to you.

20. Val and Peter want to replace their front gate with another of the same design. To have this gate made, they need to supply a diagram of it with all measurements and angles shown.

 Study the photograph of Val and Peter's gate and use it to help you answer the following questions.

 a. There are 4 different shapes formed by the metal bars of the gate. How many different types of triangles are there? Identify them.
 b. State how many types of quadrilaterals there are. Identify them.
 c. Draw a diagram of the gate showing the length measurements and the one angle that is given.
 d. Use this angle to calculate all the remaining angles in the diagram.

21. Consider the diagram.

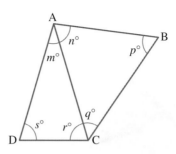

a. Prove that the sum of all angles in a quadrilateral is 360°.

b. Using a similar method, determine the sum of all angles in the following shapes. *Hint*: Draw the shape first.

i. Pentagon

ii. Hexagon

iii. Octagon

iv. Decagon

22. Determine the values of the pronumerals in the following shapes.

a.

b.

23. Calculate the values of the pronumerals in each of the following quadrilaterals.

a.

b.

c.

d.

24. Draw a quadrilateral PQRS. Label angle *P* as 100° and angle *R* as a right angle. Angle *Q* is two-thirds of angle *S*. Determine the measurement of angle *S*.

LESSON
11.8 Review

11.8.1 Topic summary

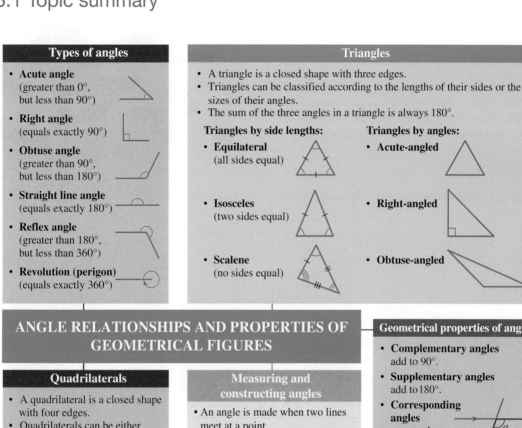

Types of angles

- **Acute angle**
  (greater than 0°,
  but less than 90°)

- **Right angle**
  (equals exactly 90°)

- **Obtuse angle**
  (greater than 90°,
  but less than 180°)

- **Straight line angle**
  (equals exactly 180°)

- **Reflex angle**
  (greater than 180°,
  but less than 360°)

- **Revolution (perigon)**
  (equals exactly 360°)

Triangles

- A triangle is a closed shape with three edges.
- Triangles can be classified according to the lengths of their sides or the sizes of their angles.
- The sum of the three angles in a triangle is always 180°.

Triangles by side lengths:
- **Equilateral**
  (all sides equal)
- **Isosceles**
  (two sides equal)
- **Scalene**
  (no sides equal)

Triangles by angles:
- **Acute-angled**
- **Right-angled**
- **Obtuse-angled**

ANGLE RELATIONSHIPS AND PROPERTIES OF GEOMETRICAL FIGURES

Quadrilaterals

- A quadrilateral is a closed shape with four edges.
- Quadrilaterals can be either convex or non-convex.
- The sum of the four angles in a quadrilateral is always 360°.
- Quadrilaterals can be divided into two major groups: **parallelograms** and **other quadrilaterals**.

Parallelograms:
- rectangle
- square
- rhombus

Other quadrilaterals:
- kite
- trapezium

Measuring and constructing angles

- An angle is made when two lines meet at a point.
- The point where the two lines meet is called the *vertex* and the lines are called the *arms* of the angle.
- Angles are measured in degrees (°).
- There are four important angles used in estimating angles: 90°, 180°, 270° and 360°.
- Circular and semicircular protractors can be used for measuring angles.

The angle is 65°
(The other arm is at 0° on the inside scale, so we use the inside scale)

Geometrical properties of angles

- **Complementary angles** add to 90°.
- **Supplementary angles** add to 180°.
- **Corresponding angles** are equal (F-shaped).
  $a = b$
- **Co-interior angles** add to 180° (C-shaped).
  $x + y = 180°$
- **Alternate angles** are equal (Z-shaped).
  $a = b$
- **Vertically opposite angles** are equal (X-shaped).
  $a = b$
- Two lines are **perpendicular** if they meet at right angles (90°).

11.8.2 Project

Tangrams

A tangram is an ancient Chinese puzzle game, also called *ch'i ch'ae pan* or Seven-board of Cunning. The first tangram books were printed in 1813, but the word *ch'i ch'ae* dates back to 740–330 BCE.

According to one Chinese legend:

> A servant of a Chinese emperor was carrying a very expensive square ceramic tray and, when he tripped and fell, it was shattered into seven pieces (called *tans*). He was unable to arrange the pieces back into the original shape, but he realised that there were many other shapes that could be built from the pieces.

Instructions

Trace the following tangram pieces onto cardboard and cut out each piece, then follow the instructions below.

1. Using only the two small triangles, make and draw the following shapes, showing the joins in your book.

   a.
      Square

   b.
      Triangle

   c.
      Parallelogram

2. Using the square, the two small triangles, the parallelogram and the medium triangle, make and draw the following shapes.

   a.
      Square

   b.
      Hexagon

c.
Trapezium

d.
Parallelogram

3. Using all the pieces, make and draw the following shapes.

a.

Triangle

b.

Rectangle

c.

Hexagon

4. Use all of the pieces to make the following shapes and all of the other shapes from the previous questions.
   a. A rocket   **b.** The letter E

 Resources

Interactivities   Crossword (int-2592)
                      Sudoku puzzle (int-3166)

Exercise 11.8 Review questions

learnon

Fluency

1. State the type of each of the angles shown here.

a.

b.

c.

2. State the type of each of the angles shown here.

a.

b.

c.

3. **MC** An acute angle has a magnitude that is:
   A. more than 90°.   **B.** less than 90°.
   C. exactly 90°.   **D.** more than 90° but less than 180°.

4. **MC** A straight angle has a magnitude of:

    A. 90°   **B.** 270°   **C.** 360°   **D.** 180°

Questions 5 and 6 refer to the following diagram.

5. **MC** The angle shown in the diagram has a magnitude of:

    A. 50°   **B.** 40°   **C.** 90°   **D.** 100°

6. **MC** The angle shown in the diagram is:

    A. a right angle.   **B.** a straight angle.

    C. an acute angle.   **D.** an obtuse angle.

7. Name the angle shown.

8. **MC** Select which of the following terms describes the angles a and b.

    Hint: There may be more than one correct answer.

    A. Equal

    B. Corresponding

    C. Co-interior

    D. Alternate

9. **MC** Select which of the following terms describes the angles a and b.

    Hint: There may be more than one correct answer.

    A. Vertically opposite

    B. Corresponding

    C. Alternate

    D. Equal

10. **MC** Select which of the following terms describes the angles a and b.

    Hint: There may be more than one correct answer.

    A. Corresponding

    B. Co-interior

    C. Alternate

    D. Equal

11. **MC** Select which of the following terms describes the angles *a* and *b*.
    Hint: There may be more than one correct answer.
    A. Vertically opposite
    B. Alternate
    C. Co-interior
    D. Supplementary

Understanding

12. For each diagram:
    a. state the type of angle relation that you use
    b. calculate the value of the pronumeral.

    i.

    ii.

    iii.

13. Determine the value of each of the pronumerals, giving reasons for your answer.

    a.

    b.

    c.

14. Name the following triangles according to the lengths of their sides.

    a.

    b.

    c.

15. Name the following triangles according to the sizes of their angles.

    a.

    b.

    c.

16. Calculate the value of the pronumeral in each of the following.

    a.

    b.

    c.

17. Calculate the value of the pronumeral in each of the following.

a.

b.

c.

18. Name the following quadrilaterals, giving reasons for your answers.

a.

b.

c.

19. Name the following quadrilaterals, giving reasons for your answers.

a.

b.

c.

20. Calculate the value of the pronumeral in each of the following quadrilaterals.

a.

b.

c.

21. Calculate the value of the pronumeral in each of the following quadrilaterals.

a.

b.

c.

Communicating, reasoning and problem solving

22. The teepee shown has an angle of 46° at its peak. What is the size of the angle, *w*, that the wall makes with the floor?

23. A circus trapeze is attached to a rope as shown in the diagram. Calculate the size of angle *t*.

24. Triangles can be classified according to the number of sides of equal length (scalene, isosceles or equilateral) and by the type of angles (acute, obtuse or right). Decide which of the following combinations are possible. Draw examples of the combinations that are possible.
    • Right–scalene, right–isosceles, right–equilateral
    • Acute–scalene, acute–isosceles, acute–equilateral
    • Obtuse–scalene, obtuse–isosceles, obtuse–equilateral

25. The following triangles are both equilateral triangles, but the side length of triangle ABC is 3 units and the side length of triangle DEF is 6 units. Determine how many of the smaller triangles would you need to fill the larger triangle.

26. If angle AOE is obtuse, name all of the acute angles in the diagram below.

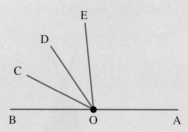

27. An equilateral triangle EDC is drawn inside a square ABCD as shown below. Determine the size of ∠AED.

28. WXYZ is a rectangle as shown here. Points P and Q are drawn so that XY = PY and ∠XPQ = 100°. Determine the size of ∠PQX.

on To test your understanding and knowledge of this topic, go to your learnON title at www.jacplus.com.au and complete the **post-test**.

Answers

Topic 11 Angle relationships and properties of geometrical figures

11.1 Pre-test

1. C
2. 135°
3. 90°
4. D
5. D
6. C
7. ∠ADB
8. A
9. D
10. a. Scalene, right-angled triangle
    b. Isosceles, acute-angled triangle
    c. Scalene, obtuse-angled triangle
    d. Isosceles, right-angled triangle
11. 59°
12. 52°
13. 68°
14. D
15. 72°

11.2 Measuring angles

1. a. B   b. B
2. a. 40°   b. 81°   c. 75°   d. 13°
3. a. 142°   b. 107°   c. 162°   d. 103°
4. a. 240°   b. 201°   c. 316°   d. 356°
    e. The accuracy of your measurement will depend on the thickness of your pencil, the position of the protractor and the side from where the angle is read.
5. a. 45°   b. 155°   c. 180°   d. 63°
6. a. Difficult, as angle is approximately 7°
    b. As recommended
7. a. 57°, 105°, 33°, 75°
    b. 57° and 33°
    c. 105°, and 75°
    d. Multiple correct answers: four 57°, and four 33°, two 105° and two 75°, two 57°, two 33°, one 105° and one 75°
8. a. Design 4. 45°
    b. Design 1. 15°
    c. Design 1, Design 2 and Design 3
    d. Design 1
    e. Design 3
    f. Design 1

9. a, b, d, e, f, g. — Responses will vary depending on the estimations made in part a.
   c.

Diagram	Measured angle
1	190°
2	130°
3	110°
4	40°
5	300°
6	200°
7	140°
8	290

10. a. As the kite moves from point A to point B, the angle increases from 0° to 90°
    b. As the kite moves from point B to point C, the angle decreases from 90° to 0°.
11. a. i. 210° (reflex angle) and 150° (obtuse angle)
        ii. 180° (straight angle) and 180° (straight angle)
       iii. 150° (obtuse angle) and 210° (reflex angle)
        iv. 15° (acute angle) and 345° (reflex angle)
         v. 108° (obtuse angle) and 252° (reflex angle)
        vi. 114° (obtuse angle) and 246°(reflex angle)
    b. 360° ÷ 60 minutes = 6°; 360° ÷ 12 hours = 30°
12. a. Triangle 1: 45°, 45°, 90°, Triangle 2: 80°, 40°, 60°, Triangle 3: 70°, 60°, 50°
    b. 180°, 180°, 180°
    c. Brinda is correct, there is a pattern. The pattern is that the angles within a triangle always add to 180°.

11.3 Constructing angles with a protractor

1. a.   b.
   c.   d.

2. a.   b.

   c.   d.

3. a. 96°

b. 92°

c. 165.2°

d. 143°

4. a. 140° **b.** 156°

c. 127° **d.** 149°

5. a. 185° **b.** 210.5°

c. 235° **d.** 260°

6. a. 243° **b.** 192°

c. 249.8° **d.** 214°

7. a. 295° **b.** 269°

c. 307° **d.** 349°

8. a. 328° **b.** 300°

c. 345° **d.** 358°

9. a. i. 45°

ii. 90°

iii. 180°

iv. 270°

v. 360°

b. 45°, 90°, 180°, 90°, 45°

10. a. 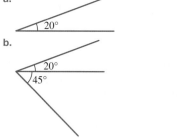 20°

b. 20° 45°

c. 20° 45° 108°

d. 187°

e. 360°

11. Construct the corresponding acute angle and then label the reflex angle.

12. a.

400°

b. 40°

13. a.

b.

c.

14. a.

60°

60°   60°

3 cm

b.

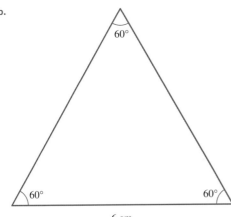

60°

60°   60°

6 cm

c. Sample response: Similarities include angles, ratios of sides, same shape. Differences include sizes, areas, accuracy.

d. 60°

15. a.

A   B

b. All triangles should be identical.

c. If two equal angles are drawn at the ends of a straight line of set length, there is only one size triangle that forms.

16. Draw any triangle on an A4 sheet of paper.

    a. Student activity; responses will vary.

    b. Despite different triangles drawn by students, the sum of the angles of a triangle is always 180°.

11.4 Types of angles and naming conventions

1. a. Acute   **b.** Reflex   **c.** Obtuse
    d. Straight   **e.** Reflex

2. a. Right   **b.** Acute   **c.** Revolution
    d. Acute   **e.** Reflex

3. a. 4; 3°, 45°, 65°, 69°
    b. 2; 123°, 165°
    c. 4; 234°, 300°, 270°, 210°

4. a.

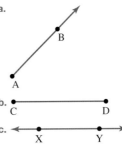

B

A

b. C   D

c. X   Y

5. a. Right   **b.** Straight   **c.** Obtuse
    d. Acute   **e.** Reflex   **f.** Revolution

6. a. ∠XUV or ∠VUX   **b.** ∠PQR or ∠RQP
    c. ∠SDX or ∠XDS   **d.** ∠GPZ or ∠ZPG
    e. ∠PFH or ∠HFP

7. In each case the middle letter should be at the vertex of the angle.

    a.

C

D   E

    b.

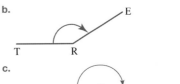

E

T   R

    c.

Q   A   S

8. a.

In each case the middle letter should be at the vertex of the angle.

b.

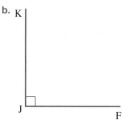

In each case the middle letter should be at the vertex.

c.

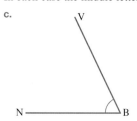

9. Acute

10. Revolution

11. Obtuse

12. Acute: ∠NOM, ∠MOY
Obtuse: ∠MOX

13. Acute: ∠PQS, ∠RQO
Obtuse: ∠SQR, ∠SQO, ∠RQP

14. a. Acute: ∠MLN, ∠NLO, ∠OLP, ∠PLK, ∠NLP, ∠OLK
Obtuse: ∠MLO, ∠MLP, ∠NLK

b. Acute: ∠YOX, ∠WOQ, ∠POQ
Obtuse: ∠XOW, ∠YOP, ∠YOW, ∠XOP, ∠YOQ, ∠POW

15. a. Acute **b.** Right **c.** Obtuse

16. a. Straight **b.** Reflex **c.** Revolution

17. a. Acute **b.** Obtuse
c. Straight **d.** Right
e. Obtuse

18. a. Sample responses:
1 o'clock, 10 o'clock
b. Sample responses:
4 o'clock, 20 past 11
c. Sample responses:
9 o'clock, 3 o'clock

19. In each case the middle letter should be at the vertex of the angle.
a. ∠DEF
b. ∠ABC, ∠XWY
c. ∠PQR
d. Sample responses:
outside angle ∠DEF,
outside angle ∠RQP

20. a. This statement is true if the sum of the two angles is less than 90°.
b. This statement is true if the sum of the two angles is equal to 90°.
c. This statement is true if the sum of the two angles is greater than 90°.
d. This statement is untrue because two acute angles can't add up to more than 180°.

21. a. Yes
b.

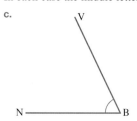

22. 145°

23. It is the naming custom to always put the letter representing the vertex in the middle of the angle's name because this allows anyone to understand the name of the angle.

24. a. Right angle **b.** Straight angle **c.** Reflex angle
d. Acute angle **e.** Obtuse angle

25. a. 25 including revolutions, 20 if revolutions not included
b. 165°

11.5 Geometrical properties of angles

1. 65°

2. a. 19° **b.** 45°

3. 136°

4. a. 10° **b.** 60° **c.** 13°

5. a. 38° **b.** 110° **c.** 22°

6. a. 19°
b. $e = 39°$, $f = 141°$, $g = 141°$
c. $i = 13°$, $h = 15°$, $k = 152°$, $j = 152°$

7. a. C **b.** D **c.** B **d.** A

8. a. i. Corresponding angles
ii. Alternate angles
iii. Vertically opposite angles
iv. Corresponding angles
v. Co-interior angles
vi. Co-interior angles
b. i. $p = 60°$
ii. $q = 45°$
iii. $s = 65°$
iv. $t = 72°$
v. $m = 110°$
vi. $n = 48°$

9. a. $y = 116°$, corresponding
b. $z = 38°$, alternate
c. $b = 136°$, supplementary

10. a. 68°, vertically opposite
b. 90°, corresponding

11. c, f

12. b, e, f

13. a. 25°   **b.** 57°

14. a. 45°   **b.** 55°

15.

Diagram	Type of angle relation	Associated shape	Rule
	Corresponding	F	Are equal
	Alternate	Z	Are equal
	Co-interior	C	Add to 180°

16. a. C   **b.** D   **c.** A   **d.** B

17. a. a and d, c and b, e and h, f and g
   b. a and e, c and g, b and f, d and h
   c. c and e, d and f
   d. c and f, d and e

18. a. 123°   **b.** 43°   **c.** 45°

19. a. 28°   **b.** 50°   **c.** 70°

20. No, because 48° + 133° ≠ 180°. For parallel lines, co-interior angles need to add to 180° (or alternate angles must be equal).

21. a. 52°. The three angles forming the straight line add to 180°.
   b. Yes. 52° + 128° = 180°, so lines parallel as co-interior angles add to 180°.
   c. 52°, 55°. Exterior angle: 125°

22. a. The vertical lines are parallel, the horizontal lines are parallel and the oblique lines are parallel because they are equally distant from each other.
   b. The vertical lines are parallel and the horizontal lines are parallel because they are equally distant from each other.

23. 60°

24. a. 130°
   b. 50°

25. Acute ∠PQR = 67° and reflex ∠PQR = 293°.

11.6 Triangles

1. a. ΔABC Scalene
   b. ΔLMN Equilateral
   c. ΔXYZ Isosceles
   d. ΔPQR Isosceles
   e. ΔRST Scalene

2. a. Right-angled triangle
   b. Acute-angled triangle
   c. Acute-angled triangle
   d. Acute-angled triangle
   e. Obtuse-angled triangle

3. A

4. B

5. D

6. A

7. a. $x = 57°$ **b.** $g = 125°$ **c.** $t = 44°$
   d. $k = 60°$ **e.** $f = 93°$ **f.** $z = 90°$

8. a. $d = 45°$ **b.** $b = 65°$ **c.** $a = 50°$

9. a. $c = 70°$ **b.** $e = 76°$ **c.** $n = 52°$

10. a. $u = 66°$ **b.** $k = 45°$ **c.** $d = 76°$
   d. $t = 74°$ **e.** $f = 57°$ **f.** $p = 70°$

11. a. $p = 50°; n = 70°$ **b.** $a = 22°; b = 28°$
   c. $x = 40°; y = 50°$ **d.** $t = 35°; s = 55°$

12. a. $b = 46°$ **b.** $n = 21°$
   c. $m = 60°$ **d.** $t = 28°$

13. a.

   b.

   c.

14. a.

   b.

   c.

15. c. Please see the worked solutions; the longest side is 10 cm in length. (**a**, **b**, **d**; Please see the worked solutions.)

16. a. $b = 20°$ **b.** $k = 30°$

17. a. ∠MPN = 72° and ∠NMP = 36°
   b. ∠MPN = 54° and ∠MNP = 54°

18. a. The sum of the angles will be greater than 180°.
   b. Start with an equilateral triangle ABC, equal sides and 3 equal angles. If point A and C are fixed and point B is extended, length AB increases. This in turn means

that length BC increases and therefore ∠C increases in size. The longest side (in this case AB) will be opposite the largest angle (in this case ∠C). This is true for all triangles, with the longest side always being opposite the largest angle.

c. Imagine that the base of a triangle is the longest side. The other two side lengths must be attached to opposite vertices of the base of the triangle. If the sum of these two lengths was shorter than the base length then these two sides could not meet and complete the triangle. Similarly, if the sum of these two lengths was equal to the base length then they could only meet in a straight line along the base, and a triangle would not be created. Therefore the sum of the two lengths must be longer than the base length to create a triangle.

19. a.

b.

c. 40°

d.

e. 20°

f.

∠F = 10°

The angle is halved every time a new triangle is constructed.

20. a. 70° **b.** 156° **c.** 50° **d.** 64°

21. a. Yes, angles add up to 180°.
  b. No, angles add up to 175°.
  c. No, angles add up to 185°.
  d. Yes, angles add up to 180°.

22. a. $a = 69°, k = 10, b = 114°$
  b. $a = 60°, c = 120°$

11.7 Quadrilaterals and their properties

1. a. Non-convex
  b. Convex
  c. Convex
  d. Non-convex

2. a. **i.** Square
  ii. Trapezium
  iii. Kite
  iv. Rhombus
  v. Rectangle
  vi. Irregular quadrilateral

  b. **i.** Square: 4 sides equal, 4 angles equal
  ii. Trapezium: one pair of parallel sides
  iii. Square: 4 equal sides and angles
  iv. Kite: adjacent sides equal

3. D

4. C

5. A

6. a. $t = 138°$ **b.** $t = 90°$ **c.** $t = 202°$
  d. $p = 92°$ **e.** $c = 66°$ **f.** $k = 67°$

7. a. $m = 78°$ (opposite angles of parallelogram equal)
  b. $u = 75°$ (opposite angles of rhombus equal)
  c. $t = 132°$ (opposite angles of a kite equal)
  d. $f = 108°$ (opposite angles of a kite equal)
  e. $p = 63°$ (opposite angles of rhombus equal)
  f. $z = 73°$ (opposite angles of parallelogram equal)

8. a. $a = 129°, b = 51°, c = 129°$
  b. $h = 130°$

9. a. $m = 113°$
  b. $c = 98°, d = 82°$
  c. $t = 116°$
  d. $p = 52°, m = 106°$
  e. $t = 103°, x = 91°$
  f. $m = 65°, p = 115°, n = 65°$

10. D

11. C

12. a. True **b.** True
  c. False **d.** True
  e. True

13. D

14. $p = 61°$

15. 115°

16. 45° and 135°

17.

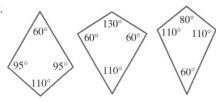

18. a. Yes, angles add up to 360°.
   b. No, angles add up to 270°.

19. a. 130°
   b. Infinite number
   c.

20. a. Two smaller obtuse-angled isosceles triangles, one larger size obtuse-angled isosceles triangle and two right-angled scalene triangles.
   b. Two types: rectangles (of different sizes) and a rhombus.
   c.

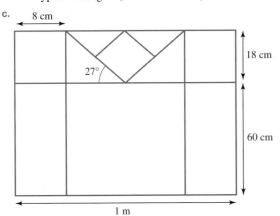

   d. In both right-angled triangles, the angles are: 27°, 63°, 90°; in all isosceles triangles, angles are: 126°, 27°, 27°; in the rhombus, angles are: 126°, 54°, 126°, 54°; in all rectangles, all angles are 90° each.

21. a. Consider the sum of the angles of the two triangles ABC and ACD.

   In $\triangle ABC : n + p + q = 180°$
   In $\triangle ACD : m + r + s = 180°$
   In quadrilateral $ABCD : n + p + q + m + r + s = 360°$

   b. i. 540°
   ii. 720°
   iii. 1080°
   iv. 1440°

22. a. $a = 70°, b = 125°, c = 125°$
   b. $x = 78°, k = 102°, m = 74°$

23. a. $a = 108°, b = 72°, c = 90°$
   b. $g = 100°, h = 80°, i = 50°$
   c. $d = 115°, e = 106.5°, f = 82°$
   d. $j = 45°, k = 135°$

24. 102°

Project

1. a.

   b.

   c.

2. a.

   b.

   c.

   d.

3. a.

b.

c.

4. a.

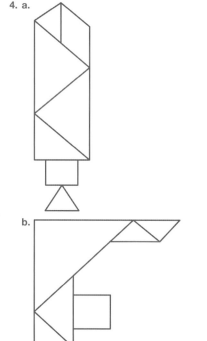

b.

11.8 Review questions

1. a. Revolution

b. Right angle

c. Straight angle

2. a. Obtuse   **b.** Reflex   **c.** Acute

3. B

4. D

5. C

6. A

7. $\angle EDF$ or $\angle FDE$

8. A, B

9. A, D

10. C, D

11. C

12. a.   **i.** Alternate angles

         ii. $x = 60°$

    b.   **i.** Vertically opposite angles

         ii. $y = 135°$

    c.   **i.** Vertically opposite angles, then co-interior angles, then vertically opposite angles. Another solution is supplementary angles, followed by corresponding angles.

         ii. $t = 50°$

13. a. $x = 15°$ because vertically opposite angles are equal.

    b. $x = 50°$ because corresponding angles are equal and the angle sum of a straight line is $180°$,
$y = 25°$ because corresponding angles are equal.

    c. $y = 100°$ because co-interior angles add to $180°$, $x = 8°$
The sum of the interior angles of a triangle equals $180°$.
Since $90 + 50 = 140$ and $180 - 140 = 40$, $5x = 40$.
Therefore $x = 8$.

14. a. Equilateral triangle

    b. Scalene triangle

    c. Isosceles triangle

15. a. Obtuse-angled triangle

    b. Acute-angled triangle

    c. Right-angled triangle

16. a. $t = 67°$ **b.** $b = 28°$ **c.** $x = 75°$

17. a. $p = 56°$ **b.** $m = 64°$ **c.** $n = 60°$

18. a. Rectangle   **b.** Parallelogram   **c.** Trapezium

19. a. Kite   **b.** Square   **c.** Rhombus

20. a. $75°$ **b.** $55°$ **c.** $42°$

21. a. $66°$ **b.** $112°$ **c.** $126°$

22. $67°$

23. $115°$

24. Because an equilateral triangle has 3 congruent angles, right-equilateral and obtuse-equilateral are not possible. Examples of the other triangles:

   i.

Right–scalene

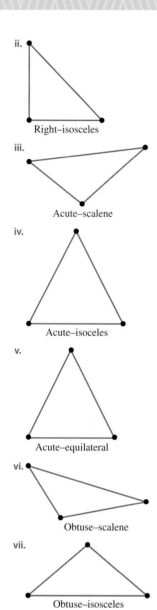

ii.

Right–isosceles

iii.

Acute–scalene

iv.

Acute–isoceles

v.

Acute–equilateral

vi.

Obtuse–scalene

vii.

Obtuse–isosceles

25. 4

26. ∠BOE, ∠BOD, ∠BOC, ∠COE, ∠COD, ∠DOE

27. 75°

28. 35°

12 Data classification and visualisation

LESSON
12.1 Overview

Why learn this?

Every day we see graphs, charts and statistics in the media. These graphs, charts and statistics all serve to condense massive amounts of information into a few simple numerical facts. This is the beauty of statistics. It turns complicated data into simpler forms that, with the knowledge gained throughout this topic, you will be able to interpret and understand.

You may have seen graphs or charts of housing prices in the news or online. One house price report stated that in March 2023 the median house price was $1 006 923 in Sydney, $743 554 in Melbourne and $694 495 in Brisbane. What is the median house price? Why are house prices discussed in terms of the median and not the mean? This topic will help you to understand the answers to these questions, and apply your understanding to other contexts.

In every business it is vital to keep a close eye on finances. Statistics can be used to determine the economic health of a business by looking at changes in revenue, costs, customer satisfaction, market share and many other factors over time. Line graphs are often used to show changes over time, and pie charts are often used to display market share or customer satisfaction.

Data is also extremely important in the medical field. Statistics are used to collate data from around the world to better understand the rates of disease, effectiveness of different medicines or treatments, and many other things. In 2020, when the Covid-19 pandemic hit the world, graphs and charts started appearing in the news and on social media displaying the case numbers for different states and countries. The data collected by each state and country was displayed graphically, enabling residents to be informed and educated on the latest developments.

1. **MC** Select the data category that suits the number of players in a netball team.
   A. Categorical and discrete
   B. Numerical and discrete
   C. Numerical and continuous
   D. Categorical and continuous

2. **MC** Select the data category that suits the amount of rainfall.
   A. Numerical and continuous
   B. Categorical and discrete
   C. Numerical and discrete
   D. Categorical and continuous

Use the following information to answer questions **3** and **4**.

The following frequency table represents the heights of 20 students in a Mathematics class.

Height group (cm)	Frequency
120–129	2
130–139	8
140–149	7
150–159	

3. Determine the number of students in the 150–159 cm group.

4. **MC** Choose the most common height group in the class.
   A. 120–129 cm
   B. 130–139 cm
   C. 140–149 cm
   D. 150–159 cm

5. A book store sold the following number of books each day for the last week. Calculate the daily average number of books sold.
   Book sales: 23, 34, 18, 28, 36, 45, 33

6. Calculate the median of the following data set.

   3, 3, 7, 8, 9, 9, 9, 11, 12, 13, 13, 17, 18, 21

7. Buddy kicks a ball 7 times. His kicks are 49, 52, 57, 43, 55, 62 and 59 metres in distance. Calculate the range of Buddy's 7 kicks.

8. The following data represents the sizes of Nike runners sold over a week. State the most common shoe size.
   Nike shoe sizes: 7, 7, 8, 9, 9, 9, 9, 10, 10, 11, 11, 11, 12, 12

9. Determine the median value in the following dot plot.

10. The following graph shows the preferred sports of a group of students.

Sport preferences

To the nearest per cent, determine the percentage of students who preferred swimming.

11. The following stem-and-leaf plot represents the ages of a group of golfers. Determine the number of golfers who are older than 20 years of age.

Key: 2|4 = 24

Stem	Leaf
1	7 8 8 9 9
2	2 4 7 9
3	1 3 3 8
4	0 2 2 2 6 6
5	5 7
6	4

12. The following back-to-back stem-and-leaf plot shows the heights of Year 7 students. Determine the difference between the height of the tallest boy and the shortest girl.

Key: 13|7 = 137 cm

Leaf (boys)	Stem	Leaf (girls)
9 8	13	7 8
9 8 8 7 6	14	3 5 6
9 8 8	15	1 2 3 7
7 6 6 5	16	3 5 6
8 7 6	17	1

13. The data presented in a divided bar graph shows the preferred Australian holiday destinations.

Preferred holiday destination

45	18	75	12

☐ Sunshine Coast
☐ Sydney
☐ Uluru
☐ Canberra

a. Based on this graph, calculate the number of people who were interviewed about their favourite Australian holiday destination.
b. Identify the number of people who preferred the Sunshine Coast.
c. Calculate the percentage of the total number of people who preferred Canberra.
d. Calculate the percentage of the total number of people who preferred Uluru.

14. A survey asked people for their favourite brand of car. The following results were found:

Toyota = 34, Ford = 28, Holden = 30, Mazda = 18

If a sector graph was constructed using these results, determine the angle used for Toyota. Give your answer to the nearest degree.

15. The following sector graph is based on the way students travel to school. Determine the percentage, to the nearest whole per cent, of students who get to school by bus.

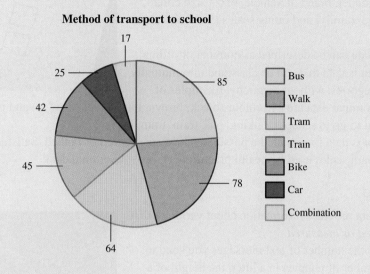

Method of transport to school

- Bus
- Walk
- Tram
- Train
- Bike
- Car
- Combination

LESSON
12.2 Classifying data

LEARNING INTENTION

At the end of this lesson you should be able to:
- classify data as numerical (discrete or continuous) or categorical (nominal or ordinal)

▶ 12.2.1 How to classify data

eles-4716

- Statistics is the study of collecting, organising, presenting, analysing and interpreting data.
- **Data** is information that is collected about a certain group.
- A **variable** is a particular characteristic that can be observed, counted or measured, such as eye colour or height. A variable can change over time or between individual observations.
- Data can be classified as **categorical data** or **numerical data**.

Categorical data

- Categorical data provides information about variables that can be *grouped into categories*.
- Take, for example, recycling bins. The different categories for recycling can include organic, paper, plastic, glass and metal.

- Categorical data can take many forms (some of which are numbers). These numbers generally represent a group or category, such as a rating scale (for example, 1 = dislike, 2 = don't care, 3 = like).
- Other common examples of categorical data include hair colour, gender, brand of vehicle, pizza size (small, medium, large, family) and rating scales (low, medium, high).
- Categorical data can be described as nominal or ordinal.
- **Nominal data** is data that can be classified into mutually exclusive categories, without hierarchy. Examples of categorical nominal data are eye colour (black, brown, blue, green, grey ...) and preferred mode of transportation to go to school (walking, bus, train, tram, bicycle etc.).
- **Ordinal data** is data for which the possible values are ordered or ranked. Satisfaction ratings in the service industry, or pain scales used by health practitioners, provide ordinal data.

Numerical data

- Numerical data provides information about variables that can be *counted or measured*.
  For example, the number of text messages you send in a day, the number of puppies in a litter, the height of a tree at your school, and the daily pollution levels on a particular highway are all examples of data that can be counted or measured.
- Numerical data can be further classified as **discrete** (able to be counted), or **continuous** (able to be measured). For example, *number of goals scored* is a discrete variable, while *height of tree* is a continuous variable.

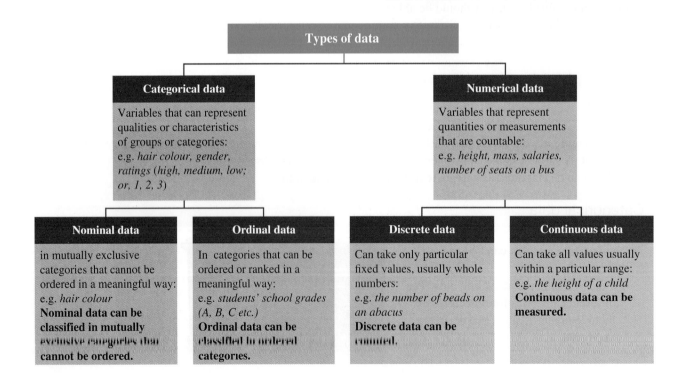

Types of data	
Categorical data	**Numerical data**
Variables that can represent qualities or characteristics of groups or categories: e.g. *hair colour, gender, ratings (high, medium, low; or, 1, 2, 3)*	Variables that represent quantities or measurements that are countable: e.g. *height, mass, salaries, number of seats on a bus*

Nominal data	**Ordinal data**	**Discrete data**	**Continuous data**
in mutually exclusive categories that cannot be ordered in a meaningful way: e.g. *hair colour* **Nominal data can be classified in mutually exclusive categories that cannot be ordered.**	In categories that can be ordered or ranked in a meaningful way: e.g. *students' school grades (A, B, C etc.)* **Ordinal data can be classified in ordered categories.**	Can take only particular fixed values, usually whole numbers: e.g. *the number of beads on an abacus* **Discrete data can be counted.**	Can take all values usually within a particular range: e.g. *the height of a child* **Continuous data can be measured.**

Numerical or categorical?

Deciding whether data is numerical or categorical is not always easy. Two things that can help your decision-making are the following.

1. Numerical data can always be used to perform arithmetic computations. This is not the case with categorical data. This is a good test to apply when in doubt.
   For example, it makes sense to calculate the average weight of a group of individuals, but not the average hair colour of a class of students.
2. It is not the variable alone that determines whether data is numerical or categorical — it is also the way the data is recorded.
   For example, if the data for the variable *weight* is recorded in kilograms, it is numerical. However, if the data is recorded using labels such as 'underweight', 'normal weight' and 'overweight', it is categorical.

WORKED EXAMPLE 1 Classifying data

Classify the following data using the following descriptive words: 'categorical' (nominal or ordinal) or 'numerical' (discrete or continuous).
a. The number of students absent from school
b. The education level of individuals in a group
c. The eye colour of each student in your class
d. The room temperature at various times during a particular day

THINK	WRITE
a. 1. Determine whether the data is categorical or numerical. 2. Determine whether the data is discrete (countable) or continuous (measurable).	a. The data is *numerical* because the number of students can be counted or measured. The data is *discrete* because the number of absences can be counted rather than measured.
b. 1. Determine whether the data is categorical or numerical. 2. Determine whether the data is nominal (cannot be ordered in any meaningful way) or ordinal (there is a natural order).	b. The data is *categorical* because the education level of individuals can be placed into groups or categories. The data is *ordinal* because the education level of individuals can ordered in a meaningful way (high school, bachelor, master, PhD).
c. 1. Determine whether the data is categorical or numerical. 2. Determine whether the data is nominal or ordinal.	c. The data is *categorical* because the colour of a person's eyes fits into groups (for example, brown, blue, green). The data is *nominal* because eye colour cannot be meaningfully ranked or ordered.
d. 1. Determine whether the data is categorical or numerical. 2. Determine whether the data is discrete (countable) or continuous (measurable).	d. The data is *numerical* because the room temperature can be counted or measured. The data is *continuous* because temperature can be measured rather than counted.

Exercise 12.2 Classifying data

| 12.2 Quick quiz on | 12.2 Exercise |

Individual pathways

■ PRACTISE	■ CONSOLIDATE	■ MASTER
1, 2, 5, 9, 11	3, 6, 7, 10, 12	4, 8, 13

Fluency

1. Match each word with its correct meaning.

   a. discrete
   b. categorical
   c. continuous
   d. numerical

   i. placed in categories or classes
   ii. data in the form of numbers
   iii. used for measurements
   iv. counted in exact values

2. **WE1** Classify each of the following data as 'categorical' or 'numerical' (discrete or continuous).

   a. The population of your town or city
   b. The types of motorbikes in a car park
   c. The heights of people in an identification line-up
   d. The weights of babies in a group
   e. The languages spoken at home by students in your class
   f. The time spent watching TV

3. Classify each of the following numerical data as either 'discrete' or 'continuous'.

   a. The number of children in the families in your suburb
   b. The air pressure in your car's tyres
   c. The number of puppies in a litter
   d. The times for swimming 50 metres
   e. The quantity of fish caught in a net

4. Classify each of the following data as 'categorical' or 'numerical' (discrete or continuous).

   a. The types of shops in a shopping centre
   b. The football competition ladder at the end of each round
   c. The lifetime of torch batteries
   d. The number of people attending the Big Day Out
   e. Final Year 12 exam grades
   f. Hotel accommodation ratings

Understanding

5. Data representing shoe (or rollerblade) sizes can be classified as categorical. Identify whether this statement is true or false.

6. Data representing the points scored in a basketball game can best be described as discrete. Identify whether this statement is true or false.

7. **MC** Select an example of categorical data.
   A. Heights of buildings in Sydney
   B. Number of pets in households
   C. Types of pets in households
   D. Birthday year of students in Year 7

8. Write a sentence explaining the difference between discrete and continuous data. Give an example of each.

Communicating, reasoning and problem solving

9. A Likert scale is used to represent people's attitudes to a topic. A five-level Likert scale could be as follows:
   1. Strongly disagree
   2. Disagree
   3. Neither agree nor disagree
   4. Agree
   5. Strongly agree

The data collected from a Likert scale is recorded as a numerical value. Explain why this data is categorical instead of numerical.

10. Explain why data such as postal codes, phone numbers and drivers' licence numbers is not numerical data.

11. The following questions would collect categorical data. Rewrite the questions so that you could collect numerical data.
    a. Do you read every day?
    b. Do you play sport every day?
    c. Do you play computer games every day?

12. The following questions would collect numerical data. Rewrite the questions so that you could collect categorical data.
    a. On average, how many minutes per week do you spend on Maths homework?
    b. How many books have you read this year?
    c. How long does it take you to travel to school?

13. A fisheries and wildlife officer released 200 tagged trout into a lake. A week later, the officer took a sample of 50 trout and found that 8 of them were tagged. The officer can use this information to estimate the population of trout in the lake. Calculate the number of trout that are in the lake. Explain how you got this answer.

LESSON
12.3 Displaying data in tables

LEARNING INTENTION

At the end of this lesson you should be able to:
- construct a frequency distribution table for a set of data
- analyse and report on the information in a frequency distribution table.

▶ 12.3.1 Frequency distribution tables

eles-4718

- Data can be displayed in a variety of forms.
- In general, data is first organised into a table, and then a suitable graph is chosen as a visual representation.
- One way of presenting data is by using a frequency distribution table.
- **Frequency** is the number of times a result or piece of data occurs.
- A **frequency distribution table** consists of three columns: score, tally and frequency.
- The following frequency table is the result of a survey of 30 students, and it shows the number of family members who live at home with them.

Score (number of family members living at home)	Tally	Frequency
2					3
3	ЖІ	5
4	ЖІ||||	9
5	ЖІ|	7
6						4
7			1
8			1
	Total	30

WORKED EXAMPLE 2 Constructing and interpreting a frequency table

A particular class was surveyed to determine the number of pets per household, and the data was recorded. The raw data was:

0, 3, 1, 2, 0, 1, 0, 1, 2, 4, 0, 6, 1, 1, 0, 2, 2, 0, 1, 3, 0, 1, 2, 1, 1, 2

a. **Organise the data into a frequency distribution table.**
b. **Calculate the number of households that were included in the survey.**
c. **Determine the number of households that have fewer than 2 pets.**
d. **State the most common number of pets.**
e. **Determine the number of households that have 3 or more pets.**
f. **Determine the fraction of households surveyed that had no pets.**

THINK

a. 1. Draw a frequency distribution table comprising three columns, headed Number of pets, Tally and Frequency.

   2. In the first column list the possible number of pets per household (from 0 to 6).

   3. Place a stroke in the tally column each time a particular score is noted.
   Hint: A score of 5 is shown as a 'gate post' (four vertical strokes and one diagonal stroke:).

WRITE

a.

Number of pets	Tally	Frequency
0	ЖІІ	7
1	ЖІ|||	9
2	ЖІ	6
3				2
4			1
5		0
6			1
	Total	26

---

4. Write the total number of tally strokes for each pet in the frequency column.

5. Calculate the total of the frequency column.

b. The total of the frequency column gives the number of households surveyed.

b. 26 households were surveyed.

c. 1. Calculate the number of households that have fewer than two pets.
Hint: Fewer than two means zero pets or one pet.

c. Fewer than two pets $= 7 + 9$
$\qquad = 16$

2. Write the answer.

16 households have fewer than two pets.

d. 1. Make a note of the highest value in the frequency column and check which score it corresponds to.

d. The score with the highest frequency (9) corresponds to one pet.

2. Write the answer.

The most common number of pets is one.

e. 1. Calculate the number of households that have 3 or more pets.
Hint: 3 or more means 3, 4, 5 or 6.

e. 3 or more pets $= 2 + 1 + 0 + 1$
$\qquad = 4$

2. Write the answer.

Four households have 3 or more pets.

f. 1. Write the number of households with no pets.

f. Households with no pets $= 7$

2. Write the total number of households surveyed.

Total number of households surveyed $= 26$

3. Define the fraction and substitute the known values into the rule.

$$\frac{\text{Households with no pets}}{\text{Total number of households surveyed}} = \frac{7}{26}$$

4. Write the answer.

Of the households surveyed, $\dfrac{7}{26}$ have no pets.

Class intervals

- Sometimes data may contain too many numerical values to list them all individually in the 'score' column. In this case, we use a range of values, called a *class interval*, as our category. For example, the range 100–104 may be used to cater for all the values that lie within the range, including 100 and 104.

WORKED EXAMPLE 3 Using class intervals in a frequency table

The data below shows the ages of a number of mobile phone owners.

12, 11, 21, 12, 30, 26, 13, 15, 29, 16, 17, 17, 17, 21, 19, 12, 14, 16, 43, 18, 51, 25, 30, 28, 33, 62, 39, 40, 30, 18, 19, 41, 22, 21, 48, 31, 33, 33, 34, 41, 18, 17, 31, 43, 42, 17, 46, 23, 24, 33, 27, 31, 53, 52, 25

a. **Construct a frequency table to classify the given data. Use a class interval of 10, that is, ages 11–20, 21–30 and so on, as each category.**

b. **Calculate the number of people who were surveyed.**

c. **State the age group that had the most people with mobile phones.**

d. **State the age group that had the fewest people with mobile phones.**

e. **Determine the number of people in the 21–30 age group who own a mobile phone.**

THINK

a. 1. Draw a frequency distribution table.

   2. In the first column, list the possible age groups (11–20, 21–30 and so on).

   3. Systematically go through the results and place a stroke in the tally column each time a particular age group is noted.

   4. Write the total tally of strokes for each age group in the frequency column.

   5. Calculate the total of the frequency column.

b. The total of the frequency column gives us the number of people surveyed.

c. 1. Make note of the highest value in the frequency column and check which age group it corresponds to.

   2. Write the answer.

d. 1. Make note of the lowest value in the frequency column and check which age group it corresponds to.
Note: There may be more than one answer.

   2. Write the answer.

e. 1. Check the 21–30 age group in the table to see which frequency value corresponds to this age group.

   2. Write the answer.

WRITE

a.

Age group	Tally	Frequency
11–20	卌 卌 卌 IIII	19
21–30	卌 卌 卌	15
31–40	卌 卌	10
41–50	卌 II	7
51–60	III	3
over 60	I	1
	Total	55

b. A total of 55 people were surveyed.

c. The 11–20 age group has the highest frequency: that is, a value of 19.

The 11–20 age group has the highest number of people with mobile phones.

d. The over-60 age group has the lowest frequency: that is, a value of 1.

The over-60 age group has the lowest number of people with mobile phones.

e. The 21–30 age group has a corresponding frequency of 15.

15 people in the 21–30 age group own a mobile phone.

 Resources

   Interactivity   Frequency distribution tables (int-4051)

12.3 Quick quiz on	12.3 Exercise

Individual pathways

■ PRACTISE	■ CONSOLIDATE	■ MASTER
1, 3, 6, 7, 10, 13	2, 4, 8, 11, 14	5, 9, 12, 15

Fluency

1. **WE2** The number of children per household in a particular street is surveyed and the data recorded as follows.

   0, 8, 6, 4, 0, 0, 0, 2, 1, 3, 3, 3, 1, 2, 3,
   2, 3, 2, 1, 2, 1, 3, 0, 2, 2, 4, 2, 3, 5, 2

   a. Organise the data into a frequency distribution table.
   b. Calculate the number of households that are included in the survey.
   c. Determine the number of households that have no children.
   d. Determine the number of households that have at least 3 children.
   e. State the most common number of children.
   f. Determine the fraction of households surveyed that have 4 children.

2. **WE3** The following data shows the incomes of a sample of doctors.

   $100 000, $105 000, $110 000, $150 000, $155 000, $106 000, $165 000, $148 000, $165 000,
   $200 000, $195 000, $138 000, $142 000, $153 000, $173 000, $149 000, $182 000, $186 000

   a. Construct a frequency table to classify the given data. Use a class interval of 10 000 to organise incomes into brackets of $100 000 to $109 999, $110 000 to $119 999, and so on.
   b. Calculate the number of people who were surveyed.
   c. State the income bracket that was most common.
   d. State the income bracket that was least common.
   e. Determine the number of doctors who earned $140 000 to $149 999.

3. Use this frequency distribution table to answer the following questions.

Score	Tally	Frequency
0	||	2
1	|||||	5
2	|||	3
3	||||| ||||||	11
4	||||| |||	8
5	||||	4
	Total	

a. Calculate the number of people who participated in the survey.
b. State the most frequent score.
c. Determine the number of people who scored less than 3.
d. Determine the number of people who scored 3 or more.
e. Calculate the fraction of the people surveyed who scored 3.

4. Draw a frequency distribution table for each of the following data sets.
   a. Andrew's scores in Mathematics tests this semester are:
   $$6, 9, 7, 9, 10, 7, 6, 5, 8, 9.$$
   b. The number of children in each household of a particular street are:
   $$2, 0, 6, 1, 1, 2, 1, 3, 0, 4, 3, 2, 4, 1, 0, 2, 1, 0, 2, 0.$$
   c. The weights (in kilograms) of students in a Year 7 class are:
   $$46, 60, 48, 52, 49, 51, 60, 45, 54, 54, 52, 58, 53, 51,$$
   $$54, 50, 50, 56, 53, 57, 55, 48, 56, 53, 58, 53, 59, 57.$$
   d. The heights of students in a Year 7 class are:
   $$145, 147, 150, 150, 148, 145, 144, 144, 147, 149, 144, 150, 150, 152,$$
   $$145, 149, 144, 145, 147, 143, 144, 145, 148, 144, 149, 146, 148, 143.$$

5. Rosemary decided to survey the participants of her local gym about their preferred sport. She asked each participant to name one preferred sport. Her results are below:

   hockey, cricket, cricket, tennis, scuba diving, netball, tennis, netball, swimming, netball, tennis, hockey, cricket, lacrosse, lawn bowls, hockey, swimming, netball, tennis, netball, cricket, tennis, hockey, lacrosse, swimming, lawn bowls, swimming, swimming, netball, netball, tennis, golf, hockey, hockey, lacrosse, swimming, golf, hockey, netball, swimming, scuba diving, scuba diving, golf, tennis, cricket, cricket, hockey, lacrosse, netball, golf

   a. State any problems with the way Rosemary has displayed the data.
   b. Organise Rosemary's results into a frequency table to show the participants' preferred sports.
   c. From the frequency table, determine:
      i. the most preferred sport
      ii. the least preferred sport.
   d. State whether any sport(s) have the same frequency.

Understanding

6. A random sample of 30 families was surveyed to determine the number of children in high school in each family. The following is the raw data collected.
   $$2, 1, 1, 0, 2, 0, 1, 0, 2, 0, 3, 1, 1, 0, 0,$$
   $$0, 1, 4, 1, 0, 0, 1, 2, 1, 2, 0, 3, 2, 0, 1$$
   a. Organise the data into a frequency distribution table.
   b. Determine the number of families that have no children of high school age.
   c. Determine the number of families that have 2 or more children of high school age.
   d. State the score that has the highest frequency.
   e. From the 30 families surveyed, state the highest number of children in high school from one family.
   f. Calculate the fraction of families that had 2 children of high school age.

7. Draw a frequency table to classify the following data on students' heights. Use a range of values (for example, 140–144, 145–149, and so on) as each category. The values are listed below.

   168 cm, 143 cm, 145 cm, 151 cm, 153 cm, 148 cm, 166 cm, 147 cm, 160 cm, 162 cm,
   175 cm, 168 cm, 143 cm, 150 cm, 160 cm, 180 cm, 146 cm, 158 cm, 149 cm, 169 cm,
   167 cm, 167 cm, 163 cm, 172 cm, 148 cm, 151 cm, 170 cm, 160 cm

8. A real estate agent has listed all the properties sold in the area in the last month. She wants to know what has been the most popular type of property from the list of sold properties:

2-bedroom house, 4-bedroom house, 3-bedroom house, 2-bedroom unit, 4-bedroom house, 1-bedroom unit, 3-bedroom house, 2-bedroom unit, 3-bedroom house, 1-bedroom unit, 2-bedroom unit, 3-bedroom house, 3-bedroom house, 3-bedroom house, 2-bedroom unit, 1-bedroom unit

a. Complete a frequency table for the list of properties and work out which type of property was most popular.

b. **MC** Select the least popular type of property.

   A. 1-bedroom unit
   C. 2-bedroom house
   B. 2-bedroom unit
   D. 3-bedroom house

c. **MC** Choose the type of property that is half as popular as a 2 bedroom unit.

   A. 4-bedroom house
   C. 2-bedroom house
   B. 3-bedroom house
   D. 1-bedroom unit

9. **MC** Select the statement(s) about the frequency column of a frequency table that are true.

   A. It adds up to the total number of categories.
   B. It adds up to the total number of results given.
   C. It adds up to the total of the category values.
   D. It displays the tally.

Communicating, reasoning and problem solving

10. Explain why tallies are drawn in batches of four vertical lines crossed by a fifth one.

11. Discuss what you need to consider when selecting a class interval for a frequency distribution table.

12. Explain the reasoning behind using class intervals and state the circumstances in which they should be used.

13. The following frequency distribution table displays the number of pets in the families of the students of a class.

Number of pets	Frequency	% Frequency
0	8	
1	11	
2	3	
3	2	
4	0	
5	1	

a. Determine the number of students in the class.
b. Calculate the percentage frequency for each row of the table.

Another class containing 25 students creates their own version of this frequency distribution table. When the data in both tables is combined, the percentage frequency of having 1 pet is 34%.

c. Determine the number of students in the other class with one pet in their family.

14. This frequency table shows the percentage occurrence of the vowels in a piece of writing. Two pieces of data are missing — the percentage occurrence of the letters O and U. The occurrence of O is 2.6 times that of U. Determine the two missing values in this table. Show your working.

Vowel	Percentage frequency
A	22.7
E	27.6
I	19.9
O	
U	

15. A survey was conducted on 45 households to identify the number of children living in each household. Use the following information to determine the results and complete the table.
   • The number of households with no children is equal to the number of households with 3 children.
   • The number of households with 1 child was 4 times the number of households with no children.
   • The number of households with 2 children was 5 times the number of households with no children.
   • There was one household with 4 children.

Score	Frequency
0	
1	
2	
3	
4	
Total	

LESSON
12.4 Measures of centre and spread

LEARNING INTENTION

At the end of this lesson you should be able to:
   • calculate the mean, median and mode of a set of data
   • calculate the range of a set of data.

12.4.1 Measures of centre

eles-4719

   • Measures of centre are statistical values that give us an idea about the centre of a data set. The measures of centre are **mean**, **median** and **mode**.

The mean

   • The mean of a set of data is the average of the values.
   • The mean is denoted by the symbol $\bar{x}$ (a lower-case x with a bar on top).
   • The average is found by adding up all of the values in the data set, then dividing by the number of values in the set.
   • The mean is not necessarily a whole number, or a number that was in the original set of data.

Calculating the mean

- The formula for the mean is:

$$\bar{x} = \frac{\text{sum of data values}}{\text{total number of data values}}$$

- In most cases the mean is thought of as the centre of a set of data.

WORKED EXAMPLE 4 Calculating the mean

For each of the following sets of data, calculate the mean, $\bar{x}$.
Give your answers correct to 2 decimal places where appropriate.

a. 5, 5, 6, 4, 8, 3, 4

b. 0, 0, 0, 1, 1, 1, 1, 1, 4, 4, 4, 4, 5, 5, 5, 7, 7

THINK	WRITE
a. 1. Calculate the sum of the given values.	a. Sum of data values $= 5 + 5 + 6 + 4 + 8 + 3 + 4$ $= 35$
2. Count the number of values.	Number of data values $= 7$
3. Define the rule for the mean.	$\bar{x} = \dfrac{\text{sum of data values}}{\text{total number of data values}}$
4. Substitute the known values into the rule and evaluate.	$= \dfrac{35}{7}$ $= 5$
5. Write the answer.	The mean (or average) of the set of data is 5.
b. 1. Calculate the sum of the given values. Take note of the number of times each value occurs. 0 occurs 3 times (3×0), 1 occurs 5 times (1×5), 4 occurs 4 times (4×4), 5 occurs 3 times (5×3), 7 occurs 2 times (7×2)	b. Sum of data values $= (3 \times 0) + (5 \times 1) + (4 \times 4) + (3 \times 5) + (2 \times 7)$ $= 0 + 5 + 16 + 15 + 14$ $= 50$
2. Count the number of values.	Number of data values $= 17$
3. Define the rule for the mean.	$\bar{x} = \dfrac{\text{sum of data values}}{\text{total number of data values}}$
4. Substitute the known values into the rule and evaluate.	$= \dfrac{50}{17}$ $= 2.941\,176\,47...$
5. Round the answer to 2 decimal places.	$= 2.94$
6. State the answer.	The mean (or average) of the set of data is 2.94.

The median

- The median represents the middle score when the data values are ordered from **smallest** to **largest** (also known as 'numerical order' or 'ascending order').
- An equal number of data values can be found below the median and above it.
- For a set with an even number of data values, there are two 'middle numbers' when the data are in ascending order, and the median is their average.

Determining the median

When determining the median, follow these steps.

1. **Arrange the data values in ascending order (smallest to largest).**
2. **If the total number of data values is odd, locate the middle value (ensure that there are as many values above the median as there are below it).**

Lower half   Upper half

Median
(middle value)

3. **If the data set contains an even number of data values there will be two 'middle' numbers. The median is the average of these numbers.**

$$1 \quad 1 \quad 4 \quad 5 \quad 5 \quad \boxed{6 \quad 8} \quad 8 \quad 11 \quad 12 \quad 12 \quad 13$$

Lower half   Middle values   Upper half

$$\frac{6+8}{2} = 7$$

Median

Alternatively:

The *position* of the median is the $\left(\dfrac{n+1}{2}\right)$ th data value, where n is the total number of data values.

WORKED EXAMPLE 6 Determining the median

Determine the middle value (median) for the following sets of data.

a. 5, 4, 2, 6, 3, 4, 5, 7, 4, 8, 5, 5, 6, 7, 5   **b.** 8, 2, 5, 4, 9, 9, 7, 3, 2, 9, 3, 7, 6, 8

THINK	WRITE
a. 1. Arrange the data values in ascending order (smallest to largest).	**a.** 2, 3, 4, 4, 4, 5, 5, 5, 5, 5, 6, 6, 7, 7, 8
2. Locate the middle value by inserting a vertical line that splits the data into two equal sections. The pink line inserted through the third number 5 splits the data into two equal sections that each have 7 data values.	2, 3, 4, 4, 4, 5, 5, ⎮, 5, 5, 5, 6, 6, 7, 7, 8
3. Write the answer.	The median of the data set is 5.
b. 1. Arrange the data values in ascending order (smallest to largest).	**b.** 2, 2, 3, 3, 4, 5, 6, 7, 7, 8, 8, 9, 9, 9

2. Locate the middle value by inserting a vertical line that splits the data into two equal sections. The pink line inserted in between the 6 and 7 splits the data into two equal sections that each have 6 data values.

2, 2, 3, 3, 4, 5, 6, | 7, 7, 8, 8, 9, 9, 9

3. Since the median (or middle position) occurs between 6 and 7, we need to determine the average of these two numbers. The median of the data lies directly between 6 and 7.

$$\text{Median} = \frac{6+7}{2}$$
$$= \frac{13}{2}$$
$$= 6.5$$

4. The number directly between 6 and 7 is 6.5. Write the answer.

The median of the data set is 6.5.

Alternative method:

a. 1. Arrange the data values in ascending order (smallest to largest).

a. 2, 3, 4, 4, 4, 5, 5, 5, 5, 5, 6, 6, 7, 7, 8

2. Count the number of data values in the set.

$n = 15$

3. Determine the $\left(\frac{n+1}{2}\right)^{\text{th}}$ data value.

$\left(\frac{n+1}{2}\right) = \frac{15+1}{2} = 8$

4. The median is located at the $\left(\frac{n+1}{2}\right)^{\text{th}}$ data value.

The median is the 8th value in the set of data. The 8th number in the set of data is 5.

5. Write the answer.

The median of the data set is 5.

b. 1. Arrange the data values in ascending order (smallest to largest).

b. 2, 2, 3, 3, 4, 5, 6, 7, 7, 8, 8, 9, 9, 9

2. Count the number of data values in the set.

$n = 14$

3. Determine the $\left(\frac{n+1}{2}\right)^{\text{th}}$ data value.

$\left(\frac{n+1}{2}\right) = \frac{14+1}{2} = 7.5$

4. The two middle values are the 7th and 8th values.

The middle values are 6 and 7

5. Calculate the average of the middle values.

$\frac{6+7}{2} = \frac{13}{2} = 6.5$

6. Write the answer.

The median of the data set is 6.5.

- As the median is the middle value, it is a common measure of centre, especially for large data sets or data sets where the average is skewed by a small number of extreme (outlier) values.

Median as a measure of centre

The median, as the middle value in a data set when all the values are arranged in numerical order, is the best measure of the centre of the data because it is not affected by outliers.

The mode

- The mode is the most common value in a set of data. It is the value that occurs most frequently.
- Some sets of data have more than one mode.
- Some sets of data have no mode at all.

Calculating the mode

When determining the mode:
1. arrange the data values in ascending order (smallest to largest) — this step is optional but does help
2. look for the value that occurs most often (the one that has the highest frequency).

Important notes about the mode

- If a set of data has no mode, your answer should be the words 'no mode', not '0'. This is what would happen when no value occurs more than once in a set of data.
  For example, in the set of data 1, 2, 3, 4, 6, 7, 10, 12, 13, no value is repeated, so there is no mode.

- If a set of data has one mode, write this value as your answer.
  For example, for the set of data 15, 31, 12, 47, 21, 65, 12, the mode is 12 because it appears twice — each other value only appears once.

- If a set of data has more than one mode, write all of the values that occur the most as your answer.
  For example, for the set of data 1, 1, 2, 2, 2, 4, 5, 6, 6, 6, 9, 9, 12, 12, 12, the modes are 2, 6 and 12.

WORKED EXAMPLE 6 Identifying the mode

Identify the mode of the following scores.
a. 2, 3, 4, 5, 5, 6, 6, 7, 8, 8, 8, 9
b. 12, 18, 5, 17, 3, 5, 2, 10, 12
c. 42, 29, 11, 28, 21

THINK	WRITE
a. 1. Look at the set of data and highlight any values that have been repeated.	a. 2, 3, 4, 5, 5, 6, 6, 7, 8, 8, 8, 9
2. Choose the value that has been repeated the most.	The numbers 5 and 6 occur twice. However, the number 8 occurs three times.
3. Write the answer.	The mode for the given set of values is 8.
b. 1. Look at the set of data and highlight any values that have been repeated.	b. 12, 18, 5, 17, 3, 5, 2, 10, 12
2. Choose the value(s) that have been repeated the most.	The number 5 occurs twice. The number 12 occurs twice.
3. Write the answer.	The modes for the given set of values are 5 and 12.
c. 1. Look at the set of data and highlight any value(s) that have been repeated.	c. 42, 29, 11, 28, 21
2. Write the answer.	No values have been repeated. The set of data has no mode, since each of the numbers occurs once only.

- The mode is generally not used as a measure of centre, rather as a measure of the most common, or popular, value.

12.4.2 Measures of spread

eles-4720

- A measure of spread indicates how far data values are spread from the centre, or from each other.
- Looking at the following two graphical displays, the pink graph on the left is more 'spread out' compared to the green graph on the right. That is, the pink graph has a larger spread (from 0 to 22) compared to the green graph (from 6 to 18).
- There are several measures of spread, but at this stage we will only discuss the **range**.
- The range of a set of values is the difference between the largest and smallest values.

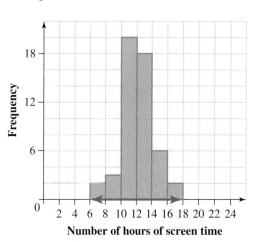

Calculating the range

The range of a set of data is the difference between the largest and smallest values in that set.

range = (largest data value) – (smallest data value)

WORKED EXAMPLE 7 Determining the range

Determine the range of the following data.

12, 76, 35, 29, 16, 45, 56

THINK	WRITE
1. Identify the largest and smallest values.	Largest value $= 76$ Smallest value $= 12$
2. State the formula for the range and substitute the values in to evaluate it.	Range = largest value – smallest value $= 76 - 12$ $= 64$
3. Write the answer.	The range of the data set is 64.

  Resources

Interactivities Measures of centre (int-4052)
Mean (int-4053)
Median (int-4054)
Mode (int-4055)
Range (int-4056)

Exercise 12.4 Measures of centre and spread

learnon

12.4 Quick quiz on	12.4 Exercise

Individual pathways

■ PRACTISE	■ CONSOLIDATE	■ MASTER
1, 4, 8, 10, 11, 15, 16, 21, 22, 27, 28, 29	2, 5, 6, 9, 12, 17, 18, 23, 24, 30, 31, 32	3, 7, 13, 14, 19, 20, 25, 26, 33, 34, 35

Fluency

1. **WE4** For each of the following sets of data, calculate the mean.

    a. 3, 4, 5, 5, 6, 7
    b. 5, 6, 7, 5, 5, 8
    c. 4, 6, 5, 4, 2, 3
    d. 3, 5, 6, 8, 7, 7
    e. 5, 4, 4, 6, 2, 3

2. For each of the following sets of data, calculate the mean.

    a. 2, 2, 2, 4, 3, 5
    b. 12, 10, 13, 12, 11, 14
    c. 11, 12, 15, 17, 18, 11
    d. 12, 15, 16, 17, 15, 15
    e. 10, 14, 12, 12, 16, 14

3. For each of the following sets of data, calculate the mean.
   Hint: Use the grouping of values to help you.

   a. 2, 2, 2, 4, 4, 4
   d. 9, 9, 8, 8, 7, 1, 1, 1, 1

   b. 1, 2, 2, 4, 4, 5
   e. 3, 3, 3, 1, 1, 1, 2, 2, 2

   c. 1, 2, 2, 5, 5, 6, 7
   f. 2, 2, 2, 3, 3, 3, 3, 6

4. **WE5a** Determine the median (middle value) for the following sets of data.

   a. 3, 3, 4, 5, 5, 6, 7
   c. 2, 2, 2, 3, 3, 4, 5

   b. 1, 2, 2, 3, 4, 8, 9
   d. 5, 5, 6, 6, 7, 7, 8, 9, 9

5. Determine the median (middle value) for the following sets of data.

   a. 7, 7, 7, 10, 11, 12, 15, 15, 16
   c. 1, 2.5, 5, 3.4, 1, 2.4, 5

   b. 4, 3, 5, 3, 4, 4, 3, 5, 4
   d. 1.2, 1.5, 1.4, 1.8, 1.9

6. **WE5b** Determine the median (middle value) for the following sets of data.

   a. 1, 1, 2, 2, 4, 4
   c. 4, 5, 7, 7, 8, 9

   b. 4, 5, 5, 5, 6, 7
   d. 1, 2, 2, 3, 3, 4

7. Determine the median (middle value) for the following sets of data.

   a. 2, 4, 4, 6, 8, 9
   c. 2, 4, 5, 7, 8, 8, 9, 9

   b. 1, 5, 7, 8
   d. 1, 5, 7, 8, 10, 15

8. **WE6** Identify the mode for each of the following sets of data.

   a. 3, 3, 4, 4, 4, 5, 6
   d. 4, 6, 4, 2, 7, 3

   b. 2, 9, 8, 8, 4, 5
   e. 2, 4, 3, 6, 2, 4, 2

   c. 1, 1, 2, 2, 2, 3

9. Identify the mode for each of the following sets of data.

   a. 4, 8, 8, 3, 3, 4, 3, 3
   c. 7, 9, 4, 6, 26, 71, 3, 3, 3, 2, 4, 6, 4, 25, 4
   e. 3, 7, 4, 5, 22

   b. 6, 2, 12, 10, 14, 14, 10, 12, 10, 12, 10, 12, 10, 12
   d. 2, 2, 3, 4, 4, 9, 9, 9, 6

10. **WE7** Determine the range of the following sets of data:

    a. 15, 26, 6, 38, 10, 17
    b. 12.8, 21.5, 1.9, 12.0, 25.4, 2.8, 1.3

Understanding

11. **MC** Select the method(s) that can be used to calculate the mean for the data 5, 5, 6, 7, 2.

    A. Adding all the results and multiplying by the number of results
    B. Adding all the results and dividing by the number of results
    C. Adding all the results
    D. Choosing the middle result

12. **MC** Select statement(s) that are true about calculating the mean of a set of data.

    A. Zeroes do not matter at all.
    B. Zeroes must be counted in the number of results.
    C. Zeroes must be added to the total as they will change it.
    D. Zeroes will make the mean zero.

13. **MC** Choose the statement that is true about the set of data 2.6, 2.8, 3.1, 3.7, 4.0, 4.2.

    A. The mean value for the data will be above 4.2.
    B. The mean value for the data will be below 2.6.
    C. The mean value for the data will be between 2.6 and 3.0.
    D. The mean value for the data will be between 3.0 and 4.0.

14. **MC** Choose the correct statement.
   A. The mean, median and mode for any set of data will always be the same value.
   B. The mean, median and mode for any set of data will never be the same value.
   C. The mean, median and mode for any set of data must always be close in value.
   D. The mean, median and mode for any set of data are sometimes close in value.

15. **MC** Select the range of the set of numbers: 16, 33, 24, 48, 11, 30, 15
   A. 48   B. 59   C. 37   D. 20

16. Dari wanted to know what her Mathematics test average was. Dari's teacher told her that she had used the mean of Dari's test results to calculate the end-of-year mark. Dari's test results (written as percentages) are:
   $$89, 87, 78, 75, 89, 94, 82, 93, 78$$
   Calculate Dari's mean Mathematics test score.

17. A cricketer had scores of 14, 52, 35, 42 and 47 in her last 5 innings. Determine her mean score.

18. Eliza works in a shoe factory as a shoe inspector. The number of shoes Eliza inspected in an hour was counted over a number of days' work. The following list contains the results.

   105, 102, 105, 106, 103, 105, 105,
   102, 108, 110, 102, 103, 106, 107,
   108, 102, 105, 106, 105, 104, 102,
   99, 98, 105, 102, 101, 97, 100

   Calculate the mean number of shoes checked by Eliza in one hour. Round your answer to the nearest whole number.

19. The number of students in the canteen each lunchtime was surveyed for 2 weeks. The following are the results.

   52, 45, 41, 42, 53, 45, 47, 32, 52, 56

   Calculate the mean number of students in the canteen at lunchtime in that fortnight. Round your answer to the nearest whole number.

20. Answer the following questions.
   a. Explain what 'the mean number of goals scored in a soccer match' means.
   b. Explain what 'the median house price' means.
   c. Explain what 'the mode of shoe-size data' means.

Communicating, reasoning and problem solving

21. Suggest why we summarise data by calculating measures of centre and spread.

22. The local football team has been doing very well. They want to advertise their average score to attract new club members. You suggest that they use the mean of their past season's game scores. They ask you to find it out for them. Their results are listed below (scores are given as total points for each game).

   110, 112, 141, 114, 112, 114, 95, 75, 58, 115, 116, 115, 75,
   114, 78, 96, 78, 115, 112, 115, 102, 75, 79, 154, 117, 62

   a. Calculate the team's mean score. Show your working.
   b. Identify whether the mode or median is the 'better' value to use to attract new club members. Justify your answer.

23. A clothing company wanted to know the size of jeans that should be manufactured in the largest quantities. A number of shoppers were surveyed and asked their jeans size. The results are given below.

13, 12, 14, 12, 15, 16, 14, 12, 15, 14, 12,
14, 13, 14, 11, 10, 12, 13, 14, 14, 10, 12,
14, 12, 12, 10, 8, 16, 17, 12, 11, 13, 12,
15, 14, 12, 17, 8, 16, 11, 12, 13, 12, 12

a. Determine the mode of the data.
b. Explain why the company would be more interested in the mode than the mean or median values.

24. Explain whether we are able to calculate the mean, median, mode and range of categorical data.

25. Tom thinks that the petrol station where he buys his petrol is cheaper than the one where his friend Sarah buys her petrol. They begin to keep a daily watch on the prices for 4 weeks and record the following prices (in dollars per litre).

Tom: 1.32, 1.37, 1.39, 1.39, 1.40, 1.42, 1.41, 1.38, 1.34, 1.30,
       1.29, 1.28, 1.27, 1.26, 1.25, 1.25, 1.24, 1.25, 1.24, 1.24,
       1.25, 1.25, 1.26, 1.27, 1.28, 1.28, 1.30, 1.32

Sarah: 1.41, 1.37, 1.32, 1.31, 1.29, 1.27, 1.24, 1.22, 1.21, 1.21,
        1.20, 1.19, 1.20, 1.21, 1.22, 1.23, 1.24, 1.24, 1.25, 1.26,
        1.27, 1.28, 1.28, 1.28, 1.29, 1.30, 1.32, 1.31

a. Calculate the mean petrol prices for Tom and Sarah.
b. Identify the petrol station that sells cheaper petrol on average. Justify your answer.
c. Explain why Tom might have been wrong about thinking his petrol station was cheaper.

26. Aniya wants to make sure the mean height of her jump in the high jump for 10 jumps is over 1.80 metres.
    a. Calculate her current mean if her jumps so far have been (in metres) 1.53, 1.78, 1.89, 1.82, 1.53, 1.81, 1.75, 1.86, 1.82. Show your working.
    b. Determine what height she needs to jump on the tenth jump to achieve a mean of 1.80. Show your working.
    c. Discuss whether this is likely, given her past results.

27. Kim has an average (mean) score of 72 in Scrabble. He has played 6 games. Determine what he must score in the next game to keep the same average.

28. Vassilis has calculated his mean score for history to be 89%, based on 5 tests. If he scores 92% in the sixth test, calculate his new mean score.

29. The club coach at a local cycling track was overheard saying that he thought at least half of the cyclists using the track were cycling at speeds of 30 km/h or more. The speeds (in km/h) of the club cyclists are listed as follows.

31, 22, 40, 12, 26, 39, 49, 23, 24, 38, 27, 16, 25, 37, 19, 25, 45, 23, 17, 20, 34, 19, 24, 15, 40, 39, 11, 29, 33, 44, 29, 50, 18, 22, 51, 24, 19, 20, 30, 40, 49, 29, 17, 25, 37, 25, 18, 34, 21, 20, 18

Discuss whether the coach's statement is correct.

30. Identify five whole numbers that have a mean of 10 and a median of 12.

31. The mean of five different test scores is 15. All test scores are whole numbers. Determine the largest and smallest possible test scores, given that the median test score is 12.

32. Gavin records the amount of rainfall in millimetres each day over a two-week period. Gavin's results are:

$$11, 24, 0, 6, 15, 0, 0, 0, 12, 0, 0, 127, 15, 0.$$

   a. Determine the mean rainfall for the two-week period.
   b. Determine the median rainfall.
   c. Identify the mode of the rainfall.
   d. State which of the mean, median and mode is the best measure of the typical rainfall. Explain your choice.

33. A group of 3 children have a mean height of 142 cm. The middle height is the same as the mean. The tallest child leaves the group, and is replaced by a child with the same height as the shortest child.
    The mean height of this group of three children is now 136 cm. Calculate the heights of the 4 children. Explain how you got your answer.

34. The mean of 5 different test scores is 10. All test scores are whole numbers. Determine the largest and smallest possible values for the median.

35. The mean of 9 different test scores ranging from 0 to 100 is 85. The median is 80. All test scores are whole numbers. Determine the greatest possible range between the highest and lowest possible test scores.

LESSON
12.5 Column graphs and dot plots

LEARNING INTENTION

At the end of this lesson you should be able to:
- display data as a column or bar graph
- display data as a dot plot
- analyse and interpret column and bar graphs and dot plots.

▶ 12.5.1 Column and bar graphs

eles-4721

Features of graphs

- Graphs are a useful way of displaying data, or numerical information. TV, websites, newspapers and magazines frequently display data as graphs.
- All graphs should have the following features:
  - a *title* — to tell us what the graph is about
  - clear *labels* for the axes — to explain what is being shown
  - *evenly scaled axes* — if the graph has numerical axes, they must have a scale that is constant for the length of the axes, with a clear indication of the units being used
  - **legends** — these are not always necessary, but they are necessary when any symbols or colours are used to illustrate some element of the graph.
  - the source of the data, indicating where (and sometimes when) the data was collected. (Note that in this topic, a source for the data is not given when the data has been created as an illustrative example.)
- Different graphical displays are used depending whether the data is categorical or numerical.

COMMUNICATING — COLLABORATIVE TASK: Features of graphs

1. Observe the graph below. Which of the features of graphs are present?
2. Using the features of the graph, identify and interpret the data displayed.
3. Write a short paragraph to summarise your findings about this data.
4. Share your answers with a partner. Discuss any differences between your answers.

Average maximum temperatures and average minimum temperatures for Sydney

Source: Australian Government, Bureau of meteorology, period 1858–2020

Using column graphs and bar graphs

- Column and bar graphs can be used to represent categorical data.
- The frequency for each category determines the length of the bar or the height of the column.
- It is easiest to graph the data from a frequency table.
- The advantage of column and bar graphs is that they are usually easy to prepare and interpret.
- The disadvantages of column and bar graphs are:
  - if they are displaying a large number of categories they become difficult to understand
  - they don't provide a visualisation of what proportion each category is of the whole.

Column graphs

- **Column graphs** should be presented on graph paper and have:
  - a title
  - labelled axes that are clearly and evenly scaled
  - columns of the same width
  - an even gap between each column
  - a space between the vertical axis and the first column.

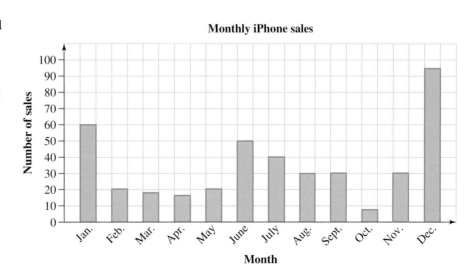

Monthly iPhone sales

Beth surveyed the students in her class to find out their preferences for the school uniform. Her results are shown in the following table. Construct a column graph to display the results.

Type of uniform	Tally	Frequency
White shirt and black skirt/trousers	ЖІ |||	8
Blue shirt and black skirt/trousers	||||	4
Blue shirt and navy skirt/trousers	ЖІ ЖІ ||	12
White shirt and navy skirt/trousers	ЖІ	5
	Total	29

THINK

1. Rule a set of axes on graph paper. Give the graph a title. Label the horizontal and vertical axes.
2. Scale the horizontal and vertical axes.
   Hint: Leave an interval at the beginning and end of the graph by placing the first column a half or one unit from the vertical axis, and making sure the horizontal axis is a half or one unit longer than needed to fit all of the columns.
3. Draw the first column so that it reaches a vertical height corresponding to 8 people. Label the section of the axis below the column as 'White shirt and black skirt/trousers'.
4. Leave a gap (measuring one column width) between the first column and the second column.
5. Repeat steps 3 and 4 for each of the remaining uniform types.

DRAW

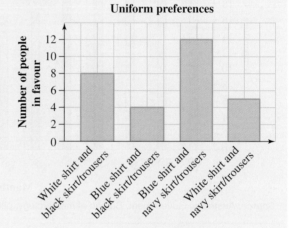

Bar graphs

- Column graphs that are drawn horizontally are known as **bar graphs**. Bar graphs can be used when category names are too long to fit at the bottom of a column.
- Bar graphs should have the same features as a column graph; that is:
  - a title
  - labelled and evenly scaled axes
  - bars of the same width
  - even gaps between bars.

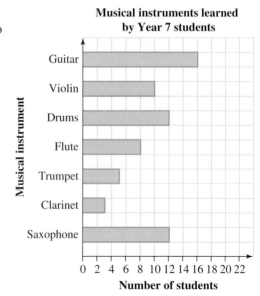

▶ 12.5.2 Dot plots

eles-4723

- **Dot plots** are usually used to represent numerical data. They use dots on a number line to show individual data values.
- If a data value appears more than once, the dots are stacked vertically.
- Dot plots give a quick overview of distribution. They show **clustering** (groups of dots) and **outliers** (values that are extremely small or large compared to the majority of values).

Number of hours of screen time each day

- The dot plot shown here displays the number of hours of screen time each day for a group of people. Each dot represents the number of hours of screen time per day for each person. Three people had 1 hour of screen time on a particular day.

WORKED EXAMPLE 9 Constructing a dot plot

Over a 2-week period, the number of packets of potato chips sold from a vending machine each day was recorded. The results are listed below.

$$10, 8, 12, 11, 12, 18, 13, 11, 12, 11, 12, 12, 13, 14$$

a. Draw a dot plot of the data.
b. Comment on the distribution.

THINK

a. 1. Use a number line to include the full range of data recorded.

2. Place a dot above the number for each value recorded.

b. Comment on interesting features of the dot plot, for example the range, clustering, extreme values or any practical information about the situation.

WRITE

a.

Number of packets of chips sold

b. The scores on this dot plot extend from 8 to 18. That gives a range of 10. On most days between 11 to 13 packets of chips were sold. There were very few days on which 8 and 18 packets of chips were sold. Making sure that there are enough chips to sell 20 packets per day should cover even the most extreme demand for chips from this machine.

---

on Resources

Interactivities Representing data graphically (int-4057)
Column and bar graphs 1 (int-4058)
Column and bar graphs 2 (int-4059)

| 12.5 Quick quiz on | 12.5 Exercise |

Individual pathways

| ■ PRACTISE | ■ CONSOLIDATE | ■ MASTER |
| 1, 4, 6, 9, 12 | 2, 5, 7, 10, 13 | 3, 8, 11, 14 |

Fluency

1. **WE8** Beth surveyed the students in her class to find out their method of travelling to school. Her results are shown in the following table. Construct a column graph to display the data.

Transport	Tally	Frequency
Car	⅟⅟⅟ ⅟⅟⅟ ⅟⅟⅟	15
Tram	⅟⅟⅟ IIII	9
Train	⅟⅟⅟ ⅟⅟⅟ ⅟⅟⅟ III	18
Bus	⅟⅟⅟ III	8
Bicycle	III	3
Total		53

2. Construct a column graph to display the data in the table shown, displaying the mean daily maximum temperatures for each month in Cairns, Queensland.

Month	Mean daily maximum temperature (°C)
January	31.8
February	31.5
March	30.7
April	29.3
May	27.6
June	25.9
July	25.6
August	26.4
September	27.9
October	29.7
November	30.8
December	31.8

3. The data in the table displays the number of students absent from school each day in a fortnight. Construct a bar graph to display the data.

Day	Number of students absent
Monday	15
Tuesday	17
Wednesday	20
Thursday	10
Friday	14
Monday	16
Tuesday	14
Wednesday	12
Thursday	5
Friday	14

4. **WE9** Over a 2-week period, the number of packets of potato chips sold from a vending machine each day was recorded as follows: 15, 17, 18, 18, 14, 16, 17, 6, 16, 18, 16, 16, 20, 18.

   a. Draw a dot plot of the data.
   b. Comment on the distribution.

5. Draw a dot plot for each of the following sets of data.

   a. 2, 0, 5, 1, 3, 3, 2, 1, 2, 3
   b. 18, 22, 20, 19, 20, 21, 19, 20, 21
   c. 5.2, 5.5, 5.0, 5.8, 5.3, 5.2, 5.6, 5.3, 6.0, 5.5, 5.6
   d. 49, 52, 60, 55, 57, 60, 52, 66, 49, 53, 61, 57, 66, 62, 64, 48, 51, 60

Understanding

6. Phone bills often include a graph showing your previous bill totals. Use the phone bills column graph shown to answer the following questions.

Phone bills April 2020–April 2021

a. State the horizontal axis label.
b. State the vertical axis label.
c. Determine how often this person receives a phone bill.
d. State in which month their phone bill was the highest.
e. Explain why it would be useful to receive a graph like this with your phone bill.
f. If the next bill was for $240.09, explain whether this would be expected.

7. An apple grower records his sales for a 12-week period.

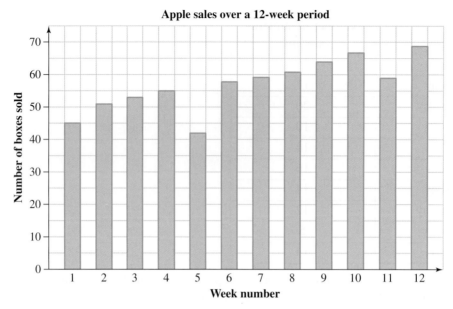

Apple sales over a 12-week period

a. Determine the number of boxes sold:

   i. in the first week
   ii. in the fifth week
   iii. in the eighth week.

b. The values for some weeks might be unusual. Determine which values these ones are.
c. Does the graph indicate that apple sales are improving? Explain your answer.

8. The students in a Year 7 physical education class were asked to sprint for 10 seconds. The teacher recorded their results on 2 different days. The graph shown displays the results collected.

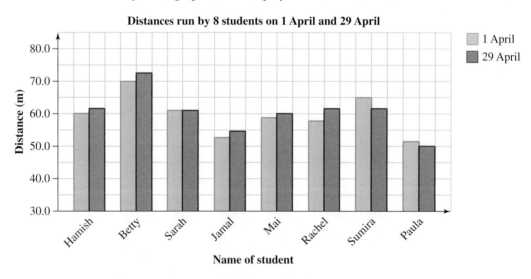

Distances run by 8 students on 1 April and 29 April

a. State who was the fastest student on each day.
b. Identify how far the fastest student ran on each day.
c. Determine who improved the most.
d. Determine which students did not improve, if any.
e. Explain whether this graph could be misleading in any way.
f. Why might the graph's vertical axis start at 30 m?

Communicating, reasoning and problem solving

9. Explain why it is important for the data points in a dot plot to be neatly aligned above the value they are representing.

10. Explain why column and bar graphs are more often used to represent categorical data rather than numerical data.

11. Explain and give an example of the effect that outliers in a set of data have on the:
    a. mean
    b. median
    c. mode
    d. range.

12. Loic surveys his class to determine which eye colour is most common and obtains the following results.
    Blue, Brown, Brown, Green, Brown, Blue, Hazel, Brown, Brown, Hazel, Brown, Brown, Blue, Brown, Brown, Brown, Hazel, Brown, Brown, Blue, Brown, Blue, Blue

    a. Represent this data as a bar graph.
    b. Determine the median of this sample.
    c. Identify the mode of this sample.
    d. If a student is selected at random from the class, calculate the probability that they don't have brown eyes.

13. Ten randomly chosen students from Class A and Class B each sit for a test in which the highest possible mark is 10. The results of the ten students from the two classes are:

Class A	1	2	3	4	5	6	7	8	9	10
Class B	1	2	2	3	3	4	4	5	9	10

    a. Display the data on a dot plot.
    b. Calculate measures of centre and spread.
    c. Explain any similarities or differences between the results of the two classes.

14. A new whitegoods store opened up in town five weeks ago. The column graph shows the number of fridges they have sold over their first five weeks.

    a. Identify the number of fridges they sold in week 2.
    b. Describe how the number of fridges sold changed from week to week.
    c. Write a formula to describe the number of fridges sold each week over the first five weeks. Let F = the number of fridges sold, and w = the week number.
    d. Using your answer to part **b** or **c**, predict the number of fridges they will sell in week 10.

Number of fridges sold per week

LESSON
12.6 Stem-and-leaf plots

LEARNING INTENTION

At the end of this lesson you should be able to:
- display data as a stem-and-leaf plot
- analyse and interpret stem-and-leaf plots.

▶ 12.6.1 Stem-and-leaf plots

eles-4724

- A **stem-and-leaf plot** may be used as an alternative to a frequency distribution table.
- Stem-and-leaf plots are used to display numerical data.
- Sometimes the term 'stem-and-leaf plot' is shortened to 'stem plot'.
- Each piece of data in a stem plot is made up of two components: a stem and a leaf.
  For example, the value 28 is made up of 20, a tens component (the stem), and 8, the units component (the leaf).
  In a stem plot, 28 would be written as:
  Key: $2|8 = 28$

Stem	Leaf
2	8

- It is important to provide a key when drawing up stem plots to make sure the data can be read and interpreted correctly.
- **Ordered stem plots** have the leaf part written in ascending order from left to right.
  Key: $6|3 = 63$

Stem	Leaf
4	3 4 7 8
5	2 4 6 9
6	1 3 5 5 5 9
7	0 0 3 6

WORKED EXAMPLE 10 Constructing an ordered stem-and-leaf plot

Prepare an ordered stem-and-leaf plot for each of the following sets of data.
a. 129, 148, 137, 125, 148, 163, 152, 158, 172, 139, 162, 121, 134
b. 1.6, 0.8, 0.7, 1.2, 1.9, 2.3, 2.8, 2.1, 1.6, 3.1, 2.9, 0.1, 4.3, 3.7, 2.6

THINK

a. 1. Rule two columns with the headings 'Stem' and 'Leaf'.

   2. Include a key to the plot that tells the reader the meaning of each entry.

   3. Make a note of the smallest and largest values of the data (that is, 121 and 172 respectively). List the stems in ascending order in the first column (that is, 12, 13, 14, 15, 16, 17). *Note:* The hundreds and tens components of the number represent the stem.

WRITE

a. Key: $12|1 = 121$

Stem	Leaf
12	9 5 1
13	7 9 4
14	8 8
15	2 8
16	3 2
17	2

---

4. Systematically work through the given data and enter the leaf (unit component) of each value in a row beside the appropriate stem. *Note:* The first row represents the interval 120–129, the second row represents the interval 130–139 and so on.

5. Redraw the stem plot so that the numbers in each row of the leaf column are in ascending order.

Key: $12|1 = 121$

Stem	Leaf
12	1 5 9
13	4 7 9
14	8 8
15	2 8
16	2 3
17	2

b. 1. Rule the stem and leaf columns and include a key.

2. Include a key to the plot that tells the reader the meaning of each entry.

3. Make a note of the smallest and largest values of the data (that is, 0.1 and 4.3 respectively). List the stems in ascending order in the first column (that is, 0, 1, 2, 3, 4). *Note:* The units components of the decimal represent the stem.

b. Key: $0|1 = 0.1$

Stem	Leaf
0	8 7 1
1	6 2 9 6
2	3 8 1 9 6
3	1 7
4	3

4. Systematically work through the given data and enter the leaf (tenth component) of each decimal in a row beside the appropriate stem. *Note:* The first row represents the interval 0.1–0.9, the second row represents the interval 1.0–1.9 and so on.

5. Redraw the stem plot so that the numbers in each row of the leaf column are in ascending order.

Key: $0|1 = 0.1$

Stem	Leaf
0	1 7 8
1	2 6 6 9
2	1 3 6 8 9
3	1 7
4	3

- The advantage of using a stem plot instead of a grouped frequency distribution table is that all the original pieces of data are retained. This means it is possible to identify the smallest and largest values. It is also possible to identify any repeated values. This makes it possible to calculate measures of centre (such as mean, median and mode) and spread (range). These identifications and calculations cannot be done when values are grouped in class intervals.
- The disadvantages of stem plots are:
  - they are not very informative for small sets of data or data that is tightly clustered together
  - they are not very manageable for data that has a large range.

Back-to-back stem-and-leaf plots

- Two sets of data can be displayed using just one stem, by placing the leaf parts for each set on either side of the stem. This is called a **back-to-back stem-and-leaf plot**.

- Back-to-back stem-and-leaf plots are used to display numerical data for two different groups. For example, test scores for two classes, Class A and Class B, can be displayed as shown here:

Key: 5|6 = 56%

Leaf (Class A)	Stem	Leaf (Class B)
4 3 2	5	6
8 5 3 4 1	6	0 0 2 4
9 6 4 4 3 1 0	7	3 5 6 7 8 8 8 9
8 8 7 5 3	8	0 1 4 6
8 6 3	9	2 3 5 6 7 8

- In a back-to-back stem-and-leaf plot, the values in the leaf are written in ascending order, working outwards from the stem.

WORKED EXAMPLE 11 Constructing a back-to-back stem-and-leaf plot

The ages of male and female groups using a ten-pin bowling centre are listed.

Males: 65, 15, 50, 15, 54, 16, 57, 16, 16, 21, 17, 28, 17, 27, 17, 22, 35, 18, 19, 22, 30, 34, 22, 31, 43, 23, 48, 23, 46, 25, 30, 21

Females: 16, 60, 16, 52, 17, 38, 38, 43, 20, 17, 45, 18, 45, 36, 21, 34, 19, 32, 29, 21, 23, 32, 23, 22, 23, 31, 25, 28

Display the data as a back-to-back stem plot and comment on the distribution.

THINK	WRITE
1. Rule three columns, headed 'Leaf (female)', 'Stem' and 'Leaf (male)'.	Key: 1|5 = 15
2. Make a note of the smallest and largest values across both sets of data (15 and 65). List the stems in ascending order in the middle column.	*see plot below*
3. Beginning with the males, work through the given data and enter the leaf (unit component) of each value in a row beside the appropriate stem.	
4. Repeat step 3 for the females' set of data.	

Leaf (female)	Stem	Leaf (male)
9 8 7 7 6 6	1	5 5 6 6 6 7 7 7 8 9
8 5 3 2 3 3 1 9 1 0	2	1 8 7 2 2 2 3 3 5 1
1 2 2 4 6 8 8	3	5 0 4 1 0
5 5 3	4	3 8 6
2	5	0 4 7
0	6	5

THINK	WRITE
5. Include a key that tells the reader the meaning of each entry.	Key: 1 | 5 = 15
6. Redraw the stem plot so that the numbers in each row of the leaf columns are in ascending order. *Note:* The smallest values are closest to the stem column and increase as they move away from the stem.	*see plot below*

Leaf (female)	Stem	Leaf (male)
9 8 7 7 6 6	1	5 5 6 6 6 7 7 7 8 9
9 8 5 3 3 3 2 1 1 0	2	1 1 2 2 2 3 3 5 7 8
8 8 6 4 2 2 1	3	0 0 1 4 5
5 5 3	4	3 6 8
2	5	0 4 7
0	6	5

THINK	WRITE
7. Comment on any interesting features.	The youngest male attending the ten-pin bowling centre is 15 and the oldest is 65. The youngest female attending the centre is 16 and the oldest is 60. Ten-pin bowling is most popular for men in their teens and 20s, and for females in their 20s and 30s.

As a class, construct a stem-and-leaf plot using data supplied by each student.

1. Write a whole number between 170 and 250 on a piece of paper.
2. As a class, decide what the stems will be for this data and come up with a key that matches this.
3. One student should volunteer to draw two columns headed 'Stem' and 'Leaf' on the left-hand side of the board and list the stems in ascending order in the first column. Remember to also include the key that you came up with in step 2.
4. Each student should then come to the board one at a time to write the leaf part of their own number in the appropriate row of the stem-and-leaf plot. Show a classmate your number and have them check that you have written your piece of data in the correct row.
5. Is the stem-and-leaf plot drawn on the board an ordered stem-and-leaf plot? How can you tell? If it is not, how can it be redrawn to make it ordered?
6. Several students should volunteer to use the answers from step 5 to draw the final version of the stem-and-leaf plot on the right-hand side of the board.
7. As a class, make some observations about the distribution of the data. Did any students have the same secret number?
8. Explain whether there is a type of graph that looks similar to a stem-and-leaf plot? Discuss the advantages and disadvantages of using this kind of graph compared with using a stem-and-leaf plot for displaying data.

On **Resources**

Interactivity   Stem-and-leaf plots (int-4060)

Exercise 12.6 Stem-and-leaf plots

learnon

| 12.6 Quick quiz **on** | 12.6 Exercise |

Individual pathways

■ PRACTISE	■ CONSOLIDATE	■ MASTER
1, 5, 6, 9, 12	2, 3, 7, 10, 13	4, 8, 11, 14

Fluency

1. **WE10** The following data gives the number of fruit that have formed on each of 40 trees in an orchard.

   29, 37, 25, 62, 73, 41, 58, 62, 73, 67, 47, 21, 33, 71,
   92, 41, 62, 54, 31, 82, 93, 28, 31, 67, 29, 53, 62, 21,
   78, 81, 51, 25, 93, 68, 72, 46, 53, 39, 28, 40

   Prepare an ordered stem plot that displays the data.

2. The number of mistakes made each week by 30 machine operators is recorded below.

12, 2, 0, 10, 8, 16, 27, 12, 6, 1, 40, 16, 25, 3, 12,
31, 19, 22, 15, 7, 17, 21, 18, 32, 33, 12, 28, 31, 32, 14

Prepare an ordered stem plot that displays the data.

3. Prepare an ordered stem plot for each of the following sets of data.

   a. 132, 117, 108, 129, 165, 172, 145, 189, 137, 116, 152, 164, 118
   b. 131, 173, 152, 146, 150, 171, 130, 124, 114
   c. 207, 205, 255, 190, 248, 248, 248, 237, 225, 239, 208, 244
   d. 748, 662, 685, 675, 645, 647, 647, 708, 736, 691, 641, 735

4. Prepare an ordered stem plot for each of the following sets of data.

   a. 1.2, 3.9, 5.8, 4.6, 4.1, 2.2, 2.8, 1.7, 5.4, 2.3, 1.9
   b. 7.7, 6.0, 9.3, 8.3, 6.5, 9.2, 7.4, 6.9, 8.8, 8.4, 7.5, 9.8
   c. 14.8, 15.2, 13.8, 13.0, 14.5, 16.2, 15.7, 14.7, 14.3, 15.6, 14.6, 13.9, 14.7, 15.1, 15.9, 13.9, 14.5
   d. 0.18, 0.51, 0.15, 0.02, 0.37, 0.44, 0.67, 0.07

5. **WE11** The number of goals scored in football matches by Mitch and Yani were recorded as follows.

Mitch	0	3	1	0	1	2	1	0	0	1
Yani	1	2	0	1	0	1	2	2	1	1

Display the data as a back-to-back stem plot and comment on the distribution.

Understanding

6. Answer the following questions for the back-to-back stem plot in question 5.

   a. Determine the number of times each player scored more than 1 goal.
   b. State who scored the greatest number of goals in a match.
   c. State who scored the greatest number of goals overall.
   d. State who is the more consistent performer.

7. The stem plot shown gives the ages of members of a theatrical group.

   a. Determine the number of people that are in the theatrical group.
   b. Identify the age of the youngest member of the group.
   c. Identify the age of the oldest member of the group.
   d. Determine the number of people who are over 30 years of age.
   e. State the most common age in the group.
   f. Determine the number of people who are over 65 years of age.

Key: 2 | 4 = 24

Stem	Leaf
1	7 8 8 9 9
2	2 4 7 9
3	1 3 3 8
4	0 2 2 2 6 6
5	5 7
6	4

8. Sprint times, in seconds, over 100 metres were recorded for a random sample of 20 runners as follows.

10.8, 11.0, 12.0, 13.2, 12.4, 13.9, 11.8, 12.8, 14.0, 15.0,
11.2, 12.6, 12.5, 12.8, 13.6, 11.5, 13.6, 10.9, 14.1, 13.9

   a. Show the data as a stem-and-leaf plot.
   b. Comment on the range of performance and other interesting points.
   c. Draw conclusions about the runners' performances.

Communicating, reasoning and problem solving

9. Explain why it is important to use a key with all stem-and-leaf plots. Give an example to illustrate your answer.

10. Explain why it is important to order and align data values when constructing a stem-and-leaf plot.

11. The scores in a Mathematics test for two classes are shown in the following table.

7F	91	77	85	82	43	84	77	79	78	92	81	80	41	88
7G	76	85	82	74	89	83	68	66	78	82	80	78	75	76

a. Display the data as a back-to-back stem-and-leaf plot.
b. Calculate the mean and median for each class.
c. Explain which of these measures of centre is the most appropriate in this situation.

12. The following stem-and-leaf plot displays the ages of the customers of a grocery store on a particular day.

a. Determine the range of their ages.
b. Determine the mode of their ages.
c. One more person in the store is added to the sample. The mean is now equal to the mode. Determine the age of this additional person. Show your working.

Key: $2\,|\,1 = 21$

Stem	Leaf
1	8
2	2 4
3	0 9
4	1 3 3 5 8
5	7
6	5
7	1 3

13. The following data shows the maximum temperature for 7 straight days in two different cities. The temperatures were recorded in the month of February.

City A	25	23	21	24	29	36	30
City B	20	17	13	13	12	12	13

a. Display the data as a back-to-back stem-and-leaf plot.
b. Calculate the range, median, mean and mode for each city.
c. One of the cities is in the northern hemisphere and the other is in the southern hemisphere. Use your answers to part b to determine which city is in the northern hemisphere and which is in the southern hemisphere.

14. Percentages in a Mathematics exam for two classes were as follows:

9A	32	65	60	54	85	73	67	65	49	96	57	68
9B	46	74	62	78	55	73	60	75	73	77	68	81

a. Construct a back-to-back stem plot of the data.
b. Determine the percentage of each group who scored above 50.
c. State which group had more scores over 80.
d. Compare the clustering for each group and comment on any extreme values.
e. Parallel dot plots are obtained by constructing individual dot plots for each set of data and positioning them on a common scale. Construct a parallel dot plot of the data (use colour).
f. Compare class performances by reference to both graphs.

LESSON
12.7 Pie charts and divided bar graphs

LEARNING INTENTION

At the end of this lesson you should be able to:
- display data as a sector graph (pie chart)
- display data as a divided bar graph
- analyse and interpret sector and divided bar graphs.

▶ 12.7.1 Sector graphs and divided bar graphs

eles-4725

- **Sector graphs** or pie charts, are made up of sectors of a circle. **Divided bar graphs** are made up of a rectangle divided into bars.
- Both of these types of graphs are used to represent categorical data — they are not suitable for numerical data.
- The sector graph and divided bar graph shown have six different sectors.

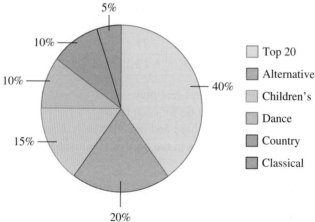

Types of music streamed on a Saturday morning

- In a sector graph the sectors should be ordered from largest to smallest in a clockwise direction (starting at 12 o'clock).
- In a divided bar graph the bars should be ordered largest to smallest from left to right.

Types of music streamed on a Saturday morning

| 40% | 20% | 15% | 10% | 10% | 5% |

☐ Top 20   ☐ Children's   ☐ Country
☐ Alternative   ☐ Dance   ☐ Classical

- Each sector or bar must be labelled appropriately either on the graph or using a legend.

Drawing sector graphs and divided bar graphs by hand

- To draw sector graphs and divided bar graphs you must determine the fraction (or percentage) of the total that is represented by each category.
- The fraction of the whole that each category represents should correspond to the fraction/percentage of the sector/divided bar graph.

- The following table shows how to draw a sector graph and a divided bar graph from data.

Category	Frequency	Fraction of total	Angle at centre of circle	Length of bar
A	3	$\dfrac{3}{12} = \dfrac{1}{4}$	$\dfrac{1}{4} \times 360° = 90°$	$\dfrac{1}{4} \times 12 = 3\,\text{cm}$
B	4	$\dfrac{4}{12} = \dfrac{1}{3}$	$\dfrac{1}{3} \times 360° = 120°$	$\dfrac{1}{3} \times 12 = 4\,\text{cm}$
C	5	$\dfrac{5}{12}$	$\dfrac{5}{12} \times 360° = 150°$	$\dfrac{5}{12} \times 12 = 5\,\text{cm}$
Total	12			

- A protractor and a ruler are required to draw these graphs by hand.

A sector graph

A divided bar graph

Drawing sector graphs and bar graphs using technology

- The quickest and easiest way to create sector graphs and divided bar graphs is to use a spreadsheet such as Microsoft Excel.
- The following worked examples provide a detailed explanation of how to create sector graphs and divided bar graphs in Microsoft Excel.

WORKED EXAMPLE 12 Constructing a sector graph

Of 120 people surveyed about where they would prefer to spend their holidays this year, 54 preferred to holiday in Australia, 41 preferred to travel overseas and 25 preferred to stay at home. Use a spreadsheet program such as Microsoft Excel to represent the data as a sector graph.

THINK

1. Input the data into a spreadsheet in order from largest category to smallest category. Put labels along the top row and numbers along the second row.

2. Select the data.

WRITE

	A	B	C	D
1	Australia	Overseas	At home	
2	54	41	25	
3				

	A	B	C	D
1	Australia	Overseas	At home	
2	54	41	25	
3				

3. Click **insert**, select the **Insert Pie or Doughnut Chart** icon, then click **Pie**.

4. The chart will appear.

Chart Title

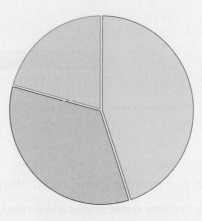

□ Australia □ Overseas ■ At home

5. Rename the chart by clicking **Chart Title** and writing a name for your chart.

Preferred holiday location

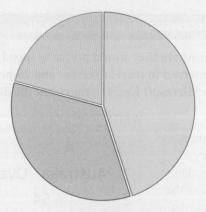

□ Australia □ Overseas ■ At home

6. Click on the **Chart Elements** in the top right corner and select **Data Labels**.

Preferred holiday location

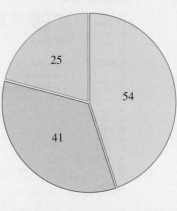

□ Australia   □ Overseas   ■ At home

WORKED EXAMPLE 13 Constructing a divided bar graph

From a community group, 90 people were surveyed about their favourite weekend activities. Of these, 43 people preferred to go to the beach, 25 preferred to watch sports and 22 preferred to go shopping. Use a spreadsheet program such as Microsoft Excel to represent the data as a divided bar graph.

THINK

1. Input the data into a spreadsheet in order from largest category to smallest category. Put labels along the top row and numbers along the second row.

2. Select the data.

3. Click **insert**, select the **Insert Column or Bar Chart** icon then click **More Column Charts**.

WRITE

	A	B	C	D
1	Beach	Sports	Shopping	
2	43	25	22	
3				

	A	B	C	D
1	Beach	Sports	Shopping	
2	43	25	22	
3				

4. From the **Bar** options, select the
   second option at the top then click
   the second graph and click **OK**.

5. The chart will appear.

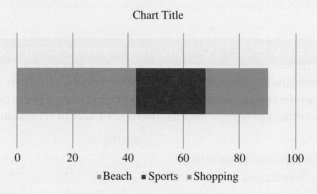

Chart Title

6. Rename the chart by clicking **Chart
   Title** and writing a name for your
   chart.

Favourite weekend activities

7. Click on the **Chart Elements** in
   the top right corner and select **Data
   Labels**.

Favourite weekend activities

Exercise 12.7 Pie charts and divided bar graphs

learn on

12.7 Quick quiz on	12.7 Exercise

Individual pathways

■ PRACTISE	■ CONSOLIDATE	■ MASTER
1, 2, 6, 9, 12, 13, 16	4, 5, 7, 10, 14, 17	3, 8, 11, 15, 18

Fluency

1. A survey was conducted of a group of students to determine their method of transport to school each day. The following sector graph displays the survey's results.

Method of transport to school

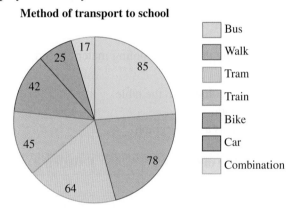

Legend:
☐ Bus
☐ Walk
☐ Tram
☐ Train
☐ Bike
☐ Car
☐ Combination

a. State the number of students surveyed.
b. State the most common method of transport to school.

2. **WE12** The following table shows the different sports played by a group of Year 7 students.

a. Complete the table.
b. Draw a sector graph to display the data.
c. Use technology to represent the data as a sector graph.

Sport	Number of students	Fraction of students	Angle at centre of circle
Basketball	55	$\dfrac{55}{180}$	$\dfrac{55}{{180}^{1}} \times \dfrac{{360}^{\circ 2}}{1} = 110°$
Netball	35	$\dfrac{35}{180}$	
Soccer	30		
Football	60		
Total	180		

3. The sector graph shown displays the candidates and the percentage of votes received in a community organisation's recent election.

   a. State who received the highest number of votes.
   b. State the percentage of votes Paula received.
   c. Can this sector graph tell us how many people voted in the election? Explain your answer.
   d. Calculate the size of the angle for the sector representing Brigid's votes. Give your answer to 2 decimal places.

Votes received

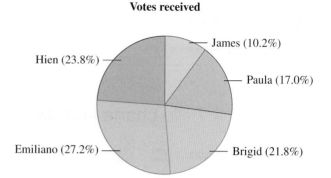

- James (10.2%)
- Hien (23.8%)
- Paula (17.0%)
- Emiliano (27.2%)
- Brigid (21.8%)

4. The sector graph shown displays the results of a student election.

   a. Identify the percentage of votes that went to Danielle.
   b. Identify the percentage of votes that went to Andrew.
   c. Calculate the percentage of votes that went to Christiana.
   d. Evaluate the total of the percentages.
   e. Using calculations, determine the size of the angle of each sectors showing the votes for each of the 3 candidates.

Election results

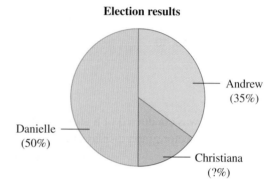

- Andrew (35%)
- Danielle (50%)
- Christiana (?%)

5. For the following table of values showing the type of roofing material used in 36 houses, answer the following questions.

   a. Complete the fraction and angle size columns in the table.
   b. Draw a sector graph to display the data.
   c. Use technology to represent the data as a sector graph.

Roofing material	Number of houses	Fraction	Angle size (°)
Concrete tiles	12		
Terracotta tiles	9		
Galvanised iron	6		
Colourbond	3		
Slate	2		
Shingles	4		
Total	36		

6. The following divided bar graph displays the types of dogs in a dog rescue centre.

Types of dogs in the rescue centre

| 18 | 12 | 6 | 6 | 3 |

- ☐ Terrier
- ☐ Rottweiler
- ☐ Labrador
- ☐ Schnauzer
- ☐ German shepherd

   a. Based on this graph, calculate the total number of dogs in the rescue centre.
   b. Identify the number of labradors that are in the centre.
   c. Identify the number of rottweilers that are in the centre.
   d. Calculate the percentage of the total number of dogs that are terriers.
   e. State which type of dog there are the fewest of.

7. **WE13** The following table of values shows the number of kilograms of meat of various types that a butcher sold in a day.

   a. Copy and complete the table of values.
   b. Construct a divided bar graph to display the data.
   c. Use technology to represent the data as a divided bar graph.

Type of meat	Amount sold (kg)	Fraction
Lamb	10	
Beef	45	
Pork	5	
Chicken	15	
Turkey	10	
Rabbit	5	
Total	90	

8. The students at Mount Birdie Secondary College counted the number of birds that landed on their school oval on one day. The results are in the following table.

Type of bird	Number of birds	Fraction
Magpie	24	
Crow	12	
Cockatoo	12	
Wattlebird	6	
Rosella	6	
Total	60	

   a. Copy and complete this table.
   b. Use technology to represent the data as:
      i. a sector graph
      ii. a divided bar graph.

Understanding

9. The following table shows the different forms of transport that 100 students use to get to school.

   a. Copy the table and complete the *Fraction* column (the first one has been done for you).
   b. Using a total bar length of 10 cm, complete the *Length of bar* column (the first one has been done for you).
   c. Represent the data as a divided bar graph.
   d. If the total bar length was 5 cm (instead of 10 cm) calculate how long the *Walk* section would be.

Transport	Number of students	Fraction	Length of bar
Walk	50	$\dfrac{50}{100} = \dfrac{1}{2}$	$\dfrac{1}{2} \times 10 = 5\,\text{cm}$
Car	20		
Bus	15		
Bike	15		
Total	100		

10. The sector graph shown represents the favourite colours of 200 students. The angle of each section is shown.

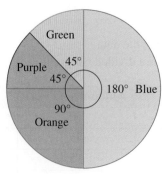

   a. Calculate the number of students who said their favourite colour was blue.
   b. Calculate the number of students who said their favourite colour was orange.
   c. Calculate the number of students who said their favourite colour was purple.

11. The sector graph shown represents the furniture sold at Devin's Furniture Land over the period of one week. The total number of sales was 90 items. Calculate the number of pieces of each furniture item that were sold. Use a protractor to measure the angles in the sector graph.

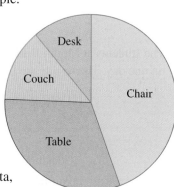

12. State which type of data is represented by the following graphs: column and bar graphs, dot plots, stem-and-leaf plots and sector graphs.

Communicating, reasoning and problem solving

13. Discuss the advantage of using a divided bar graph to display categorical data, rather than a column graph.

14. The following two sector graphs represent the sales at Charlie's Kiosk. One sector graph is from summer and one sector graph is from winter.

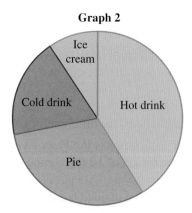

   a. State which graph represents the sales from summer.
   b. Explain your answer to part a.

15. For the table of values showing the sports played by a group of students, answer the following questions.

   a. Complete the angle size column before drawing a sector graph to display the data.
   b. In this data we are not told how many students were surveyed.

      i. Explain why the data could not represent ten students.
      ii. Explain why the data could represent twenty students.

Sport	Percentage of total	Angle size (°)
Cricket	20	
Hockey	25	
Netball	40	
Soccer	10	
Tennis	5	
Total	100	

16. Omar is working on his budget. His income is $2000 per month and the following divided bar graph shows how he is currently spending his money. Omar needs your help to increase the amount he saves.

□ Rent ■ Utilities □ Food ■ Entertainment ■ Savings

a. Omar does not want to move house and he cannot change his utility bills. Determine the sections of the bar graph that must stay the same.
b. Determine how much Omar is currently spending on food and entertainment in total.
c. If Omar halves the amount he spends on food and entertainment and puts that money into savings, determine how much he would:
   i. spend on food
   ii. spend on entertainment
   iii. put into savings in total each month.

d. Use technology to draw a new divided bar graph showing the changes to Omar's budget you made in part c.

17. Beach Brellas Inc. is designing a new beach umbrella using the design shown.

a. Determine the percentage of the umbrella that is blue.
b. Determine the angle for one section of the umbrella.
c. If it takes 1.5 m² of fabric to make the whole design, determine how much of each colour fabric is needed to make the design. Give your answer to 3 decimal places and include units.

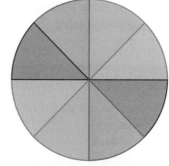

18. The sector graph shown has four sectors representing different amounts. Starting at the smallest sector, each new sector is double the size of the previous one. Without using a protractor, calculate the third sector as a percentage. Show your working.

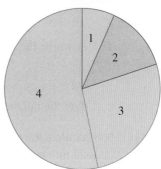

LESSON
12.8 Comparing data

LEARNING INTENTION

At the end of this lesson you should be able to:
- calculate measures of centre and spread from a graphical display
- identify the most appropriate measure of centre for a graphical display.

▶ 12.8.1 Determining measures of centre and spread from graphs
eles-4726
- We have considered the calculation of measures of centre and spread from lists of data.
- We also need to know how to calculate these measures from graphs of individual pieces of data.
- We can then make comparisons between data presented in listed form and data presented in graphical form.
- When data is displayed graphically, the spread may be obvious, but we often need to calculate the data's measures of centre so that we can understand it better.
- We also need to be able to determine which measure of centre best represents the data.

Determining the most appropriate measure of centre

- If a set of data contains outliers, the median is the most appropriate measure of centre because outliers do not affect it, whereas outliers can significantly affect the value of the mean.
- If a set of data does not contain outliers, the mean or median could be used as the measure of centre. Since the concept of the mean (or average) is more familiar to most people, it is generally used in preference to the median.
- The mode is not normally considered a good measure of centre because it tells us only about the most common score.

WORKED EXAMPLE 14 Determining measures of centre and spread from a dot plot

a. Use the dot plot shown to determine the following statistical measures.
    i. Mean   **ii. Median**   **iii. Mode**   **iv. Range**
b. Comment on the most suitable measure of centre for the data.

THINK	WRITE
a. **i.** **1.** Determine the sum of the values.	**a.** **i.** Sum of values $= 16 + 3 \times 18 + 4 \times 19$ $+ 2 \times 20 + 21 + 24$ $= 231$
2. Count the number of values.	There are 12 values.
3. Calculate the mean by dividing the sum by the total number of values.	$\text{Mean} = \dfrac{\text{sum of data values}}{\text{total number of data values}}$ $= \dfrac{231}{12}$ $= 19.25$

ii. 1. The values are already in order. The median is the middle value. There are 12 values, so the middle one is the average of the 6th and 7th values.

ii. The 6th and 7th values are both 19.

2. Calculate the average of these.

The median value is 19.

iii. The mode is the most common value. Look for the one that occurs most frequently.

iii. The mode is 19.

iv. The range is the difference between the largest value and the smallest value.

iv. Range $= 24 - 16$
$= 8$

b. The scores 16 and 24 are possible outliers. When a set of data contains outliers the best measure of centre is the median, as it is not affected by outliers as much as the mean.

b. Since there may be outliers in the data, the median is the most suitable measure of centre in this case.

WORKED EXAMPLE 15 Determining measures of centre and spread from a stem plot

a. Use the stem plot to determine the following measures of the data.
 i. Mean
 ii. Median
 iii. Mode
 iv. Range
b. Comment on the most suitable measure of centre for the data.

Key: $1|8 = 18$

Stem	Leaf
1	8 9
2	2 2 5 7 7 8
3	0 1 4 6 7
4	0 5

THINK

WRITE

a. i. 1. Determine the sum of the values.

a. i. Sum of values $= 18 + 19 + 22 + 22 + 25$
$+ 27 + 27 + 28 + 30 + 31$
$+ 34 + 36 + 37 + 40 + 45$
$= 441$

2. Count the number of values.

There are 15 values.

3. Calculate the mean by dividing the sum by the total number of values.

$$\text{Mean} = \frac{\text{sum of data values}}{\text{total number of data values}}$$
$$= \frac{441}{15}$$
$$= 29.4$$

ii. The values are already in order. The median is the middle value. There are 15 values, so the middle one is the 8th value. Locate this.

ii. The middle position of the 15 values is the 8th value. This is 28. The median value is 28.

iii. The mode is the most common value. Look for the one that occurs most frequently.

iii. There are two modes (the data set is bimodal): 22 and 27.

iv. The range is the difference between the largest value and the smallest value.

iv. Range $= 45 - 18$
$= 27$

b. Look at the measures of mean, median and mode to see which best represents the values in terms of their closeness to the centre.

b. Because there do not appear to be any outliers, the mean or median could be used to represent the centre of this set of data.

 Resources

Interactivity Determining measures of centre and spread from graphs (int-4062)

Exercise 12.8 Comparing data   learn on

12.8 Quick quiz on	12.8 Exercise

Individual pathways

■ PRACTISE	■ CONSOLIDATE	■ MASTER
1, 4, 6, 7, 13, 16	2, 5, 8, 9, 10, 14, 17	3, 11, 12, 15, 18

Fluency

1. **WE14** Consider the dot plot shown.

   a. Use the dot plot to determine the following measures of the data.

   i. mean   ii. median   iii. mode   iv. range

   b. Comment on the most suitable measure of centre for the data.

   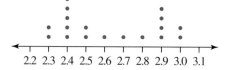

2. Consider the dot plot shown.

   a. Use the dot plot to determine the following measures of the data.

   i. mean   ii. median   iii. mode   iv. range

   b. Comment on the most suitable measure of centre for the data.

3. Consider the dot plot shown.

   a. Use the dot plot to determine the following measures of the data.

   i. mean   ii. median   iii. mode   iv. range

   b. Disregard the score of 10 and recalculate each of these values.

   c. Discuss the differences/similarities in your two sets of results.

4. **WE15** Consider the stem-and-leaf plot shown.

   a. Use the stem plot to determine the following measures of the data.
   i. mean
   ii. median
   iii. mode
   iv. range

   b. Comment on the most suitable measure of centre for the data.

   Key: 6.1|8 = 6.18

Stem	Leaf
6.1	8 8 9
6.2	0 5 6 8
6.3	0 1 2 4 4

5. Consider the stem-and-leaf plot shown.

a. Use the stem plot to determine the following measures of the data.
   i. mean
   ii. median
   iii. mode
   iv. range
b. Disregard the score of 44 and recalculate each of these values.
c. Discuss the differences/similarities in your two sets of results.

Key: 1|10 = 10

Stem	Leaf
1	0 2
2	1 3 3 5
3	
4	4

Understanding

6. **MC** Select the score that shows up most often.

A. median
B. mean
C. mode
D. average

7. **MC** Select the best term for *average* in everyday use.

A. the mean
B. the mode
C. the median
D. the total

8. **MC** Choose the measure affected by outliers (extreme values).

A. the middle
B. the mode
C. the median
D. the mean

The back-to-back stem plot shown displays the heights of a group of Year 7 students. Use this plot to answer questions **9** and **10**.

Key: 13 | 7 = 137cm

Leaf (boys)	Stem	Leaf (girls)
98	13	7 8
9 8 8 7 6	14	3 5 6
9 8 8	15	1 2 3 7
7 6 6 5	16	3 5 6
8 7 6	17	1

9. **MC** Select the total number of Year 7 students.

A. 13
B. 17
C. 30
D. 36

10. **MC** Choose the tallest male and shortest female heights.

A. 186 cm and 137 cm
B. 171 cm and 148 cm
C. 137 cm and 188 cm
D. 178 cm and 137 cm

11. A survey of the number of people in each house in a street produced the following data:

$$2, 5, 1, 6, 2, 3, 2, 1, 4, 3, 4, 3, 1, 2, 2, 0, 2, 4$$

a. Prepare a frequency distribution table with an $f \times x$ column and use it to calculate the average (mean) number of people per household.

b. Draw a dot plot of the data and use it to find the median number per household.

c. Calculate the modal number per household.

d. State which measure of centre (mean, median or mode) would be most useful to:

   i. real estate agents renting out houses

   ii. a government population survey

   iii. an ice-cream truck owner.

12. The mean of 12 scores is 6.3. Calculate the total of the scores.

Communicating, reasoning and problem solving

13. A class of 26 students had a median mark of 54 in Mathematics, even though no-one actually scored this result.

a. Explain how this is possible.

b. Explain how many students must have scored below 54.

14. A clothing store records the dress sizes sold during one day in order to work out the most popular sizes. The results for a particular day are given below.

$$12, 14, 10, 12, 8, 12, 16, 10, 8, 12, 10, 12, 18, 10, 12, 14$$
$$16, 10, 12, 12, 12, 14, 18, 10, 14, 12, 12, 14, 14, 10$$

Rebecca is in charge of marketing and sales. She uses these figures when ordering future stock. From these figures she decides on the following ordering strategy. She will order:
- the same number of size 8, 16 and 18 dresses
- three times this number of size 10 and size 14 dresses
- five times as many size 12 dresses as size 8, 16 and 18.

Comment on Rebecca's strategy.

15. A small business pays these salaries listed below (in thousands of dollars) to its employees.

$$18, 18, 18, 18, 26, 26, 26, 35$$

The business owner's salary is 80 thousand dollars.

a. Determine the wage earned by most workers. Make sure to include the business owner's salary in this calculation.

b. Calculate the average wage. Show your working.

c. Calculate the median of the distribution.

d. Explain which measure of centre might be used in salary negotiations by:

   i. the union that represents the business's employees

   ii. the business owner.

16. Five scores have an average of 8.2. Four of those scores are 10, 9, 8 and 7. Determine the fifth score.

17. A tyre manufacturer selects 48 tyres at random from their production line for testing. The total distance travelled during the safe life of each tyre is shown in the following table.

Distance in km ('000)	46	50	52	56	78	82
Number of tyres	4	12	16	10	4	2

a. Calculate the mean, median and mode.
b. Explain which measure best describes the average safe life of these tyres.
c. Recalculate the mean with the 6 longest-lasting tyres removed. Determine how much it is lowered by.
d. If you selected a tyre at random, determine how long it would most likely last.
e. In a production run of 10 000 tyres, determine how many could be expected to last for a maximum of 50 000 km.
f. If you were the tyre manufacturer, explain what kind of distance guarantee you would be prepared to offer on your tyres.

18. A soccer team had averaged 2.6 goals per match after 5 matches. After their 6th match the average dropped to 2.5 goals per match. Determine the number of goals they scored in their 6th match. Show your working.

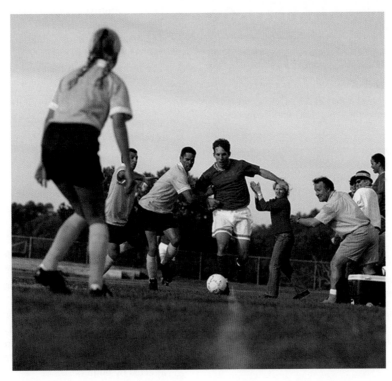

LESSON
12.9 Review

12.9.1 Topic summary

Types of data

- There are two main types of data: *categorical* data and *numerical* data.

Categorical data	Numerical data
Data that can be placed into a group or category • Nominal categorical data cannot be ordered in a meaningful way e.g. Eye colour, gender • Ordinal categorical data can be ordered/ ranked in a meaningful way e.g. Shoe size, rating scale (high, medium, low)	Data that can be counted or measured • **Discrete** numerical data can be counted. e.g. The number of people in a room • **Continuous** numerical data can be measured. e.g. A person's height

DATA CLASSIFICATION AND VISUALISATION

Frequency distribution tables

- Frequency is the number of times a result or piece of data occurs.
- A frequency table shows the score, tally and frequency in a tabular form.

e.g.

Age group (score)	Tally	Frequency
11–20	IIII	4
21–30	II	2
31–40	I	1
	Total	7

Representing data graphically

- There are many types of graphical displays to represent data.
- For *categorical* data the most appropriate graphical displays are:
  - frequency table
  - column or bar graph
  - sector graph (pie chart)
  - divided bar graph.
- For *numerical* data the most appropriate graphical displays are:
  - frequency table
  - dot plot
  - stem-and-leaf plot.

Measures of centre

- A measure of centre gives us some idea about the location of the centre of the data.
- The three measures of centre are *mean*, *median* and *mode*.
- The **mean** is commonly referred to as the *average*. The symbol used to represent the mean is $\bar{x}$. The formula used to calculate the mean is:

$$\bar{x} = \frac{\text{sum of data values}}{\text{total number of data values}}.$$

- The **median** represents the middle score when the data is ordered from smallest to largest.

e.g. 1, 1, 3, 4, **4**, 10, 12, 25, 41
Median = 4

- The **mode** is the most commonly occurring value in a data set. Data sets might have one mode, no mode or more than one mode.

Comparing data

- By calculating measures of centre, we can make comparisons between data sets.
  e.g. Mean battery life is higher for brand A batteries than for brand B so, on average, brand A batteries last longer.
- When extreme values (outliers) are present in a data set, the best measure of centre is the median as it is least affected by outliers.

Measures of spread

- A measure of spread tells us how spread out the data is.
- The **range** is one measure of spread; it is the difference between the largest and smallest values in that set.

Range = largest value − smallest value

12.9.2 Project

Families with children

Surveys help us to find out about lots of different things. They can be conducted by collecting information from an entire population or smaller groups. The information collected in a survey, known as data, can be examined and presented in many different ways. Graphs are commonly used to present the results of a survey.

The graph shown displays the percentage of families in Australia with children under the age of 15. This data is taken from the 2011 census.

Family composition

1. State the type of graph that has been used to display the information.
2. State the most common category of Australian families with children under 15.
3. Give an example of another type of graph that could be used to present this information.

Conduct your own survey of the number of children under 15 in your classmates' families. Compare your results with the results of the 2011 census.

4. Record your survey results in the following frequency distribution table.

Number of children	Tally	Frequency	Percentage
1			
2			
3			
4 or more			
	Total		

5. Is the survey you conducted an example of a *census* (contains the entire population) or a *sample* (contains only a portion of the entire population)? Explain your answer.
6. Explain whether the data you collected is classified as categorical data or numerical data.
7. Present the information from your survey as a column graph. Use the percentage values on the vertical axis and number of children on the horizontal axis.
8. Explain how the results of your class compare with the results obtained in the 2011 census. Discuss the major difference between these two sets of results.
9. Design and conduct a new survey on a topic of interest (choose the topic as a class). You can survey members of your class on this topic or expand it to include a larger target audience. Present your data as a poster that displays the findings of your survey using an appropriate style of graph.

 Resources

Interactivities Crossword (int-2608)
Sudoku puzzle (int-3174)

Fluency

1. Identify whether the following statement is true or false.
   The classification of data describing the number of iPhones sold during the year is continuous.

2. State whether the data that describes a person's mass (in kilograms) is categorical, discrete or continuous.

3. State whether the data that describes sandwich types at a takeaway outlet is categorical, discrete or continuous.

4. Eighty students in a school of 800 participated in a survey to find their favourite drinks. The results were: Coke 20, Pepsi 14, other soft drink 10, milk/milk-based 16, coffee 9, tea 4 and fruit juice 7.
   a. Identify the most popular beverage.   b. Identify the least popular beverage.

5. Use the frequency distribution table shown to answer the following questions.

Score	Frequency
13	2
14	9
15	3
16	5
17	6
18	1

   a. Calculate the number of students in the class.
   b. State the most frequent score.
   c. State the least frequent score.
   d. Determine the number of students who scored 15 or less.
   e. Determine the number of students who scored at least 16.

6. A random sample of 24 families was surveyed to determine the number of vehicles in each household. The following list is the raw data.

   2, 0, 3, 2, 1, 0, 2, 3, 4, 2, 2, 1, 0, 1, 3, 2, 1, 0, 0, 0, 2, 2, 3, 3

   a. Organise the data into a frequency distribution table.
   b. Identify the number of families who have no vehicles in their household.
   c. Determine the number of families who have two or more vehicles in their household.
   d. Identify the score that has the highest frequency.
   e. State the highest score.
   f. Calculate the fraction of families who had two vehicles in their household.

7. For the following sets of data, determine the following.

   i. the mean   **ii.** the mode   **iii.** the median

   a. 4, 4, 4, 3, 2, 2, 3   **b.** 2, 4, 2, 4, 3, 7, 4   **c.** 4, 5, 6, 6, 5, 5, 3

   d. 1, 2, 1, 4, 1, 5, 1   **e.** 5, 8, 1, 7, 7, 5, 2, 7, 5, 5   **f.** 1, 5, 8, 7, 4, 8, 5, 6, 8

Understanding

8. The graph shown displays the test marks of a group of students.

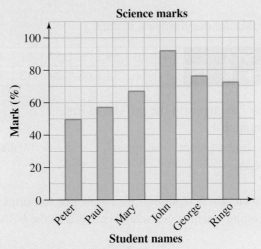

   a. State the label on the horizontal axis.

   b. State the scale used on the vertical axis.

   c. State the name of the student with the highest mark.

   d. Determine which student(s) would have failed if the pass mark was 50%.

   e. Identify Paul's mark.

   f. Identify George's mark.

9. Look at the following graph, showing air quality.

   a. State what the horizontal axis represents.

   b. State what the vertical axis represents.

   c. Identify the region with a rating that ranged from very good to good every day.

   d. Determine the day when the Western region had a fair to poor air quality rating.

   e. Determine the region with a rating that ranged from good to fair every day.

10. a. The following data gives the speed of 30 cars recorded by a roadside speed camera along a stretch of road in a rural area where the speed limit is 100 km/h.

75, 90, 83, 92, 103, 96, 110, 92, 102, 93, 78, 94, 104, 85, 88,
82, 81, 115, 94, 84, 87, 86, 96, 71, 91, 91, 92, 104, 88, 97

Present the data as an ordered stem-and-leaf plot.
b. Comment on the data recorded by the roadside speed camera.

11. Complete the specified tasks for the given sets of data.

Set A	64	30	59	1	57	13	45	28
Set B	3	41	38	68	29	32	30	31

a. Construct a back-to-back stem plot of the two sets of data.
b. For each data set, determine the:
   i. lowest score   **ii.** highest score   **iii.** mean   **iv.** median   **v.** range.
c. Compare the two data sets.

12. Consider the dot plot shown, which represents the number of soft drinks sold at lunchtime over a 20-day period from a vending machine in a school cafeteria in the USA.

a. Use the dot plot to determine the following.
   i. mean   **ii.** median   **iii.** mode   **iv.** range
b. Comment on the distribution of the data.

13. Consider the stem plot shown, which represents the ages of participants in an aerobics class at a gym.

Key: $1|7 = 17$

Stem	Leaf
1	4 7 8 8 9
2	0 1 3 5 6 7 9
3	2 4 5 6 7 8
4	2 5
5	3

a. Use the graph to determine the:
   i. mean age
   ii. median age
   iii. modal age
   iv. range of ages.
b. Comment on the distribution of the data.

14. A group of 150 people were surveyed about what they had for breakfast that morning. Of the people surveyed, 62 had cereal, 70 had toast, 12 had a cooked breakfast and 6 had no breakfast. Use technology to represent the data as a sector graph.

15. Use technology to draw a divided bar chart to display the following data.

Hobby	Percentage of total
Stamp collecting	20
Photography	15
Sailboarding	20
Skateboarding	30
Model making	15
Total	100

Communicating, reasoning and problem solving

16. This table shows the maximum and minimum daily temperatures in a city over a one-week period.

Day	1	2	3	4	5	6	7
Maximum (°C)	12	13	10	11	9	10	8
Minimum (°C)	3	3	2	1	0	4	2

Use the table to answer the following questions.
a. State the maximum temperature on day 3.
b. State the day that had the lowest minimum temperature.
c. Determine the coldest day.
d. Identify the day that had the warmest overnight temperature.
e. Calculate the temperature range (variation) on day 2.
f. Determine the day that had the smallest range of temperatures.

The table shown is part of a teacher's mark book. Use it to answer questions **17** and **18**.

Name	Test 1	Test 2
Oisìn	85	94
Alfonso	85	63
Mark	95	58
James	82	67
David	76	95
Rachel	62	85
Mary	87	75
Bahar	94	63
Esther	68	68

17. a. Calculate the class mean for test 1.
b. Calculate the class mean for test 2.
c. Identify the test in which the class received better marks.
d. Determine the number of students who scored above the mean in test 1.
e. Determine the number of students who scored above the mean in test 2.

18. **a.** Determine the mode score for test 1.
    b. Determine the mode score for test 2.
    c. Determine the median score for test 1.
    d. Determine the median score for test 2.
    e. Calculate the mean score for the two tests for each student and list them.
    f. Determine the range of values for these means.

19. The results of height measurements (in cm) for the students in class 9A are given below.

    145, 152, 148, 152, 163, 148, 165, 158, 159, 162, 145, 153, 156, 158, 157,
    159, 169, 156, 156, 156, 152, 154, 128, 141, 154, 153, 156, 156, 165, 168

    a. Calculate the range of heights.
    b. Calculate the mean height of the class.
    c. Determine the mode for the heights.
    d. Determine the median height of the class.
    e. Explain whether these three values (mean, mode, median) are similar.

20. Use the table of values shown to answer the following questions.

Month	Number of flies (thousands)
January	24
February	28
March	26
April	20
May	18
June	15
July	12
August	10
September	11
October	12
November	15
December	19

a. State what would be on the horizontal axis in drawing a column graph using this data.
b. State what would be on the vertical axis in drawing a column graph.
c. Identify the minimum value required on the vertical axis.
d. Identify the maximum value required on the vertical axis.
e. Determine the scale that should be used for the vertical axis.
f. Draw a column graph to display the data.

21. The number of pets cared for by each of 20 families was surveyed, giving the following data.

    1, 2, 3, 2, 0, 1, 0, 2, 5, 3, 2, 1, 2, 0, 2, 0, 1, 3, 2, 1

   a. Organise the data into a frequency distribution table.
   b. Draw a dot plot of the data and comment on the distribution.
   c. Calculate the mode, median and mean.

22. a. For the given set of scores 9, 3, 8, 5, 6, 9, 4, 5, 5, 25, determine the following.
       i. mode
      ii. median
     iii. mean
      iv. outliers (if any)
       v. mean when outliers are omitted
      vi. mean and median when 10 is added to *each* of the given scores
   b. Comment on any discoveries from part a.

23. This back-to-back stem-and-leaf plot shows the heights of a group of boys and girls.

Key: 13 | 7 = 137 cm

Leaf (boys)	Stem	Leaf (girls)
98	13	78
98876	14	356
988	15	1237
7665	16	356
876	17	1

Give a full description of the heights of the boys and the girls.

To test your understanding and knowledge of this topic, go to your learnON title at www.jacplus.com.au and complete the **post-test**.

Answers

Topic 12 Data classification and visualisation

12.1 Pre-test

1. B
2. A
3. 3
4. B
5. 31
6. 10
7. 19 m
8. 9
9. 2.55
10. 20%
11. 17
12. 41 cm
13. a. 150   b. 45
    c. 8%   d. 50%
14. 111
15. 24%

12.2 Classifying data

1. a. iv   b. i
   c. iii   d. ii
2. a. Numerical (discrete)
   b. Categorical
   c. Numerical (continuous)
   d. Numerical (continuous)
   e. Categorical
   f. Numerical (continuous)
3. a. Discrete   b. Continuous   c. Discrete
   d. Continuous   e. Discrete
4. a. Categorical
   b. Categorical
   c. Numerical (continuous)
   d. Numerical (discrete)
   e. Categorical
   f. Categorical
5. True
6. True
7. C
8. a. Discrete data deals with values that are exact and must be counted, for example the number of people at a football match. Continuous data deals with values that are measured and may be written in decimal form, for example the length of each football quarter.
   b. Primary data is data collected by me, for example a survey I conduct, and secondary data is data collected by someone else, for example AFL/NRL data.
9. In a Likert scale the data is split into categories, so even though these categories can be represented by numbers, the data is still categorical.

10. Although represented by digits, these types of data are categorical because they do not represent numerical ordered values.
11. Possible answers:
    a. How many minutes, on average, do you read per day?
    b. How many times a week do you play sport?
    c. How many times a week do you play computer games?
12. Possible answers:
    a. Do you do Maths homework every night?
    b. What books have you read this year?
    c. What suburb do you live in?
13. 1250

12.3 Displaying data in tables

1. a.

Score	Tally	Frequency
0	$\cancel{IIII}$	5
1	IIII	4
2	$\cancel{IIII}$ IIII	9
3	$\cancel{IIII}$ II	7
4	II	2
5	I	1
6	I	1
7		0
8	I	1
	Total	30

b. 30

c. 5

d. 12

e. 2

f. $\dfrac{2}{30} = \dfrac{1}{15}$

2. a.

Value (thousand dollars)	Tally	Frequency
100–109	III	3
110–119	I	1
120–129		0
130–139	I	1
140–149	III	3
150–159	III	3
160–169	II	2
170–179	I	1
180–189	II	2
190–199	I	1
200–209	I	1
	Total	18

b. 18

c. $100 000 - $109 999, $140 000 - $149 999 and $150 000 - $159 999

d. $120 000 - $129 999

e. 3

3. a. 33   b. 3   c. 10

   d. 23   e. $\frac{1}{3}$

4. a.

Score	Tally	Frequency
5	|	1
6	||	2
7	||	2
8	|	1
9	|||	3
10	|	1
	Total	10

b.

Score	Tally	Frequency
0	卌	5
1	卌	5
2	卌	5
3	||	2
4	||	2
5		0
6	|	1
	Total	20

c.

Score	Tally	Frequency
45	|	1
46	|	1
47		0
48	||	2
49	|	1
50	||	2
51	||	2
52	||	2
53	||||	4
54	|||	3
55	|	1
56	||	2
57	||	2
58	||	2
59	|	1
60	||	2
	Total	28

d.

Score	Tally	Frequency
143	||	2
144	卌|	6
145	卌	5
146	|	1
147	|||	3
148	|||	3
149	|||	3
150	||||	4
151		0
152	|	1
	Total	28

5. a. The list is messy. It is difficult to see how many different sports there are or to tell how many people prefer a particular sport.

b.

Score	Tally	Frequency
Hockey	卌|||	8
Cricket	卌|	6
Tennis	卌||	7
Netball	卌||||	9
Swimming	卌||	7
Golf	||||	4
Scuba diving	|||	3
Lacrosse	||||	4
Lawn bowls	||	2
	Total	50

c. i. Netball

   ii. Lawn bowls

d. Yes, Tennis and swimming had a frequency of 7, and golf and lacrosse had a frequency of 4.

6. a.

Score	Tally	Frequency
0	卌 卌 |	11
1	卌 卌	10
2	卌 |	6
3	||	2
4	|	1
	Total	30

b. 11

c. 9

d. 0

e. 4

f. $\dfrac{1}{5}$

7.

Height	Tally	Frequency
140–144	||	2
145–149	ЖЖ |	6
150–154	||||	4
155–159	|	1
160–164	ЖЖ	5
165–169	ЖЖ |	6
170–174	||	2
175–179	|	1
180–184	|	1
	Total	28

8. a.

Type	Tally	Frequency
1-bedroom unit	|||	3
2-bedroom unit	||||	4
2-bedroom house	|	1
3-bedroom house	ЖЖ |	6
4-bedroom house	||	2
	Total	16

A 3-bedroom house was the most popular.

b. C

c. A

9. B

10. Because it makes it easier to collect and read the data.

11. When selecting a class interval we need to select ranges that are narrow enough to highlight variations in the data, and that are also wide enough to ensure the amount of data we are looking at is manageable.

12. Class intervals can be used to group continuous numerical data so that it can be represented in tables and other graphical forms.

13. a. 25

b.

Number of pets	Frequency	% Frequency
0	8	32
1	11	44
2	3	12
3	2	8
4	0	0
5	1	4

c. 6

14. O: 21.5%, U: 8.3%

15.

Score	Frequency
0	4
1	16
2	20
3	4
4	1
Total	45

12.4 Measures of centre and spread

1. a. 5   b. 6   c. 4
   d. 6   e. 4

2. a. 3   b. 12   c. 14
   d. 15   e. 13

3. a. 3   b. 3   c. 4
   d. 5   e. 2   f. 3

4. a. 5   b. 3   c. 3   d. 7

5. a. 11   b. 4   c. 2.5   d. 1.5

6. a. 2   b. 5   c. 7   d. 2.5

7. a. 5   b. 6   c. 7.5   d. 7.5

8. a. 4   b. 8   c. 2
   d. 4   e. 2

9. a. 3   b. 10 and 12   c. 4
   d. 9   e. no mode

10. a. 32   b. 24.1

11. B

12. B

13. D

14. D

15. C

16. 85%

17. 38

18. 104 shoes

19. 47 students

20. a. The average number of goals scored in each match, calculated by summing the total number of goals scored and dividing by the number of matches played.

   b. The average house price, calculated by listing all of the house prices and finding the value that has the same number of houses being more expensive and less expensive than it.

   c. The most popular shoe size.

21. By calculating measures of centre and spread we can see how data is grouped around a central point. This allows us to see any patterns in the data and interpret those patterns.

22. a. 101.9

   b. Both the mode (115) and median (112) give a better impression of how the team has performed, even though they could give a 'misleading' impression of the team's performance.

23. a. Mode = 12

   b. The mode shows which size to order more of. The mean and median would not show the more common sizes and would give only an indication of the middle of the range.

24. We cannot calculate the mean, median and range of categorical data, because these values are all calculated using the numerical values of the variables. We can calculate the mode of categorical data, which is the most popular or frequency category.

25. a. Tom: $1.30, Sarah: $1.27

b. Sarah's sells cheaper petrol.

c. Tom may have seen the price on a very expensive day. There's a greater range of values at Sarah's petrol station.

26. a. 1.75 m

b. 2.21 m

c. Based on her past performances and her past range of values, she cannot jump this high.

27. 72

28. 89.5%

29. The coach's statement was not correct.

30. One possible answer is 5, 6, 12, 13, 14. If the 5 numbers are in ascending order, the 3rd number must be 12 and the other 4 numbers must add up to 38.

31. The highest score is 49 and the lowest score is 0. The scores would be 0, 1, 12, 13, 49

32. a. 15

b. 3

c. 0

d. The median is the best measure because the mode is the lowest of all scores and the mean is inflated by one much larger figure.

33. 133 cm, 133 cm, 142 cm, 151 cm

34. Largest value for the median is 15. Scores would be 0, 2, 15, 16, 17 or 0, 1, 15, 16, 18. Smallest value for the median is 2. Scores would be 0, 1, 2, a, b, where $a + b = 47$

35. 43; highest score: 100; lowest score: 57. The scores would be 57, 77, 78, 79, 80, 97, 98, 99, 100

12.5 Column graphs and dot plots

1.

Method of transport used to go to school

2.

Mean daily maximum temperatures, Cairns

3.

Students absent

4. a.

b. The scores lie between and include 6 to 20; that is, a range of fourteen.
Mostly 16 to 18 packets were sold. Sales of 6 and 20 packets of chips were extremely low. A provision of 20 packets of chips each day should cover the most extreme demands.

5. a.

b.

c.

d.

6. a. Months

b. Bill total

c. Every 3 months (quarterly)

d. January 2021

e. A graph like this is useful to monitor the spending pattern over time and to decide whether there have been any unusual increases.

f. It would not be normal, but there might be reasons to explain why it was so much higher, for example making a lot of overseas phone calls.

7. a. i. 45

ii. 42

iii. 61

b. 5 and 11

c. It does indicate an improvement, despite these two low weeks, because there is an overall upward trend throughout the time period.

8. a. Betty

b. 1 April: 70 m, 29 April: 73 m

c. Rachel

d. Yes. Sarah remained the same. Sumira and Paula sprinted at a slower pace.

e. Yes, the graph could be misleading because the vertical scale begins at 30 m and not at 0 m. If the vertical scale was not read properly, it might appear that the students covered a greater distance in a shorter period of time.

f. The graph's vertical axis starts at 30 m because no-one ran less than that distance and the teacher wanted to use all the space to show small differences in the distances.

9. By aligning the data points neatly above the value they are representing, these data points cannot be mistakenly seen as representing any other value. A dot plot is also neater and easier to interpret.

10. Column and bar graphs have discrete columns and bars, which are perfect for representing categorical data. There are often better graphs for displaying numerical data.

11. a–d. Generally, outliers affect the mean and range, have little effect on the median, and have no effect on the mode.

12. a.

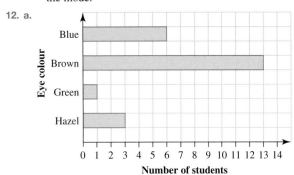

b. The data is categorical, not numerical. Therefore there is no median.

c. Brown

d. $\dfrac{10}{23} \approx 43\%$

13. a.

Test results

Class A
0 1 2 3 4 5 6 7 8 9 10
Class B

b. Class A: Mean 5.5, median 5.5, no mode, range 9; Class B: Mean 4.3, median 3.5, modes 2, 3 and 4, range 9

c. Class A has a higher mean and median than Class B, while the range is the same for both classes. The results for Class A are scattered throughout the whole range, while those for Class B are concentrated more towards the lower end of the range.

14. a. 35

b. Each week the number of fridges sold increases by 5.

c. $F = 25 + 5w$

d. $F = 25 + 5(10)$
$= 25 + 50$
$= 75$

12.6 Stem-and-leaf plots

1. Key: 2|7 = 27

Stem	Leaf
2	1 1 5 5 8 8 9 9
3	1 1 3 7 9
4	0 1 1 6 7
5	1 3 3 4 8
6	2 2 2 2 7 7 8
7	1 2 3 3 8
8	1 2
9	2 3 3

2. Key: 3|6 = 36

Stem	Leaf
0	0 1 2 3 6 7 8
1	0 2 2 2 2 4 5 6 6 7 8 9
2	1 2 5 7 8
3	1 1 2 2 3
4	0

3. a. Key: 12|7 = 127

Stem	Leaf
10	8
11	6 7 8
12	9
13	2 7
14	5
15	2
16	4 5
17	2
18	9

b. Key: 13|2 = 132

Stem	Leaf
11	4
12	4
13	0 1
14	6
15	0 2
16	
17	1 3

c. Key: 23|7 = 237

Stem	Leaf
19	0
20	5 7 8
21	
22	5
23	7 9
24	4 8 8 8
25	5

d. Key: $65|2 = 652$

Stem	Leaf
64	1 5 7 7
65	
66	2
67	5
68	5
69	1
70	8
71	
72	
73	5 6
74	8

4. a. Key: $1|7 = 1.7$

Stem	Leaf
1	2 7 9
2	2 3 8
3	9
4	1 6
5	4 8

b. Key: $6|7 = 6.7$

Stem	Leaf
6	0 5 9
7	4 5 7
8	3 4 8
9	2 3 8

c. Key: $13|7 = 13.7$

Stem	Leaf
13	0 8 9 9
14	3 5 5 6 7 7 8
15	1 2 6 7 9
16	2

d. Key: $1|5 = 0.15$

Stem	Leaf
0	2 7
1	5 8
2	
3	7
4	4
5	1
6	7

5.

Key: $0|2 = 2$

Leaf (Mitch)	Stem	Leaf (Yani)
3 2 1 1 1 1 0 0 0 0	0	0 0 1 1 1 1 1 2 2 2

Mitch scored between 0 and 3 goals inclusive.
Yani scored between 0 and 2 goals inclusive.

6. a. Mitch: 2, Yani: 3

b. Mitch

c. Yani: 11

d. Yani, as he scored goals in more games.

7. a. 22   **b.** 17

c. 64   **d.** 13

e. 42   **f.** 0

8. a. Key: $10|8 = 10.8$

Stem	Leaf
10	8 9
11	0 2 5 8
12	0 4 5 6 8 8
13	2 6 6 9 9
14	0 1
15	0

b. Range $= 4.2$ There is a reasonably large range of sprint times in this sample.

c. The majority of these runners completed the 100 m sprint between 11 and 13.9 seconds. The average sprint time was 12.68 s and the distribution is approximately evenly spread around the middle.

9. Stem plots need a key because these plots can show a variety of data, ranging from whole numbers to decimals. A key helps to show what the data represents. For example, a stem plot key showing data could be $15|1 = 151$ while a stem plot key showing decimals could be $11|1 = 11.1$. In these examples, the number on the leaf represents units (in the first example) and decimal values (in the second example).

10. By ordering a dot plot it is easy to identify key values, for example the median, the lowest value and the highest value. By aligning the data values we can easily see how many values lie in each class interval. We can also see the shape of the data.

11. a.

Test scores (%)
Key: $5|6 = 56\%$

Leaf (Class 7F)	Stem	Leaf (Class 7G)
3 1	4	
	5	
	6	6 8
9 8 7 7	7	4 5 6 6 8 8
8 5 4 2 1 0	8	0 2 2 3 5 9
2 1	9	

b. 7F: mean $= 77$, median $= 80.5$
7G: mean $= 78$, median $= 78$

c. Since class 7F has two outliers, the best measure of centre is the median, not the mean.

12. a. 55

b. 43

c. 26

13. a.

Temperatures (°C)
Key: 2 | 1 = 21°C

Leaf (City A)	Stem	Leaf (City B)
	1	2 2 3 3 3 7
9 5 4 3 1	2	0
6 0	3	

b. City A: range = 15, median = 25, mean = 26.9, there is no mode.
City B: range = 8, median = 13, mean = 14.3, mode = 13.

c. In February the northern hemisphere is in winter and the southern hemisphere is in summer. Looking at the answers to part **b** City A has a higher mean and median temperature, so is most likely in the southern hemisphere. City B has a lower mean and median, so is in the northern hemisphere.

14. a.

Key: 3|2 = 32

Leaf (9A)	Stem	Leaf (9B)
2	3	
9	4	6
7 4	5	5
8 7 5 5 0	6	0 2 8
3	7	3 3 4 5 7 8
5	8	1
6	9	

b. 9A: 83%; 9B: 92%

c. 9A: 2

d. For class 9A, exam marks are quite spread out overall, but there is a cluster of students scoring in the 60% range. For class 9B, the spread of exam marks is smaller — they are generally clustered in the 70% range.

e. See the dot plot at the foot of the page.*

f. The dot plot tells a slightly different story than the stem-and-leaf plot. For class 9A there appears to be 3 outliers (32%, 85% and 96%). This was not apparent in the stem-and-leaf plot. The results do not seem clustered around any particular percentage range. For class 9B there is one clear outlier (there are possibly more) at 46%. The results still appear clustered in the 70% range.

12.7 Pie charts and divided bar graphs

1. a. 356   **b.** Bus

2. a.

Sport	Number of students	Fraction of students	Angle at centre of circle
Basketball	55	$\dfrac{55}{180}$	$\dfrac{55}{^1\cancel{180}} \times \dfrac{^2\cancel{360°}}{1} = 110°$
Netball	35	$\dfrac{35}{180}$	$\dfrac{35}{\cancel{180}} \times \dfrac{^2\cancel{360}}{1} = 70°$
Soccer	30	$\dfrac{30}{180}$	$\dfrac{30}{\cancel{180}} \times \dfrac{^2\cancel{360}}{1} = 60°$
Football	60	$\dfrac{60}{180}$	$\dfrac{60}{\cancel{180}} \times \dfrac{^2\cancel{360}}{1} = 120°$
Total	180		

b, c.

Sports played by Year 7 students

Football, Basketball, Netball, Soccer

3. a. Emiliano

b. 17%

c. No, this sector graph gives percentages only, not actual values or numbers of people who voted.

d. 78.48°

4. a. 50%

b. 35%

c. 15%

d. 100%

e. i. 180°   **ii.** 126°   **iii.** 54°

5. a.

Roofing material	Number of houses	Fraction	Angle size (°)
Concrete tiles	12	$\dfrac{1}{3}$	120
Terracotta tiles	9	$\dfrac{1}{4}$	90
Galvanised iron	6	$\dfrac{1}{6}$	60
Colourbond	3	$\dfrac{1}{12}$	30
Slate	2	$\dfrac{1}{18}$	20
Shingles	4	$\dfrac{1}{9}$	40
Total	36	1	360

***14. e.**

32 34 36 38 40 42 44 46 48 50 52 54 56 58 60 62 64 66 68 70 72 74 76 78 80 82 84 86 88 90 92 94 96
33 35 37 39 41 43 45 47 49 51 53 55 57 59 61 63 65 67 69 71 73 75 77 79 81 83 85 87 89 91 93 95

Percentages

b.

Roofing materials

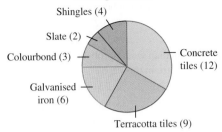

- Shingles (4)
- Slate (2)
- Colourbond (3)
- Galvanised iron (6)
- Terracotta tiles (9)
- Concrete tiles (12)

6. a. 45

   b. 6

   c. 12

   d. 40%

   e. German Shepherd

7. a.

Type of meat	Amount sold (kg)	Fraction
Lamb	10	$\frac{1}{9}$
Beef	45	$\frac{1}{2}$
Pork	5	$\frac{1}{18}$
Chicken	15	$\frac{1}{6}$
Turkey	10	$\frac{1}{9}$
Rabbit	5	$\frac{1}{18}$
Total	90	1

   b, c. See the figure at the foot of the page.*

8. a.

Type of bird	Number of birds	Fraction
Magpie	24	$\frac{2}{5}$
Crow	12	$\frac{1}{5}$
Cockatoo	12	$\frac{1}{5}$
Wattlebird	6	$\frac{1}{10}$
Rosella	6	$\frac{1}{10}$
Total	**60**	

b. i.

Birds seen on school oval

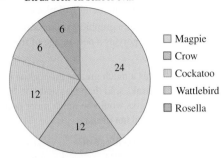

- Magpie
- Crow
- Cockatoo
- Wattlebird
- Rosella

ii.

Birds seen on school oval

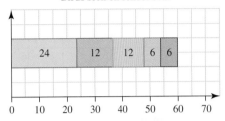

- Magpie
- Crow
- Cockatoo
- Wattlebird
- Rosella

9. a, b.

Transport	Number of students	Fraction	Length of bar
Walk	50	$\frac{50}{100} = \frac{1}{2}$	$\frac{1}{2} \times 10 = 5\,cm$
Car	20	$\frac{20}{100} = \frac{1}{5}$	$\frac{1}{5} \times 10 = 2\,cm$
Bus	15	$\frac{15}{100} = \frac{3}{20}$	$\frac{3}{20} \times 10 = 1.5\,cm$
Bike	15	$\frac{15}{100} = \frac{3}{20}$	$\frac{3}{20} \times 10 = 1.5\,cm$
Total	**100**		

c.

Modes of transport to school

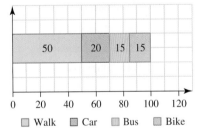

- Walk
- Car
- Bus
- Bike

   d. 2.5 cm

10. a. 100   **b.** 50   **c.** 25

11. Chair: 40; Table: 28; Couch: 12; Desk: 10.

---

*7b, c.

Type of meat sold

45	15	10	10	5	5

- Beef
- Chicken
- Lamb
- Turkey
- Pork
- Rabbit

12. Categorical data: column and bar graphs and sector graphs
Numerical data: dot plots, stem-and-leaf plots

13. A divided bar graph gives a graphical representation of the proportion of each category as a percentage of the whole.

14. a. Graph 1

b. In graph 1 there was a much greater proportion of ice creams and cold drinks sold than pies and hot drinks. Ice creams and cold drinks would be much more popular in summer than pies and hot drinks.

15. a.

Sport	Percentage of total	Angle size (°)
Cricket	20	72
Hockey	25	90
Netball	40	144
Soccer	10	36
Tennis	5	18
Total	100	360

Sports

Tennis (5)

Soccer (10) — Cricket (20)

Hockey (25)

Netball (40)

b. i. If there were only 10 students in the class, each student would represent 10% of the class total. It would therefore be impossible for only 5% of the class to play tennis.

ii. If there were 20 students in the class, each student would represent 5% of the class total. Since all of the percentages for the sports are multiples of 5%, this is a plausible scenario.

16. a. Rent and utilities

b. $800

c. i. $200 **ii.** $200 **iii.** $500

Omar's new budget

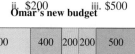

| 700 | 400 | 200 | 200 | 500 |

☐ Rent ☐ Utilities ☐ Food ☐ Entertainment ☐ Savings

17. a. 25% **b.** 45° **c.** 0.375 m²

18. $26\frac{2}{3}\%$

12.8 Comparing data

1. a. i. 82.4 **ii.** 82 **iii.** 81 **iv.** 9

b. The mean, median and mode are all quite close, so any could be used as a measure of centre.

2. a. i. 2.63 **ii.** 2.55 **iii.** 2.4 **iv.** 0.7

b. This distribution is quite spread out, with a significant number at the lower end and towards the top. For this reason the mode is probably the best measure of centre.

3. a. i. 4.5 **ii.** 5 **iii.** 5 **iv.** 10

b. i. 4.1 **ii.** 5 **iii.** 5 **iv.** 7

c. The mean and range are affected by the outlier. The median and mode have not been affected.

4. a. i. 6.26 **ii.** 6.27
iii. 6.18 and 6.34 **iv.** 0.16

b. The mean or median would probably be the best measures of centre in this case.

5. a. i. 22.6 **ii.** 23 **iii.** 23 **iv.** 34

b. i. 19 **ii.** 22 **iii.** 23 **iv.** 15

c. The mean and range are most affected by the outlier. The median is affected slightly, while the mode is not affected.

6. C

7. A

8. D

9. C

10. D

11. a.

Score x	Frequency f	Freq. × score $f \times x$
0	1	0
1	3	3
2	6	12
3	3	9
4	3	12
5	1	5
6	1	6
	$n = 18$	$\sum fx = 47$

Mean ≈ 2.6

b.

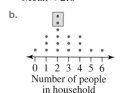

Number of people in household

c. 2

d. i. Median **ii.** Mean **iii.** Mode

12. 75.6

13. a. The median was calculated by taking the average of the 2 middle scores.

b. 13

14. Rebecca's strategy seems reasonable. Sample responses can be found in the worked solutions in your online resources.

15. a. $18 000

b. $29 444

c. $26 000

d. i. Mode **ii.** Mean

16. 7

17. a. 55 250 km, 52 000 km, 52 000 km

b. The median would be a better measure of centre because it will be less affected by the outliers in the data than the mean.

c. 51 810 km; It is reduced by 3440 km

d. 52 000 km

e. 3333

f. 50 000 km; 92% last that distance or more.

18. 2

Project

1. Sector graph

2. Families with no children in this age range is the most common category.

3. Column or bar graph

4. Responses will vary depending on the students surveyed. Here is a sample response.

Number of children	Tally	Frequency	Percentage
1	|||||	6	24%
2	||||||||||	11	44%
2	|||||	5	20%
4 or more	|||	3	12%
	Total	25	

5. Sample (the entire population was not surveyed)

6. Numerical data

7. Responses will vary depending on the students surveyed. The following is a sample response.

Percentage of households with *x* number of children under the age of 15

8. Responses will vary depending on the students surveyed. One major difference will be that none of the households surveyed have 0 children under 15 years as they all contain at least 1 Year 7 student. This is a major variation from the census data.

9. Responses will vary depending on the survey undertaken by the student.

12.9 Review questions

1. False

2. Continuous

3. Categorical

4. a. Coke   b. Tea

5. a. 26   b. 14   c. 18   d. 14   e. 12

6. a.

Cars	Frequency
0	6
1	4
2	8
3	5
4	1
Total	24

b. 6

c. 14

d. 2

e. 4

f. $\dfrac{1}{3}$

7. a. i. 3.14   ii. 4   iii. 3

b. i. 3.71   ii. 4   iii. 4

c. i. 4.86   ii. 5   iii. 5

d. i. 2.14   ii. 1   iii. 1

e. i. 5.2   ii. 5   iii. 5

f. i. 5.78   ii. 8   iii. 6

8. a. Names   b. 1 unit = 10%

c. John   d. Peter

e. 58%   f. 77%

9. a. Days of week   b. Air quality rating

c. Northern region   d. Wednesday

e. Eastern region

10. a.   Key: 3|6 = 36

Stem	Leaf
7	1 5 8
8	1 2 3 4 5 6 7 8 8
9	0 1 1 2 2 2 3 4 4 6 6 7
10	2 3 4 4
11	0 5

b. Only 6 cars were travelling over the 100 km/h speed limit. The slowest speed recorded was 71 km/h, while the fastest speed recorded was 115 km/h. Most of the recorded speeds were in the 90–99 km/h class interval. The most common (modal) speed recorded was 92 km/h.

11. a.   Key: 3|1 = 31

Leaf Set A	Stem	Leaf Set B
1	0	3
3	1	
8	2	9
0	3	0 1 2 8
5	4	1
9	5	
4	6	8

b. Set A
  i. 1 ii. 64 iii. 37.1 iv. 37.5 v. 63

  Set B
  i. 3 ii. 68 iii. 34 iv. 31.5 v. 65

c. The two sets have similar minimum and maximum values, giving them a similar range. The mean of Set A is slightly higher than that of Set B. The scores of Set A are spread out more evenly than those in Set B, which tend to be clustered towards the middle.

12. a. i. 6.55 ii. 6 iii. 6 iv. 9

b. There appear to be two outliers (11 and 12). The mean, median and mode values are quite close, indicating that it is common for 6 soft drinks to be sold quite frequently. The minimum number sold over the period is 3, while the maximum is 12.

13. a. i. 29 years ii. 27 years
  iii. 18 years iv. 39 years

b. The ages of the group range from 14 years to 53 years, with most of the group being under 40. The mean or median values would be a good representation of the age of the group.

14.
People who ate different types of breakfast

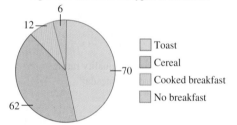

- Toast
- Cereal
- Cooked breakfast
- No breakfast

15. See the figure at the foot of the page.*

16. a. 10 °C b. Day 5 c. Day 7
  d. Day 6 e. 10 °C f. Days 6 and 7

17. a. 81.56 b. 74.22 c. Test 1
  d. 6 e. 4

18. a. 85
  b. 63
  c. 85
  d. 68
  e. Oisìn 89.5, Alfonso 74, Mark 76.5, James 74.5, David 85.5, Rachel 73.5, Mary 81, Bahar 78.5, Esther 68
  f. 21.5

19. a. 41 cm b. 154.8 cm c. 156 cm
  d. 156 cm e. Yes

20. a. Months
  b. Number of flies
  c. 10 000

d. 28 000

e. 1 cm = 5000 flies (if beginning from 0 flies) or
  1 cm = 2000 flies (if beginning from 10 000 flies)

f.
Fly population

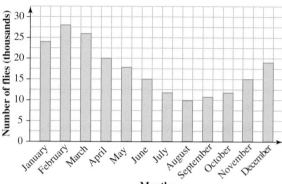

21. a.

x	f
0	4
1	5
2	7
3	3
4	0
5	1
Total	20

b.

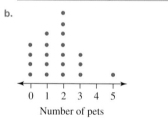

Number of pets

c. 2, 2, 1.65

22. a. i. 5 iii. 7.9 v. 6
  ii. 5.5 iv. 25 vi. 16.9, 15.5

b. When 10 was added to each score, both the mean and median increased by exactly 10.

23. The boys have a mean height of 158 cm, with a median height of 158 cm and three modes (148 cm, 158 cm and 166 cm). They range in height from 138 cm to 178 cm. The girls have a mean height of 153 cm, with a median height of 152 cm and no mode. They range in height from 137 cm to 171 cm.
The shortest person is a girl, while the tallest person is a boy. On the whole, the boys are taller than the girls.

*15

Hobbies				
30	20	20	15	15

- Skateboarding
- Stamp collecting
- Sailboarding
- Photography
- Model making

13 Probability

LESSON
13.1 Overview

Why learn this?

Probability helps us make informed decisions because it measures the likelihood of different events occurring and therefore allows us to predict the future. As with many different areas of mathematics, probability has its own terminology or language. To understand the chances of an event happening, it is important to understand the language of probability.

Whether you realise it or not, you use and consider probabilities every single day. For example, the weather report may say that there is a high chance of rain tomorrow. When you decide what to wear each day, you are taking into account the likelihood of it raining, or the likelihood of the sun shining all day.

Many board games depend on probability. Have you ever played a board game with 2 dice where you wanted to throw a double 6? The chance, or probability, of that occurring is 1 in 36. Lotteries use probability to determine the likelihood of a winning ticket being bought, and set the prize money based on these calculations. To win first prize in TattsLotto, you need to have selected the correct 6 balls from the 45 balls in the barrel. The chance of this happening is 1 in 8 145 060. This means that you would be likely to win TattsLotto approximately once per 8 million tickets you buy! In Oz Lotto you need to select 7 balls from a 45 ball barrel, with the chance of winning division 1 dropping lower, to 1 in 45 379 620!

It is widely used in many industries, including finance. Insurance brokers look at the chance, or likelihood, of an event occurring, and set their rates accordingly. Many different professions require knowledge of probability, including science, market research analysis, meteorology, financial analysis, statistics and many others.

Hey students! Bring these pages to life online

▶ **Watch videos**

🧩 **Engage with interactivities**

A+ **Answer questions and check solutions**

Find all this and MORE in jacPLUS ▶

Reading content and rich media, including interactivities and videos for every concept

Extra learning resources

Differentiated question sets

Questions with immediate feedback, and fully worked solutions to help students get unstuck

1. **MC** Select the option that best describes the probability of rolling an odd number with a die.
   A. Impossible   B. Unlikely   C. Even   D. Likely

2. **MC** Select the probability that is best linked to a likely probability.
   A. 0.1   B. 0.3   C. 0.5   D. 0.7

3. A bag contains 3 blue balls, 5 red balls and 2 yellow balls. Determine the probability of randomly selecting a yellow ball from the bag.

4. **MC** A ball is selected from a bag. Determine the sample space of the bag that contains 5 blue balls, 8 red balls and 2 yellow balls. (B = blue, R = red, Y = yellow)
   A. $S = \{5B, 8R, 2Y\}$ B. $S = \{B, R, Y\}$
   C. $S = \{B, R, W, Y\}$ D. $S = \{BBBBB, RRRRRRRR, YY\}$

5. A card is selected from a standard deck of cards and the suit is noted. Determine the number of elements in the sample space.

6. Two dice are rolled and the sum of the two dice is 5. Determine the number of ways the sum of 5 can be obtained.

7. **MC** If a fair die is rolled, the probability of rolling a number less than 5 is:
   A. $\dfrac{2}{3}$ B. $\dfrac{1}{6}$ C. $\dfrac{1}{3}$ D. $\dfrac{2}{5}$

8. **MC** If a 12-sided die is rolled, the probability of rolling a multiple of 4 is:
   A. $\dfrac{1}{2}$ B. $\dfrac{1}{4}$ C. $\dfrac{1}{3}$ D. $\dfrac{5}{12}$

9. **MC** A card is chosen at random from a standard deck of cards. The probability of selecting a red picture card is:
   A. $\dfrac{2}{13}$ B. $\dfrac{3}{26}$ C. $\dfrac{3}{28}$ D. $\dfrac{3}{13}$

10. A bag contains 5 blue balls, 7 red balls and 3 green balls. If a ball is selected at random, calculate the probability that neither a red or blue ball was selected.

11. **MC** Lauren does not know when Emily's birthday is. Determine which of the following months it is most likely to be in.
    A. April   B. June   C. September
    D. All options are equally likely.

12. Australia Post delivers parcels during business hours (9 am to 5 pm). Sally has ordered a parcel that will arrive next week. If each hour has equal chance, determine the probability that the parcel arrives before midday on Tuesday.

13. **MC** Identify which of the following options best describes the probability of randomly choosing a prime number out of the first 100 whole numbers.
    A. Impossible   B. Unlikely   C. Even chance   D. Likely

▶

14. **MC** Identify which of the following options would best simulate selecting a ball from a bag of 6 differently coloured balls.
    A. Spinning a square spinner
    B. Spinning a 4-sector circular spinner
    C. Spinning a 5-sector circular spinner
    D. Rolling a standard die

15. A box contains 3 red dice and 4 blue dice. Determine the smallest number of dice needed to be taken from the box to be *sure* you have a die of each colour.

LESSON
13.2 The language of chance

LEARNING INTENTION

At the end of this lesson you should be able to:
- understand that a probability is a number between 0 and 1 that describes the possibility of an event
- classify the chance of an event occurring using words such as *certain, likely, unlikely, even chance* or *impossible*.

13.2.1 The language of chance

eles-4710

- An **event** is one or more outcomes of a chance experiment.
  - When classifying the chance of an event occurring, we use words such as *certain, likely, even chance, unlikely* and *impossible*.

  - Probabilities can be expressed as decimals, fractions and percentages. For example, $0.5 = \dfrac{1}{2} = 50\%$.

  - The following **probability** scale associates important words used to describe probability with their approximate corresponding numerical values.
  - An **outcome** in a chance experiment refers to any possible result.
  - An event can describe either one outcome (for example, rolling a 1 on a die) or a collection of outcomes (for example, rolling an even number on a die).

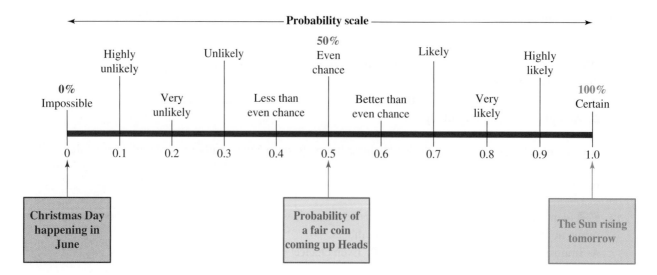

Discuss other examples of events that have a probability of 0, 0.5 or 1.

WORKED EXAMPLE 1 Classifying an event occurring

For each of the given statements, specify whether the chance of the following events occurring is _certain, likely, even chance, unlikely_ or _impossible_.
a. **You will compete in the next Olympics.**
b. **Every student in Year 7 will obtain 100% in their next mathematics test.**
c. **Each person in your class has been to the zoo.**
d. **You flip a coin and Tails comes up.**
e. **March is followed by April.**

THINK	WRITE
a. The chance of a person competing in the next Olympics is very small; however, it could happen.	a. It is _unlikely_ that this event will occur.
b. Due to each student having different capabilities and the number of students involved, this situation will almost never happen.	b. The chance of this event occurring is _very unlikely_.
c. The chance that each student in your class has been to the zoo, either with their family or primary school, is probable.	c. It is _likely_ this event has occurred.
d. When you flip a coin in a chance experiment, there are only two possibilities, a Head or a Tail. So there is a 50% chance of Tails coming up.	d. There is an _even chance_ this event will occur.
e. April always follows the month of March. This is a true statement.	e. It is _certain_ this event will occur.

Assign a number between and including 0 and 1 to represent the estimated probability of the following events, depending on how likely they are.

a. One of two evenly matched tennis players will win the next game.

b. You will guess the correct answer on a multiple choice question with four options.

c. Rolling a fair die and obtaining a number less than 6.

THINK	WRITE
a. 1. Determine the likelihood of an event occurring, with reasoning.	a. Since the two players are evenly matched, one does not have an advantage over the other. Therefore, they each have an equal chance of winning the next game.
2. Express the answer as a decimal.	The probability that one player wins the game is $\frac{1}{2}$ or 0.5.
b. 1. Determine the likelihood of an event occurring, with reasoning.	b. When guessing an answer on a multiple choice question with 4 options, 1 out of the 4 possibilities will be correct. One out of 4 may be expressed as a fraction.
2. Express the answer as a decimal.	The probability of guessing the correct answer is $\frac{1}{4}$ or 0.25.
c. 1. Determine the likelihood of an event occurring, with reasoning.	c. In the chance experiment of rolling a die, there are six possibilities. They are 1, 2, 3, 4, 5, 6. A number less than 6 includes 1, 2, 3, 4, 5. Therefore, five out of the six possibilities may be rolled. Five out of six may be expressed as a fraction.
2. Express the answer as a decimal, correct to 2 decimal places.	The probability of obtaining a number less than six is $\frac{5}{6}$ or approximately 0.83.

COMMUNICATING — COLLABORATIVE TASK: Draw a spinner

Create spinners with the following probabilities.

a. $P(\text{blue}) = \frac{1}{3}$ and $P(\text{white}) = \frac{2}{3}$

b. $P(\text{blue}) = \frac{1}{2}$, $P(\text{white}) = \frac{1}{4}$, $P(\text{green}) = \frac{1}{8}$ and $P(\text{red}) = \frac{1}{8}$

c. $P(\text{blue}) = 0.75$ and $P(\text{white}) = 0.25$

Once you have created your spinners, test them out with your classmates. How accurately do your results match the probabilities for which the spinners were designed?

Exercise 13.2 The language of chance

learn on

13.2 Quick quiz **on**	13.2 Exercise

Individual pathways

■ PRACTISE	■ CONSOLIDATE	■ MASTER
1, 3, 5, 9, 12, 15	2, 6, 7, 10, 13, 16	4, 8, 11, 14, 17

Fluency

1. **WE1** For each of the given statements, specify whether the chance of the following events occurring is certain, likely, even chance, unlikely or impossible.

   a. New Year's Day will be on 1 January next year.
   b. You roll a fair die and obtain a number less than 5.
   c. Water will boil in the fridge.
   d. There will be snow on the ski fields this winter.
   e. You will grow 18 cm taller this year.

2. For each of the given statements, specify whether the chance of the following events occurring is certain, likely, even chance, unlikely or impossible.

   a. You will win first prize in Tattslotto.
   b. You roll a fair die and obtain an odd number.
   c. The year 2024 will be a leap year.
   d. You choose a white ball from a bag that contains only white balls.
   e. You choose a yellow ball from a bag containing 4 red balls and 4 yellow balls.

3. **WE2** Assign a number between and including 0 and 1 to represent the estimated probability of the following events, depending on how likely they are.

   a. You flip a coin and obtain a Tail.
   b. You choose a red ball from a bag containing only 8 white balls.
   c. You guess the correct answer in a multiple choice question with 5 options.
   d. You roll a die and obtain a number greater than 4 on a fair die.

4. Assign a number between and including 0 and 1 to represent the estimated probability of the following events, depending on how likely they are.

   a. You flip a coin and obtain a Head.
   b. You choose a green ball from a bag containing only four green balls.
   c. You have science classes this year.
   d. Australia will win the Boxing Day cricket test.

5. **MC** The word that has the same meaning as *improbable* is the word:

   A. impossible   **B.** unlikely   **C.** uncertain   **D.** likely

6. **MC** The word that has the same meaning as *certain* is the word:

   A. definite   **B.** possible   **C.** likely   **D.** probable

Understanding

7. Compare the given events A, B, C, D and order them from least to most likely.

   A. It will be sunny in Queensland most of the time when you visit.
   B. Melbourne Cup Day will be on the first Tuesday in November next year.
   C. You will be the next Australian Eurovision contestant.
   D. Saturn will be populated next year.

8. List five events that are:

   a. impossible   **b.** unlikely to happen
   c. likely to happen   **d.** sure to happen.

For questions 9 to 11, match the word with the number that best represents the probability associated with the word. Use only the following numbers: 0, 0.25, 0.5, 0.75, 1.

9. **a.** Certain   **b.** Likely   **c.** Unlikely   **d.** Probable   **e.** Improbable

10. **a.** Slim chance   **b.** Sure   **c.** Doubtful
    d. Not able to occur   **e.** More than likely

11. **a.** Definite   **b.** Impossible   **c.** Fifty-fifty   **d.** Fair chance

Communicating, reasoning and problem solving

12. Explain your answers to the following questions using the language learned in this lesson.

    a. If today is Monday, what is the chance that tomorrow is Thursday?
    b. If today is the 28th of the month, what is the chance for tomorrow to be the 29th?
    c. If you toss a coin, what is the chance it will land Heads up?

13. 'Fifty-fifty' is an expression commonly used in probability. Explain the meaning of this expression, giving its fractional value and its decimal number form, as well as expressing it as a percentage.

14. Five balls numbered 1, 2, 3, 4 and 5 are placed in a bag. You draw a ball out of the bag at random and check whether it is odd or even. Explain why the probability of drawing an odd ball from the bag is not 0.5.

15. Anthony has 10 scrabble pieces, as shown. His friend Lian is blindfolded and is asked to pick a piece at random during a chance experiment. Determine the chance she will pick:

    a. an I   **b.** an A   **c.** a U   **d.** an E.

16. On 1 January, two friends, Sharmela and Marcela, were chatting with each other. Sharmela commented, 'It is very likely that tomorrow the temperature will be around 35 °C.' Marcela replied, 'It is very likely that tomorrow it is going to snow.' They were both correct.

   a. Explain how this is possible.

   b. Think and discuss any other situations like the one described in this question.

17. You roll a regular 6-sided die. Identify an event that could be described by the following words:

   a. Impossible   **b.** Very unlikely   **c.** Even chance   **d.** Very likely   **e.** Certain

LESSON
13.3 The sample space

LEARNING INTENTION

At the end of this lesson you should be able to:
- determine the outcomes and sample space of a chance experiment.

⊙ 13.3.1 The sample space

eles-4711

- The **sample space** refers to the list of all possible outcomes of a chance experiment. It is represented by the letter S.
  For example, the sample space for the spinner shown is $S = \{$blue, pink, green, orange$\}$.
  Note: Each outcome is only listed once in the sample space (for example, even if there are multiple sectors with red on the spinner, red is only listed once).
- Each outcome in the sample space is called an element of the sample space. For example, green is an outcome of the chance experiment and an element of the sample space.
- The number of elements in the sample space is denoted $n(S)$.
- The number of elements in the sample space of the spinner shown is 4: $n(S) = 4$.

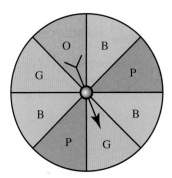

The sample space

The sample space is the list of all possible outcomes.

For example, for a 6-sided die, the sample space (S) is:

$$S = \{1, 2, 3, 4, 5, 6\}$$

WORKED EXAMPLE 3 Listing the sample space

A card is drawn from a standard deck. The suit of the card is then noted.
List the sample space for this chance experiment.

THINK	WRITE
Although there are 52 cards in the deck, we are concerned only with the suit. List each of the four suits as the sample space for this chance experiment.	$S = \{$clubs, spades, diamonds, hearts$\}$

WORKED EXAMPLE 4 Identifying the number of elements in a sample space

A die is rolled and the number on the uppermost face is noted. Identify the number of elements in the sample space.

THINK	WRITE
A die has six outcomes: $S = \{1, 2, 3, 4, 5, 6\}$	$n(S) = 6$ There are 6 elements in the sample space.

On Resources

Interactivities The sample space (int-4027)

Exercise 13.3 The sample space

learn on

13.3 Quick quiz on	13.3 Exercise

Individual pathways

■ PRACTISE	■ CONSOLIDATE	■ MASTER
1, 2, 3, 9, 10, 13, 16	4, 6, 7, 11, 14, 17	5, 8, 12, 15, 18, 19

Fluency

WE3 For each of the chance experiments in questions **1** to **6**, list the sample space.

1. A spinner with equal sectors labelled 1 to 10 is spun.

2. A coin is tossed.

3. A multiple choice question has five alternative answers: A, B, C, D and E.

4. A soccer team plays a match and the result is noted.

5. A card is selected from the four aces in a deck.

6. An exam paper is given the grade A to F.

7. A card is selected from a standard deck. List the sample space if we are interested in:
   a. the suit of the card chosen
   b. the colour of the card chosen.

8. A bag contains 8 red marbles, 9 green marbles and 2 orange marbles. A marble is selected from the bag. List the sample space.

Understanding

9. **WE4** A coin is tossed. Identify the number of elements in the sample space.

10. In each of the following, state the number of elements in the sample space.

    a. A card selected from a standard deck.
    b. The first ball drawn in the Tattslotto draw. (Balls are numbered from 1 to 45.)
    c. The winner of the NRL premiership. (There are 17 teams.)
    d. A day of the year is selected.

11. In each of the following, state the number of elements in the sample space.
    a. A letter from the alphabet is selected at random.
    b. The first prize in the lottery is chosen from tickets numbered 1 to 180 000.
    c. A term is selected from a school year.
    d. You win a medal at your chosen event at the world swimming championships.

12. **MC** From the following list, select the event that has the most elements in the sample space.
    A. Selecting a card from a standard deck
    B. Selecting a page at random from this book
    C. Selecting a student at random from your class
    D. Selecting a video game from an online digital video game platform such as Steam or Epic Games Store

Communicating, reasoning and problem solving

13. Explain the difference between a chance experiment, an outcome, an event and the sample space.

14. Are all elements in a sample space equally likely to occur? Justify your answer.

15. Determine the number of different ways in which change can be given for a 50 cent coin using only 20 cent, 10 cent and 5 cent coins. Justify your answer.

16. Michelle studies elective music. Her assignment this term is to compose a piece of music using as many instruments as she chooses, but only those that she can play. Michelle plays the acoustic guitar, the piano, the double bass and the electric bass. Determine the different combinations Michelle could choose.

17. Alex has one brother, one sister, a mother and a father. The family owns a 5-seat car. When the family goes out, the parents always sit in the front two seats. There are three seats behind that. Determine the different seating arrangements that are there.

18. If you had any number of ordinary six-sided dice, determine the number of different ways you could roll the dice and obtain a total of 6.

19. Four students, Aimee, Ben, Carla and Donald, are standing in a line waiting to enter the classroom. Determine the number of possible ways these 4 students can be arranged if Ben refuses to stand at the front of the line.

LESSON
13.4 Theoretical probability

LEARNING INTENTION

At the end of this subtopic you should be able to:
- calculate theoretical probabilities.

▶ 13.4.1 Theoretical probability

eles-4713

- An outcome is a particular result of a chance experiment.
- A **favourable outcome** is one that we are looking for.
- **Equally likely outcomes** have an equal chance of occurring.
- For the spinner shown:
  - there are three blue sectors out of eight possible outcomes (sectors), so
    $P(\text{blue}) = \dfrac{3}{8}$
  - pink and green are equally likely outcomes as they both are two out of the
    eight sectors: $P(\text{pink}) = \dfrac{2}{8} = \dfrac{1}{4}$ and $P(\text{green}) = \dfrac{2}{8} = \dfrac{1}{4}$
  - orange is the least likely outcome as it is one sector out of the eight, so
    $P(\text{orange}) = \dfrac{1}{8}$
  - the total of the probabilities of all outcomes of an event is 1. In the example
    shown, $\dfrac{3}{8} + \dfrac{2}{8} + \dfrac{2}{8} + \dfrac{1}{8} = 1$

Theoretical probability

- **The theoretical probability of a particular result or event is defined as:**

$$P(\text{event}) = \frac{\text{number of favourable outcomes}}{\text{total number of outcomes}}$$

DISCUSSION

Design a spinner that has four colours: red, green, blue and yellow.

Red should be twice as likely to occur as green and three times as likely to occur as blue. Yellow should be half as likely to occur as green.

Are there multiple designs that meet the description above?

Discuss with your classmates and try to come up with multiple designs.

WORKED EXAMPLE 5 Determining outcomes of chance experiments

State how many possible outcomes there are for each of the following chance experiments and specify what they are. Discuss whether the outcomes are equally likely or not equally likely to occur.
a. Tossing a coin
b. Spinning a circular spinner with 9 equal sectors labelled from a to i
c. Drawing a picture card (jack, queen, king) from a standard pack of cards

THINK	WRITE
a. 1. Make a note of how many sides the coin has and what each side represents.	**a.** The coin has 2 sides, a Head and a Tail.
2. State the outcomes and specify whether they are equally likely or not equally likely to occur.	When tossing a coin, there are two possible outcomes: Head or Tail. These outcomes are equally likely to occur.
b. 1. Make a note of how many sectors are in the circular spinner and what each one represents.	**b.** The circular spinner has 9 sectors labelled a to i.
2. State the outcomes and specify whether they are equally likely or not equally likely to occur.	When spinning the circular spinner, there are 9 possible outcomes; they are a, b, c, d, e, f, g, h or i. These outcomes are equally likely to occur.
c. 1. State the possible outcomes.	**c.** There are 2 possible outcomes, drawing a picture card or drawing a non-picture card.

2. Consider the possibility of the outcomes and specify whether they are equally likely or not equally likely to occur.

There are 52 cards in a standard pack. There are 3 picture cards in each of the four suits: $3 \times 4 = 12$ picture cards in a pack. Therefore the probabilities are:

$$P(\text{picture card}) = \frac{12}{52}$$

$$P(\text{non-picture card}) = \frac{52 - 12}{52} = \frac{40}{52}$$

The probability of drawing a picture card is less than the probability of drawing a non-picture card.

WORKED EXAMPLE 6 Calculating probability

Christopher rolls a fair 6-sided die.
a. State all the possible results that could be obtained.
b. Calculate the probability of obtaining:
  i. a 4
  ii. a number greater than 2
  iii. an odd number.

THINK	WRITE
a. Write all the possible outcomes.	**a.** There are six possible outcomes: $1, 2, 3, 4, 5, 6$.
b. i. 1. The number 4 occurs once.	**b. i.** Number of favourable outcomes $= 1$ Total number of outcomes $= 6$
2. Write the rule for probability.	$$P(\text{event}) = \frac{\text{number of favourable outcomes}}{\text{Total number of outcomes}}$$
3. Substitute the known values into the probability formula and evaluate the probability of obtaining a 4.	$$P(4) = \frac{1}{6}$$
4. Write the answer.	The probability of obtaining a 4 is $\frac{1}{6}$.
ii. 1. 'A number greater than 2' implies: $3, 4, 5, 6$.	**ii.** Number of favourable outcomes $= 4$ Total number of outcomes $= 6$
2. Write the rule for probability.	
3. Substitute the known values into the probability formula and evaluate the probability of obtaining a number greater than 2.	$$P(\text{greater than 2}) = \frac{4}{6}$$
4. Simplify the fraction.	$$= \frac{2}{3}$$
5. Write the answer.	The probability of obtaining a number greater than two is $\frac{2}{3}$.

| iii. | 1. | 'An odd number' implies 1, 3, 5. | iii. | Number of favourable outcomes = 3 |
| | | | | Total number of outcomes = 6 |

2. Write the rule for probability.

3. Substitute the known values into the probability formula and evaluate the probability of obtaining an odd number.

$$P(\text{an odd number}) = \frac{3}{6}$$

4. Simplify the fraction.

$$= \frac{1}{2}$$

5. Write the answer.

The probability of obtaining an odd number is $\frac{1}{2}$.

 Resources

 Interactivity Simple probability (int-4028)

Exercise 13.4 Theoretical probability

 learn on

| **13.4 Quick quiz** on | **13.4 Exercise** |

Individual pathways

■ PRACTISE	■ CONSOLIDATE	■ MASTER
1, 4, 7, 11, 14, 18, 21	2, 5, 8, 10, 13, 16, 19, 22	3, 6, 9, 12, 15, 17, 20, 23

Fluency

For each of the chance experiments in questions **1** to **5**, state the number of possible outcomes and specify what they are.

1. Rolling a 12-sided die, numbered 1 to 12 inclusively

2. Spinning a spinner for a game that has 5 equal-sized sections, numbered 1 to 5 inclusively

3. Choosing a consonant from the word *cool*

4. Rolling an even number greater than 2 on a fair 6-sided die

5. Choosing an odd number from the first 20 counting numbers

WE5 For each of the chance experiments in questions **6** to **10**, state the possible outcomes and specify whether each outcome is equally likely to occur or not.

6. Rolling a fair 6-sided die

7. Spinning a spinner where half is white and half is black

8. Spinning a spinner where half is white, a quarter is blue and a quarter is red

9. Rolling a 6-sided die that has the numbers 1, 2, 3, 4, 5, 5 on it

10. Choosing a vowel in the word *mathematics*

Understanding

11. **WE6** Christina rolls a fair 10-sided die with faces numbered from 1 to 10.
    a. State all the possible results that could be obtained.
    b. Calculate the probability of obtaining:
        i. a 9
        ii. a number less than 7
        iii. a prime number
        iv. a number greater than 3
        v. a multiple of 3.

12. Leo has been given a bag of marbles to play with. Inside the bag, there are 3 blue, 6 red, 4 green and 7 black marbles. If Leo takes out one marble from the bag, calculate:

    a. P(getting a red marble)
    b. P(getting a green marble)
    c. P(getting a black marble)
    d. P(getting a blue or black marble)
    e. P(getting a green, red or blue marble)
    f. P(getting a green, red, blue or black marble).

13. There is a valuable prize behind 2 of the 5 doors in a TV game show. Determine the probability that a player choosing any door will win a valuable prize.

14. **MC** A circular spinner is shown. Select the probability of obtaining an orange sector.

    A. $\dfrac{4}{7}$　　B. $\dfrac{1}{7}$　　C. $\dfrac{3}{7}$　　D. $\dfrac{1}{2}$

15. **MC** For an octagonal spinner with equal sectors numbered 1 to 8, select the chance of getting a number *between* 2 and 7.

    A. $\dfrac{5}{8}$　　B. $\dfrac{3}{8}$　　C. $\dfrac{3}{4}$　　D. $\dfrac{1}{2}$

16. A pack of playing cards is shuffled and a card is chosen at random (in no particular order or pattern). Calculate the probability that the card chosen is:

    a. a black card (that is, spades or clubs)
    b. an ace
    c. a diamond
    d. a picture card (that is, jack, queen or king)
    e. the queen of hearts.

17. A pack of playing cards is shuffled and a card is chosen at random (in no particular order or pattern). Calculate the probability that the card chosen is:

    a. a diamond or a black card   **b.** not a king   **c.** a club, diamond, heart or spade

    d. not a spade   **e.** red and a ten.

Communicating, reasoning and problem solving

18. If we know the probability of an event occurring, explain how we can work out the probability of it not occurring.

19. In a raffle where there is only 1 prize (a car), 100 000 tickets have been sold, at a cost of $5.00 each.

    a. Determine the chance of winning the prize for a buyer who:

      i. purchases only 1 ticket   **ii.** purchases 20 tickets

      iii. purchases 50 tickets   **iv.** purchases all the tickets.

    b. If someone bought all the tickets in the raffle, explain whether they had made a wise purchase.

20. Answer the following questions and justify your answer.

    a. If you had only one pair of shoes, determine what the probability would be that you would wear that pair of shoes on any given day.

    b. If you had two pairs of shoes, determine what the probability would be that you would wear a certain pair of shoes on any given day.

    c. If you had seven pairs of shoes, determine what the probability would be that you would wear a certain pair of shoes on any given day.

    d. If you had seven pairs of shoes but two pairs were identical, state what the probability would be that you would wear one of the two identical pairs of shoes on any given day.

    e. Explain what happens to the probability when the number of pairs of shoes increases.

    f. Explain what happens to the probability when the number of identical pairs of shoes increases.

21. Melbourne City FC is a soccer team in the Australian A-league. Over its history the team has won 60% of its matches, and lost half as many as it has won. If one of the team's past matches is selected at random, determine the probability that the match was a draw.

22. At Jaca college there are 220 Year 7 students, with a gender split of 55% girls and 45% boys. Every morning, 59 Year 7 students catch a bus to school. If one third of Year 7 boys catch a bus to school every morning, determine the probability that a randomly selected Year 7 student is a girl who does not catch a bus to school.

23. An apple, a banana, an orange, a peach and a bunch of grapes are in May's fruit basket. If she chooses two fruits to take to school with her lunch, determine the probability that one of the fruits is a banana.

LESSON
13.5 Observed probability

LEARNING INTENTION

At the end of this lesson you should be able to:
- calculate observed probabilities from data and by performing repeated trials of a chance experiment.

▶ 13.5.1 Observed probability

eles-4714

- In real life, the chance of something occurring may be based on factors other than the number of favourable and possible outcomes.

  For example, the chances of you beating your friend in a game of tennis could theoretically be $\frac{1}{2}$, as you are one of 2 possible winners. In practice, however, there are other factors (such as experience and skill) that would influence your chance of winning.
- A **trial** is one performance of an experiment to collect a result.
- An **experiment** is a process that allows us to collect data by performing trials. In experiments with repeated trials, it is important to keep the conditions for each trial the same.
- A **successful trial** is one that results in the desired outcome.
- The **observed probability** of an event is found by conducting an observation and counting the number of times the event occurs.

Observed probability

The observed probability of a particular result or event is defined as:

$$P(\text{event}) = \frac{\text{number of successful trials}}{\text{total number of trials}}$$

WORKED EXAMPLE 7 Calculating observed probability

A coin is flipped 10 times and the results are seven Heads and three Tails. Calculate the observed probability of obtaining a Tail.

THINK	WRITE
1. Obtaining a Tail is considered a success. Each flip of the coin is a trial.	$P(\text{success}) = \dfrac{\text{number of successful trials}}{\text{total number of trials}}$
2. Tails was flipped three times, so there were three successful trials out of a total of 10 trials.	$P(\text{Tail}) = \dfrac{3}{10}$ $= 0.3$

Digital technology

Scientific calculators have the ability to be random number generators in two different features.

- **Ran#** (SHIFT + •): Ran# generates a 3-decimal random number between 0 and 1. Each time you press = another random number is displayed.

- **RanInt** (ALPHA + •): RanInt generates a random number between a certain interval. For example, RanInt#(1, 6) will randomly generate a number from 1 to 6 (which can simulate a die). Each time you press = another random number in the interval will be displayed.

▶ 13.5.2 Long-term trends

eles-4715

- The more times an experiment is performed, the more accurate the observed probability becomes.
- In fair experiments, the **long-term trend** (that is, the trend observed for results from a very large number of trials) shows that the results obtained through observed probability will closely match those of theoretical probability.

WORKED EXAMPLE 8 Investigating observed probability

a. **Copy the following table. Toss a coin 10 times and record the result in row 1 of the table.**

Experiment number	Heads		Tails	
	Tally	**Count**	**Tally**	**Count**
1				
2				
3				
4				
5				
6				
	Total		**Total**	

b. **Calculate the probability of obtaining a Head from your experiment.**
c. **Calculate the probability of obtaining a Tail from your experiment.**
d. **Discuss how these values compare with the theoretical results.**
e. **Repeat step a another 5 times and combine all of your results.**
f. **Explain how the combined result compares with the theoretical results.**

	THINK		WRITE

THINK

a. Toss a coin 10 times and record the results in the first row of the table.

Notes: (a) Place a stroke in the appropriate tally column each time an outcome is obtained. Five is denoted by a 'gatepost': that is, 4 vertical strokes and 1 diagonal stroke $\left(\cancel{||||}\right)$.

(b) The same coin must be used throughout the experiment. The style of the toss and the surface the coin lands on must be the same.

WRITE

a. Sample answer:

	Heads		Tails	
Exp. No.	**Tally**	**Count**	**Tally**	**Count**
1	||||	4	$\cancel{||||}$ |	6
2	||||	4	$\cancel{||||}$ |	6
3	$\cancel{||||}$ ||	7	|||	3
4	|||	3	$\cancel{||||}$ ||	7
5	||||	4	$\cancel{||||}$ |	6
6	$\cancel{||||}$ |||	8	||	2
	Total	30	Total	30

b. 1. For experiment 1, calculate the probability of obtaining a Head using the rule.

b. $P(\text{event}) = \dfrac{\text{number of favourable outcomes}}{\text{number of possible outcomes}}$

$P(\text{Heads}) = \dfrac{\text{number of Heads obtained}}{\text{total number of tosses}}$

2. Substitute the given values into the rule.

$P(\text{Heads}) = \dfrac{4}{10}$

3. Evaluate and simplify.

$= \dfrac{2}{5}$

4. Convert the fraction to a percentage by multiplying by 100%.

As a percentage, $\dfrac{2}{5} = \dfrac{2}{5} \times 100\%$

$= \dfrac{200}{5}\%$

$= 40\%$

5. Answer the question.

The probability of obtaining a Head in this experiment is $\dfrac{2}{5}$ or 40%.

c. 1. For experiment 1, calculate the probability of obtaining a Tail for this experiment.

c. $P(\text{Tails}) = \dfrac{\text{number of Tails obtained}}{\text{total number of tosses}}$

2. Substitute the given values into the rule and simplify.

$P(\text{Tails}) = \dfrac{6}{10}$

$= \dfrac{3}{5}$

3. Convert the fraction to a percentage by multiplying by 100%.

As a percentage $\dfrac{3}{5} = \dfrac{3}{5} \times 100\%$

$= \dfrac{300}{5}\%$

$= 60\%$

The probability of obtaining a Tail in this experiment is $\dfrac{3}{5}$ or 60%.

d. Compare the results obtained in parts **b** and **c** with the theoretical results.

d. The observed value obtained for P(Heads) is $\frac{2}{5}$ (or 40%) and P(Tails) is $\frac{3}{5}$ (or 60%).

The theoretical value of these probabilities is $\frac{1}{2}$ (or 50%). Therefore, the observed probabilities differ from the theoretical probabilities by 10%.

e. 1. Repeat the procedure of part **a** 5 times.

2. Calculate the total number of Heads and Tails obtained and enter the results in the table.

e. Refer to the results in the table in part **a**.

f. 1. Calculate the probability of obtaining a Head for all 10 experiments combined.

f. $P(\text{Heads}) = \dfrac{\text{number of Heads obtained}}{\text{total number of tosses}}$

2. Substitute the given values into the rule and simplify.

$P(\text{Heads}) = \dfrac{30}{60}$

$= \dfrac{1}{2}$

3. Convert the fraction to a percentage by multiplying by 100%.

As a percentage $\dfrac{1}{2} = \dfrac{1}{2} \times 100\%$

$= \dfrac{100}{2}\%$

$= 50\%$

The probability of obtaining a Head in all 10 experiments is $\dfrac{1}{2}$ or 50%.

4. Calculate the probability of obtaining a Tail for all 10 experiments combined.

$P(\text{Tails}) = \dfrac{\text{number of Tails obtained}}{\text{total number of tosses}}$

$P(\text{Tails}) = \dfrac{30}{60}$

$= \dfrac{1}{2}$

As a percentage $\dfrac{1}{2} = \dfrac{1}{2} \times 100\%$

$= \dfrac{100}{2}\%$

$= 50\%$

The probability of obtaining a Tail in all 10 experiments is $\dfrac{1}{2}$ or 50%.

5. Compare the combined result obtained with the theoretical results.

The combined results in this experiment produced probability values that were equal to the theoretical probability values.

Therefore, the long-term trend of obtaining a Head or Tail when tossing a coin is equal to $\dfrac{1}{2}$.

Digital technology

Probability observations can be simulated using simple devices (such as coins or dice) or by using technology. Examples include random number generators in Excel spreadsheets or online adjustable spinner simulators, as shown. These simulators allow you to select the number of sectors and the number of spins (trials) and let the simulation run. You will notice that the observed probabilities get closer to the theoretical probabilities as you increase the number of spins.

COMMUNICATING — COLLABORATIVE TASK: Experimental versus theoretical probability!

1. Roll a die 20 times and record how many even numbers you roll.
2. Calculate the number of even numbers you should have rolled. How does your number compare to the theoretical number you would expect from 20 rolls?
3. Compare your results with your classmates' results. How do your classmates' results compare to the expected number?
4. Collate all of your classmates' results together. How does the total number of evens compare to the theoretical expected number? Are the class results closer to the expected results than each student's individual results?

 Resources

Interactivities Experimenting with chance (int-4030)
Observed probability versus theoretical probability (int-4031)
Generating random numbers on an Excel spreadsheet (int-4032)
Simulations (int-5352)

Exercise 13.5 Observed probability

| 13.5 Quick quiz on | 13.5 Exercise |

Individual pathways

■ PRACTISE	■ CONSOLIDATE	■ MASTER
1, 3, 4, 9, 10, 13	2, 5, 7, 11, 14	6, 8, 12, 15

Fluency

1. **WE8** Teagan was playing the game Trouble and recorded the number of times she rolled a 6. During the game, she was successful 5 times out of the 25 times she tried. Calculate the observed probability of rolling a 6 in the game.

2. In the 2020 AFL season, the Brisbane Lions won 14 out of 17 games. Determine the observed probability of the Lions winning a game in 2020.

3. **WE9**

Experiment number	Heads		Tails	
	Tally	Count	Tally	Count
1				
2				
3				
4				
5				
6				
	Total		Total	

a. Copy the table provided. Toss a coin 10 times and record the results in the first row of the table.
b. Calculate the observed probability of obtaining a Head.
c. Calculate the observed probability of obtaining a Tail.
d. Discuss how these values compare with the theoretical results.
e. Repeat step a another 5 times and combine all your results.
f. Explain how the combined results compare with the theoretical results.

4. If you wanted to create a device that would give a theoretical probability of achieving a particular result as $\frac{1}{4}$, state how many sections a spinner such as this would need to be divided into.

5. Determine how you would divide or colour a spinner if you wanted to achieve the probability of a success equal to $\frac{3}{10}$.

6. Two students conducted a chance experiment using a standard 6-sided die. The results were recorded in the following table.

Experiment number	Even number		Odd number	
	Tally	Count	Tally	Count
1					3				2
2						4			1
3				2					3

a. Describe in words the experiment that was conducted by the two students.
b. State the number of times they conducted the experiment.

c. Calculate the theoretical probability of rolling an even number.

d. Compare the observed probability with the theoretical probability of rolling an even number.

e. Compare the observed probability with the theoretical probability of rolling an odd number.

Understanding

7. Toss a coin 60 times, tallying up the number of Heads and Tails that you toss in trials of 10 tosses at a time. (If you have already completed question **3**, you may use the results you obtained from that observation.)

   a. The long-term trend of the probability of obtaining a Head on the toss of a coin is the P(Heads) from your experiment.
   Determine the long-term trend of the probability after:
   i. 10 tosses of the coin
   ii. 20 tosses of the coin
   iii. 30 tosses of the coin
   iv. 60 tosses of the coin.

   b. Obtain a classmate's 60 results. Combine these with yours. State the long-term trend of P(Heads) obtained.

   c. Combine your pair's 120 results with those of another pair. State the long-term trend of P(Heads) obtained.

   d. Finally, count the results obtained by the whole class for this experiment. (Make sure nobody's results are counted twice.) You should have 60 tosses per person. State the long-term trend of P(Heads) obtained.

   e. Copy and complete the following table.

Number of tosses	Heads		Tails	
	P(Heads)	P(Heads) as percentage	P(Tails)	P(Tails) as percentage
10				
20				
30				
60				
120				
240				
Whole class (specify number of tosses)				

   f. Comment on the changes of the long-term trend value of P(Heads) as you toss the coin more times.

8. Use a random number generator to simulate a 5-colour spinner.

   a. Determine the chance of getting any one of the five colours when you spin the spinner (theoretically).

   b. Spin the spinner 10 times and, using a table such as the following one, record your results.

Colour	1	2	3	4	5
Number of times it occurs					

   c. From your results, list the probabilities of obtaining each colour. For example, divide the number of times a particular colour was obtained by the total number of spins (that is, 10).

   d. Explain why these probabilities might not be the same as the theoretical probability would suggest.

   e. Spin your spinner and record the results for another 10 spins.

   f. Spin your spinner so that you have 100 results. Is the observed probability closer to the pure probability? Why might this be? Discuss.

Communicating, reasoning and problem solving

9. Cory records the fact that it has rained on 65 out of 365 days in a year.

   a. Write the number of days that it has rained as a simple fraction.

   b. Karen says that since any day can be wet or dry, the probability of rain on any day is $\frac{1}{2}$. Explain whether Karen is correct.

   c. Determine the observed probability of rain on any given day, expressed as a decimal, correct to 2 decimal places.

10. a. Conduct the following observations:

    i. Toss a coin 10 times and record the number of Heads and Tails that occur.

    ii. Toss a coin 25 times and record the number of Heads and Tails that occur.

    iii. Toss a coin 50 times and record the number of Heads and Tails that occur.

    iv. Toss a coin 100 times and record the number of Heads and Tails that occur.

    b. Calculate the observed probability for each experiment.

    c. Compare these values and explain your findings.

11. A die is rolled 30 times, giving the following results.

4	3	5	4	3	5	2	1	1	5
3	2	2	4	1	3	1	6	1	3
2	1	6	6	3	5	1	3	5	3

    a. Display these results in a frequency table.

    b. Calculate the probability of obtaining a 6 when you roll a die.

    c. Calculate how many times you would expect to obtain a 6 in 30 rolls of a die.

    d. Explain the difference between your expected results and the actual results shown in this question.

12. You have calculated previously that the chance of getting any particular number on a 6-sided die is $\frac{1}{6}$. You have 2 different coloured dice. Explain whether there is any difference in your dice apart from colour and whether one could be biased (more likely to give a particular result than theory says it should).

    a. Design a test to determine whether the dice you have are fair. Write down what you are going to do.

    b. Perform your test and record results.

    c. Determine the probability of getting each of the numbers on each of your dice, based on your tests and on the long-term trend you have observed.

    d. State what your test says about your two dice. Explain whether there are any things that need to be considered before giving your answer. (Perhaps your dice have slightly uneven shapes or something that might cause them to lean towards one result more than others.)

13. Inside a bag are 36 shapes that are either squares or triangles. One shape is taken out at random, its shape noted and put back in the bag. After this is repeated 72 times, it is found that a triangle was taken out 24 times. Estimate how many triangles and how many squares there are in the bag.

14. In your desk drawer, there are 5 identical red pens and 6 identical black pens. Determine the smallest number of pens you have to remove from the drawer in the dark so that you will be absolutely sure of having:

    a. 2 black pens
    c. 1 black pen and 1 red pen.

    b. 2 red pens
    d. Explain your answers to parts a, b and c.

15. Jim operates a parachute school. Being a man who is interested in statistics, he keeps a record of the landing position of each jump from first-time parachutists. With experience, parachutists are able to land on a particular spot with great accuracy. However, first-time parachutists do not possess this ability.
Jim has marked out his landing field with squares, as shown here.

Legend
Region A =
Region B =
Region C =

We are going to look at the areas of each of the regions A, B and C. To do this, we will determine each of the areas in terms of one of the small squares in region A. We will say that each small square has an area of 1 square unit.

a. Determine the area of Jim's whole landing field (in square units).
b. Determine the areas of regions A, B and C (in square units).
c. Assuming that the parachutist lands in the field, calculate the theoretical probability that the landing will occur in:
  i. region A
  ii. region B
  iii. region C.

d. Jim's records indicate that, from 5616 jumps of first-time parachutists, the landing positions were:
  i. 592 in region C
  ii. 1893 in region B
  iii. 3131 in region A.

Comment on these results in comparison with the probabilities you calculated in part **c**.

LESSON
13.6 Review

13.6.1 Topic summary

Understanding probability

- Probability looks at how **likely** something is to happen.
- Probabilities are stated as a number from 0 to 1, where 0 corresponds to *impossible* and 1 corresponds to *certain*. That is,

$$0 \leq P(\text{event}) \leq 1$$

- Different words or phrases in the English language can be associated with different probabilities, as indicated on the probability scale below.

PROBABILITY

The language of chance

- A **chance experiment** is an experiment in which there are several different possible outcomes, each having a defined chance of occurring.
  e.g. Rolling a die is a chance experiment.
- An **outcome** is a possible result of a chance experiment, such as rolling a 2.
- An **event** is either a single outcome (e.g. rolling a 4) or a collection of outcomes (e.g. rolling a 1, 2 or 3).
- The **sample space**, S, is the set of *all* possible outcomes of a chance experiment.
  e.g. The sample space of rolling a fair six-sided die is $S = \{1, 2, 3, 4, 5, 6\}$.
- The number of elements in the sample space is often important in probability. It is denoted by the symbol $n(S)$.

Theoretical probability

- A *favourable outcome* is one that we are looking for.
- *Equally likely outcomes* have an equal chance of occurring.
- The **theoretical probability** of an event occurring is calculated as:

$$P(\text{event}) = \frac{\text{number of favourable outcomes}}{\text{total number of outcomes}}.$$

e.g. Consider the following chance experiments.

Rolling a die: $P(\text{rolling a 1}) = \dfrac{1}{6}$

Flipping a coin: $P(\text{Tail}) = \dfrac{1}{2}$ (or 0.5)

Picking a card from a deck of cards:
$P(\text{diamond}) = \dfrac{13}{52} = \dfrac{1}{4}$ (or 0.25)

Observed probability

- The **observed probability** of an event is determined by the results of repeated trials of the experiment.
- $P(\text{event}) = \dfrac{\text{number of successful trials}}{\text{total number of trials}}$

e.g. If a team has won 42 out of their last 70 matches, the observed probability that they will

win a match is $\dfrac{42}{70} = \dfrac{3}{5} = 60\%$.

- The observed probability of an event becomes more accurate as the number of trials increases, and approaches the theoretical probability.

13.6.2 Project

Snakes, ladders and probability!

In the game of snakes and ladders, you roll 2 dice, add the 2 numbers that appear on the uppermost faces to get a total, and move a marker the total number around a 100-square board. If you are lucky and your marker lands on the base of a ladder, you can advance more quickly in the game by 'climbing' to the top of the ladder. However, if your marker lands on the head of a snake, you must 'slide down' the length of the snake.

1. Complete the following table to show the possible totals when 2 dice are rolled. Some of the values have been included for you.
2. List the sample space for the possible totals when 2 dice are rolled.
3. Determine which total appears the most.
4. Determine which total appears the least.
5. Calculate the probability of getting a total of 3 when 2 dice are rolled.

		Die 1					
		1	**2**	**3**	**4**	**5**	**6**
Die 2	**1**	2	3	4			
	2	3	4	5			
	3	4					
	4						
	5						
	6						

Using the table that shows the frequency of the totals, we can investigate the probabilities involved in moving around the snakes and ladders board. The following situations will enable you to investigate some of the possibilities that occur in snakes and ladders.

6. Imagine you landed on square 95 and slid down the snake to square 75. What total would you need to get to go up the ladder at square 80 on your next move? Determine the number of ways in which this total can be achieved in one turn.
7. If you slid down the snake at square 87, is it possible to move up the next ladder with your next turn? Explain.

8. Explain what would happen if you were on square 89 and rolled two 1s and rolled two 1s again with your next turn. Discuss what would be the likelihood of this happening in a game.

9. Describe how you could get from square 71 to square 78 in one turn. Work out the probability of this happening.

10. Imagine you had a streak of luck and had just climbed a ladder to square 91. Your opponent is on square 89. Explain which player has the greater chance of sliding down the snake at square 95 during the next turn.

11. Investigate the different paths that are possible in getting from start to finish in the fewest turns. For each case, explain the totals required at each turn and discuss the probability of obtaining these totals.

Play a game of snakes and ladders with a partner. Examine your possibilities after each turn, and discuss with each other the likelihood of moving up ladders and keeping away from the snakes' heads as you both move around the board.

 Resources

 Interactivities   Crossword (int-2596)
Sudoku puzzle (int-3171)

Exercise 13.6 Review questions

learn**on**

Fluency

1. For each of the given statements, specify whether the chance of the following events occurring is certain, likely, even chance, unlikely or impossible.
   a. Australia is in the Southern Hemisphere.
   b. You will still be alive in the next century.
   c. You obtain an even number on a circular spinner numbered from 1 to 16.

2. For each of the given statements, specify whether the chance of the following events occurring is certain, likely, even chance, unlikely or impossible.
   a. England is in the Southern Hemisphere.
   b. You roll a fair die and obtain a number less than or equal to 4.
   c. Humans can survive without water indefinitely.

3. List one event for which the chance of it occurring would be:
   a. impossible   **b.** certain   **c.** unlikely

4. Copy the number line and place the following words on it to indicate what sort of chance each number would represent: *certain, very unlikely, unlikely, likely, highly likely, highly unlikely, even chance, impossible, very likely*

$$0 \quad 0.1 \quad 0.2 \quad 0.3 \quad 0.4 \quad 0.5 \quad 0.6 \quad 0.7 \quad 0.8 \quad 0.9 \quad 1.0$$

5. List the sample space for the following chance experiments.
   a. Tossing a coin
   b. Drawing a marble out of a bag containing red, green and blue marbles
   c. Answering a true/false question in a test

▶

6. List the sample space for the following chance experiments.
   a. Rolling a 6-sided die
   b. Spinning a circular spinner numbered from 1 to 10
   c. Competing in a race in which there are 9 runners

7. For each of the following, state the number of elements in the sample space, $n(S)$.
   a. The first prize in a raffle is chosen from tickets numbered from 1 to 5000.
   b. A card is selected from the red cards in a standard deck.
   c. A circular spinner numbered 1 to 5 is spun.
   d. A day is selected in the month of July.

Understanding

8. A six-sided die labelled 1, 2, 3, 3, 4, 5 is rolled.
   State:
   a. P(getting a 5)   **b.** P(getting a 3)
   c. P(getting an even number)   **d.** P(getting an odd number).

9. A six-sided die labelled 1, 2, 3, 3, 4, 5 is rolled.
   State:
   a. P(getting a number greater than or equal to 3)   **b.** P(getting a number less than 3)
   c. P(getting a 0)   **d.** P(getting a 6).

10. A person has a normal pack of cards and draws one out.
    Calculate:
    a. P(picking a heart)
    b. P(picking a spade)
    c. P(picking a picture card)
    d. P(picking a card with a number less than 5 on it). Do not count an ace as a number.

11. Ten cards are numbered from 1 to 10, shuffled and placed face down on a table. If a card is selected at random, calculate the probability that the card selected is:
    a. 4   **b.** an even number
    c. divisible by 3   **d.** an even number and divisible by 3.

12. Ten cards are numbered from 1 to 10, shuffled and placed face down on a table. If a card is selected at random, calculate the probability that the card selected is:
    a. an even number or divisible by 3   **b.** not divisible by 3
    c. greater than 8   **d.** 12.

13. A sample of 250 students at a particular school found that 225 of them had access to the internet at home. Given this sample is a good representation for the entire school, calculate the probability that a student selected at random in the school will have internet access at home.

Communicating, reasoning and problem solving

14. In your sock drawer there are 4 identical blue socks and 5 identical black socks. Determine the smallest number of socks you can take from the drawer in the dark so that you will be absolutely sure of having:
    a. a pair of black socks
    b. a pair of blue socks
    c. a black pair and a blue pair of socks.

15. The arrows on Spinner A and Spinner B are spun. (If an arrow lands on a line, the spinner is spun again.) The two numbers are added to get a score. For example, in the diagram the score is 10.

Spinner A   Spinner B

    a. Determine the highest total score possible.
    b. Determine the possible total scores.
    c. List all the ways to get a total score of 8.
    d. Determine the probability of getting a total of 9.
    e. Determine the probability of getting a total score of 10.

16. Charlotte and Rhianna have a flower garden. They have 17 red flowers, 12 pink flowers and 13 yellow flowers. Charlotte picks some flowers for her vase. She selects 6 red flowers, 3 pink and 3 yellow flowers. If Rhianna picks a flower at random for her hair, determine the probability that it is yellow.

17. A bowl contains blue marbles and white marbles. If there are twice as many blue marbles as white marbles, determine the probability that a blue marble is selected.

18. Chloë is a contestant on a game show. There are five sealed cases on the podium and each of the cases contains one of the following amounts: $1, $50, $500, $1000 and $5000. The game show host offers her a deal of $400, or she can choose a case and keep the amount of money in it, instead. Determine the probability she will win more than the $400 the host is offering.

19. Rebecca plays a sideshow game where she puts four balls into a clown's mouth, and the balls then fall into slots numbered 1 to 6. To win, one of her balls must go into slot 6.
She has noticed that the numbers 2 and 5 come up 4 times as often as 1 and 6, and the numbers 3 and 4 come up 5 times as often as 1 and 6. Determine the probability that she will get a 6.

20. You have a spinner. Draw a pie or circle graph to represent the following list of colours and their associated sectors of the circle, in degrees.

Discuss the chances of landing on each of the colours. List the likelihoods from smallest to largest.

Colour	Degree
Yellow	25°
Orange	40°
Green	60°
Pink	100°
Purple	15°
Blue	120°

on To test your understanding and knowledge of this topic, go to your learnON title at www.jacplus.com.au and complete the **post-test**.

Answers

Topic 13 Probability

13.1 Pre-test

1. C
2. D
3. $\frac{1}{5}$
4. B
5. 4
6. 4
7. A
8. B
9. B
10. $\frac{1}{5}$
11. D
12. $\frac{11}{40}$
13. B
14. D
15. 5

13.2 The language of chance

1. a. Certain   b. Likely
   c. Impossible   d. Likely
   e. Unlikely
2. a. Unlikely   b. Even chance
   c. Certain   d. Certain
   e. Even chance
3. a. $\frac{1}{2}$ (or 0.5)   b. 0
   c. $\frac{1}{5}$ (or 0.2)   d. $\frac{1}{3}$
4. a. $\frac{1}{2}$ (or 0.5)
   b. 1
   c. 1
   d. 0.3–0.7. Any value within this range is reasonable.
5. B
6. A
7. D, C, A, B
8. Sample responses can be found in the worked solutions in the online resources.
9. a. 1   b. 0.75   c. 0.25
   d. 0.75   e. 0.25
10. a. 0.25   b. 1   c. 0.25
    d. 0   e. 0.75
11. a. 1   b. 0   c. 0.5   d. 0.5

12. a. It is impossible because Tuesday always follows Monday.
    b. It is certain because the 29th always follows the 28th day of the month. However, if the month is February, depending on whether the year is a leap year or not, the chance is either impossible or certain.
    c. It has an equal chance or a fifty-fifty chance if it is a coin with a Head on one side and a Tail on the other side.
13. Equal chance: 50% or $\frac{1}{2}$ or 0.5.
14. There are 3 odd-numbered balls (1, 3, 5) and only 2 even-numbered balls (2, 4). Therefore you are more likely to draw an odd-numbered ball from the bag than an even-numbered ball.
15. a. Highly unlikely because there is only one piece with the letter I. The chance is $\frac{1}{10}$ or 0.1.
    b. Equal chance because there are five pieces with the letter A and five pieces with other vowels.
    c. Impossible because there are no pieces with the letter U.
    d. Unlikely because the chance is $\frac{3}{10}$ or 0.3.
16. a. The two girls are living in two different countries.
    b. Time zones in different places, night-time and daytime in different places in the world.
17. a. Rolling a 7
    b. Rolling a 1
    c. Rolling an even number
    d. Rolling a number that is greater than 1
    e. Rolling a number that is less than 10

13.3 The sample space

1. $S = \{1, 2, 3, 4, 5, 6, 7, 8, 9, 10\}$
2. $S = \{\text{Heads, Tails}\}$
3. $S = \{A, B, C, D, E\}$
4. $S = \{\text{win, loss, draw}\}$
5. $S = \{\text{ace of clubs, ace of spades, ace of hearts, ace of diamonds}\}$
6. $S = \{A, B, C, D, E, F\}$
7. a. $S = \{\text{clubs, spades, diamonds, hearts}\}$
   b. $S = \{\text{red, black}\}$
8. $S = \{\text{red, green, orange}\}$
9. 2
10. a. 52   b. 45
    c. 17   d. 365 (or 366 in a leap year)
11. a. 26   b. 180 000   c. 4   d. 3
12. D
13. A chance experiment is an experiment in which the outcome is left to chance. An outcome is any possible result of the chance experiment. An event can describe either one outcome or a collection of outcomes. The sample space is a list of all possible outcomes.
14. It depends on whether each element has an equal chance of occurring. In the sample space for rolling a die, each element is equally likely to occur. In more complex experiments, each event may not be likely to occur.

15. 12 different ways:
20, 20, 10; 20, 20, 5, 5; 20, 10, 10, 10;
20, 10, 10, 5, 5; 20, 10, 5, 5, 5, 5; 20, 5, 5, 5, 5, 5, 5;
10, 10, 10, 10, 10; 10, 10, 10, 10, 5, 5;
10, 10, 10, 5, 5, 5, 5; 10, 10, 5, 5, 5, 5, 5, 5;
10, 5, 5, 5, 5, 5, 5, 5, 5; 5, 5, 5, 5, 5, 5, 5, 5, 5, 5;

16. 15 choices

17. 12

18. 11

19. 18 ways

13.4 Theoretical probability

1. $12 : 1, 2, 3, 4, 5, 6, 7, 8, 9, 10, 11, 12$

2. $5 : 1, 2, 3, 4, 5$

3. $2 : C, L$

4. $2 : 4, 6$

5. $10 : 1, 3, 5, 7, 9, 11, 13, 15, 17, 19$

6. $1, 2, 3, 4, 5, 6$. All equally likely.

7. White, black. Each is equally likely.

8. White, blue, red. It is more likely you will get white rather than red or blue, because white takes up a bigger area.

9. $1, 2, 3, 4, 5$. It is more likely that a five will be rolled because it appears twice on the die.

10. a, e, i. It is more likely that a consonant would be obtained as there are 7 consonants and 4 vowels.

11. a. $1, 2, 3, 4, 5, 6, 7, 8, 9, 10$
    b. i. $\frac{1}{10}$ ii. $\frac{3}{5}$ iii. $\frac{2}{5}$ iv. $\frac{7}{10}$ v. $\frac{3}{10}$

12. a. $\frac{3}{10}$ b. $\frac{1}{5}$ c. $\frac{7}{20}$
    d. $\frac{1}{2}$ e. $\frac{13}{20}$ f. 1

13. $\frac{2}{5}$

14. C

15. D

16. a. $\frac{1}{2}$ b. $\frac{1}{13}$ c. $\frac{1}{4}$ d. $\frac{3}{13}$ e. $\frac{1}{52}$

17. a. $\frac{3}{4}$ b. $\frac{12}{13}$ c. 1 d. $\frac{3}{4}$ e. $\frac{1}{26}$

18. An event must either occur or not occur — there is no other option. If we know the probability of an event occurring, the probability it will not occur is $1 - P(\text{event})$.

19. a. i. $\frac{1}{100\,000}$
    ii. $\frac{1}{5000}$
    iii. $\frac{1}{2000}$
    iv. 1
    b. It would not be a very wise purchase because the total cost would be \$500 000 unless of course the car was worth more than \$500 000.

20. a. 100% b. 50%
    c. $\frac{1}{7}$ d. $\frac{2}{7}$
    e. Decrease f. Increase

21. $0.1 = 10\%$

22. $\frac{95}{220} = \frac{19}{44} \approx 43\%$

23. The probability of taking a banana is $\frac{2}{5}$.

13.5 Observed probability

1. $\frac{1}{5}$

2. $\frac{14}{17} \approx 82\%$

3. a.

Exp. No.	Heads		Tails	
	Tally	Count	Tally	Count
1	IIII	4	JHT I	6
2	I	1	JHT IIII	9
3	JHT I	6	IIII	4
4	JHT III	8	II	2
5	IIII	4	JHT I	6
6	JHT II	7	III	3
	Total	30	Total	30

Note: This is only one possible solution. Answers will differ each time.

b. $\frac{2}{5}$ (40%)

c. $\frac{3}{5}$ (60%)

d. The theoretical value for both results is $\frac{1}{2}$ (50%). They differ by 10%.

e. Refer to table.

f. The combined result equals the theoretical value.

4. 4

5. Divide the spinner into 10 sections; 3 of these sections will be shaded in 1 colour and 7 sections will be shaded in another colour.

6. a. The students rolled a die 5 times and recorded whether the result was an odd number or an even number.
   b. 3 times
   c. 0.5
   d. The observed probability was 0.6; this is 0.1 higher than the theoretical probability.
   e. The observed probability was 0.4, this is 0.1 lower than the theoretical probability.

7. Values will differ for each group. As you complete more trials, you will probably notice the values you get observationally are closer to those you would expect from

theoretical probability $\left(\dfrac{1}{2} \text{ or } 50\%\right)$. If this is not occurring, you will probably need more trials.

8. a. $\dfrac{1}{5}$

   b.
Colour	1	2	3	4	5
Number of times it occurs		II	II	IIII	II

   c. $P(1) = 0$, $P(2) = \dfrac{1}{5}$, $P(3) = \dfrac{1}{5}$, $P(4) = \dfrac{2}{5}$, $P(5) = \dfrac{1}{5}$

   Note: This is only one possible solution. Answers will differ each time.

   d. Your spinner may not be evenly balanced and this may lead to an increased likelihood of getting one result or another.

   e-f. Values will differ for each group. As you complete more trials, you will probably notice the values you get observationally are closer to those you would expect from theoretical probability $\left(\dfrac{1}{5} \text{ or } 20\%\right)$. If this is not occurring, you will probably need more trials.

9. a. $\dfrac{13}{73}$

   b. Karen is not correct, as a wet day (65 out of 365) and a dry day (300 out of 365) are not equally likely.

   c. 0.18

10. The general trend as the number of trials increases should show that the probability of getting a Head or a Tail gets closer to 50%.

11. a.
Score	Frequency
1	7
2	4
3	8
4	3
5	5
6	3

    b. $P(6) = \dfrac{1}{6}$

    c. 5

    d. The result differs by pure chance. The more often we roll the die, the closer we can expect the results to match the expected number of sixes.

12. a. Each die would need to be thrown about 120 times for us to get some impression of whether it is biased or not. The same person would need to roll the die each time in the same manner and onto the same surface.

b.
Number on die	Black	White
1	20	19
2	19	21
3	20	20
4	21	20
5	19	20
6	21	20
Total	120	120

Note: These values will differ each time the experiment is conducted.

c.
Number on die	Probability of black	Probability of white
1	$\dfrac{20}{120} = \dfrac{1}{6}$	$\dfrac{19}{120}$
2	$\dfrac{19}{120}$	$\dfrac{21}{120} = \dfrac{7}{40}$
3	$\dfrac{20}{120} = \dfrac{1}{6}$	$\dfrac{20}{120} = \dfrac{1}{6}$
4	$\dfrac{21}{120} = \dfrac{7}{40}$	$\dfrac{20}{120} = \dfrac{1}{6}$
5	$\dfrac{19}{120}$	$\dfrac{20}{120} = \dfrac{1}{6}$
6	$\dfrac{21}{120} = \dfrac{7}{40}$	$\dfrac{20}{120} = \dfrac{1}{6}$
Total	1	1

The long-term trend suggests that the probability of obtaining each value on either die will be $\dfrac{1}{6}$.

d. The 2 dice appear to be fair as each value occurred approximately 20 times, which is what we would expect in 120 throws. More trials, however, could be conducted. It is very important when conducting an experiment such as this that the devices used are even in shape and size and that one doesn't have an advantage over the other.

13. 12 triangles, 24 squares

14. a. 7

    b. 8

    c. 7

    d. In part a there are 5 red pens, so, to make sure that you pick 2 black pens, you have to pick $5 + 2 = 7$.
    In part b there are 6 black pens, so, to make sure that you pick 2 red pens, you have to pick $6 + 2 = 8$.
    In part c there are 5 red pens and 6 black pens, so, to make sure that you pick 1 black pen and 1 red pen, you have to pick $6 + 1 = 7$.

15. a. 36 square units

    b. A = 20 square units, B = 12 square units, C = 4 square units

    c. i. $\dfrac{5}{9}$ ii. $\dfrac{1}{3}$ iii. $\dfrac{1}{9}$

d. i. $\dfrac{37}{351}$, which is approximately 0.11. This is very close to the theoretical probability.

   ii. $\dfrac{631}{1872}$, which is approximately 0.34. This is very close to the theoretical probability.

   iii. $\dfrac{3131}{5616}$, which is approximately 0.56. This is very close to the theoretical probability.

Project

1.

Die 2 \ Die 1	1	2	3	4	5	6
1	2	3	4	5	6	7
2	3	4	5	6	7	8
3	4	5	6	7	8	9
4	5	6	7	8	9	10
5	6	7	8	9	10	11
6	7	8	9	10	11	12

2. $\{2, 3, 4, 5, 6, 7, 8, 9, 10, 11, 12\}$

3. 7

4. 2, 12

5. $\dfrac{1}{18}$

6. 5, four different ways.

7. You end up at square 36. The next ladder is at square 51. It is not possible to obtain a 15 when you roll 2 dice.

8. At the end of these two turns, you would end up at square 73. P(two ones in a row) $= \dfrac{1}{36} \times \dfrac{1}{36} = \dfrac{1}{1296}$.

9. You need to roll a total of 7. P(7) $= \dfrac{1}{6}$.

10. Your opponent has a better chance, because the chance of scoring a 6 is higher than that of scoring a 4.

11. There are many different paths that get from start to finish in 5 turns.

13.6 Review questions

1. a. Certain   b. Unlikely   c. Even chance

2. a. Impossible   b. Likely   c. Impossible

3. a. The month of January having 29 days
   b. Tuesday coming after Monday
   c. Rolling a 3 on a regular die

4.

Impossible	Highly unlikely	Very unlikely	Unlikely	Even chance	Likely	Very likely	Highly likely	Certain
0	0.1	0.2	0.3	0.4 0.5 0.6	0.7	0.8	0.9	1.0

5. a. {Head, Tail}
   b. {red marble, green marble, blue marble}.
   c. {true, false}

6. a. $\{1, 2, 3, 4, 5, 6\}$
   b. $\{1, 2, 3, 4, 5, 6, 7, 8, 9, 10\}$
   c. $\{1, 2, 3, 4, 5, 6, 7, 8, 9\}$

7. a. 5000   b. 26   c. 5   d. 31

8. a. $\dfrac{1}{6}$ b. $\dfrac{1}{3}$ c. $\dfrac{1}{3}$ d. $\dfrac{2}{3}$

9. a. $\dfrac{2}{3}$ b. $\dfrac{1}{3}$ c. 0   d. 0

10. a $\dfrac{1}{4}$ b $\dfrac{1}{4}$ c $\dfrac{3}{13}$ d $\dfrac{3}{13}$

11. a. $\dfrac{1}{10}$ b. $\dfrac{1}{2}$ c. $\dfrac{3}{10}$ d. $\dfrac{1}{10}$

12. a. $\dfrac{7}{10}$ b. $\dfrac{7}{10}$ c. $\dfrac{1}{5}$ d. 0

13. $\dfrac{9}{10}$

14. a. 6 socks. Worst case scenario is first picking 4 blue socks, then 2 black socks.
   b. 7 socks. Worst case scenario is first picking 5 black socks then 2 blue socks.
   c. 7 socks. Minimum needed to guarantee a pair of black socks is 6 (from part a). Picking one more guarantees a pair of blue socks will be included (from part b).

15. a. 12   b. 4, 6, 8, 10, 12   c. $2 + 6, 6 + 2, 4 + 4$
   d. 0   e. $\dfrac{2}{9}$

16. $\dfrac{1}{3}$

17. $\dfrac{2}{3}$

18. $\dfrac{3}{5}$

19. $\dfrac{29\,679}{160\,000}$

20. Likelihood smallest to largest: purple, yellow, orange, green, pink, blue

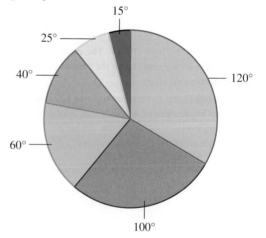

Semester review 2

The learnON platform is a powerful tool that enables students to complete revision independently and allows teachers to set mixed and spaced practice with ease.

Student self-study

Review the **Course Content** to determine which topics and lessons you studied throughout the year. Notice the green bubbles showing which elements were covered.

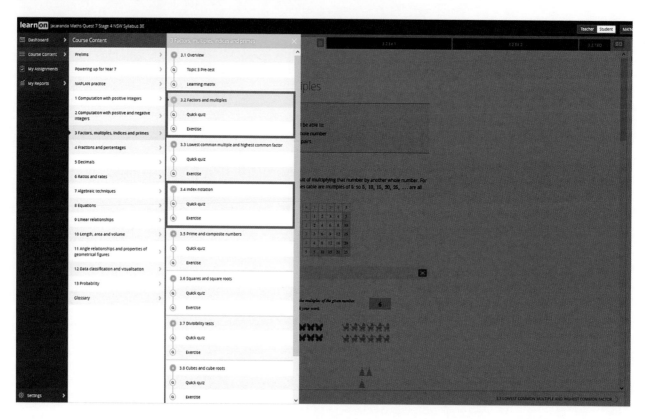

Review your results in **My Reports** and highlight the areas where you may need additional practice.

Use these and other tools to help identify areas of strengths and weakness and target those areas for improvement.

Teachers

It is possible to set questions that span multiple topics. These assignments can be given to individual students, to groups or to the whole class in a few easy steps.

Go to **Menu** and select **Assignments** and then **Create Assignment**. You can select questions from one or many topics simply by ticking the boxes as shown below.

Once your selections are made, you can assign to your whole class or subsets of your class, with individualised start and finish times. You can also share with other teachers.

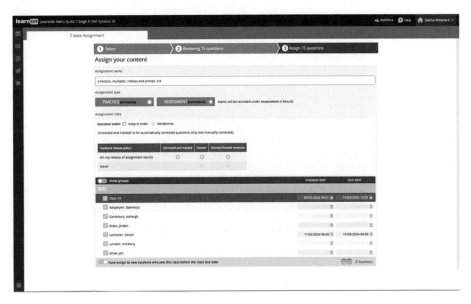

More instructions and helpful hints are available at www.jacplus.com.au.

GLOSSARY

3-dimensional object shape that has depth, as well as length and width; also called a polyhedron

acute angle angle greater than 0° but less than 90°

acute-angled triangle triangle where each angle is acute; that is, less than 90°

adjacent angles angles lying next to another angle

algorithm a step-by-step set of tasks to solve a particular problem. A program is an implementation of an algorithm.

allied angles *see* **co-interior angles**

alternate angles angles 'inside' the parallel lines, on alternate sides of a transversal; think of them as Z-shaped

arithmagon triangular figure in which the two numbers at the ends of each line add to the number along the line

ascending order in order from smallest to largest

Associative Law a number law that refers to the order in which three numbers may be added, subtracted, multiplied or divided, taking two at a time

Associative Law for addition a number law that refers to the order in which three numbers may be added or subtracted

back elevation the view observed from standing directly behind an object

back-to-back stem-and-leaf plot a type of table used to organise, display and compare two sets of data

backtracking process of working backwards through a flow chart. Each step is the inverse operation of the corresponding step in the flow chart. It can be used to solve equations.

back view *see* **back elevation**

bar graph graph drawn in a similar way to a column graph, with horizontal bars instead of vertical columns. Categories are graphed on the vertical axis and frequencies (numbers) on the horizontal axis.

base the number being repeatedly multiplied in a power term. The power is the number of times the base is written.

BIDMAS the order in which calculations are performed. The order is: brackets; index (or power); division; multiplication from left to right; addition; and subtraction from left to right.

bird's eye view view observed from the top of the object looking down on it

capacity maximum amount of fluid that can be contained in an object. The units may be the same as for volume or litres and millilitres may be used.

Cartesian coordinates a pair of numbers that gives the position of a point on the Cartesian plane. The first number is the x-coordinate and the second number is the y-coordinate.

Cartesian plane area formed by a horizontal line with a scale (x-axis) joined to a vertical line with a scale (y-axis). The point of intersection of the lines is called the origin.

categorical data data that involves grouping or classifying things, not numbers; for example, grouping hair colour

circumference distance around the outside of a circle, given by the rule $2\pi r$ or πD, where r is the radius and D is the diameter of the circle

clustering a grouping of values around a particular value

coefficient the number written in front of a term

co-interior angles angles on the same side of the transversal positioned between the parallel lines. These angles are supplementary.

column graph graph in which equal width columns are used to represent the frequencies (numbers) of different categories

common factor a factor common to more than one number

common multiple a number that is a multiple of two or more numbers

Commutative Law a rule for a particular operation showing that the result is the same regardless of the order of the numbers

Commutative Law for addition a number law that refers to the order in which two numbers may be added or subtracted

complementary angles angles that sum (add) to 90°

complete revolution an angle of 360° (a full circle)

composite number number with more than two factors

composite shape figure made up of more than one basic shape

constant a value that is not a variable

continuous describes data that take on any value that is measurable; for example, the heights of people in a class

conversion factor used to convert one unit of measurement to another

convert to change a measurement or value to different units

convex a shape with all its interior angles less than 180°

corresponding angles angles on the same side of the transversal that are both either above or below the parallel lines. The angles are equal in size.

cross-section the two-dimensional shape we can see when we cut straight through a three-dimensional object

cube number a number that can be arranged in a cube

cube root number that multiplies by itself three times to equal the original number. Finding a cube root is the opposite of cubing a number.

data information collected from surveys

decimal places digits after the decimal point

decimal point a dot, placed after a whole number, used to represent parts of a decimal fraction

decomposition method method of subtraction. The larger number is decomposed by taking 10 from the tens column and adding it to the units column. It may also involve taking 100 from the hundreds column and adding it to the tens column.

denominator bottom term of a fraction. It shows the total number of parts into which the whole has been divided.

descending order numbers in order from largest to smallest

diameter straight-line distance across a circle though its centre

difference the answer obtained when two numbers are subtracted

directed number a number with a positive or negative sign indicating that it is a certain direction from the origin along a number line

discrete describes data that takes on exact values that may be counted; for example, the number of televisions sold each week

Distributive Law a rule stating that each term inside a pair of brackets is to be multiplied by the term outside the brackets

divided bar graph a representation of data whereby the different sectors of a bar represent different groups of data

dividend the quantity in a division that is to be divided

divisibility test a series of rules that tests for factors of a number

division the sharing of a number into equal parts; the opposite of multiplication

divisor the quantity in a division by which the dividend is divided

dot plot consists of a horizontal axis labelled and evenly scaled, with each data value represented by a dot

equally likely outcomes two or more outcomes that have an identical chance of occurring

equation mathematical statement containing a left-hand side, a right-hand side and an equals sign between them

equilateral triangle triangle with all sides equal in length. All angles will be equal too.

equivalent expressions two or more expressions that are identical but displayed in different ways

equivalent fractions two or more fractions that are equal in value but have different numerators and denominators; for example, $\frac{1}{2} = \frac{2}{4}$

event a possible outcome of a trial in a probability experiment

experiment the process in probability of performing repeated trials of an activity for obtaining data in order to be able to predict the chances of certain things happening

exponent *see* **power**

expression mathematical statement made up of pronumerals (letters) and numbers. An expression does not contain an equals sign.

exterior angle angle formed when any side of a closed shape is extended outwards. The exterior angle and the interior angle adjacent to it are supplementary (add up to 180°).

factor a whole number that divides exactly into another whole number, without any remainder

factor tree diagram that displays the prime factors of a composite number. Each branch shows a factor of all numbers above it. The last numbers at the bottom of the tree are the prime factors of the original number.

favourable outcome the outcome we are looking for

fraction used to describe a part of a whole number

frequency number of times a result or piece of data occurs.

frequency distribution table table used to organise data by recording the number of times each data value occurs

front elevation view observed from directly in front of an object

front view *see* **front elevation**

highest common factor (HCF) largest factor that is common to all the numbers given

improper fraction fraction whose numerator is larger than its denominator; for example, $\dfrac{7}{4}$

index *see* **power**

input number the first number in a flow chart

inspection determining the answer by examining the equation

integers positive whole numbers, negative whole numbers and zero

interior angle an angle on the inside edges of a shape

interval a segment that can be labelled with two capital letters, one at each end

inverse operation an operation that 'undoes' a previous operation. Addition and subtraction are inverse operations; and multiplication and division are inverse operations.

is greater than more or higher in value, expressed by the symbol >

is less than less or lower in value, expressed by the symbol <

isometric drawing a 2-dimensional drawing of a 3-dimensional object

isosceles triangle triangle that has two sides of equal length. The two angles opposite the equal sides will also be equal.

legend lists definitions of different symbols or colours used in a graph

like terms terms that contain exactly the same pronumeral (letter) part; for example, $5a$ and $33a$ are like terms but $4ab$ is not

long-term trend in probability, the trend observed for results from a very large number of trials. This tends to match the theoretical probability.

lowest common multiple (LCM) lowest multiple that two or more numbers have in common

magnitude the size of a number, relative to its distance from zero; also called absolute value

mean a measure of the centre of a set of data. If the total value of all the observations is shared equally among all of the values in the data set, the mean is the amount that each observation would receive.

median the middle value of a set of data when the data is arranged in order from smallest to largest

mixed number number that is made up of a whole number and a fraction; for example, $2\dfrac{3}{4}$

mode the most common value in a set of data

multiple the result of multiplying a number by another whole number

multiplication a quick, shortened process of adding a number to itself a specific number of times

negative integers whole numbers less than zero

nominal data data that can be classified into mutually exclusive categories without hierarchy

number plane *see* **Cartesian plane**

numerator top term of a fraction. It shows how many parts there are.

numerical data data that can be measured with a numerical value

observed probability determined by conducting an experiment and gathering data

obtuse angle angle greater than 90° but less than 180°

obtuse-angled triangle triangle that has one angle greater than 90°

opposite integers numbers equidistant from zero, but one is positive and one is negative

opposite operations *see* **inverse operations**

ordered stem plot a type of table used to organise and display data, with the leaf part written in ascending order from left to right

order of operations set of rules that determines the order in which mathematical operations are performed

ordinal data data for which the possible values are ordered or ranked

outcome result of an experiment

outlier an extreme value in a data set

output number the last number in a flow chart

parallel lines lines that run in the same direction and never meet

parallelogram quadrilateral with both pairs of opposite sides parallel to each other. Rectangles, squares and rhombuses are parallelograms.

per cent out of 100

perfect square a number that can be arranged in a square; a number that is the square of another number

perigon an angle of 360° (a full circle)

perimeter distance around the outside (border) of a shape

perpendicular describes two objects with a right angle (90°) between them

place value the position of a digit in a number

power the number of times that the base is written in the expanded form of a power term; also known as an exponent or index

prime factors factors that are only prime numbers

prime number counting number that has exactly two factors, itself and 1

prism a solid object with identical parallel ends, and with the same cross-section along its length

probability likelihood or chance of a particular event (result) occurring

product result of a multiplication

pronumeral letter used in place of a number

proper fraction fraction whose numerator is smaller than its denominator; for example, $\frac{3}{4}$

protractor instrument used to measure angles in degrees

quadrant one of four regions of the Cartesian plane produced by the intersection of the x- and y-axes

quadrilateral 2-dimensional, closed shape formed by four straight lines

quotient result of a division

radius the distance from the centre of a circle to the circumference

range the set of all y-values in a relationship between x and y (algebra); the difference between the largest and smallest data value (measure of spread)

rate a ratio that compares two measurements with different units; for example, km/h or $/g

ratio a comparison of two or more quantities of the same kind

ray a fixed starting point but no end point with an arrow on one end

reciprocal the reciprocal of a number is obtained by first expressing it as a fraction then tipping the fraction upside down; for example, $\frac{4}{3}$ is the reciprocal of $\frac{3}{4}$. A mixed number must be converted to an improper fraction first and then tipped.

recurring decimal decimals that have one or more digits repeated continuously; for example, 0.999... They can be expressed exactly by placing a dot or horizontal line over the repeating digits, as in this example: 8.343 434... = 8.3̇4̇ or 8.$\overline{34}$.

reflex angle an angle greater than 180° but less than 360°

remainder number left over after numbers have been divided

right angle angle that equals 90°

right-angled triangle triangle that has one of its angles equal to 90° (a right angle). A small square is placed in that angle to mark the right angle.

sample space in probability, the complete set of outcomes or results obtained from an experiment. It is shown as a list enclosed in a pair of braces (curled brackets) { }.

scalene triangle triangle that has no equal sides. The angles will all be unequal too.

sector graph a representation of data whereby the different sectors of a circle represent different groups of data; also known as a pie graph

side elevation view observed when standing directly to one side of an object. The left view or the right view of the object can be drawn.

side view *see* **side elevation**

square number a numbers that can be arranged in a square

square root number that multiplies by itself to equal the original number. Finding a square root is the opposite of squaring a number; for example, $\sqrt{49} = 7$.

squaring a number multiplying a number by itself; for example, $2 \times 2 = 2^2 = 4$

stem-and-leaf plot a type of table used to organise and display data

substitution process by which a number replaces a pronumeral in a formula

successful trial a trial that results in the outcome you wanted

sum answer obtained when numbers are added together

supplementary angles angles that sum (add) to 180°

term part of an equation or expression. A term can be expressed as one or more than one pronumeral, or a number only. Examples include $5x$, $3ab$, xyz, 8 and so on.

terminating decimal a decimal number that has a fixed number of decimal places; for example, 0.6 and 2.54

theoretical probability probability determined by predicting outcomes rather than by gathering data

top view *see* **bird's eye view**

transversal a line that intersects a pair (a set) of parallel lines

trial number of times a probability experiment is conducted

unit price the price for a single unit of the item in question. This can be the price of a specified quantity (e.g. 1 kg or 100 mL) or of a single unit (e.g. 1 banana).

unlike terms terms that do not contain exactly the same pronumeral (letter) part; for example, $5a$ and $33a$ are like terms but $4ab$ is an unlike term

variable another word for a pronumeral (algebra); a symbol in an equation or expression that may take many different values (data)

vertex the point at which two line s meet

vertically opposite angles special angles formed when two straight lines intersect. The two non-adjacent angles are called vertically opposite angles. These angles are equal in size.

vinculum horizontal bar separating the numerator from the denominator

volume amount of space that a 3-dimensional object occupies. The units used are cubic units, such as cubic metres, m^3, and cubic centimetres, cm^3. A volume of $1\ m^3$ is the space occupied by a cube of side length 1 m.

whole numbers counting numbers: 0, 1, 2, 3, 4, …

INDEX